INTERNATIONAL SERIES OF MONOGRAPHS IN
PURE AND APPLIED BIOLOGY

Division: **ZOOLOGY**

GENERAL EDITOR: G. A. KERKUT

VOLUME 57

THE BIOLOGY OF THE MOLLUSCA

THE BIOLOGY OF THE MOLLUSCA

SECOND EDITION

BY

R. D. PURCHON, D.Sc.

*Professor of Zoology in the University of London,
at Chelsea College of Science and Technology*

*(formerly: First Raffles Professor of Zoology
in the University of Malaya, 1950–60;
Professor of Zoology, University of Ghana, 1961–2)*

PERGAMON PRESS

OXFORD · NEW YORK · TORONTO · SYDNEY
PARIS · FRANKFURT

U.K.	Pergamon Press Ltd., Headington Hill Hall, Oxford OX3 0BW, England
U.S.A.	Pergamon Press Inc., Maxwell House, Fairview Park, Elmsford, New York 10523, U.S.A.
CANADA	Pergamon of Canada Ltd., 75 The East Mall, Toronto, Ontario, Canada
AUSTRALIA	Pergamon Press (Aust.) Pty. Ltd., 19a Boundary Street, Rushcutters Bay, N.S.W. 2011, Australia
FRANCE	Pergamon Press SARL, 24 rue des Ecoles, 75240 Paris, Cedex 05, France
FEDERAL REPUBLIC OF GERMANY	Pergamon Press GmbH, 6242 Kronberg-Taunus, Pferdstrasse 1, Federal Republic of Germany

First edition 1968

Second edition 1977

Reprinted 1978

Library of Congress Cataloging in Publication Data

Purchon, R. D.

The Biology of the Mollusca

(International series in pure and applied biology:

Division, zoology; v. 57)

1. Mollusks. I. Title

QL403.P8 1976   594   76–10804

ISBN 0–08–021028–7

Printed in Great Britain by Biddles Ltd., Guildford, Surrey

TO

MY WIFE

CONTENTS

LIST OF TABLES

LIST OF ILLUSTRATIONS

ACKNOWLEDGEMENTS

I HAVE been greatly aided over a long period of time by many authors who have kindly sent me reprints of their publications, thereby enabling me to read far more widely than would otherwise have been possible, and I would like to record my grateful thanks for all this help.

I would like to thank Mrs. R. Helyar and Miss H. Haworth for their help in the preparation of the manuscript, and Miss E. Turner who prepared the microscope slide from which Fig. 106 was drawn. I have greatly appreciated the excellent service provided by the library of Chelsea College during the whole period of preparation of this book, and I would like to acknowledge my gratitude to the Librarian, Mr. Teague, and to the Assistant Staff of the library.

I have to thank Dr. Ronald G. Evans, Department of Biology, University of Keele, who very kindly identified the species of *Patella* which were collected in the Azores by the Chelsea College Expedition to the Azores, 1965; I am also deeply grateful to Dr. Evans for information he placed at my disposal regarding the distribution of species of *Patella* on British and European shores.

First-class illustrations are more informative and more stimulating than lengthy descriptions, and I have drawn freely on the rich supply of excellent illustrations in scientific journals; I wish to thank the Editors, Boards of Management, and Proprietors of many scientific journals, and all other holders of copyright, who kindly gave me permission to reproduce the large numbers of beautiful illustrations which had previously been published elsewhere. For each illustration I have cited in the legend the full details of its source.

I have greatly appreciated constant encouragement from my wife in the course of drafting the manuscript for this book, and assistance in correction of the proofs and preparation of the index.

I am particularly indebted to Professor Gareth Owen, Department of Zoology, the Queen's University of Belfast, who read the first draft of the whole manuscript, and who made many valuable suggestions and criticisms. The book is undoubtedly much the better for Professor Owen's advice, but I remain responsible for all residual errors and shortcomings.

INTRODUCTION

THE principal purpose served by this volume is to make readily available the interesting arguments brought forward, the illustrations, and the conclusions reached in a large number of authoritative papers on molluscs which have been published during the last 40 or 50 years. Outstanding among the pioneer works on the functional biology of molluscs are the papers of the late J. H. Orton on *Crepidula* in 1912 and 1914, and of C. M. Yonge on *Mya* in 1923 and on *Ostrea* in 1926. Subsequent to these there has been an ever-increasing flow of publications of absorbing interest on the biology of various molluscs, the principal contributors being C. M. Yonge, J. Z. Young, A. Graham, V. Fretter, and J. E. Morton. Important papers from these and many other authors are scattered widely in zoological journals where they are only likely to be studied by specialists who are pursuing a particular research interest. Yet many zoologists, conchologists, and naturalists who lack the compulsion to work their way laboriously through the *Zoological Record* in search of reading material will find the main conclusions of these papers to be of absorbing interest. For their benefit I have gathered together in this book the essential features of several hundreds of original publications in the form of eight essay reviews, using as many as possible of the original illustrations, and endeavouring to represent as fairly as possible the views of the original authors. The book is liberally studded with references to the original papers, so that any reader who wishes can check the validity of any statement or can pursue in more detail any subject in which his interest has been aroused.

According to the *Zoological Record* about a thousand scientific articles are published each year on molluscs, including those published in foreign languages. Thus in the period of time covered by this book about 50,000 papers have been published on the mollusca alone. It is obvious that I could not attempt to consult more than a small fraction of these. Many of these papers are not relevant in this particular context, e.g. papers describing new species, locality records, and the constituents of geological faunas. I have been obliged to exclude from consideration many papers which did not fit into my chosen essay topics and, in particular, I have not attempted to review papers which are essentially physiological in content. Except for a few very important papers in foreign languages, e.g. the paper by L. Tinbergen on the behaviour of *Sepia*, for which there is no equivalent in the English language, I have generally confined my attention to papers in English: this enabled me to work more quickly in preparing the manuscript, and has the advantage that the reader can follow up any point by reference to the original literature without difficulty. I do not doubt that as a result of this policy I have missed many interesting and important publications on molluscs, but perhaps the general reader will agree that I have already provided more than sufficient references to support my subject-matter.

I have another important incentive in drafting this book. If a three-year undergraduate course in zoology is to include a general review of the animal kingdom together with introductory courses in genetics, physiology, ecology, etc., then the amount of time which can be allocated even to such an important phylum as the Mollusca is pitiably small. In such a short course it may be possible to include a few exercises on living material, but in general one is obliged to give adequate coverage to basic knowledge of classification and comparative anatomy without which any consideration of special cases would be pointless. Having completed these foundation studies, all too little time remains for consideration of adaptive radiation within the major classes, and yet it is here that the subject becomes so deeply fascinating. This is where the real interest lies, in the arena of current scientific investigation. I confess that I have no great interest in anatomy as such, and the anatomy of one particular species is no more important than that of any other example in the same group. The serious student should not set out to learn the anatomy of a selected and approved type species—and then be apprehensive of tackling the anatomy of another, fairly closely related example. Surely, instead of learning the anatomy of a selected type species he should learn how to investigate the anatomy of an unknown animal. After all, this is how the great anatomists of a previous century had to work. The best authority on the anatomy of a species is the animal itself, so I have offered in Appendix B a number of exercises in which the student investigates the anatomy of an animal, aided by a narrative description, or a questionnaire. In some cases I have deliberately provided descriptions of exotic species, so that the student has to look for points where the description fails to accord with his specimen, e.g. *Aspatharia brumpti* instead of *Anodonta cygnea*; *Aplysia winneba* instead of *A. punctata*; and, going a little further, *Achatina fulica* instead of *Helix pomatia*. By this device I am attempting to make the student think during his practical class work instead of abandoning himself to a solely mechanical activity; in using these descriptive accounts, the student should test the validity of each sentence—true or false?—and not take it all for granted. At the same time he should consider these descriptions as models of brief, lucid, scientific prose which he could profitably emulate in his own written work—it is not easy to write concise, meaningful descriptions of the anatomies of animals.

Finally, I venture to hope that this book may be of interest to many amateur shell collectors who would like to know something about the living animals which occupy the shells which they seek for their cabinets. One cannot do entirely without technical terms when writing on the biology of molluscs, but I have tried to keep these to a minimum, and to explain those which have to be used. To understand something of the processes of feeding and digestion, of reproduction, and of the adaptations to special habitats, and to know something of the problems which have to be met by a variety of molluscs should add greatly to the pleasures of possession of a fine collection of rare and beautiful shells. Modern students of the functional biology of molluscs owe a great deal to the patient work of amateur conchologists and naturalists of previous centuries, and it is desirable that the amateur should be able to reap the benefits of the endeavours of the professional. Moreover, much valuable work remains to be done by field workers in recording the feeding and reproductive habits of molluscs, and it is hoped that this volume will encourage such field-work.

I have not attempted to modernise or standardise Latin names used in this book, being doubtful whether those names in current usage will be any more lasting than those used 10 or 20 years ago. I have generally employed the name used in the publication cited, which has the slight disadvantage that one particular species may be referred to under different names in different places in the book; it has the advantage that a reader will not experience confusion when consulting original literature. In some places I have indicated a synonym where I have thought this to be helpful.

R. D. PURCHON

Otford, Kent
October, 1966

PREFACE TO THE SECOND EDITION

CHANGES incorporated in this second edition include the following recent developments: consideration of the time factor in investigations into feeding and digestion in bivalves, and recognition of the possible occurrence of cyclical patterns in the digestive process which may be linked to environmental variables, or to behavioural rhythms; development of the carnivorous habit in the Verticordiidae (Bivalvia) and the bearing of this on the origin of the Septibranchia; patterns of distribution of terrestrial molluscs on Pacific islands; the opportunity has also been taken to correct minor errors in the original text.

R.D.P.

Otford, Kent

CHAPTER 1

FORM AND FUNCTION OF THE MANTLE CAVITY AND ASSOCIATED ORGANS

Synopsis

The mantle cavity and its associated organs, together with the buccal mass and the radula, are basic molluscan features although each has been lost in certain lineages.

In the Aplacophora there is a posterior mantle cavity which houses the ctenidia where such are present.

In the Monoplacophora the mantle cavity is lateral in position, between the foot and the mantle; six pairs of nephridia discharge into the mantle cavity, which contains a metameric series of ctenidia.

In the Polyplacophora the mantle cavity is lateral, and contains numerous true ctenidia which are secondarily reduplicated, not metamerically segmented. An excretory duct and a gonoduct open into the mantle cavity posteriorly on each side. The nature of the respiratory currents and the cleansing mechanisms in the mantle cavity are described.

In the Scaphopoda the tubular mantle cavity is very simple and contains no ctenidia. The paired excretory organs open into the mantle cavity on either side of the anus.

In the Bivalvia the ctenidia have become concerned in the process of feeding, only to a limited extent in the Protobranchia, but very extensively in all other types of bivalves. There is a tendency for fusion of the left and right mantle lobes with the formation of siphons and with progressive enclosure of the mantle cavity in higher forms. The mantle cavity is most extensively enclosed in this way in deep-burrowing members of the in-fauna and in the wood-boring shipworms. The respiratory and feeding water current in the mantle cavity is typically maintained by the lateral cilia of the ctenidial filaments, but in the Nuculanidae and in the Septibranchia there is a muscular ctenidial pumping mechanism. Vigorous clapping of the shell valves by contractions of the posterior adductor muscle may serve to rid the mantle cavity of sediment in the free-living epifauna such as *Pecten*, or may serve to embed the animal more deeply in the substratum in certain members of the infauna such as *Pinna*. This mechanism has provided the basis for the development of free swimming in some of the Pectinidae and Limidae. In *Lima hians* the swimming action has been applied to darting into crevices between stones prior to nest building. The sea water in the mantle cavity is used as a hydro-skeleton in certain examples where the mantle cavity can be effectively sealed; e.g. *Hiatella* which uses this principle to grip the walls of its burrow by the swollen base of the siphonal process; e.g. *Mya*, which uses fluid pressure in the mantle cavity to extend its siphons.

In the Gastropoda the mantle cavity has been moved by the process of torsion from its primitive posterior position to a new, anterior position above the head. This change may have enabled the animal to withdraw its head quickly into the protection afforded by the shell when alarmed. It would also be advantageous for the sense organs (osphradia) in the mantle cavity to be sited anteriorly. In prosobranchs each individual undergoes torsion during its larval development. The anterior position of the mantle cavity poses important sanitary problems which have been solved in a variety of ways in different lineages of prosobranchs. The most successful solution of these problems, as judged by the development of genera and species, is that adopted by the Mesogastropoda and Neogastropoda in which there is a single pectinibranch ctenidium on the left (post-torsional) side of the mantle cavity, and the respiratory stream of water passes through the mantle cavity from left to right. The greater efficiency of this arrangement enabled these lineages to colonise soft sedimentary deposits without the risk of the mantle cavity becoming · clogged with silt.

The reduction of the right side of the mantle cavity and loss of the right ctenidium and auricle was accompanied also by the loss of the right excretory organ. The right excretory duct became incorporated into the single gonoduct. Since the respiratory current passes out of the mantle cavity on its right side, and the single gonoduct opens on this side, it became possible for a penis to develop on the right side of the male, and for the development of copulation, and internal fertilisation. The mantle cavity thus becomes involved in the process of reproduction by sheltering the penis and the female genital opening, by contributing a distal section to the genital duct, and in some cases by harbouring the developing embryos in their egg capsules.

In some cases, e.g. *Crepidula*, the ciliary cleansing mechanisms of the ctenidia and of the mantle surface have become modified to constitute a novel process of feeding on minute particles suspended in the inhalant water stream—an interesting example of convergent evolution with respect to the filibranch bivalves.

Various prosobranchs have invaded fresh waters, e.g. *Theodoxus*, and some are amphibious and can breathe either fresh water or air, e.g. *Pila*. Many little-known prosobranchs have acquired the capacity to breathe air, have lost all trace of the ctenidia, and have invaded dry land, e.g. *Pomatias*.

The Opisthobranchia exhibit the consequences of detorsion, and in extreme examples, e.g. the Nudibranchia, the mantle cavity and ctenidia have been completely lost. The penis, situated on the right side of the head, was presumably left behind as the mantle cavity moved backwards along the right side of the body during the process of detorsion.

The Pulmonata occur on dry land, in fresh water, and also on the sea shore. Some freshwater forms, e.g. *Lymnaea stagnalis* and *Physa fontinalis*, are facultative air or water breathers; so also are some primitive marine pulmonates such as *Trimusculus*, but *Siphonaria*, which also occurs inter-tidally on the sea shore, apparently only breathes sea water.

In the Cephalopoda the mantle cavity is posterior in position, and this has been considered to be the primitive condition in the phylum. The respiratory demands of large and active predacious animals could not be maintained by ciliary means, and the respiratory flow is here effected by the extremely powerful pallial muscles. The mantle cavity is concerned with the reproductive process since it houses the nidamental and accessory nidamental glands in the female and since the hectocotylised arm of the male is inserted into the mantle cavity of the female in order to transfer spermatophores during copulation. The muscular mantle not only supplies a highly efficient respiratory mechanism, but it also serves for locomotion, a powerful exhalant jet of water being directed in the appropriate direction by the manœuverable funnel. The mantle cavity of cephalopods is concerned with defence, since the duct of the ink sac discharges into the mantle cavity, and the emission of ink serves to distract any potential predator which alarms the cephalopod.

In decapods the striated pallial muscles are under dual nervous control; nerve impulses arriving in the normal nerve fibres in the pallial nerves maintain the gentle movements required for respiration and for slow cruising motion, while nerve impulses travelling in the giant fibres in the pallial nerves mediate the powerful all-or-nothing escape reaction. There are no such giant fibres in octopods.

The mantle cavity, with the organs contained in it, contributes more to the molluscan mode of life than does any other organ complex save, perhaps, the buccal mass and the radula. Although not actually in view, it undoubtedly makes a major contribution to the "molluscan facies". It is in the three "major" classes—the Bivalvia, the Gastropoda, and the Cephalopoda—that the possibilities of the mantle cavity are exploited to the full.

Serving primarily for respiration, in certain cases the mantle cavity has assumed additional functions including the collection and sorting of food, incubation of early developmental and larval stages, and even locomotion. Conversely, a minority of molluscs have lost the mantle cavity.

Prior to the discovery of the class Monoplacophora, the diversity of form of the mantle cavity in modern molluscs was "explained" as having been derived from a hypothetical ancestral form in which the mantle cavity was posterior in position and contained two "aspidobranch" or bi-pectinate ctenidia. We now have to consider how to assimilate the Monoplacophora into this picture, for this class has great antiquity with fossil representatives from the Cambro-Silurian, while the modern representative has a lateral mantle cavity and metamerically segmented ctenidia.

Aplacophora

These small, worm-like and carnivorous animals are comparatively simple in form; the mantle cavity is posterior in position and it houses the anus, the urino-genital apertures, and also the gills in such forms as possess these. In *Chaetoderma* there are two aspidobranch gills; in *Neomenia* and some other genera there is a circlet of laminate gills, while in other cases there are no gills. It seems probable that the Aplacophora are secondarily simplified, in which case the site and form of the mantle cavity offers no reliable hint as to the remote ancestral condition.

Monoplacophora

The peripheral mantle cavity is circular in form, being bounded internally by the wall of the foot and externally by the margin of the mantle. A metameric series of ctenidia lies evenly spaced in the lateral part of the mantle cavity. Each ctenidium is suspended from the roof of the mantle cavity, and bears a single series of well-developed leaflets on the posterior border of its axis. These leaflets do not contain any skeletal supports. There are five pairs of ctenidia in *Neopilina galatheae* (Lemche and Wingstrand, 1959), and six pairs in *N. ewingi* (Clarke and Menzies, 1959). The ctenidia are covered by a ciliated epithelium, and probably serve exclusively for respiration. The mouth opens into the mantle cavity anteriorly. The anterior rim of the mouth bends backwards and expands laterally to form a pair of velar folds, which could be compared with the outer labial palps of a bivalve. The posterior border of the mouth is marked by two posterior tentacle ridges which pass backwards and outwards between the foot and the velar folds. These posterior tentacle ridges bear a series of short, dichotomously branched tentacles, and can be compared in position with the inner labial palps of a bivalve mollusc. The tentacles are poorly ciliated, and the part they might play in the feeding process is uncertain. Two short, pre-oral tentacles, of unknown function, lie lateral to the velar lobes.

Six pairs of nephridia, in metameric series, open into the mantle cavity by separate nephridiopores, most of which are sited near the bases of the ctenidia. In *N. galatheae* the third and fourth nephridia serve also as gonoducts for liberation of sperm or ova into the mantle cavity. There are no accessory genital structures. The anal papilla lies posteriorly, in the median plane (see Fig. 150, p. 456).

The peripheral mantle cavity thus serves for respiration and for excretion and defaecation; it is possible that in addition eggs may be fertilised and incubated in the mantle cavity, but this is not known.

Polyplacophora

The mantle cavity lies peripherally between the foot and the margin of the mantle, as in the Monoplacophora. Here, however, the margin of the mantle is in the form of a girdle, which does not secrete shelly material, and which exhibits a considerable degree of flexibility.

The anterior end of the mantle cavity is more or less filled by the proboscis which bears the mouth centrally, and by the pair of palps which lie one on each side of the proboscis. Posteriorly the anal papilla lies in the median line and opens into the mantle cavity, as in the Monoplacophora. Numerous bi-pectinate ctenidia lie in the posterior part of the mantle cavity; the larger the number of ctenidia present, the further forward do these extend along the sides of the foot. The number of ctenidia increases as the animal develops, additional ctenidia being added in front of the post-renal* ctenidium, and sometimes also posteriorly to this ctenidium (see Fig. 160, p. 474).

In *Lepidopleurus asellus* the ctenidia number from 11 to 13 in the adult; likewise in *Acanthochitona crinitus* there are about 15 ctenidia; in *Lepidochitona cinereus* the number varies from 16 to 19, while in *Tonicella marmorea* the ctenidia range from 19 to 26 in number. Since the number of ctenidia is variable for any one species, and since they do not conform in position with any other serially repeated structures—e.g. the eight shell valves—it is agreed that the series of ctenidia does not indicate metameric segmentation. On the contrary, it is judged that there is a secondary reduplication of the ctenidia in the Polyplacophora.

The ctenidia hang downwards from the roof of the mantle cavity. The anterior ctenidia curve inwards towards the side of the foot, and so divide the mantle cavity into an outer and an inner channel. Further posteriorly the ctenidia bend inwards and increasingly backwards, directing the inner, exhalant channel towards the posterior median line where the anal papilla lies. Ciliary currents on the ctenidia draw water through the spaces between the ctenidia and between the ctenidial filaments, from the outer to the inner channel. The ctenidial filaments do not contain internal skeletal supporting rods.

* The post-renal ctenidium is the first ctenidium to be formed as the animal develops; it lies immediately adjacent to the renal aperture and it is usually, but not always, the largest ctenidium in the series.

In order to admit water for this respiratory current, the margin of the girdle can be raised locally at any point. Elevation of the margin of the mantle for this purpose is usually anterior, provided that the animal is wholly submerged in water (Yonge, 1939). The exhalant stream of water passes out posteriorly, and carries with it any gametes (ova or sperm) emitted from the genital pores, any fluid waste matter emitted from the excretory pores, and any faecal pellets voided at the anus. The posterior positions of the excretory pores and of the anus, and the firm texture of the faecal pellets, minimise the risk of fouling the respiratory organs.

In common with all molluscs in which the mantle cavity serves for respiration and not for feeding, there is the need to ensure that the organs in the mantle cavity are not fouled by deposition of particles brought in by the inhalant current. In those species with numerous ctenidia the inhalant current is so strong that particles will not settle in the mantle cavity. In addition the roof of the mantle cavity is ciliated, the cilia beat posteriorwards, and aid in keeping the mantle cavity clean. Where the ctenidia are fewer and the inhalant current is less powerful, these ciliary tracts on the surface of the mantle cavity are more powerfully developed.

Tracts of unicellular mucus cells, interspersed with ciliated cells, occur at various points in the mantle cavity, e.g. on the sides of the foot, on the roof of the pallial groove, on the inner border of the girdle, or on the bases of the ctenidia. The extent of these tracts is greatest in species in which there are few ctenidia and in which the respiratory water current is least strong. Their function is to consolidate particles of sediment brought into the mantle cavity in the respiratory stream, and these tracts are therefore analogous with the hypobranchial glands of prosobranch gastropods.

Sense organs may be found at various points in the mantle cavity. These include the paired osphradia, which lie in the exhalant chamber between the post-renal gill and the anus. Each is an elongated area of ciliated and sensory cells, the epithelium of which is covered by a cuticle which is penetrated by the sensory "hairs". The osphradia are innervated from the pallial nerve, which lies immediately below. Similar sensory epithelial patches may occur elsewhere, e.g. on the roof of the mantle cavity anterior to the series of ctenidia, where they are innervated from the pedal cord. It is doubtful whether these osphradia can be considered to be homologous with those of the Bivalvia and of the Gastropoda. (In Bivalves the osphradia lie in the supra-branchial, or exhalant chamber, while they lie in the path of the inhalant stream in the gastropods.) It has been suggested that in all cases these sensory patches may be tactile organs, concerned with estimating the amount of sediment being brought in by the respiratory current. The older idea that they were olfactory seems less satisfactory, for it is inherently improbable that an olfactory organ would evolve in the exhalant chamber to test water which was about to leave the mantle cavity. As an alternative hypothesis, do the osphradia serve as chemo-receptors capable of detecting the presence of spermatozoa or ova of the same species? If so, their purpose might be to ensure co-ordination of spawning.

At the posterior end of the mantle cavity in some genera the girdle bears a pair of folds on its inner wall—one on each side. These girdle folds form horizontal shelves which press against the large posterior ctenidia, and so help to perfect the division of the mantle cavity into inhalant and exhalant chambers.

Two excretory ducts open into the mantle cavity posteriorly, one on each side, the excretory opening lying adjacent to the first ctenidium to be formed during the development of the individual. The gonad ducts open into the mantle cavity, one on each side and a little in front of the excretory pore. In some genera the eggs are retained in the mantle cavity, where they are fertilised and undergo the early stages of development; this, however, is not characteristic of the class as a whole.

Scaphopoda

The mantle cavity is a tubular chamber with a large anterior and a small posterior orifice, which extends through the whole length of the shell. It acquires this tubular form at an early stage in development when the left and right lobes of the mantle meet and fuse in the median ventral line. The mantle cavity lies lateral, ventral, and posterior to the visceral mass.

The mantle cavity of *Dentalium* is extremely simple, as there are no ctenidia, and respiration is presumably effected by the general surface of the mantle. There is a system of oblique ciliated ridges on the sides and floor of the mantle cavity just in front of the anus, and the ciliary currents set up by this organ will probably aid the slow inhalant water current. Particles of suspended matter which fall on these ridges are passed towards the base of the foot, whence they may be collected by the captacula and carried to the mouth.

The two excretory organs open into the mantle cavity on either side of the anus, and the gonad discharges via the right excretory pore. The mantle cavity serves solely for respiration and for elimination of the ill-formed faeces and excretory wastes.

Bivalvia

In the Bivalvia the mantle cavity lies latero-ventrally between the left and right mantle lobes, and forms a spacious chamber the principal function of which is to house the organs of feeding and respiration. The ctenidia cross the mantle cavity and divide it into a lower infrabranchial chamber and an upper supra-branchial chamber. The lateral cilia on the ctenidial filaments draw a current of water into the infra-branchial chamber and between the ctenidial filaments. This water current then passes out of the supra-branchial chamber as the exhalant current. This system first evolved to provide adequate respiration, but in the filter-feeding bivalves it became modified for feeding purposes. The excretory and genital apertures lie within the mantle cavity, on the sides of the visceral mass, and in a variety of specialised examples the eggs are ferti-lised within the mantle cavity and are retained there during the early stages of development.

Primitively the mantle cavity communicated with the exterior by a very extensive median aperture, e.g. in *Glycymeris* and *Arca*. This gape is even more extensive in monomyarian forms such as *Ostrea*, *Pinctada*, or *Pecten*, and many other members of the Anisomyaria, in which the animal lies on one side and there has been a secondary approach towards radial symmetry. In these forms the ctenidium is draped round the sub-central posterior adductor muscle, and an inhalant water current carrying food particles in suspension is drawn into the mantle cavity through 180° of arc, or more.

In more advanced forms there are points of fusion between the margins of the left and right mantle lobes. These fusions differ in number, in degree of participation by the three marginal folds of the mantle edge, and in ex-tent along the free edge of the shell. There may be a single region of union, providing a complete exhalant orifice, as in *Mytilus*, *Ostrea*, and *Astarte*, or there may be a second region of fusion providing in addition a complete inhalant aperture. The exhalant aperture is always posterior in position; the inhalant aperture is normally posterior (but not invariably). In members of the superficial infauna, e.g. the Tellinidae, the fusion below the inhalant siphon is short, and there is a very extensive ventral pedal gape through which the foot may protrude into the substratum. In the Tellinacea this ventral mantle fusion is generally strengthened by the cruciform muscle, a feature which is unique to the order. This muscle is associated with a curious tactile sense organ which is thought to serve as a proprioceptor which can record the degree to which the siphons are extended or with-drawn (Yonge, 1949) (see Fig. 1).

In deep-burrowing members of the infauna, e.g. *Mya* and *Lutraria*, as also in rock- and wood-boring bivalves, e.g. *Hiatella*, *Pholas*, and *Teredo*, the ventral fusion of the left and right mantle lobes is far more extensive and there is a greatly reduced pedal gape at the anterior end. Here the mantle cavity is tubular in form, much like that of the Scaphopoda. In some cases, e.g. *Zirfaea* among the rock-borers, the mantle cavity has been secondarily extended at its posterior end, so that in the relaxed condition the mantle cavity and the contained ctenidium extend beyond the posterior borders of the shell valves (Purchon, 1955). This is carried to an extreme in the wood-boring genera in the Teredinidae, where the greater part of the visceral mass, and almost the whole of the ctenidium, lie posterior to the shell valves in a worm-like prolongation of the mantle. This great extension of the mantle cavity in the Teredinidae permits the development of the stomach caecum as a huge food-storage organ.

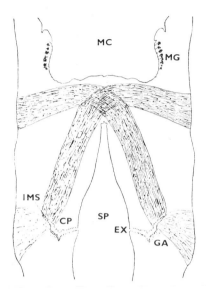

FIG. 1. *Scrobicularia plana*. Diagrammatic horizontal section through the cruciform muscle which unites the left and right mantle lobes ventrally, a little anterior to the base of the inhalant siphon. Associated with the cruciform muscle there is an unique sense organ (CP) in the form of a deep slit penetrating from the siphonal space (SP) into the substance of the muscle (IMS). An area of ciliated epithelium overlies a small ganglion (GA) and probably supplies proprioceptive information regarding the degree of extension or retraction of the siphons. These associated structures are of special interest in being restricted to the order Tellinacea, and providing firm evidence of relationship between its constituent families. For interpretation of other lettering, see pp. 38. (Originally published in Yonge, 1949, *Phil. Trans.* B, **234**, 52, fig. 18.)

The fusion of the left and right mantle lobes may concern only the two inner folds of the mantle margin—"type A", e.g. the Tellinacea. Alternatively, both inner and middle folds are concerned in the fusion—"type B", e.g. the Cardiacea and Veneracea; again, the fusion may also concern that part of the outer fold which secretes the periostracum, in which case the ventral wall of the mantle cavity is strengthened by an external coat of periostracum—"type C", e.g. the Saxicavacea (Yonge, 1957). By these mantle fusions the mantle cavity has become progressively more enclosed and more clearly defined.

The ventral fusion of the left and right mantle lobes may include a cross-fusion of muscle fibres, with the formation of an accessory ventral adductor muscle which holds together the ventral borders of the two shell valves, e.g. in the Pholadidae (Turner, 1954). The inhalant and exhalant orifices may lie flush with the margin of the shell, or they may be raised upon shorter, or longer siphons. These siphons may remain separate, and highly mobile, as in the Tellinidae, or they may be fused to form a rather rigid siphonal process, as in *Mya*, or *Zirfaea*. Where the siphons are fused the interior of the siphonal process remains divided into inhalant and exhalant channels by a continuous partition. Strictly speaking the lumina of the two siphons are not parts of the mantle cavity, but are separate manifestations. There is often a delicate membranous valve at the base of each siphon, separating its lumen from the mantle cavity.

Although there may be well-developed muscles in the mid-ventral wall of the mantle cavity where the two mantle lobes have fused, pallial muscles play very little part in the movements of water through the mantle cavity. The normal respiratory and feeding current is effected by the lateral cilia of the ctenidial filaments in most bivalves. Exceptional cases are found in the Nuculanidae where water is pumped through the mantle cavity by muscle fibres lying in the suspensory membranes of the two ctenidia, and in the Septibranchia where a similar pumping action is provided by the muscular septum. Water may be expelled suddenly from the mantle cavity by sudden contractions of the adductor muscles in order to flush out accumulations of pseudofaeces, or when the animal is alarmed. From this pre-adaptive condition there has arisen the capacity to swim, which is most clearly witnessed in *Chlamys opercularis, Pecten irradians*, and in *Lima fragilis* (Yonge, 1936). Other examples which may display comparable jet propulsion are *Solen delesserti*, which may become exposed on the surface of the sand in which it normally lies buried, and *Scintilla deshayesi*, which normally lies sheltered in crannies under coral boulders. In all these cases muscular action upon the contents of the mantle cavity is effected by the adductor muscles and not by the pallial muscles. The failure of the Bivalvia to generate any pelagic representatives can be attributed to two limitations: the inability to develop adequate pallial muscles and the inability to lighten the shell sufficiently.

There are a few unrelated examples of bivalves which enter or penetrate the substratum more deeply, in contrasted ways, with the aid of water currents raised by muscular action on the fluid-filled mantle cavity. I know of only two bivalves in which the pallial muscles have developed to any great extent for the purpose of throwing water from the mantle cavity. In the protobranch *Solemya* the shell valves are bordered by a flexible strip of periostracum, which remains attached to the mantle margin, and which can be drawn inwards by the action of powerful pallial muscles, thereby obliterating the mantle cavity and causing the ejection of water and waste matter from the mantle cavity. *Brechites* is a burrowing form which lies buried vertically in muddy sand at or below low-water mark on tropical shores. The mantle lobes are very extensively fused, there being a very small anterior pedal gape in the mature animal. The anterior end of the mantle cavity is closed by a very thick and muscular layer of mantle which contains powerful pallial muscle bundles radiating outwards to their insertions on the shelly tube from a small central valvular pedal aperture. Rhythmic contractions of this pallial partition draw water up from the substratum, through the pedal orifice, and expel it through the inhalant siphon. The purpose served by this unique organ is apparently to embed the shell more deeply in the substratum. This digging action may be evoked when the shell is partly uncovered by wave action (Purchon, 1960). The ear shell, *Pinna*, is able to embed itself more deeply in the substratum by sudden contractions of the posterior adductor muscle, causing jets of water to be directed downwards from the mantle cavity into the substrate from the anterior end of the shell. The substratum is loosened, and the animal is then able to penetrate more deeply by contracting the byssal retractor muscles (Yonge, 1953) (See Fig. 76, p. 202).

Lima hians usually lives in nests formed by stones linked together by byssus threads. When beginning to form such a nest, *L. hians* darts suddenly into a crevice between two small stones with considerable force, penetrating as far as possible. It then relaxes the adductor muscle and the opening thrust of the ligament pushes the stones slightly apart. In this way the animal embeds itself sufficiently before beginning to construct its nest. Once the nest has been constructed, *L. hians* varies its position within the nest from time to time. By sudden contractions of the posterior adductor, jets of water are shot out of the mantle cavity into the interstices between the stones, preventing these from becoming blocked with sediment. In this way easy access is maintained for the respiratory and feeding water supply (Gilmour, 1963).

In some rock-boring bivalves the water in the mantle cavity and in the lumen of the siphonal process may be used as a hydroskeleton. In *Hiatella* spp. the siphonal tentacles are turned inwards, the siphonal apertures are closed, and then the whole siphonal process is contracted without detectable leakage of water at the tip. Water is evidently also unable to escape from the mantle cavity at the small pedal orifice, and consequently the basal half of the siphonal process becomes swollen and engages with the walls of the boring. This provides a fixed point, or fulcrum, on which the shell valves can be moved with consequent abrasion of the rock and deepening of the burrow. This method of boring depends on the fact that the siphons and the mantle cavity together form a sealed water-tight unit (Hunter, 1949). Chapman and Newell (1956) have shown that in some examples, e.g. *Mya*, the siphons are not extended by blood pressure but by hydraulic pressure generated in the mantle cavity. The pedal gape is closed anteriorly, and the siphonal orifices are also closed, so that the lumina of the siphons and the mantle cavity comprise a closed system. Contraction of the adductor muscles and relaxation of the retractor muscles of the siphons causes water to pass from the mantle cavity into the siphons, and the siphons are accordingly elongated. The siphonal orifices are opened, the adductors are relaxed, and the opening thrust of the ligament causes more water to enter the system. Repetition of the process causes further expansion of the siphons.

Gastropoda

Prosobranchia

In the Prosobranchia, as in the Pulmonata, the mantle cavity occupies an anterior position and both the gut and the nervous system are twisted. The rectum opens as usual into the mantle cavity, and faeces are discharged from an anus which is sited above and behind the head. To say the least, this is a curious arrangement, and one which seems at first sight to have serious disadvantages. There must have been very powerful grounds for establishing this as the basic configuration in this large and successful class. It is agreed that originally the mantle cavity occupied a posterior position and that neither the gut nor the nervous system was twisted. It is further agreed that at a very early stage in the phylogeny of the class a process of torsion of the visceral mass occurred whereby the mantle cavity became displaced through 180° of arc in an anti-clockwise direction when viewed from above, so that henceforth the mantle cavity lay in its present position above and behind the head. In this process of torsion it is obvious that—disregarding coils and loops—the gut would be twisted into a U-shape. The nervous system includes two long connectives which pass backwards from the pleural ganglia to the visceral ganglia which lie posteriorly in the viscera. It is obvious that 180° of torsion of the visceral mass would carry the visceral ganglia from a posterior to an anterior position, and that the ganglion which previously occupied a left posterior position would now lie anteriorly and on the right. The connectives passing backwards from the left and right pleural ganglia would cross each other as the result of torsion, and the connective which arises on the right side anteriorly would pass over to the left side, and vice versa.

We cannot ever know the functional pressures which actually brought about this realignment, but we can offer some plausible speculations on the disadvantages of having a posteriorly sited mantle cavity.

If the ancestral gastropod crawled on a surface which carried a certain amount of sediment, some of this would be disturbed and raised in suspension temporarily. The water drawn into a posteriorly situated mantle cavity would then be slightly contaminated by unwanted small particles. Moreover, the opening of the mantle cavity would be sheltered behind the visceral mass where it would not get the benefit of gentle water currents caused when the snail moved about in calm water. Again, sense organs would develop within the mantle cavity in close association with the delicate ctenidia. These sense organs are the osphradia. With the mantle cavity in a posterior position it would be difficult for the snail to locate the source of contaminating material and avoid it.

In contrast, there seem to be various advantages in having the mantle cavity in an anterior position. Water entering the mantle cavity would be uncontaminated by sediment raised by the movement of the snail, while symmetrically sited osphradia in an anterior mantle cavity would detect contamination as soon as the snail entered an unsuitable area, thereby enabling the snail to take immediate avoiding action. Different views have been advanced as regards the functions of the osphradia. Judging by the indirect evidence of the occurrence and degree of development of the osphradia in gastropods occupying different habitats and feeding in contrasted ways, the osphradia are thought to be tactile organs for estimation of the turbidity of the inhalant water stream (Hulbert and Yonge, 1937; Yonge, 1947). On the other hand, experimental evidence has been offered to advance the view that the osphradia of the stenoglossan *Bullia laevissima* are chemo-receptors by means of which this predacious carnivore can detect its prey and avoid its enemies. The osphradium of *Bullia* is said to be sensitive to trimethylamine, which emanates from the normal food of the snail, and to tetra-methyl-ammonium, a common constituent of coelenterates, and both of these substances are attractive. In contrast, avoidance behaviour is elicited by substances which occur in high concentration in the flesh of elasmobranchs, which feed upon *Bullia*. If the osphradial nerve of *Bullia* is severed, the snail fails to respond to its favourite food when this is placed in close proximity (Brown and Noble, 1960; Brown, 1961). It is possible that the osphradium may have developed different sensory capacities in different phylogenies, and care should be taken not to over-generalise from the results of experiments on only a few species.

Anteriorly sited osphradia can supplement the cephalic sense organs, thereby strengthening the general organisation of the animal. With the mouth of the mantle cavity facing forwards, the movement of the animal would cause a slight rise in pressure of water entering the mantle cavity, which would augment the inflow of water caused by the lateral cilia of the ctenidial filaments. This would facilitate respiration. Most important, perhaps, is the presence of a large chamber just within the shell into which the head can be withdrawn immediately in the event of alarm. The head having been protected first, the tail is then drawn in, and finally the mouth of the shell can be sealed by the operculum which lies on the dorsal surface of the tail.

These are quite compelling arguments that it is generally more efficient for the mantle cavity to lie anteriorly, above and behind the head, provided that various contingent disadvantages could be overcome in some way. But it would be teleological to suggest that these advantages could have been anticipated, and we still do not know how the process of torsion was initiated. It is probable that torsion was originally a larval adaptation to meet the special requirements of planktonic life (Garstang, 1928), and that this is yet another example of paedomorphosis, the post-torsional larval form happening to supply the basis for a more efficient model for solving the problems of the bottom-dwelling adult. The only mutation necessary to initiate torsion would be for the production of asymmetry of the two retractor muscles. Torsion illustrates von Baer's law—"ontogeny recapitulates phylogeny"—since to the present day each individual prosobranch undergoes torsion during its own larval life. In some examples observed the process occurs in two stages. Thus in *Patella* the first stage takes about 40 hours and the final stage is then completed in a few hours (Smith, 1935); in *Haliotis* the first stage of torsion takes up to 6 hours, but the process is only completed about 5 days later, after the individual has adopted a benthic life (Crofts, 1937).

Before leaving the subject of torsion, there is one further aspect of special interest which deserves mention. Whatever may have been the environmental pressures, or the functional stresses, which caused torsion in the first place, surely this process can have occurred once and once only. It seems highly improbable that this remarkable realignment of the parts of the body could possibly have occurred on separate occasions in two or more independent lineages by parallel evolution. The chances of this selfsame change occurring twice on independent occasions seems impossibly small. If we postulate that torsion was originally caused by a repetitive single gene mutation at a particular locus on a chromosome, then this could, and presumably would, occur independently in all populations bearing that chromosome. This would seem to indicate that all the gastropods which exhibit torsion spring from one single common origin, and are monophyletic in almost the strictest sense of the word. Thus it seems that the whole vast edifice of the Gastropoda has arisen from this single aboriginal assumption of torsion, which probably met the exigencies of planktonic life in the first place, but which proved to have great pre-adaptive value for the exploitation of bottom-dwelling by the adult. The development of an anterior mantle cavity ranks high in importance in the establishment of the gastropod model and in the ensuing adaptive radiation of the class. Had torsion not occurred, the Gastropoda as we know them would never have come into existence.

The only way we can understand the anatomies of modern gastropods is by asserting that torsion occurred in some remote common ancestor and that this anatomical modification has been incorporated into the ontogeny of most modern forms. On these grounds we derive modern gastropods from a form which had a posterior mantle cavity which contained two aspidobranch, or bipectinate, ctenidia. This picture of the hypothetical ancestral gastropod has been projected still further back into the past to contribute to the conjectured form of the remote ancestor of all molluscs.

Primitively the rectum opens into the mantle cavity centrally, and with the mantle cavity in an anterior position some further modification would be necessary to make the system workable. Inhalant water currents are generated by the cilia on the ctenidia, and these enter the mantle cavity laterally and pass between the ctenidial leaflets towards the median line where faeces and excretory wastes are discharged. The exhalant current, laden with waste material, would then have to leave the mantle cavity centrally and pour over the head of the animal, but for some modification to avoid this obviously undesirable arrangement. There are, in fact, various alternative arrangements which enable the exhalant current to pass in a different direction. These are summarised in Table 1, which shows two contrasted trends in the Prosobranchia. One of these trends is reduction and disappearance of the ctenidia, while another is the development of secondary pallial gill leaflets. Individual examples can be

TABLE 1. *Diversity of respiratory arrangements in the Prosobranchia*

Five genera in the Patellacea are marked with an asterisk (*). [†]With the exception of the two genera *Valvata* and *Adeorbis*

	Two bi-pectinate ctenidia; shell slit or perforated	One bi-pectinate ctenidium; water current left to right	One pectinibranch ctenidium; water current left to right	No ctenidia
No secondary pallial leaflets	*Pleurotomaria* *Scissurella* *Emarginula* *Haliotis* *Diodora* etc.	Trochacea Neritacea *Patelloida* *Valvata* *Adeorbis*	[†]Mesogastropoda Neogastropoda	*Lepeta*
Incomplete ring of secondary pallial leaflets		*Lottia*		*Patina*
Complete ring of secondary pallial leaflets				*Patella*

selected and arranged in a series to illustrate one or other of these trends, but such arrangements are entirely artificial and without phylogenetic significance. It seems clear from Table 1 that several different possible arrangements have been attempted, with clear examples of parallel or convergent evolution. Some of these arrangements have been notably more successful than others, e.g. the Patellacea have been extraordinarily successful in colonising the exposed rocky shore. The most successful arrangement of all is that found in the Mesogastropoda and Neogastropoda, in which there is a single pectinibranch ctenidium. This has provided an efficient working model which has been applied with great success in a very extensive adaptive radiation.

Let us now examine in turn each of these trends in the organisation of the mantle cavity complex.

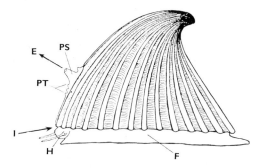

FIG. 2. The limpet *Emarginula reticulata*, seen from the left side. The anterior mantle cavity contains two symmetrically disposed ctenidia, and the inhalant water stream (I) is anterior. The shell bears a deep median anterior emargination and the exhalant water current (E) emerges forwards and upwards from the mantle cavity through a pallial siphon (PS) which lies at the summit of this cleft in the shell. For interpretation of other lettering, see p. 38. (Originally published in Yonge, 1947, *Phil. Trans.* B, **232**, 462, fig. 13.)

In *Emarginula* and in *Pleurotomaria* there is a deep median emargination of the roof of the mantle cavity and a corresponding slit in the shell above. The anus lies close to the inner end of this emargination and the exhalant current passes through the emargination, carrying faeces and excretory wastes away from the head of the animal (see Fig. 2). In *Haliotis* there is a related condition; there is a deep sub-median emargination of the roof of the mantle, but there is no corresponding slit in the shell above. Instead there is a regular series of apertures, a new aperture periodically being added anteriorly as the shell grows, and old apertures being closed posteriorly one by one in such a way that the animal may possess as many as five or six of these apertures at any one time. *Haliotis* has a greatly enlarged body whorl and a very deep mantle cavity. An extensive slit would weaken the shell too much, and the series of apertures serves the same purpose without weakening the shell. The respiratory currents pass inwards between the ctenidial leaflets and impinge on the roof of the mantle cavity where the hypobranchial glands secrete mucus to immobilise any particles of silt suspended in the water (see Fig. 3). Due to the asymmetry of the shell the right ctenidium is

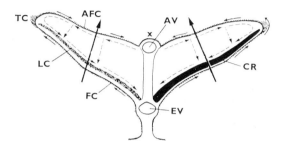

FIG. 3. *Haliotis tuberculata.* Transverse section through the aspidobranch ctenidium, showing a pair of short triangular ctenidial leaflets, one on each side of the ctenidial axis. The efferent branchial vein (EV) lies in the suspensory membrane while the afferent vein (AV) lies at the opposite side of the bases of the ctenidial leaflets. The frontal surfaces of the leaflets bear frontal cilia (FC), on either side of which lateral cilia (LC) create the respiratory water current which passes between adjacent leaflets. The frontal margin of each leaflet is supported internally by a chitinous rod (CR) which holds the leaflet out firmly in the respiratory current. The above-mentioned features are only shown on one side of the figure for clarity in presentation. It will be noted that the respiratory water currents, indicated by two large arrows, are in the opposite direction to the flow of blood through the filaments from afferent to efferent sides (broken arrows). For interpretation of other lettering see p. 38. (Originally published in Yonge, 1947, *Phil. Trans.* B, **232**, 451, fig. 4.)

slightly reduced and the right hypobranchial gland is greatly reduced in size. The exhalant current escapes from the mantle cavity through the series of shell apertures (see Fig. 4).

This line of modification is carried to its extreme form in *Diodora*, in which a single aperture is formed in the roof of the mantle cavity and in the overlying shell, early in life. This aperture remains at the apex of the shell, enlarging gradually as the animal grows, and so comes to lie at the posterior end of the mantle cavity of the adult animal. The exhalant current is ejected at the apex of the shell, far away from the animal's head (see Fig. 5). There is a secondary return to bilateral symmetry in the Fissurellidae, e.g. *Diodora*. This modification is not without disadvantage, for it lowers the resistance of the animal to desiccation, and gastropods such as *Diodora, Haliotis,* etc., are unable to colonise the

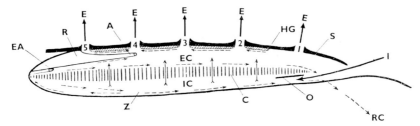

FIG. 4. *Haliotis tuberculata*, diagrammatic vertical longitudinal section through the mantle cavity and shell. The anterior mantle cavity contains two ctenidia (C) of which the right is slightly smaller than the left. The shell bears a linear series of orifices (1–5) above the mantle cavity, and the mantle is very deeply cleft along the line of these apertures. The inhalant water current (I) enters the mantle cavity anteriorly above the head, passes between the ctenidial filaments and out of the mantle cavity via the five apertures (E). Periodically a new aperture is formed anteriorly at the growing edge of the shell and old apertures, closer to the apex of the shell are obliterated. The rectum (R) lies in the posterior half of the mantle cavity, above the ctenidia, and faecal material is carried away in the exhalant current emerging from the fourth shell aperture. The inhalant current impinges on the osphradium (O) before reaching the anterior tips of the ctenidia. The orifice of the right renal organ (EA) is shown at the base of the rectum. Broken arrows represent ciliary cleansing currents on the floor of the mantle cavity. For interpretation of other lettering, see p. 38. (Originally published in Yonge, 1947, *Phil. Trans.* B, **232**, 453, fig. 5.)

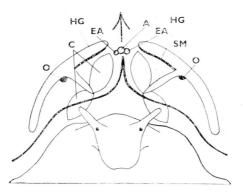

FIG. 5. Diagrammatic transverse section through the head and mantle cavity of a diotocardian prosobranch such as *Diodora*. The arrows indicate the course taken by the respiratory currents which enter symmetrically from both sides, anteriorly, and emerge dorsally through the "keyhole". On each side of the mantle cavity there is an aspidobranch ctenidium (C), an osphradium (O) and a hypobranchial gland (HG). The anus (A) and the excretory orifices (EA) open dorsally in the median line and waste products are voided into the dorsally directed exhalant water current. (Originally published in Graham, 1948, Inaugural Lecture, Birkbeck College, University of London, p. 8, fig. 2.)

middle and upper parts of the rocky shore (Graham, 1948, 1951). This is true where zonation is based primarily upon changes in tidal level. Where the splash zone is great relative to tidal variation, rocks well above "high tide level" are permanently wetted by splash, and keyhole limpets are common on upward-facing surfaces at notably higher levels on the shore, e.g. on the rocky shores of West Africa.

There is another trend in which there has been reduction and loss of the right ctenidium, followed by loss of one series of filaments in the remaining left ctenidium. Here, in order to keep the body compact, the visceral mass and the shell became coiled in a turbinate spiral, with a consequent departure from bilateral symmetry. The viscera are attached to the interior of the shell by shell muscles, and one result of this asymmetry was compression of the right side of the mantle cavity. This led first to reduction of the size of the right ctenidium, as in *Haliotis*, and then to the loss of the right ctenidium, as in the Trochacea, e.g. *Trochus*, *Gibbula*, *Calliostoma*, and *Turbo*; in the Neritacea, e.g. *Theodoxus* and *Nerita*; and certain of the Patellacea, e.g. *Lottia* and *Patelloida* (= *Acmaea*) In all these examples the one remaining ctenidium is bi-pectinate, and is turned towards the right side so that the respiratory current passes through the mantle cavity from left to right. This condition has been achieved independently by parallel evolution in the Trochacea, Neritacea, and Patellacea.

The Patellacea are interesting in exhibiting a novel further development of the situation by divesting the original mantle cavity—here termed the "nuchal" cavity—of its primary respiratory function, and transferring this function to a series of secondary gills in the newly formed marginal pallial groove. In this respect the family Lepetidae demonstrate the most reduced condition, the mantle cavity having lost both ctenidia and osphradia, and there being no secondary pallial gills. The Lepetidae are sublittoral limpets whose respiratory demands are evidently small (Yonge, 1960). In the family Acmaeidae (Fig. 6), *Patelloida* has a nuchal cavity of reduced size which contains the left ctenidium and the left osphradium. The pallial groove resembles that of *Lepeta* in lacking secondary gills. *Lottia* is basically comparable to *Patelloida*, but possesses a series of secondary pallial gills which is incomplete anteriorly. *Lottia* tends to be oriented head downwards when it is at rest on the rocks. The margin of the shell does not fit the rock surface at all well and air is able to enter the pallial groove. Even when heavily pounded by waves, the shell may be held well up from the rock and the water streaming down the rock surface swills through the pallial groove and over the secondary gills. At this time the respiratory flow is effected by the gravitational movement of the water. The family Patellidae resemble the Lepetidae in having no ctenidia, but they possess secondary pallial gills; *Patina* resembles *Lottia* in possessing an incomplete series of pallial gills, while *Patella*

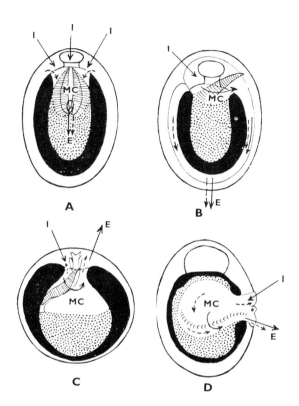

Fig. 6. Convergent evolution in the Gastropoda: diagrammatic representation of four different kinds of limpet, all seen from above. The shell muscle is shown black, and the visceral mass is stippled. The mantle cavity receives an inhalant respiratory current (I) and discharges an exhalant current (E), the positions and directions of these currents differing in the different types of limpets. Broken arrows indicate ciliary cleansing currents. A, *Diodora*, the "keyhole" limpet (Prosobranchia, Archaeogastropoda, Fissurellidae), in which the paired aspidobranch ctenidia are retained and the exhalant current passes out dorsally via the apical "keyhole". B, *Acmaea* (Prosobranchia, Archaeogastropoda, Patellacea), in which only one aspidobranch ctenidium is retained and in which the respiratory current passes transversely through the mantle cavity from left to right. In other members of the Patellacea, e.g. *Patella*, both ctenidia are lost, and secondary respiratory leaflets are found marginally on the mantle. C, *Hipponyx* (Prosobranchia, Mesogastropoda), only one, pectinibranch ctenidium is retained and the exhalant current leaves the mantle cavity anteriorly on the right side. D, *Siphonaria* (Pulmonata), a secondary gill has developed in the mantle cavity, and both inhalant and exhalant currents are on the right side of the visceral mass. (Originally published in Yonge, 1953a, *Proc. Calif. Acad. Sci.* **28,** 18, fig. 9.)

has advanced one stage further and has a complete ring of secondary gills (see Figs. 7 and 8).

Where these secondary pallial gills occur, the respiratory current is drawn centripetally into the pallial groove; it passes between the respiratory leaflets in the opposite direction to the flow of blood in the leaflets, and is then passed out centrifugally from the pallial groove. From a respiratory point of view, therefore, there has been a secondary approach to radial symmetry in these sedentary limpets.

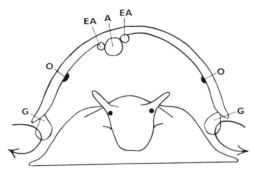

FIG. 7. Diagrammatic transverse section through the head and mantle cavity of the limpet *Patella*. The arrows indicate the respiratory water currents between the secondary gill leaflets (G). The nuchal cavity is no longer respiratory in function, the ctenidia having been lost. The osphradia (O) are still present. The anus (A) and the excretory apertures (EA) open dorsally into the nuchal cavity. (Originally published in Graham, 1948, Inaugural Lecture. Birkbeck College, University of London, p. 10, fig. 4.)

The sequence—*Patelloida, Lottia, Patina, Patella*—has no phylogenetic significance, as was emphasised above, but can be taken to signify that *Patella*, for example, has probably passed through successive "patelloidiform" "lottiform", and "patiniform" phases in its ancestry. As is indicated in Table 1, the Prosobranchia in general, and the Patellacea in particular, have made several *independent* experiments towards a solution of the problems posed by an anteriorly sited mantle cavity. These have met with various degrees of success. Those which retain two ctenidia but possess a perforated shell are generally confined to the lower levels of the rocky shore or even to the sub-littoral fringe except where there is heavy surf. The Patellacea have not been particularly effective as regards the numbers of genera and families which have been generated, but they have been exceptionally successful from the point of view of their achievement in colonisation of exposed and desiccated levels on the rocky shore. It is relevant to add in this connection that in the Patellidae the mid-gut and hind-gut have been greatly extended, and the latter carries a great accumulation of well-compacted faeces which are not discharged into the mantle cavity while the animal is exposed by the ebbing tide. Well-consolidated strings of faeces are voided when the animal is submerged, and are carried away immediately by water currents. This

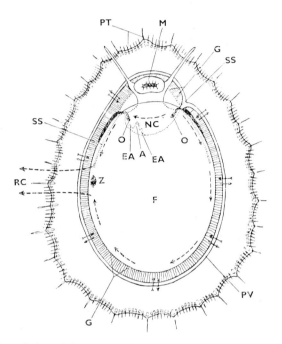

Fig. 8. Ventral view of the limpet *Patella vulgata*. The large oval foot (F) provides attachment to the substratum. Anterior to the foot lies the proboscis with the mouth (M) and the cephalic tentacles. The mantle cavity includes the nuchal cavity (NC) which lies above the head, and a wide groove which passes round the entire animal. The nuchal cavity contains the two osphradia (O) and receives the anus (A) and the two excretory apertures (EA); there are no ctenidia. A circum-pallial vein (PV) supplies blood to a complete circle of secondary, pallial gill leaflets (G). Streaks of sensory epithelial cells on the sides of the foot (SS) can detect differences in the salinity of water. Waste material which accumulates at the point Z is periodically ejected in the rejection current (RC) caused by sudden contractions of the shell muscle. For interpretation of other lettering, see p. 38. (Originally published in Yonge, 1947, *Phil. Trans.* B, **232**, 469, fig. 21.)

protection of the mantle cavity from contamination by faecal material is a major contribution to the success of the Patellidae as colonists of the exposed rocky shore (Graham, 1948).

The Trochacea and the Neritacea are comparable in their successful colonisation of the shore, and in the invasion of fresh waters by the latter order. The most outstanding success, however, was that of the pectinibranch gastropods, i.e. the Mesogastropoda and the Neogastropoda which we are now about to discuss.

In the pectinibranch condition we see the most satisfactory conclusion to the process of simplification of the respiratory mechanism. Only the left ctenidium remains, and this is monopectinate, i.e. there is only one series of filaments along one side of the ctenidial axis. The ctenidial axis is fused throughout its length to the wall of the mantle cavity. The inhalant current is drawn into the mantle cavity on the left side of the head, it passes between the ctenidial filaments from left to right, and the exhalant current is ejected on the right side of the head (see Fig. 9).

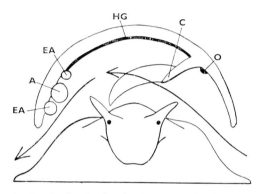

FIG. 9. Diagrammatic transverse section through the head and mantle cavity of a monotocardian prosobranch such as *Buccinum*. The arrows indicate the course taken by the respiratory current which passes through the mantle cavity from left to right. A single osphradium (O) lies at the base of the one remaining pectinibranch ctenidium (C), in the path of the inhalant water current. The hypobranchial gland (HG) covers an extensive area of the roof of the mantle cavity above the ctenidium. The anus (A) and the excretory apertures (EA) open into the mantle cavity on the right side, and discharge waste material into the exhalant water current. (Originally published in Graham, 1948, Inaugural Lecture. Birkbeck College. University of London, p. 10, fig. 3.)

There is a striking measure of uniformity in the configuration of the ctenidium and the respiratory flow of water through the mantle cavity of the pectinibranch snails; the abundance and diversity of the pectinibranchs indicates that this is an ideal solution to the respiratory and excretory problems which were posed by the adoption of torsion.

In the aspidobranch condition the ctenidium is suspended in the mantle cavity by membranes on the afferent and on the efferent side. This arrangement divides the mantle cavity incompletely into compartments and this leads to the possibility of clogging of the mantle cavity if the water were markedly turbid. Perhaps it is for this reason that aspidobranch gastropods tend to be restricted to hard, clean substrates.

In the pectinibranch condition, on the other hand, the ctenidium is attached to the mantle wall along one side only, along the ctenidial axis, and the arrangement is less susceptible to clogging by sediment. Perhaps it is in consequence of this that pectinibranch gastropods have been so successful in invading even muddy substrates.

The inhalant current may be guarded and extended in burrowing forms by the development of the left margin of the mantle cavity into an inhalant siphon. The respiratory current passes across the mantle cavity from left to right, between the ctenidial leaflets and, as always, travels in the opposite direction to that of the blood flowing in these leaflets. This is the most efficient arrangement from the point of view of respiration. So that the ctenidial leaflets are not displaced by the force of the inhalant current, each is supported internally by a skeletal rod which lies under the zone occupied by the current-producing lateral cilia. Resulting from the reduction of the right side of the mantle cavity, the rectum lies along its right hand border, and both faeces and excretory wastes are carried out of the mantle cavity by the exhalant current which passes out on the right. With the loss of the right kidney, the right excretory duct has been converted into the single gonoduct, and a well-developed penis is present in the male. Internal fertilisation is the normal practice, and thus the mantle cavity is now more actively concerned in the reproductive process. A further modification which accompanies the loss of the topographically right ctenidium is the loss of the corresponding auricle of the heart.

As measured by the numbers of families and genera involved, the "pectinibranch" solution to the respiratory and excretory problem, coupled with the highly versatile "taenioglossan" radular structure, has been remarkably successful.

In the pectinibranch snails the remaining left ctenidium is attached to the mantle wall for the whole length of its axis, and has a single series of leaflets which are usually short and comparatively broad-based. The roof of the mantle cavity bears the large, left hypobranchial gland, the function of which is to immobilise with mucus all particles of suspended matter which pass between the ctenidial leaflets and impinge on the roof of the mantle cavity. These particles, bound in mucus, are then expelled from the mantle cavity by ciliary cleansing currents, and the mantle cavity is accordingly protected from the risk of being fouled with silt. In such forms there is only one osphradium, which lies near the mouth of the mantle cavity on the left side, at the point where the inhalant water current enters the mantle cavity. The osphradium may be large and well developed, and in some cases it is a conspicuous bi-pectinate structure. In many cases the left margin of the mantle cavity is developed to form an incomplete tube, the siphon, by means of which the snail can obtain a clean inhalant water current even when the greater part of the snail is buried in the substratum. The carnivorous *Buccinum* and *Melongena* are characteristic examples of this line of modification of the mantle cavity (see Fig. 175, p. 510).

A further development of the asymmetrical mantle cavity in the Prosobranchia concerns modification of the ctenidial leaflets and of the ciliary cleansing mechanisms, with the provision of a new method of feeding on small particles suspended in the inhalant water current. The ctenidial leaflets are extended to form long, slender filaments. Suspended particles are collected by cilia on both the frontal and the abfrontal surfaces of the filaments, and also on the floor of the mantle cavity, and are gathered together in a ciliated food groove which leads to the right side of the head. The animal collects strips of mucus-bound particles from this groove by means of its radula. A full description of such a feeding mechanism has been given for *Crepidula* and also for other examples in Chapter 2 (Orton, 1912). Here the mantle cavity has acquired a new function, namely the collection and grading of food, without surrendering any of its responsibilities for respiration, excretion, and reproduction.

It should not be thought that the prosobranchs of the inter-tidal zone are necessarily only able to breathe in water. Sandison (1966) has shown that *Thais* (= *Nucella*) *lapillus* and three species of *Littorina* are able to respire in air, and have higher respiratory rates in air than in water. *L. saxatilis*, which occurs relatively high on the rocky shore, was less affected by prolonged drying than was *T. lapillus* which occurs at lower levels on the shore.

Opisthobranchia

Members of the Opisthobranchia exhibit some degree of detorsion, the mantle cavity lying on the right-hand side of the visceral mass, as in *Aplysia* in which the condition is most clearly displayed. Here the ctenidium is plicate, with alternate ridges and furrows on opposite sides of the axis, and without filaments, and this structure cannot have been derived from a pectinibranch ancestor. The occurrence of a single ctenidium, a single auricle, and a single excretory organ provides a clear indication that the Opisthobranchia have been derived from ancestors which had undergone torsion to the full extent. According to Dollo's law, the organs of the right side of the mantle cavity having been lost, these organs can never be regained. The genital aperture lies just within the anterior border of the mantle cavity, and this communicates by means of a ciliated sperm groove with the penis. The penis is situated on the right side of the head, far anterior to the mantle cavity, and has evidently been "left behind" when the mantle cavity retreated posteriorly during the process of detorsion (see Fig. 177, p. 520).

The mantle cavity in these opisthobranchs retains the functions of respiration, excretion, and defaecation, and still participates in the reproductive funcitons. In *Aplysia* the anus is situated at the posterior end of the mantle cavity, in a backwardly directed anal siphon, so that there is no danger of contamination of the mantle cavity by faecal material. In *Aplysia* the mantle cavity houses the orifice of the purple gland, which discharges a purple stain when the animal is irritated. This secretion evidently has a repugnatorial function and will discourage attack by a potential predator.

Mention has already been made of the capacity of the stenoglossan *Bullia laevissima* to detect trimethylamine and tetra-methyl-ammonium, both of which are natural substances emanating from potential food sources, by means of the osphradium (Brown and Noble, 1960; Brown, 1961).

The marine slug *Onchidella*, formerly thought to be a pulmonate, is now known to be an opisthobranch. There is a very small posterior mantle cavity, which receives the anus and the excretory aperture, and which contains no ctenidium or secondary gill. The female genital aperture lies adjacent but external to the mantle cavity, while the male genital aperture lies anteriorly, close to the head. Here the mantle cavity is greatly reduced in size, and serves solely as an excretory channel. An entirely new pulmonary chamber has developed, enabling this opisthobranch to breathe air (Fretter, 1943). On sheltered rocky shores in the tropics, e.g. in Singapore waters, *Onchidella* is extremely abundant and dominates the upper half of the beach while the tide is out. It retreats into a rock crevice when submerged by the tide. *Onchidella* glides about over the muddy rocks, discharging copious strings of faeces. Presumably these faecal strings are accumulated in the hind-gut in large quantities while the animal is submerged by the tide in order to avoid fouling the greatly reduced mantle cavity.

Many opisthobranchs, such as the dorids and the eolids, possess no mantle cavity and have lost both ctenidia and both osphradia. They are, in general, bilaterally symmetrical, this being a secondary condition. However, the penis is situated anteriorly on the right side of the head, and this can be taken as evidence of derivation of these animals from "fully torted" ancestors.

Air-breathing Forms

The mantle cavity, or a portion of it, has assumed the function of air breathing, with the acquisition of a lung, in more than one phylogeny. This is true of various small prosobranch genera which have become fully terrestrial, e.g. the large family Cyclophoridae (Prosobranchia, Mesogastropoda) with over sixty genera including *Cyclophorus*, *Alycaeus*, *Opisthostoma*, *Diplommatina*, etc. Very little is known of the modes of life of these terrestrial prosobranchs or of their respiratory processes. The ctenidium has been lost, but a sense organ remains on the left side of the mantle cavity, and this resembles the osphradium of aquatic gastropods. These terrestrial, air-breathing prosobranchs tend to be overlooked, and it is well to emphasise that the Pulmonata are by no means the only terrestrial gastropods.

One terrestrial prosobranch, *Pomatias elegans*, occurs in Britain and can readily be recognised by the large circular operculum by means of which it closes the aperture of the shell when it is inactive. Rees (1964) has shown how respiration can be maintained by means of a smaller, secondary opening to the mantle cavity when the operculum is withdrawn in certain tropical terrestrial prosobranchs. Such devices presumably serve to minimise loss of water through respiration during periods of drought (see Fig. 10).

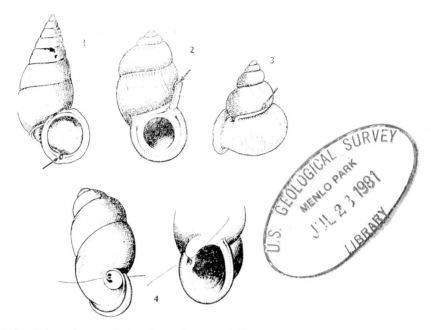

FIG. 10. Special respiratory devices in various tropical terrestrial prosobranchs in the family Cyclophoridae. Secondary orifices near the mouth of the shell lead in to the mantle cavity and permit respiration with minimal loss of water during periods of drought. 1, *Tortulosa pyramidatus*. 2, *Rhaphaulus chrysalis*. 3, *Alycaeus major*. 4, *Pupinella macgregori*, with a piece of cotton passed through the columellar notch and twisted external tube. (Originally published in Rees, 1964, *Proc. malac. Soc. Lond.* **36**, 63, figs. 5–8.)

In contrast, the freshwater prosobranch family Ampullariidae is amphibious. In *Pila* the large mantle cavity is incompletely divided into left and right compartments by an oblique ridge on its floor (see Figs. 11 and 12). A typical pectinibranch ctenidium lies on the right side, between this ridge and the rectum and genital duct which follow the right margin of the mantle cavity. The ctenidium serves for aquatic respiration in the normal manner. An incomplete partition hangs downwards across the left part of the mantle cavity, the chamber lying behind this partition serving as a lung. On the left side of the head there is a short siphon, an incomplete tube formed by an upwardly rolled layer of mantle. *Pila* may climb up to the surface,

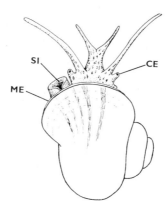

FIG. 11. Dorsal view of a living specimen of *Pila* sp. showing the head fully extended, with cephalic eyes (CE) at the bases of the tentacles. The mantle edge (ME) borders the edge of the shell, and the siphon (SI) is expanded.

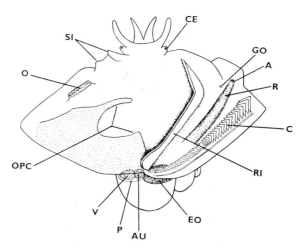

FIG. 12. Dorsal view of *Pila* sp. after the mantle cavity has been opened by a longitudinal incision along its right side, the roof of the mantle cavity having been reflected to the left. The osphradium (O) lies on the left side, in the path of the inhalant water stream. A ridge (RI) on the floor of the mantle cavity incompletely divides the mantle cavity into two parts, with the ctenidium (C), the anus (A), and the genital aperture (GO) on the right-hand side. The left side of the mantle cavity is again divided by an extensive opaque white curtain which hangs downwards from its roof. A wide opening in this curtain (OPC) provides communication between the anterior and posterior moieties of the left half of the mantle cavity. During air breathing the aperture to the posterior chamber is approximated to the base of the siphon, and air is passed into the posterior chamber. This will serve both for respiration and for provision of buoyancy. For interpretation of other lettering, see p. 38.

extend this siphon through the meniscus, and take air into the mantle cavity. The whole body of the snail is rhythmically raised and lowered several times, thereby drawing air into and expelling air from the mantle cavity. When the snail leaves the surface, the siphon is withdrawn and its ventral lip is rolled upwards. A bubble of air may remain in the posterior chamber of the mantle cavity, and this will provide the posterior part of the shell with buoyancy (personal observations). The mantle cavity thus has dual respiratory methods during at least part of the time while the snail is submerged, and this is undoubtedly advantageous when the available oxygen in the water is low. This must often be the case in standing water in the tropics. The presence of the air bubble in the lung will either minimise the effect of gravity or will provide a small degree of lift, and this new function of the mantle cavity as a buoyancy tank is a significant innovation. There is here a striking example of convergent evolution with respect to the aquatic Pulmonata, e.g. *Lymnaea*.

Pulmonata

In the order Pulmonata the primary modification is the conversion of the whole mantle cavity into an air-breathing organ with a heavily vascularised lung on its roof. The ctenidia have been lost. In typical examples such as *Helix* the greater part of the margin of the mantle has been fused to the dorsal surface of the body, the only remaining entrance to the mantle cavity being the pneumostome on the right-hand side. Air is inspired by opening the pneumostome and then lowering the muscular floor of the mantle cavity. The pneumostome is then closed and the floor of the mantle cavity is relaxed and allowed to rise. This raises the pressure within the mantle cavity, and this aids in the uptake of oxygen at the surface of the lung. When the pneumostome is opened again much of the stale air in the mantle cavity is driven out, due to the raised pressure within the chamber.

Various members of the Pulmonata have approached or have invaded fresh waters. Among the principally aquatic Basommatophora, *Lymnaea truncatula* is a marsh-dwelling and air-breathing snail. Other species of *Lymnaea*, e.g. *L. stagnalis* and *L. peregra*, are truely aquatic forms and these can breathe either air or water, as also can *Physa fontinalis*.

At inshore sites *L. peregra* makes regular migrations to the surface in order to breathe air. The snail approaches the surface gradually, one of its tentacles being bent backwards underneath the meniscus. Suddenly the shell is revolved and the pneumostome opens at the surface for about 3 seconds to refresh the air in the mantle cavity. The opening is ciliated, and is coated with a hydrofuge film of mucus which aids in breaking the meniscus and stops water running into the mantle cavity. While exchanging air the snail is quite passive, there being no active movements for ventilation, and the

rapid gaseous exchange is effected solely by diffusion. In populations of
L. peregra living in shallow water the snails tend to surface about once every
2 hours to respire air.

In populations of *L. peregra* which live in deeper water, more distant
from the shore line, all the specimens have their mantle cavity filled with
water and do not come to the surface to take in air. When such water-
breathing specimens are brought into shallower water they adopt the habit
of air breathing, presumably being stimulated in some way by the proximity of the surface of the water. When a water-breathing specimen first
surfaces in order to take in air, it behaves in a different manner from that
described above for a regular air breather. It passes up through the meniscus until about a half of the shell is exposed, and it then swings about from
side to side vigorously, during which time the water drains out of the
mantle cavity. Having filled its mantle cavity with air, it submerges again
and thereafter behaves in the same way as other air-breathing specimens.
P. fontinalis tends to occur in greater abundance in deeper water and further off-shore than does *L. peregra*. Specimens of *P. fontinalis* are similarly
air breathers when in shallow water, and water breathers when in deeper
waters. The body surface of *P. fontinalis* is augmented by finger-like processes of the mantle which extend round the exterior of the shell, and which
probably aid in respiration. The spat of both species commence life as
water breathers and remain so longer in nature than they do in the artificial conditions in an aquarium (Hunter, 1953).

In air-breathing specimens of *L. peregra* and *P. fontinalis*, the air bubble
carried in the mantle cavity will doubtless be of advantage in its capacity
as a buoyancy device.

Aquatic pulmonates differ from freshwater prosobranchs in lacking a
ctenidium. As specified by Dollo's law, the ctenidia have been irrevocably
lost. A secondary gill may develop, but this is an entirely different organ
and is a development of a lobe of the anus, which lies external to the mantle
cavity. This secondary gill is somewhat better developed in the wholly
aquatic *Planorbis* than in *Lymnaea*. In the freshwater limpets of the family
Ancylidae the mantle cavity seems to have divested itself of all its original
functions. In *Ancylus* the anal gill is relatively small, but in the related
Laevapex it is developed into a large triangular, pleated pseudobranch which
lies on the left side of the body, below the anus and external to the mantle
cavity. The excretory aperture lies laterally in the lobe of the mantle which
overlies the pseudobranch. The female genital aperture lies on the left side
immediately below the pseudobranch, while the male aperture lies anteriorly on the left side, just behind the tentacle (Basch, 1959).

Some of the pulmonata are well established on the sea shore, e.g. the
limpet genera *Siphonaria*, *Trimusculus*, and *Williamia*. *Siphonaria* closely
resembles *Patella* in its growth form, this being a striking example of convergent—or perhaps parallel—evolution. Although a pulmonate, the

mantle cavity has no lung and has acquired a secondary series of gill leaflets, the mantle cavity being filled with sea water (see Fig. 13).

The mantle cavity communicates with the exterior via a siphon which reaches to the margin of the shell on the right side. The dorsally situated mantle cavity is divided into an anterior inhalant and a posterior exhalant part by means of a ridge which crosses the floor of the mantle cavity and enters the siphon. The rectum lies between the inhalant and exhalant channels of the siphon, and the anus discharges the faeces outside the shell.

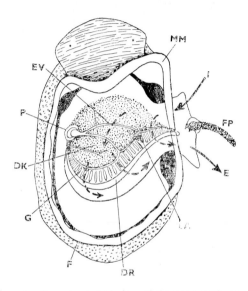

Fig. 13. *Siphonaria alternata*, dorsal view of the animal after removal of the shell. The mantle cavity lies dorsal to the visceral mass and the secondary gill (G) consists of about thirty triangular leaflets which hang downwards in the mantle cavity. The inhalant current (I) enters the mantle cavity antero-laterally and passes backwards between the gill leaflets. A dorsal ridge on the roof of the mantle cavity and a corresponding ventral ridge on the floor of the mantle cavity isolate a posterior part of the mantle cavity from which an exhalant water current (E) passes postero-laterally. An afferent branchial vein lies on the posterior side of the gill and blood flows forwards through the leaflets and via the efferent branchial veins (EV) to the heart in the pericardium (P). Faecal pellets (FP) are discharged external to the mantle cavity. The dorsal part of the excretory organ (DK) discharges into the mantle cavity via a renal pore (EA). For interpretation of other lettering, see p. 38. (Originally published in Yonge, 1952, *Proc. malac. Soc. Lond.* **29**, 193, fig. 2.)

In a large specimen of *S. alternata* there may be as many as thirty leaflets in the secondary gill and these hang down from the roof of the mantle cavity. As in the prosobranchs, the respiratory water current passes etbween the gill leaflets in the opposite direction to the flow of blood, which provides the maximum efficiency for the respiratory process (Yonge, 1952).

I have watched specimens of *S. atra* and *S. exigua* in Singapore waters on many occasions with the aid of a diving mask, and have always noted that these species are firmly clamped down on the rock when they are covered by the tide. In this they differ from the prosobranch limpets. When these two species have just been uncovered by the retreating tide they invariably rise up and leave their "homes" and move about, grazing on the wet rock surfaces. Before the rocks become dry and hot in the sun the limpets have all returned to their "homes" and have clamped down firmly on the rock again. While the limpet is moving and grazing the shell is held fairly high above the rock surface. The mantle cavity must have been completely filled with sea water at the beginning of the comparatively brief period of grazing. Some water may ooze out and moisten the surface of the head, the mantle, and the foot, but air must be excluded from the mantle cavity for it would be impossible to dislodge an air bubble from the dorsally sited mantle cavity, except on extremely steep rock surfaces.

In the vicinity of Singapore *S. exigua* occurs principally at mean sea level, but extends upwards to about high-water mark of neap tides. *S. atra* occurs at a lower level, at about low-water mark of neap tides (Purchon and Enoch, 1954). The pulmonate limpet *Williamia vernalis* occurs only at and below low-water mark of spring tides on the Pacific coast of North America (Yonge, 1960a).

Whereas *Siphonaria* is found on exposed, upward-facing rock surfaces, the pulmonate *Trimusculus* (= *Gadinia*) is invariably found on the under sides of overhanging ledges where it will tend to remain moist and cool. In *Trimusculus* the mantle cavity is sited dorso-laterally above the viscera on the left side. It does not contain any secondary gills. The mantle cavity communicates by a long and slender duct with the pneumostome, which lies anteriorly on the right side of the head. The excretory organ opens within the mantle cavity at the inner end of the duct to the pneumostome. The anus, however, lies adjacent to the pneumostome and is thus external to the mantle cavity.

Trimusculus is amphibious, being able to breathe both water and air. When the animal is under water an air bubble may sometimes be partly expelled from the pneumostome and then be withdrawn again. This process may be repeated about once every 2 minutes. Presumably oxygen and carbon dioxide are being exchanged between the air bubble and the surrounding water and thereby the air in the mantle cavity is refreshed. Alternatively the limpet may allow sea water to enter at the pneumostome, and then similar inhalant and exhalant movements may be observed. In both cases respiration is effected by muscular, not ciliary movements. Because it occupies shady, cool, and moist sites under rock ledges, and thanks to its amphibious capacities, *Trimusculus* is able to extend further up the rocky shore than *Siphonaria* can. However, *Siphonaria* is better adapted than *Trimusculus* to withstanding the rigours of exposure and desiccation.

The fact that *Siphonaria* and *Trimusculus* are classified as pulmonates does not mean that they have been derived from a terrestrial, air-breathing ancestry, and that they have secondarily returned to the sea shore and to a littoral mode of life. On the contrary, they comprise an early and aberrant offshoot from the main pulmonate stock, and never left the sea shore. They probably lost their ctenidia for the same reasons as did the Patellidae among the Prosobranchia. A number of these primitive inter-tidal pulmonate genera give evidence of their primitive origins by still possessing a free-swimming veliger larva (Morton, 1955, 1955a).

Cephalopoda

In the Cephalopoda the mantle cavity occupies a posterior position, comparable to that in the hypothetical remote ancestral mollusc. The two ctenidia are considered to have migrated and to have become re-attached to the wall of the visceral mass throughout their length by a greatly extended afferent membrane. The respiratory current is no longer produced by ciliary action, and the ctenidial filaments are not ciliated. Their respiratory surfaces are considerably increased by secondary folds. The tips of the filaments bear membranous extensions by means of which they are attached to the afferent membrane, and skeletal supporting rods have developed adjacent to the afferent side of the leaflet. These rods serve as ties for the tips of the filaments and so support them against the surge of the respiratory water currents, which may possibly pass to and fro in both directions between the filaments. Since these skeletal supports lie on the *afferent* side of the filaments, they cannot be homologous with the rods found in other molluscan ctenidia, e.g. in the aspidobranch gastropods. They serve a comparable purpose, and can be taken as examples of parallel evolution.

There are no hypobranchial glands in the Cephalopoda and this may be related to the powerful water currents passing through the mantle cavity and to the pelagic mode of life which has been adopted. Osphradia are present in *Nautilus*, but are absent in all the Dibranchia. In *Nautilus* the respiratory current is generated by muscular pulsations of the funnel. This might be inadequate for so bulky an animal, but for the presence of a second pair of ctenidia. Nothing is known of the number of ctenidia in the extinct Nautiloidea and Ammonoidea, but the orthodox view at present is that the tetrabranchiate condition in *Nautilus* is secondary, not primitive. This view fits in well with the concept of the Mollusca as being unsegmented. With the discovery of the metamerically segmented class Monoplacophora, perhaps some further thought should be directed to the possibility that *Nautilus* is metamerically segmented, with at least two body segments which bear ctenidia. I would not wish to press the point, but it is always sound scientific policy to re-examine basic assumptions from time to time, especially in the light of any new evidence available.

The mantle cavity in the Cephalopoda retains the paired excretory apertures and a single genital opening on the left side. Nidamental and accessory nidamental glands are present in the female, and the mantle cavity is further concerned in the reproductive process in that the hectocotylus of the male is inserted into the mantle cavity of the female during copulation, e.g. in *Octopus* (Young, 1962) (see Fig. 183, p. 536).

With the exception of the Pearly Nautilus, which has a heavily built external shell, the modern Cephalopoda either have a greatly reduced internal shell (Decapoda) or no shell at all (Octopoda). Correlated with the loss of an external protective shell, the wall of the mantle cavity had to be greatly strengthened in order to provide this part of the body with a firm and protective body wall. Although the octopus skulks in crevices in the rocks and lies in wait for passing prey, it is capable of the most vigorous action when the occasion demands, and the Dibranchiata as a whole can generally be regarded as actively moving, predacious carnivores with a comparatively high metabolic rate. The respiratory demands of such active animals must be high and probably could not be satisfied by inhalant currents generated by ciliary action. The development of a muscular wall to the mantle cavity was not only a necessary protective device, but it also provided an entirely new and more efficient method of breathing by rhythmic muscular contractions and relaxations of the mantle.

When the circular muscles of the mantle relax, the radial muscle fibres presumably contract and so the volume of the mantle cavity is increased, and water enters at all points along the margin of the mantle cavity. When the circular muscles of the mantle contract, the mantle cavity is compressed. The mantle wall near its margin becomes closely applied to the base of the muscular funnel, preventing leakage of water around the margin, and the exhalant water stream is concentrated into a powerful jet which leaves the funnel. The original purpose of this may have been to ensure that the exhaled water did not contaminate the water about to be inhaled at the mouth of the mantle cavity (see Fig. 14).

This muscular respiratory process is immensely powerful, and the jet of water emitted from the funnel has acquired importance in locomotion as a contributor of propulsive force. This is most striking in the escape reaction when the funnel is directed forwards and the emission of a series of rapid and powerful jets causes the animal to shoot backwards with great velocity. At other times when the animal is swimming forwards the funnel may be directed downwards, and the more gentle jets of water

emitted from the funnel will provide the body with a certain amount of
lift. With its new-found muscular process of ventilation, the mantle
cavity is now also concerned in locomotion, and is deeply involved in
such important matters as the capture of prey and escape from predators.
It is interesting to pause here to consider the means by which this muscular
system has acquired its great efficiency.

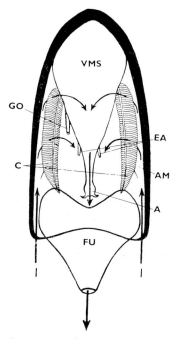

FIG. 14. Diagrammatic representation of the mantle cavity of the cuttlefish,
Sepia officinalis, showing the disposition of the organs in the mantle cavity
and the course of water currents effecting both respiration and locomotion.
These water movements are caused by contractions in the thick and highly
muscular mantle wall. For interpretation of lettering, see p. 38. (Originally published in Yonge, 1947, *Phil. Trans.* B, **232**, 502, fig. 35.)

The occurrence of two sets of antagonistic muscle fibres pervading the
same mass of tissue seems to be a characteristic of the Cephalopoda.
The circular muscle fibres are very long, and are striated; they are joined
end to end, and the whole mass is attached "dorsally" to the internal
shell. These fibres need to be stimulated individually before they can
contract, and the efficiency of the mechanism depends on the extent to
which the action of the muscle fibres can be co-ordinated. A wave of
contraction in the mantle would be less effective than a single co-ordinated
contraction in creating an exhalant jet of water. It is equally obvious
that the muscles of the left and right sides must act in unison.

The gentle pallial contractions required for respiration and for slow "cruising" motion are effected as the result of nerve impulses reaching the pallial muscle fibres from innumerable nerve fibres of normal dimensions in the pallial nerves. The degree of muscular contraction depends on the number of muscle fibres activated—an increase in the number of neurones participating will result in a more powerful jet of water being expelled from the funnel.

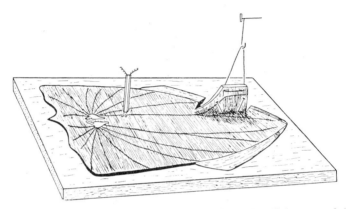

FIG. 15. Preparation of the mantle, stellar ganglia, and pallial nerves of the squid for recording contractions of the circular pallial muscle fibres after stimulation of the posterior stellar nerve. (Originally published in Young, 1938, *J. exp. Biol.* **15**, 174, fig. 3.)

The violent escape reaction of decapods is effected by an entirely different mechanism and is an all-or-nothing response to some visual, tactile, or physical disturbance. In decapods a system of giant nerve fibres passes from the central nervous system to the retractor muscles of the head and of the funnel, and also to the pallial muscle fibres. The giant fibres in the pallial nerves branch repeatedly and individually innervate large numbers of muscle fibres. Nerve impulses travel more rapidly along the giant fibres than along ordinary nerve fibres, and so reach their destination earlier; the rate of conduction is proportional to the square root of the diameter of the axon. The giant fibres in the pallial nerves are graded in size in the squid, those in the longest nerves being of greatest diameter (Pumphrey and Young, 1938). At each synapse in the giant fibre system there is extensive and intimate association between the two neurones, so that the nerve impulse cannot fail to pass the synapse.

Each impulse travelling in the giant fibre system leads to a single act of expulsion of water through the funnel in an all-or-nothing manner (see Figs. 15 and 16). Complete relaxation of the pallial muscles follows, and this allows the mantle cavity to expand and fill up with water again in anticipation of the next emission. Due to the small number of synapses and the high rate of conduction of impulses in the giant fibres, a large squid can get under way within 55 msec of receiving a visual shock. This is about twice as fast as would be possible in the absence of giant fibres (Young, 1938, 1939; Prosser and Young, 1937).

This hierarchical system of giant fibres is particularly important with respect to the escape reaction in which locomotion by jet propulsion is performed at its maximum capacity. For the more gentle movements

FIG. 16. Records on a smoked drum of the contractions of the circular pallial muscle fibres of the squid, produced experimentally by excitation of the nerve fibres in the posterior stellar nerve. The contractions of the fibres are shown by the deviations of the trace, and the calibration of the lever in grams is shown on the right of the figure. The voltage used in each individual stimulation is shown in the horizontal row of figures. The lowermost trace shows the time intervals (approximately 3 sec). The giant fibre was pricked and the left-hand half of the trace records the results of stimulation central to the prick, and concerns only the undamaged small fibres; here muscle contractions are graded in proportion to the voltage, up to a certain maximum, this being due to increase in the numbers of nerve fibres—and of muscle fibres—involved in the contraction. At "A" the electrodes were placed distal to the prick in the giant fibre and on the right of this point stimulation of the giant fibre produces only maximal contractions. (Originally published in Young, 1938, *J. exp. Biol.* **15**, 177, fig. 7.)

associated with ventilation of the mantle cavity for respiration, and with forward movement at "cruising speed", the giant fibre system is probably inactive, and the pallial muscle fibres are probably only stimulated by the ordinary nerve fibres.

In the cuttle-fish, *Sepia*, the body is much broader than is that of *Loligo*, and accordingly there is less variation in the distances along the stellar nerves to the various regions of pallial muscles. There are more than one giant fibre per stellar nerve in *Sepia*, and there is no clear relation between the diameters and the lengths of their axons. It must be concluded that the neuro-muscular organisation is more highly advanced in the pelagic, fast-swimming squid than it is in the demersal, slow-swimming cuttle-fish. There can be little doubt that the development of the giant fibre system has played an important part in the evolution of the decapod cephalopods. In contrast, the pallial musculature of the octopods is not served by giant fibres.

It is, finally, worthy of note that the mantle cavity of many cephalopods is concerned with defensive mechanisms since the duct of the ink sac opens into the mantle cavity. In certain cases at least, the ink is not emitted in the form of a cloud, or smoke screen, as is commonly believed. On the contrary, the darkly coloured animal may emit a mass of dark-stained water of about the same size as itself. At the same time the animal becomes pale and inconspicuous and darts away backwards. The would-be aggressor is deceived by this "sleight-of-hand" trick, having concentrated its attention on the dark blob of water which appeared in between it and the escaping cephalopod. The mantle cavity, in mixing the ink with the water contained therein, and in discharging the mixture in a compact mass, has played a major part in this defensive procedure (Hall, 1956). It is said that in the unlit waters of the abyss the same sleight-of-hand trick is played with a luminous secretion instead of ink.

Key to the Lettering on the Figures

A	Anus.	IMS	Intramuscular slit.
AFC	Abfrontal cilia.	LC	Lateral cilia.
AM	Afferent membrane supporting the ctenidium.	LCT	Left ctenidium.
		LO	Left osphradium.
AU	Auricle.	M	Mouth.
AV	Afferent branchial vein.	MC	Mantle cavity.
C	Ctenidium.	ME	Mantle edge.
CE	Cephalic eye.	MG	Mucous glands.
CP	Ciliated pit.	MM	Margin of mantle.
CR	Chitinous rod supporting the ctenidial leaflets.	NC	Nuchal cavity.
		O	Osphradium.
DK	Dorsal portion of the excretory organ.	OPC	Opening into pulmonary cavity.
DR	Dorsal ridge.	P	Pericardium.
E	Exhalant water current.	PS	Pallial siphon.
EA	Excretory aperture.	PT	Pallial tentacles.
EC	Exhalant part of mantle cavity.	PV	Pallial vein.
		R	Rectum.
EO	Excretory organ.	RC	Rejection current.
EV	Efferent branchial vein.	RI	Ridge on floor of mantle cavity, separating ctenidial from pulmonary portion.
EX	External opening of sense organ.		
F	Foot.	S	Shell.
FC	Frontal cilia.	SI	Siphon.
FP	Faecal pellet.	SM	Suspensory membrane of ctenidium.
FU	Funnel.		
G	Secondary gill leaflets.	SP	Siphonal space.
GA	Ganglion below sense organ.	SS	Sensory strips.
GO	Genital orifice.	TC	Terminal cilia.
H	Head.	V	Ventricle.
HG	Hypobranchial gland.	VMS	Visceral mass.
I	Inhalant water current.	Z	Region where waste material accumulates.
IC	Inhalant part of mantle cavity.		

Reference List

BASCH, P. F. (1959) The anatomy of *Laevapex fuscus*, a freshwater limpet (Gastropoda: Pulmonata), *Misc. Publ. Mus. Zool. Univ. Mich.* **108**, 1–56.

BROWN, A. C. (1961) Chemo-reception in the sandy-beach snail *Bullia*, *S. Afr. J. Lab. clin. Med.* **7**, 160.

BROWN, A. C. and NOBLE, R. G. (1960) Function of the osphradium in *Bullia* (Gastropoda), *Nature, Lond.* **188**, 1045.

CHAPMAN, G. and NEWELL, G. E. (1956) The role of the body fluid in the movement of soft-bodied invertebrates. II. The extension of the siphons of *Mya arenaria* L. and *Scrobicularia plana* (da Costa), *Proc. roy. Soc.* B, **145**, 564–580.

CLARKE, A. H. and MENZIES, R. J. (1959) *Neopilina (Vema) ewingi*, a second living species of the palaeozoic class Monoplacophora, *Science* **129**, 1026–7.

CROFTS, D. R. (1937) The development of *Haliotis tuberculata*, with special reference to organogenesis during torsion, *Phil. Trans.* B, **228**, 219–68.

FRETTER, V. (1943) Studies in the functional morphology and embryology of *Onchidella celtica* (Forbes and Hanley) and their bearing on its relationships, *J. mar. biol. Ass. U.K.* **25**, 685–720.

GARSTANG, W. (1928) The origin and evolution of larval forms, *Rep. Brit. Ass. Glasgow*, 1928, Sect. D, 77–98.

GILMOUR, T. H. J. (1963) *Studies on the Limidae*, thesis, University of Glasgow.

GRAHAM, A. (1948) *Form and Function in the Littoral Gastropod*, inaugural lecture, 20.4. 1948, Birkbeck College, London.

GRAHAM, A. (1951) *Dealing with the Raw Material (Animals)*, Linn. Soc. Lond. Taxonomy Pamphlet 1949–50, 16–24.

HALL, D. N. F. (1956) Ink ejection by cephalopoda, *Nature, Lond.* **177**, 663.

HULBERT, G. C. E. B. and YONGE, C. M. (1937) A possible function of the osphradium in gastropoda, *Nature, Lond.* **139**, 840.

HUNTER, W. R. (1949) The structure and behaviour of *Hiatella gallicana* (Lamarck) and *H. arctica* (L.), with special reference to the boring habit, *Proc. roy. Soc. Edinb.* B, **63**, 271–89.

HUNTER, W. R. (1952) A note on the amphibious snail *Succinea pfeifferi* Rossmässler, in a previously undescribed habitat, *Glasg. Nat.* **17**, 91.

HUNTER, W. R. (1953) The condition of the mantle cavity in two pulmonate snails living in Loch Lomond. *Proc. roy. Soc. Edinb.* B, **65**, 143–165.

LEMCHE, H. and WINGSTRAND, K. G. (1959) The anatomy of *Neopilina galatheae* Lemche, 1957 (Mollusca, Tryblidacea), *Galathea Rep.* **3**, 1–63.

MORTON, J. E. (1955) The functional morphology of the British Ellobiidae (Gastropoda, Pulmonata) with special reference to the digestive and reproductive systems, *Phil. Trans.* B, **239**, 89–160.

MORTON, J. E. (1955a) The evolution of the Ellobiidae, with a discussion on the origin of pulmonates, *Proc. zool. Soc. Lond.* **125**, 127–68.

ORTON, J. H. (1912) The mode of feeding of *Crepidula*, with an account of the current-producing mechanism in the mantle cavity, and some remarks on the mode of feeding in gastropods and lamellibranchs, *J. mar. biol. Ass. U.K.* **9**, 444–78.

PROSSER, C. L. and YOUNG, J. Z. (1937) Responses of muscles of the squid to repetitive stimulation of the giant nerve fibres, *Biol. Bull. Woods Hole*, **73**, 237–41.

PUMPHREY, R. J. and YOUNG, J. Z. (1938) The rates of conduction of nerve fibres of various diameters in cephalopods, *J. exp. Biol.* **15**, 453–66.

PURCHON, R. D. (1955) The structure and function of the British Pholadidae (rock-boring Lamellibranchia), *Proc. zool. Soc. Lond.* **124**, 859–911.

PURCHON, R. D. (1960) A further note on the biology of *Brechites penis* (L.) (Lamellibranchia), *Proc. malac. Soc. Lond.* **34**, 19–23.

PURCHON, R. D. and ENOCH, I. (1954) Zonation of the marine fauna and flora on a rocky shore near Singapore, *Bull. Raffles Mus.* **25**, 47–65.

REES, W. J. (1964) A review of breathing devices in land operculate snails, *Proc. malac. Soc. Lond.* **36**, 55–67.

SANDISON, E. E. (1966) The oxygen consumption of some intertidal gastropods in relation to zonation, *J. Zool.* **149**, 163–73.

SMITH, F. G. W. (1935) The development of *Patella vulgata*, *Phil. Trans.* B, **225**, 95–125.

TURNER, R. D. (1954) The family Pholadidae in the Western Atlantic and the Eastern Pacific. Part I. Pholadinae, *Johnsonia* **3**, 1–64.

YONGE, C. M. (1936) The evolution of the swimming habit in the Lamellibranchia, *Mém. Mus. r. Hist. nat. Belg.*, ser. 2, **3**, 78–100.

YONGE, C. M. (1939) On the mantle cavity and its contained organs in the Loricata (Placophora), *Quart. J. micr. Sci.* **81**, 367–90.

YONGE, C. M. (1947) The pallial organs in the aspidobranch gastropoda and their evolution throughout the mollusca, *Phil. Trans.* B, **232**, 443–518.

YONGE, C. M. (1949) On the structure and adaptations of the Tellinacea, deposit-feeding Eulamellibranchia, *Phil. Trans.* B, **234**, 29–76.

YONGE, C. M. (1952) The mantle cavity in *Siphonaria alternata* Say, *Proc. malac. Soc. Lond.* **29**, 190–9.

YONGE, C. M. (1953) Form and habit in *Pinna carnea*, *Phil. Trans.* B, **237**, 335–74.

YONGE, C. M. (1953a) Observations on *Hipponix antiquatus* (Linnaeus). *Proc. Calif. Acad. Sci.* **28**, 1–24.

YONGE, C. M. (1956) Water circulation in the mantle cavity of the owl limpet, *Lottia gigantea* Gray, *Nautilus* **69**, 79–87.

YONGE, C. M. (1957) Mantle fusion in the Lamellibranchia, *Publ. Staz. zool. Napoli* **29**, 151–71.

YONGE, C. M. (1958) Observations in life on the pulmonate limpet *Trimusculus (Gadinia) reticulatus*, *Proc. malac. Soc. Lond.* **33**, 31–37.

YONGE, C. M. (1960) Mantle cavity, habits, and habitat in the blind limpet, *Lepeta concentrica* Middendorff, *Proc. Calif. Acad. Sci.* **31**, 103–10.

YONGE, C. M. (1960a) Further observations on *Hipponyx antiquatus* with notes on north pacific pulmonate limpets, *Proc. Calif. Acad. Sci.* **31**, 111–19.

YOUNG, J. Z. (1938) The functioning of the giant nerve fibres of the squid, *J. exp. Biol.* **15**, 170–85.

YOUNG, J. Z. (1939) Fused neurons and synaptic contact in the giant nerve fibres of cephalopods. *Phil. Trans.* B, **229**, 465–503.

YOUNG, J. Z. (1962) Courtship and mating by a coral reef octopus *(O. horridus)*, *Proc. zool. Soc. Lond.* **138**, 157–162.

CHAPTER 2

FEEDING METHODS AND ADAPTIVE RADIATION IN THE GASTROPODA

Synopsis

The adaptive radiation of the Gastropoda is principally based on adaptations of the buccal mass and radula to meet the problems faced in feeding in various ways on a wide variety of food substances. The main features of interest in this radiation are the retention of the buccal mass and radula and of the habit of ingesting only small particles—with only few exceptions—the plasticity in form and function of the radula and associated musculature, and participation in the feeding mechanism by various other organ systems, e.g. the ctenidium and the pedal mucous gland. Diversification of the feeding mechanism has assisted gastropods to colonise almost all conceivable habitats in the sea, in fresh water, and on dry land, and to exploit almost every possible mode of life—as herbivores, as deposit feeders, as plankton feeders, as scavengers, as parasites, and as predacious carnivores. The Prosobranchia, the Opistho-branchia, and the Pulmonata are considered separately.

Among the Prosobranchia the Archaeogastropoda tend to exhibit rather primitive feeding methods such as browsing and grazing on algae, e.g. *Haliotis* and *Diodora*; rasping algae off rock surfaces, e.g. *Patella*. Some archaeogastropods have colonised fresh waters, e.g. *Theodoxus* and *Nerita*, while some are terrestrial, e.g. *Hydrocena*. The Mesogastropoda show extraordinary diversification in feeding methods including rasping algae off rock surfaces, e.g. *Littorina*; cutters of algae, e.g. *Strombus*; collectors of organic deposits, e.g. *Hipponyx* and *Hydrobia*: it has been shown that *Hydrobia* does not feed directly on the organic debris but it digests various micro-organisms which gradually reduce the organic matter in the sediment. Other feeding mechanisms exhibited in the Mesogastropoda include collecting plankton by interaction of the cteni-dium and the radula, e.g. *Crepidula*—in some respects this represents convergence on the ctenidial ciliary feeding mechanisms of filibranch bivalves; in members of the Vermetidae the feeding process is aided by mucus secreted by the pedal gland and in *Vermetus gigas* planktonic food is trapped in mucous threads which are extended far into the surrounding calm sea water. In contrast to the above, some mesogastropods feed on colonial animal growths such as sponges, e.g. *Cerithiopsis*, or on com-pound ascidians, e.g. *Trivia*.

Some mesogastropods are benthic hunters, e.g. *Natica*, which drills a hole through the shell of its bivalve prey and then excavates the flesh within. Some mesogastropods have become planktonic and are carnivorous, e.g. *Ianthina* which feeds on *Velella*, and the Heteropoda are other planktonic hunters. Some mesogastropods are ectoparasitic on anemones or on echinoderms, sucking up the body fluids of the host, while others are endo-parasitic and may have lost many of the features characteristic of gastropods. Some mesogastropods have invaded fresh waters, where they may feed on suspended particles, e.g. *Viviparus*, or on water weeds, e.g. *Pomacea*. Finally, many mesogastropods, all of which are herbivorous, have colonised dry land in competition with stylommatophoran pulmonates.

The Neogastropoda mostly exhibit rather advanced feeding mechanisms such as scavenging, e.g. *Buccinum* and *Bullia*; and as benthic hunters, e.g. *Purpura* (= *Nucella*) and *Urosalpinx*—these drill holes through the shell of their molluscan prey partly mechanically by the radula and partly chemically by secretions from a special glandular organ. The predacious mode of life reaches a climax in the Toxoglossa where a single radular tooth is used as a harpoon or as an arrow by species of *Conus*. Species of *Conus* are stenophagous, and feed either on polychaets, on enteropneusts, on other gastropods or on fish. The molluscivorous and the piscivorous species of *Conus* kill their prey with a very powerful poison which in extreme cases is lethal to man. In contrast to the above, a few neogastropods are herbivorous, e.g. *Columbella*; one genus, *Antenome*, has colonised fresh water where it preys upon fish and on other snails.

The details of many of the above feeding mechanisms are fully described, and the adaptive radiation of the Prosobranchia is expressed diagrammatically on the basis of feeding habits.

Turning to the sub-class Opisthobranchia, we find that the adaptive radiation in feeding habits is rather less diversified than is that of the Prosobranchia; this is due to the failure of the Opisthobranchia to colonise fresh waters and dry land. Among the Opisthobranchia in fact, only one group, the herbivorous Succineidae, have established themselves in fresh water. Some of the more primitive of the opisthobranch feeding methods include browsing and grazing on algae, e.g. *Akera* and *Onchidella*; feeding on organic deposits, e.g. *Actaeon*; and cutting of algae, e.g. *Aplysia*. More advanced feeding methods include the collecting of plankton, e.g. the pteropod *Limacina*; sucking the contents of plant cells, e.g. *Elysia*; feeding on colonial animal growths, e.g. the dorid *Jorunna* and the eolid *Doto*; feeding on sedentary animals, e.g. *Pleurobranchus* which feeds on ascidians, and *Aeolidia* which feeds on sea anemones. An extreme example is *Calma*, which feeds on fish eggs.

Some opisthobranchs are benthic hunters, e.g. *Philine* and *Scaphander*, which may swallow small molluscs whole and crush them in the gizzard—this is a notable exception to the general molluscan principle of feeding on comparatively small particles. Some of the Opisthobranchia have become planktonic hunters, e.g. the nudibranch *Fiona* which feeds on the siphonophore *Velella*, and the gymnosomatous pteropods such as *Clione* which probably feeds on *Limacina*. The eolid *Melibe* has evolved a remarkable cephalic veil which it uses as a casting net to trap the copepods on which it feeds. Finally, the family Pyramidellidae, recently transferred from the Prosobranchia to the Opisthobranchia, live as ecto-parasites on various polychaets and bivalves.

Certain of the above-mentioned feeding habits have been described in detail, and the adaptive radiation of the Opisthobranchia is expressed diagrammatically on the basis of feeding habits. It is suggested that the Opisthobranchia are polyphyletic, the term "opisthobranch" representing a structural and functional stratum which has been attained independently by a number of phylogenies originating in the Prosobranchia. The schemes of classification of the Opisthobranchia by various authorities are compared in tabular form.

The sub-class Pulmonata comprise the terrestrial Stylommatophora, the freshwater Basommatophora, and also various primitive phylogenies of problematical status such as *Siphonaria* and *Trimusculus*, which feed by rasping fine algal growths off rocks in much the same way as do various primitive prosobranchs. Both freshwater and terrestrial pulmonates are generally herbivorous, some specialising on particular types of plant food, but some are facultative carnivores, while a few are obligatory carnivores, e.g. the Streptaxacea (*Gonaxis*) and Oleacinacea (*Testacella*). In terms of diversity of feeding habit the adaptive radiation of the Pulmonata may appear rather restricted, but the Pulmonata have colonised all conceivable habitats in fresh water and on dry land, even including arid areas and the islands of Oceania.

The radula has been found in all classes of Mollusca except the Bivalvia, and must have been the original molluscan feeding organ. From the structure and mode of action of the radulae of a great variety of modern forms we can conclude that the food of the remote common ancestor of modern molluscs must have been comparatively small particles which were collected in some way by the radula.

Modern gastropods exhibit an extraordinary diversity in both their feeding methods and in the nature of the food material (a summary of the food taken by a wide variety of gastropods is given by Graham, 1955). In certain cases the feeding methods are obviously highly specialised and the source of food is entirely surprising, while in other cases the feeding methods and the sources of food seem to be primitive and may scarcely have changed since the earliest origins of the class. Gastropods with such primitive feeding mechanisms are sometimes demonstrably primitive in other respects; it is therefore possible to select certain modern gastropods and to describe their organs of feeding, and their feeding methods, as probably being characteristic of the primitive condition in the ancestral gastropod stock. It is logical to start on these lines and then to proceed to study more advanced and more specialised feeding methods.

Arranging a sequence of this sort is very helpful in enabling us to envisage how novel feeding mechanisms *could have come into existence* and how new sources of food may have become exploitable. We must remember, however, that the actual course of events in the evolution of a particular feeding mechanism may have been quite otherwise, and our logical sequence cannot claim to be an evolutionary series.

There have been two phases in the evolution of the class Gastropoda: firstly, the mega-evolutionary phase which led to the emergence of the biological model which we define as "gastropod"; secondly, the elaboration of this model in a variety of ways, with the production of innumerable variants each of which is able to function efficiently under restricted and specialised circumstances. This second phase of "adaptive radiation" of the gastropods is heavily based on modifications, developments, and transformations of the original feeding mechanisms, so enabling representatives of the class to invade a wide variety of habitats and to exploit a surprising diversity of sources of food.

It is important that in our survey we should not become lost in a vast array of petty detail; these details are intended to illustrate a few major points of principle, e.g. that the adaptive radiation of the gastropods was mainly brought about by diversification of feeding methods in which process the following considerations are important: firstly, the retention of an extremely primitive feeding organ, the radula; secondly, the tendency to retain as a general rule—but not invariably—the ancient function of ingesting food in the form of comparatively small particles; thirdly, the great plasticity in form and function that has been displayed by the radula and its associated musculature; fourthly, the role played by pre-adaptation of other organ systems and their ultimate incorporation into the feeding mechanism. The effectiveness of these factors in the adaptive radiation of the Gastropoda, coupled with significant collateral developments of the digestive processes, is witnessed in the ubiquity of the class and the wide range of their modes of life. With the sole exception of aerial locomotion, the Gastropoda have *successfully* undertaken every possible mode of life; within the class may be found herbivores, deposit feeders, plankton feeders, scavengers, predacious carnivores, and parasites. It is difficult to think of any habitat in the sea, in fresh water, or on dry land which does not harbour a gastropod. This extraordinary and unparalleled success is largely attributable to the four features listed above, as we hope to illustrate sufficiently in the following review of feeding mechanisms throughout the class.

The Gastropoda comprise three great divisions—the Prosobranchia, the Opisthobranchia, and the Pulmonata. Each of these has given rise to a broad adaptive radiation based largely on feeding methods. We shall consider each of these three adaptive radiations in turn.

Prosobranchia

This sub-class is composed of three orders, of which the Archaeogastropoda largely retain primitive, ancestral features, the Mesogastropoda display a remarkable diversity of modes of life, while the Neogastropoda are mostly predacious carnivores and scavengers and may be regarded as the most advanced order.

Order Archaeogastropoda

Browsers and grazers on algae

The most primitive of these can be recognised by the presence of two ctenidia in the anterior mantle cavity, e.g. *Haliotis, Diodora, Fissurella, Emarginula*, and *Pleurotomaria*. In all such cases there is an extremely large number of comparatively slender teeth in each transverse tooth row on the radula. This is also true for several genera in which there is only one ctenidium and which are therefore less primitive, e.g. *Trochus, Turbo, Monodonta, Gibbula, Nerita*, and *Theodoxus* (Whitaker, 1951). Such a radula is described as "rhipidoglossan". With the exception of some species of *Theodoxus* which are fluviatile, all these genera live on rocky shores either on the upward facing rock surfaces or perhaps on the under surfaces of boulders; they feed on the filamentous algae and other fine algal growths on the rock surfaces, browsing on these growths by protruding the radula through the mouth and by withdrawing it again with a licking action. When protruded, the radula passes over a "bending plane", at which point the sub-radular membrane is flattened and the marginal teeth swing outwards. The radula is curled backwards under the proboscis and its toothed surface is pressed against the algal food source. As the radula is drawn backwards into the mouth the rows of teeth are erected one by one and scrape against the substratum, removing fine particles of vegetable material. The marginal teeth brush scattered fragments of potential food material towards the median line as they swing inwards prior to returning over the bending plane into the mouth. This description is true also for taenioglossan and stenoglossan radular types, but it is not applicable to docoglossan limpets, to dorids, or to pulmonates. On returning through the mouth the radula becomes compressed laterally and rolled transversely to form a shallow gutter in which the lateral teeth point inwards and grip the small food particles. In this inwardly directed position the radular teeth will not scrape the lining of the oesophagus, which is further protected by lubricating secretions from the salivary glands.

Fine particles of vegetable matter, so collected and admixed with secretions from the salivary glands, will then pass along the oesophagus and be subjected to the processes of digestion. In other words, the feeding mechanism is basically the same as that employed by the Polyplacophora, and this may be regarded as the most primitive feeding mechanism in the Gastropoda.

Raspers of algae off rock surfaces

Limpets in the family Patellacea are adapted for life on exposed rocky shores and to rasping on the surface of the rocks themselves, feeding on fine algal growths. The radula is robustly built for this heavy work, with only twelve or thirteen teeth in each transverse row, each tooth being large and heavily built. Examples are *Patella* (Graham, 1931, 1964) *Cellana* and *Acmaea*, and also the more delicate *Patina* (Graham and Fretter, 1947) which lives under more sheltered circumstances on the fronds and holdfasts of laminarian weeds, and which feeds by rasping the comparatively soft tissues of the seaweed. This type of radula is known as "docoglossan". In a docoglossan such as *Patella vulgata* the buccal mass is much more robust than are those of other types of prosobranch, and much more work is done by the radula in rasping food from the surface of the rocks. The buccal mass is strengthened by five pairs of cartilages, and is manipulated by muscles inserted into the body wall of the head, the latter being made rigid by blood pressure in the cephalic haemocoele. The buccal mass is pushed forwards out of the mouth and the radula is protruded and laid against the substratum. Where the sub-radular membrane passes over the bending plane the membrane is flattened, but the teeth are not erected and the marginal teeth are not swung out laterally. The buccal mass may continue to move forwards, thereby causing the radula to be dragged forwards, several rows of powerful teeth engaging with the substratum simultaneously at the "working plane" which is below and behind the bending plane. The radula is finally drawn back over the bending plane and returns into its resting position, particles of plant material being carried in at the same time. The radular teeth press against the crescentic jaw and it is possible that this may serve partly as a cutting device (Graham, 1964).

The genus *Neritina* and some species of *Theodoxus* have invaded fresh waters, while the families Helicinidae and Hydrocenidae are terrestrial.

Order Mesogastropoda

In this very large order the number of teeth in each transverse row on the radula has been reduced still further, usually to seven. There is normally one central tooth and three pairs of lateral teeth in each transverse tooth row. The lateral teeth are hook-shaped and curve inwards towards the central tooth. This radular form, which is described as "taenioglossan", is evidently very efficient and highly adaptable, for it occurs in over fifty families which exhibit many contrasted methods of feeding.

Browsers and grazers on algae

Omalogyra scrapes the surface cells of the green weed, *Ulva*, and sucks up the cell contents (Fretter, 1948).

Raspers of algae off rock surfaces

In some genera, e.g. *Littorina* and *Nodilittorina* (= *Tectarius*), fine algal particles are rasped directly from rock surfaces. Thus *L. neritoides* in European waters, and *N. malaccensis* in Malayan waters, feed on more or less bare rock surfaces at and above high-water mark on rocky shores. In contrast, *L. littoralis* feeds on the soft fronds of *Fucus* and of *Ascophyllum* on European shores, and *L. melanostoma* feeds on the leaves of mangrove plants in Malaya, at about mid-tide level. Bakker (1959) subjected *L. littoralis* (= *L. obtusata*) to carefully controlled food preference tests under laboratory conditions, first testing only two seaweeds at a time and finally testing four seaweeds simultaneously. It was found that the most favoured weed was *F. vesiculosus*, this being followed by *F. spiralis* (which occurs slightly higher up the shore) and then by *A. nodosum* (which occurs a little lower down the shore), and *F. serratus* was the least acceptable of the four species of weed. Although *L. littoralis* does not occur on *Pelvetia canaliculata* on the shore, laboratory tests showed that it was able to feed on this weed, which was chosen by some individuals in preference to *F. spiralis*. It seems probable that some factor other than food preference prevents *L. littoralis* from extending upwards into the *Pelvetia* zone.

Some prosobranchs are so minute that they have to use their jaws to grip the substratum in order to obtain sufficient purchase to rasp off algal cells or diatoms, e.g. *Skeneopsis* and *Rissoella* (Fretter, 1948).

Cutters of algae

Pterocera (= *Lambis*) lives on the reef flat of coral reefs, on areas of sand intermixed with stones, and boulders covered with a variety of seaweeds. In this genus there are two laterally placed jaws and a small and delicate radula. When feeding, the proboscis is everted, thus exposing the radula which is then used to nibble off the finest epiphytic algae (Yonge, 1932). *Strombus* feeds in a similar fashion (Robertson, 1961).

Collectors of organic deposits

In *Hipponyx* the radula has assumed an entirely different function, for which the curved lateral teeth are admirably suited. *H. antiquatus* is a limpet-shaped creature which has secreted a "second shell valve" ventral to the foot, and lives permanently fixed on the floor of crevices or in depressions in the surface of rocks where sediment is liable to accumulate. The long proboscis is thrust out and the radular teeth grasp any particulate matter that has drifted into the crevice, more or less regardless of the size of the particles. *Hipponyx* feeds to a great extent on fragments of calcareous algae. This material is passed back to the stomach when the radula is withdrawn. *H. australis* feeds in the same way, but this species is attached to the shells of species of *Turbo*, usually close to the exhalant aperture, and it feeds on the faecal pellets of the *Turbo* (Yonge, 1953, 1960). Thus in one species of *Hipponyx* the radula has been turned to an entirely novel technique of feeding on organic matter in sediment, while in another species of the same genus, due to the adoption of a special epizoic posture, deposit feeding has been further specialised to become coprophagy.

Various accounts have been given of the diet of *Hydrobia ulvae*. A stimulating investigation by Newell (1965) indicates that in all probability *H. ulvae* feeds principally on micro-organisms which may multiply on various substrates which are themselves largely indigestible by the snail. Micro-organisms thrive in "detritus", utilising some of the available carbon compounds as a source of energy, and obtaining nitrogen either from the sea water or from the atmosphere. When detritus is passed through the gut of the snail the micro-organisms are digested, but the finely divided organic matter which comprises the bulk of the detritus is indigestible. This investigation throws a new light on "detritus-feeding" wherever it occurs.

H. ulvae occurs intertidally on sheltered estuarine beaches, particularly favouring that part of the beach which is the upper limit of the zone of muddy sand, where populations of over 10,000 per square metre may be encountered. When the tide is out *H. ulvae* tends to burrow into the muddy sand, taking up a vertical position just below the surface, with the tips of the tentacles exposed, and there it feeds on the organic matter in the surface layers of the mud by means of its protrusible proboscis. Of all the material ingested, only the micro-organisms will be digested. When the tide rises the snails emerge from the substratum and launch themselves upside down on the surface of the water, hanging there by the broad, creeping sole of the foot, aided by a small sheet of mucus which is up to 0·5 cm in diameter. The snails float in this manner for some hours until they are deposited once more on the beach by the receding tide. It has been found that while floating the snails consume portions of the mucus float and, with it, the small particles of water-borne matter which become stuck to it. Once again, the beneficial food material will consist solely of micro-organisms, and "detritus" will pass through the gut undigested. In this way *H. ulvae* exhibits a remarkable diurnal rhythmic change of feeding habits, feeding on organic deposits while lying buried on the foreshore, and feeding on suspended particles while it is hanging from the meniscus (Newell, 1962) (see Fig. 127, p. 378).

Collectors of plankton

The primary functions of the complex of organs in the mantle cavity are to effect respiration and to eliminate faecal and nitrogenous wastes. This requires the passage of water through the mantle cavity and between the ctenidial leaflets. This inhalant water current necessarily carries particles of suspended matter into the mantle cavity, and it is essential to the proper functioning of the mantle cavity that these should not be allowed to accumulate and so to clog the ctenidial leaflets. This is fully discussed in the chapter on the mantle cavity. It is only necessary here to summarise that discussion by observing that the ctenidial filaments are typically short and triangular with a large respiratory surface, and that ciliary cleansing currents dispose of particulate waste material, bound in mucus secreted by the hypobranchial gland, at two sites: relatively large particles tend to accumulate at the left side, and smaller particles at the right side of the orifice of the mantle cavity. This particulate matter which is voided from the mantle cavity is partly organic in content, and is therefore a potential source of food. In certain taenioglossans, e.g. *Hipponyx*, described above, the radula is protruded and sedimentary material is grasped between the radular teeth and is carried back into the mouth as food. If excessive quantities of waste material were accumulating at the mouth of the mantle cavity its presence there would be detected by epithelial sense organs and presumably the radula would be used to drag away the offending material and so clear the orifice of the mantle cavity. Such a train of circumstances would inevitably lead in due course to utilisation of the "waste" material as a new source of food, and as soon as this became an established practice the normal evolutionary forces would begin to operate on the mantle cavity and its contained organs. The ctenidial filaments would serve not only for respiration but also for the collection of food material; elongation of the leaflets into long slender filaments would strengthen the inhalant current and would provide a more effective meshwork to sieve the entire inhalant water current; the increased areas of ciliated frontal surfaces would be more effective in trapping food material. The binding of this material into a mucous cord, and the guidance of this food cord to a selected point within reach of the radula would complete the working details of an entirely new type of feeding mechanism.

Such a feeding mechanism has been described for *Crepidula fornicata* in a classic account by Orton (1914) to which interesting further details have been supplied by Werner (1951, 1953). In *C. fornicata* the mantle cavity is particularly large and stretches over almost the entire dorsal surface of the animal. The single, pectinibranch ctenidium is attached at its base to the mantle on the left side, and its filaments extend nearly horizontally across the mantle cavity and divide this into a left, lower, inhalant chamber, and a right, upper, exhalant chamber. Due to the action of the cilia on the ctenidial filaments, fresh sea water is drawn into the left lower part of the mantle cavity and upwards between the ctenidial filaments into the right, upper exhalant chamber, where it passes out on

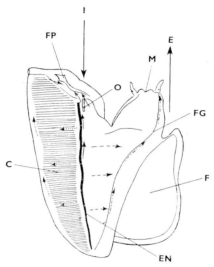

Fig. 17. Ciliary feeding mechanisms in the mantle cavity of the prosobranch *Crepidula fornicata*. The mantle cavity has been opened by a longitudinal incision along its right side and the roof has been reflected to the left. The inhalant water current (I) impinges on the osphradium (O). The endostyle (EN) passes a sheet of mucus over the frontal surfaces of the ctenidial filaments. A food groove (FG) passes over the floor of the mantle cavity towards the head, and food is collected from this groove periodically by means of the radula. For interpretation of other lettering, see pp. 93–4. (Originally published in Yonge, 1938a, *J. mar. biol. Ass. U.K.* 22, 458, fig. 3.)

the right-hand side. When *Crepidula* is feeding a mucous filter covers the entrance to the mantle cavity and this collects the larger particles borne in the inhalant water stream. Any large particles remaining in the inhalant stream after passing the filter are liable to drop on the anterior floor of the mantle cavity, whence they are passed by ciliary currents into a food pouch on the left side of the head, which also receives mucus-bound material from the filter. Material in this food pouch may be ingested by the radula or it may be discarded as pseudofaeces (see Fig. 17).

Slightly smaller particles may pass deeper into the mantle cavity before falling on to its floor; these are conveyed to the right by ciliary acceptance currents and are guided into a ciliated gutter, or food groove, which passes down the right side of the floor of the mantle cavity and terminates on the right side of the head. The smallest water-borne particles will be carried upwards in the water current which passes between the ctenidial filaments; these particles are collected mainly on the lower, frontal surfaces of the filaments and also, to a lesser extent, on the upper, abfrontal surfaces. Werner observed that a second mucous net is formed from mucus secreted by the endostyle and spread over the frontal surfaces of the ctenidial filaments. This mucous net passes from the tips of the filaments into the ciliated food groove on the floor of the mantle cavity. The tips of the ctenidial filaments form a temporary roof to this groove.

The hypobranchial gland is reduced in size. The "endostyle", which supplies mucus on to the frontal surface of the ctenidium, lies alongside the base of the ctenidium on the floor of the mantle cavity. A comparable endostyle has been found in other members of the Calyptraeidae (Yonge, 1938), and also in *Struthiolaria* (Morton, 1951).

The single ctenidium thus serves both for respiration and for the collection of food material, and the particles carried suspended in the inhalant current are sorted into two fractions: larger particles accumulate in a mucus-bound mass in a pocket on the left side of the head, while smaller particles are presented in a mucus-bound food cord on the right side of the head. The development of an oralward ciliary current along the free edge of the ctenidium assists in the passage of food towards the head and is a significant contribution to the feeding mechanism. The radula is everted periodically, the lateral teeth diverging as it is protruded. The radula is applied to the food cord on the right side of the head, the lateral teeth converge and grasp the food cord, and a small portion of the cord is torn off and carried back into the mouth. Typically, the entire food cord is gradually taken as food in this way. Occasionally material is also taken into the mouth from the accumulation of larger particles on the left side of the head. Thus the ciliary mechanisms in the mantle cavity collect and sort suspended material into two grades and the animal then, using its radula, deliberately selects the finer of these grades as its basic source of food. A new feeding organ has been created—the ctenidium —and this functions in much the same way as it does in a filibranch bivalve. It is to be noted, however, that the radula has not been rendered obsolete. As in *Hipponyx* it serves not as a rasping, but as a grasping organ.

A further point of interest in this new feeding mechanism is that *Crepidula* retains the ancient method of feeding on minute particles, but these are made up into mucoid packages of a size which is convenient for manipulation by the radula.

In *Capulus* the food material collected in the mantle cavity is passed to the centre of a flat propodial extension of the foot, anterior to the head (see Fig. 18). From this point food material is passed partly by ciliary and partly by muscular action up a groove in the dorsal surface of a short proboscis, and so reaches the mouth where it is grasped by the radula and is ingested (Yonge, 1938). Sharman (1956) observed that *C. ungaricus* may often be found on the shell of a living bivalve mollusc such as *Chlamys opercularis*, *Pecten maximus*, or *Modiolus modiolus*, close to the free ventral margin. In such a position the animal is capable of taking pseudo-

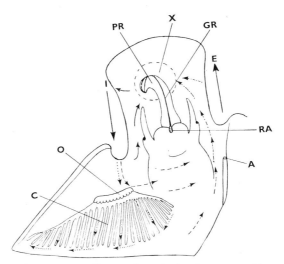

FIG. 18. *Capulus ungaricus*. The mantle cavity has been opened by an incision along the right side and the roof of the mantle cavity has been reflected to the left, inverting the ctenidium (C) and the osphradium (O) and exposing the dorsal surface of the head. Ciliary currents are represented by dotted arrows and by broken arrows. Food material accumulates in a hollow on the upper surface of the propodium (X) from which it is collected periodically by the proboscis (PR). For interpretation of other lettering, see pp. 93–4. (Originally published in Yonge, 1938a, *J. mar. biol. Ass. U.K.* **22**, 460, fig. 5.)

faeces from the mantle cavity of the bivalve and this has actually been observed. Feeding mechanisms similar to that of *Crepidula* have been described for *Turritella communis* (Graham, 1938), and for the freshwater snail, *Viviparus viviparus* (Cook, 1949) (see Fig. 19).

Both *Aporrhais* and *Struthiolaria* lie buried in the surface layers of marine deposits of firm muddy gravel or of firm muddy sand. Both genera are deposit feeders, but they have contrasted feeding mechanisms. In *Struthiolaria* the feeding mechanism is basically the same as that of *Crepidula*, the main supply of mucus to trap food material on the ctenidium being similarly provided by an endostyle, but the main source of food being particles stirred up from the surface of the muddy sand Morton, 1951). *Aporrhais*, on the other hand, employs the more archaic method of foraging for organic debris, bits of algae, and diatoms, by protruding its proboscis and grasping prospective food material by its radular teeth in the same way as does *Hipponyx* on rocky shores (Yonge, 1937).

Studies on the feeding mechanisms of species of *Vermetus* reveal yet another method of feeding, and point to the pre-adaptive features from which it evolved. *V. novae hollandiae* occurs only in the most turbulent

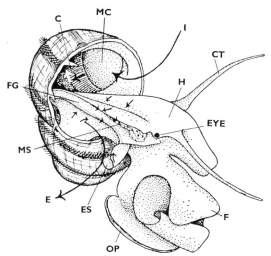

Fig. 19. A young specimen of *Viviparus viviparus* seen with its body fully extended and illustrating the ctenidial ciliary feeding mechanism which is sometimes employed. An inhalant current (I) bearing particles of silt, infusorians, etc., enters the mantle cavity (MC) on its left side. Small particles collected on the elongated ctenidial filaments, and also on the floor of the mantle cavity, enter a ciliated food groove (FG) passing obliquely forwards on the floor of the mantle cavity towards the right side of the head. The animal periodically turns its head and takes this food material into its mouth. For interpretation of other lettering, see pp. 93–4. (Originally published in Cook, 1949, *Proc. malac. Soc. Lond.* **27**, 266, fig. 2.)

water at the reef edge, e.g. on the Great Barrier Reef of Australia. This species has elongated ctenidial filaments and feeds in the same way as does *Crepidula*. In addition, however, mucous balls are formed by the pedal gland and are added to the anterior end of the food cord to increase its bulk, so that it can be readily grasped by the radula. In contrast to this, *V. gigas* is to be found only in calm water; its shell is coiled with the orifice raised up above the substratum where the animal's feeding processes will receive no interference from water currents produced by other reef organisms nearby. The ctenidium is of the typical pectinibranch respiratory type, with short triangular leaflets, and there is no oralward ciliary current along the tips of the filaments, i.e. the ctenidium is not used to collect food material. Such particulate material *as may happen* to fall on the floor of the mantle cavity and become bound with mucus, and particles which become immobilised by secretions from the hypobranchial gland, are compacted into mucous balls which pass to the right side of the head and are ingested by use of the radula as in *Crepidula*. This, however, does not form the whole, or even the major part of the food supply of *V. gigas*; an additional feeding mechanism has sprung into existence. The pedal gland and the pedal tentacles are greatly enlarged and the latter are capable of great extension. Mucus secreted by the pedal gland is passed into a longitudinal groove in each of the pedal tentacles. The pedal tentacles are extended far out into the calm water while the sticky mucous cord is formed in the groove. The tentacles are then disengaged from the mucous cords which are left hanging in the water. Three or four mucous threads, each as long as 30 cm, may be in action at a time. The presence of living zooplankton in the water is said to stimulate the production of mucous threads. In due course plankton and inanimate particles become stuck to the feeding-threads, which are eventually drawn back to the mouth and are consumed.

Both the pedal tentacles and the pedal gland have become involved in this novel feeding mechanism, and the use of secretions from the pedal gland to increase the bulk of the food cord in *V. novae hollandiae* can be seen as an essential pre-adaptation towards the new feeding method of *V. gigas*. A number of other factors are contributory, and are worthy of mention. The feeding mechanism can only occur in calm water where the threads can hang undisturbed for some time while gathering their cargo of food material. The mechanism would be impossible—and, of course, entirely unnecessary—in rough water which would bring a plentiful supply of plankton to be collected by the ctenidium as in *V. novae hollandiae*. It could only develop in an animal in which the water currents entering and leaving the mantle cavity were too weak to disturb the mucous threads; presumably as this new feeding mechanism began to evolve the ctenidial filaments gradually diminished in size and reverted to the ancestral short triangular leaflets, which are once again solely respiratory. It could only develop in an animal which was sufficiently raised up above the substratum for the mucous threads to be undisturbed by water currents produced by other organisms on the coral reef (Yonge, 1932, 1938; Yonge and Iles, 1939). Clearly the pre-adaptive utilisation of mucus from the pedal gland, as in *V. novae hollandiae*, enabled the genus to invade an environment—calm coastal waters—that would have otherwise been unattainable.

Morton (1951) has shown that *Serpulorbis zelanicus* and *Novastoa lamellosa* resemble *V. novae hollandiae* in several respects; the ctenidial leaflets are short and triangular, and feeding is predominantly by ciliary ctenidial sorting mechanisms. This may be supplemented in *S. zelanicus* by use of sticky mucoid threads emitted by the pedal gland, so *Serpulorbis* may occupy a position intermediate between *V. gigas* and *V. novae hollandiae*.

Morton has studied various other Vermetidae from New Zealand waters and finds that *Stephopoma* has long slender ctenidial filaments and feeds principally in the manner of *Crepidula* (as described above). In addition, however, *Stephopoma* was occasionally seen to employ an unique feeding method which suggests convergent evolution with respect to the Cirripedia and to the Serpulidae (Polychaeta, Sedentaria), and which is doubtless advantageous to a very crowded colony of encrusting organisms. The anteriormost ctenidial filaments project like a fan from the mouth of the mantle cavity, mucus is secreted by an "endostyle" at the ctenidial axis and is passed up the ctenidial filaments and there serves to trap planktonic organisms. These projecting ctenidial filaments were periodically drawn through the water like the limbs of a barnacle, only rather more slowly (Morton, 1951). It becomes evident that the Vermetidae, exploiting the encrusting mode of life, are exhibiting a minor phase of adaptive radiation in feeding methods, and further interesting minutae of feeding habit may still remain to be discovered in this family.

Feeders on colonial animal growths

Certain mesogastropods with taenioglossan radulae have adopted carnivorous habits. In the case of *Cerithiopsis* and of *Triphora*, which feed on sponges, this is of more significance in the consideration of digestion than of feeding methods, for the animal is feeding on an inactive substratum. In these two genera the mouth is at the tip of a long slender proboscis. When feeding, the proboscis is thrust through the osculum of a sponge and an attack is made on the deeper tissues. Thus these predators have adopted an advantageous pattern of behaviour and so avoid ingesting the superficial layers of the sponge which are heavily loaded with indigestible spicules. The dorsal border of the mouth carries a pair of jaws, each of which comprises a pad of about forty spikes. These jaws serve to loosen the tissues of the sponge, and the radula then rakes up the loosened pieces of tissue and conveys them to the buccal mass. The jaws and the radula are lubricated by secretions of the salivary glands. In *Cerithiopsis* the saliva may contain a digestive enzyme for extra-alimentary digestion of the tissues of the sponge (?), derived from special glandular cells found only in the enlarged left salivary gland (Fretter, 1951a). Although these animals have adopted a carnivorous diet we should note that the food material presented to the alimentary canal for digestion conforms approximately to tradition, consisting of comparatively small particles. The need for an extracellular protease, and the loss of the crystalline style in carnivores, are discussed elsewhere.

Such carnivores tend to adopt stenophagous habits, each species of snail restricting its feeding to a particular substratum. Thus *Cerithiopsis tubercularis* in the adult stage apparently feeds on the sponge *Hymeniacidon sanguinea* (young specimens were observed to feed on diatoms and detritus in the crevices of shells of *Chlamys opercularis*).

Cowries, also, are stenophagous carnivores (Lebour, 1932). *Trivia* feeds on compound ascidians, preferring *Diplosoma* to *Botryllus* and to *Botrylloides*. Again, *Simnia patula* feeds on either *Alcyonium digitatum* or on

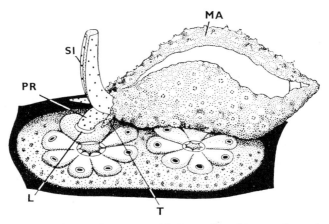

FIG. 20. *Erato voluta* feeding on a zooid of *Botryllus schlosseri*. The proboscis (PR) of the cowrie is thrust down the mouth of the tunicate. The shell of the cowrie is almost completely covered by the expanded lobes of the mantle (MA). For interpretation of other lettering, see p. 94. (Originally published in Fretter, 1951b, *Proc. malac. Soc. Lond.* **29**, 15, fig. 1.)

Eunicella verrucosa and is restricted to these substrates. *Trivia* feeds indiscriminately on the ascidian *Diplosoma*, consuming both the test and the zooids within, although it is unable to digest the tunicin which makes up the test. The test is eaten first, thus exposing the zooid which is then eaten whole. *Erato voluta*, on the other hand, is more selective. Having tested the water current at the mouth of a zooid, *Erato* thrusts its proboscis down the mouth of the ascidian, bites off the flesh of the zooid with its jaws, and uses its radula to convey the flesh up to the stomach. In this way only the digestible flesh is taken and the indigestible tunicin of the test is avoided. It may take 20 minutes for the snail to eat one zooid, and either one or two zooids may be eaten in one meal (Fretter, 1951b) (see Fig. 20).

Benthic hunters

Some mesogastropods are adapted for life as predacious carnivores on sandy shores and plough forwards just below the surface of the sand in search of small bivalves for food. The foot is greatly developed for pushing through the sand and is equipped with pores by which sea water can be admitted to an "aquiferous system". By this means the foot can be dilated, and quickly deflated, and the aquiferous system acts as a hydraulic skeleton. This is particularly striking in *Natica didyma* and also in *Sinum neritoideum* in Malayan waters. In the last named example the enormous foot cannot be withdrawn into the shell even after maximum deflation; it serves as a very powerful burrowing organ, by which surprisingly rapid progress is made. *N. alderi* feeds on a wide variety of bivalves which lie buried in the sand, e.g. species of *Nucula*, *Thyasira*, *Montacuta*, *Venus*, *Tellina*, etc. The prey is grasped by special folds of the foot and this determines the normal position at which the shell of the bivalve is drilled, and the size range of the prey which are attacked. Pelseneer (1924) reported that *Natica alderi* drilled *Donax* and *Macoma* in the vicinity of the umbo and never near the free ventral margin, but Ansell (1960) found that *Venus striatula* was normally attacked near the ventral margins. Ansell showed that there was a close correlation between the size of the prey and the diameter of the hole drilled in its shell, i.e. small *Natica* fed upon small *V. striatula*, and larger *Natica* fed on larger specimens of *V. striatula*. Ansell listed fifteen species of bivalves attacked by *N. alderi* at Kames Bay.

There has been controversy as to the method by which *Natica* drills the hole in the shell of its prey. Treating the shells of bivalves with a coat of paraffin wax did not stop *Natica* from boring, which suggested that acid alone could not be responsible (Turner, 1953). In all probability boring is partly by the rasping action of the radula—which would quickly cut through an acid-resistant periostracum or through an experimental layer of paraffin wax, and partly by the use of an acid secretion or an enzyme. The source of the secretion, which has been claimed to be sulphuric acid, is a disc-shaped gland which lies on the ventral side of the proboscis, near its tip (Ankel, 1937; Graham, 1953). *Pollinices duplicata* feeds similarly on *V. mercenaria* on American shores (Carriker, 1951).

Planktonic hunters

Four families in the Mesogastropoda have taken to pelagic life, and these are all predacious carnivores. *Ianthina* has no eyes, and no powers of directive swimming, and can only feed when it drifts by chance against one of a variety of other pelagic animals (see Fig. 21). It feeds principally on the siphonophore *Velella*. *Ianthina* has been observed to abandon its float when feeding and to browse on the underside of the siphonophore, periodically liberating a quantity of purple dye which may perhaps serve to anaesthetise the tentacles of the *Velella* (Wilson and Wilson, 1956). When *Ianthina* is feeding the buccal mass is extruded through the mouth apparently by a rise in blood pressure in the cephalic haemocoele. When extruded, the left and right halves of the odontophore are spread apart, and their hemispheric surfaces are covered with erected radular teeth. When the buccal mass is withdrawn the two halves of the odontophore are drawn together, meeting in the median line and grasping food material between them. Fragments of food are torn off the prey by the radula, the jaws apparently playing little part in the process (Graham, 1965).

The remaining three families are collectively known as the Heteropoda, and include *Atlanta*, *Carinaria*, and *Pterotrachea*. The two last-named ge-

FIG. 21. *Ianthina janthina*, lateral view of a living specimen in a glass aquarium vessel. The snout and tentacles are not quite fully extended. The colourless translucent epipodium of the right side is well displayed. The contracted podium is pressed against the anterior end of the float. The posterior end of the float is disintegrating. Note the hydroid attached to the shell of the snail near its apex. (Originally published in Wilson and Wilson, 1956, *J. mar. biol. Ass. U.K.* **35**, plate 1.)

nera are elongate, actively swimming forms in which the visceral mass is greatly reduced. The shell is greatly reduced in size and weight in *Carinaria* and is lacking in *Pterotrachea*. *Pterotrachea* swims with its ventral surface uppermost; its foot has been reduced and modified to form a median fin, the original ventral creeping sole having been modified to form a small sucker which may be used to grasp its prey such as medusae, small fish, etc. *Pterotrachea* is predacious and uses its radula to grasp and to swallow its prey whole.

Ectoparasites

Opalia crenimarginata is a temporary ectoparasite on the anemone *Anthopleura xanthogrammica* (see Fig. 22). *Balcis devians* (= *Eulima distorta*) lives on *Antedon rosacea*, its proboscis being deeply embedded in a pinnule. It has no radula and no jaws. Food material is drawn up into the

FIG. 22. The prosobranch gastropod *Opalia crenimarginata* is a temporary parasite on the anemone *Anthopleura xanthogrammica*. The body wall of the anemone is perforated by the proboscis of the snail, and the snail then sucks nourishment from the tissues of the anemone. (Originally published as Dr. Gunnar Thorson's Christmas card for 1956.)

stomach by a suction pump formed by the buccal mass. The gut is more highly modified to suit the parasitic mode of life than it is in the opisthobranchiate Pyramidellidae (for which see below) (Fretter, 1955). The host-relationships of most of these small ectoparasites have yet to be discovered and recorded, but Habe (1952) has prepared a useful list of host-relationships for twenty-two species of Eulimidae and of Styliferidae living in Japanese waters.

The capulid genus *Thyca* is ectoparasitic on echinoderms and is somewhat comparable with *Balcis*. In *Thyca* the true foot is much reduced and a "false foot" surrounds the mouth and serves as an organ of attachment. It is thought that *Thyca* feeds on the tissues of its host probably after predigestion by salivary secretions. A further adaptation to the parasitic mode of life in *Thyca* is sexual dimorphism, the males being much smaller than the females (Baer, 1952). In *Mucronalia*, which crawls on the body of *Synapta*, and which may enter the intestine and feed on the intestinal wall, the proboscis is surrounded by a sheath-shaped pseudopallium. It is possible that this pseudopallium serves to localise digestive juices when these are applied to the epithelium when *Mucronalia* begins to feed. After the perforation has been made the proboscis is inserted and then the pseudopallium is reversed in position and now forms an upwardly projecting collar which surrounds the base of the proboscis.

More advanced ectoparasites such as *Stylifer* become almost completely embedded in the dermal tissues of the host sea urchin on which they live. The foot is reduced in size, and a pseudopallium arises from the base of the proboscis and forms an upwardly projecting collar of considerable size, which encloses the entire animal and isolates it from the flesh of the host. The shell and the contained viscera have therefore come to lie in a pseudopallial chamber which communicates with the exterior in the region of the spire of the shell. We can envisage the creation of this condition by development from the condition shown in *Mucronalia* by an increase in the burrowing action caused by feeding and by hypertrophy of the pseudopallial sheath on the base of the proboscis. The ctenidium lies in the mantle cavity as usual, and this opens into the pseudopallial chamber. Sea water is drawn into the latter for respiration either by contractions and expansions of the pseudopallium itself, or by piston-like movements of the proboscis (Baer, 1952).

The condition found in *Gasterosiphon* could have been derived from a condition comparable to that now found in *Stylifer*, by yet deeper invasion of the host's body, while retaining contact with the sea water outside for respiratory purposes by means of the pseudopallial chamber. This, however, is not necessarily the correct interpretation. *Gasterosiphon* is an elongate, worm-like animal which has been found in the body cavity of a holothurian, still being attached to the body wall at one end. A swelling close to the point of attachment indicates the position of the visceral mass, which has lost its protective shell. As in *Stylifer*, the visceral mass lies in a pseudopallial chamber which communicates with the exterior via the point of attachment to the body wall of the host for respiratory purposes. The animal is hermaphrodite, and eggs are retained and are incubated for a period in the pseudopallial chamber.

Endoparasites

Other genera are still more highly specialised and, having lost connection with the outside environment, have become truly endoparasitic, e.g. *Entoconcha* and *Enteroxenos*. *Entoconcha* occurs in the body cavity of a holothurian and feeds by drawing blood directly from a blood vessel on the wall of the gut of the host. *Enteroxenos* has lost all traces of the alimentary canal, and has even lost its mouth. It is therefore even more highly modified for parasitic life than the Cestoda, which are now believed never to have had a gut. The genus is solely recognisable as a mollusc through retention of a veliger larva. The adult presumably feeds by absorbtion of nourishment through its body wall, as does a cestode.

The development of dwarf males—possibly neotenic larvae—which live in the brood pouch of the female in both *Entoconcha* and *Enteroxenos* are major adaptations without which this extreme degree of parasitism could not have been reached (Baer, 1952).

No account of the Mesogastropoda would be complete without reference to their successful invasion of fresh waters and also of dry land.

Freshwater herbivores

There are several families of freshwater mesogastropods including the Ampullariidae, Valvatidae, Hydrobiidae, Viviparidae, Melaniidae, etc. These can easily be distinguished from freshwater pulmonates by the presence of an operculum, and they are often referred to as "operculates" for that reason. The feeding methods of *Viviparus* have been mentioned above. The freshwater prosobranchs of Lake Tanganyika have attracted particular attention. Due to the great size and depth of this lake a greater diversity of habitat is offered than in most bodies of fresh water. Due to the long period over which this body of water has been isolated, a minor phase of adaptive radiation has occurred among the prosobranchs which were restricted to the lake. The resulting forms showed striking resemblances to various marine prosobranchs, and for this reason it was first thought that the prosobranchs of Lake Tanganyika represented a relict marine fauna. This has now been shown not to be the case. The snails in question had first invaded fresh waters via estuaries, and had become fully adapted to the freshwater environment. Subsequently they became isolated in Lake Tanganyika and became adapted to a variety of habitats and different modes of life. Such an adaptive radiation is characteristic of a siuation in which a small group of animals is subjected over a very long perod to isolation, freedom from predation, and freedom from competition for food, and are presented at the same time with a diversity of unoccupied ecological niches. *Nassopsis* and *Tanganyicia* occur on rocky sites in the lake and have short blunt radular teeth which are adapted for rasping algae off the surfaces of rocks. *Tiphobia*, on the other hand, has a radula with long delicate marginal teeth which may serve to pick up detritus from a muddy bottom (Yonge, 1938). *Pomacea canaliculata* feeds on freshwater plants such as *Pistia* and *Vallisneria*, grasping a leaf between the sensory palps on each side of the mouth, perforating the leaf by the radular teeth, and tearing off relatively large portions of the leaf by action of the jaws (Andrews, 1965).

Terrestrial herbivores

Similarly, a number of families of Mesogastropoda have invaded dry land, including the family Cyclophoridae, with the interesting genera

Cyclophorus, *Alycaeus*, *Diplommatina*, and *Opisthostoma*, which are discussed elsewhere in this book. All these terrestrial mesogastropods are herbivorous.

Order Neogastropoda

This order of prosobranchs consists largely of predacious carnivores. In many the number of radular teeth in each transverse row is reduced to three, each of these being strongly built and bearing several cusps. Such a radula is known as "rachiglossan"; it can be used to drill through thick shells, sometimes being assisted by a chemical softening agent. Some of these neogastropods have elaborate methods of tracking down their prey and have adopted complex and tedious methods of opening their shells, thus indicating the high degree of development of their sense organs and their nervous systems. Some, however, have lost all trace of radular teeth, this condition being known as "aglossan". This is true for *Coralliophila* and for *Magilus*, and for one member of the Turridae (Smith, 1964) and it is not yet known how these animals feed or what is the nature of their food.

Scavengers

Some of the Neogastropoda have adopted a scavenging habit and do not drill holes (Graham, 1953). They are attracted to dead and decaying flesh presumably by its scent. *Buccinum undatum* is such an animal, and advantage is taken of the scavenging habits of this sub-littoral species to catch it commercially in baited traps. This is one of the largest of the British marine snails, and it is an important source of food collected principally on the east coast of Britain (Simpson, 1960).

Brown has described the feeding habits of two species of *Bullia* which occupy slightly different habitats. Both are scavengers and feed on medusae, dead crabs, dead fish, etc., to which they are attracted by emanations from the decaying flesh. *B. laevissima* occurs sub-littorally or in very sheltered inter-tidal situations where it is not disturbed by surf; it only feeds when under water. *B. digitalis*, on the other hand, favours surf-beaten inter-tidal situations. During the lower half of the tidal cycle *B. digitalis* lies in the water with its foot fully expanded, and it is swept to and fro in the swash zone. When the tide rises above mid-tide level *B. digitalis* buries itself in the sand and cannot be induced to leave the sand to feed until the tide has fallen below mid-tide level again. When the tide is below mid-tide level, specimens of *B. digitalis* are quickly attracted by dead fish placed on the surface of the wet sand *above* the swash zone. If they are still buried in the sand they quickly emerge and allow the surf to carry them passively inshore, and as soon as they emerge from the water they move actively over the surface of the wet sand towards the bait. *B. digitalis* never feeds under water in nature, but only on the surface of the wet sand or in the surf.

Detection of food material by *Bullia* was shown to be effected by the osphradium. Substances emanating from natural foods of the snail, such as trimethylamine, cause a positive response by the snail even when the dilution is very great. Snails which had been operated upon by sectioning the osphradial nerve failed to respond to comparable stimulation (Brown, 1961, 1961a).

Freshwater carnivores

Among the Gastropoda the only predacious carnivore or scavenger to become established in fresh waters is the genus *Antenome* (Buccinidae), which occurs in Siam, Cambodia, Malaya, Sumatra, Borneo, and Java. *A. helena* is always found in fresh waters, e.g. in lakes, fish ponds, and in rivers. Specimens kept under observation in aquaria were seen to consume the snails *Indoplanorbis exusta* and *Bellamya javanica*, an earthworm, and the fish *Panchax panchax*, but *Antenome* was never seen to feed on the algae in the aquarium (Jutting, 1956).

Marine benthic hunters

Melongena is such a predacious carnivore, which feeds on bivalve molluscs. In America *M. coronata* prefers to feed on species such as *Tagelus divisus*, which have permanently gaping shell valves and consequently have no form of protection against the predator. *M. coronata* is also able to attack clams which can shut their shell valves tightly, e.g. *Mercenaria mercenaria*, taking an opportunity to slip its proboscis between the margins of the shell valves when these are gaping slightly. The proboscis must be toughly constructed to withstand the inevitable pinching which it will receive from the edges of the shell valves of the prey. Under aquarium conditions a number of *Melongena coronata* tend to co-operate in feeding together on one clam, and there can be little doubt but that this is the normal practice. Once a *Melongena* has begun to feed on a clam other *Melongena* will join it, even if they have to pass clams which have been broken open and on which they could feed more easily. This pattern of behaviour will be advantageous in reducing the pressure exerted on any one proboscis by the pinching action of the shell valves of the clam, and by accelerating the final relaxation of the shell valves (Turner, 1959).

Fasciolaria hunteri is a normal member of the oyster bed community in the vicinity of Beaufort, Carolina; it may attack a wide variety of bivalves, but it prefers oysters and it selects *Crassostrea virginica* in preference to *Ostrea equestris*. When feeding on a bivalve it slips the outer edge of its shell between the shell valves of its prey when they are slightly parted before inserting its proboscis and commencing to feed. *F. hunteri* also feeds on gastropods such as *Nassarius* spp. and on *Urosalpinx cinerea*, the well-known predator of oysters. *F. hunteri* prefers to feed on *U.*

cinerea rather than upon *Crassostrea* if *Urosalpinx* is abundant and easily obtained. Since *Urosalpinx* prefers to feed on small oysters and eats these in large numbers, the presence of *F. hunteri* on oyster beds may therefore be advantageous to the oysters although it may consume some oysters itself (Wells, 1958).

Murex fulvescens also feeds on a variety of bivalves, preferring *C. virginica* and apparently finding *O. equestris* comparatively unattractive. This large snail mounts the upper, flat shell of an oyster and drags the shell valves of its prey apart, holding the lowermost valve down by wedging the outer lip of its own shell against it. When feeding on *Mercenaria* or on *Modiolus*, *Murex fulvescens* holds the prey with its foot and drags the shell against the lip of its own shell. The tangential force so applied scrapes and chips away fragments from the edges of the two shell valves until a gap is formed, through which the proboscis is then thrust. It is not known whether small specimens of *M. fulvescens* feed in the same way (Wells, 1958a).

Busycon similarly likes to feed on *Tagelus*, but will feed on *Chione* or on *Venus* or on oysters if *Tagelus* is not available. When feeding on oysters, *Busycon* may wait for the oyster to open its shell valves and then slip the outer lip of its shell between the gaping valves, so wedging them apart. The proboscis can then be inserted and the flesh consumed. Alternatively, *Busycon* may grasp the bivalve firmly with its foot, bring the outer lip of its shell to bear against the ventral margin of the shell valves of the prey and, by repeated pressure through contractions of its columella muscle, chip fragments off the edge of the clam shell. As soon as a sufficiently large gap has been made in this manner, the lip of the snail shell is inserted as a wedge to hold the clam shell valves apart (Magalhaes, 1948). *Busycon* is attracted by the exhalant current emitted by bivalves, and by this means it is able to find bivalves which are quite deeply buried, such as "Quahogs" (*Venus mercenaria*). *Busycon* can find a quahog, dig it up, mount it, and attack the ventral shell valve as described above. When the shell valve is too thick to be broken in this way *Busycon* may leave it after a sustained attack. Generally, however, *Busycon* prefers to feed on thin-shelled forms such as *Mytilus edulis* rather than on the more troublesome thick-shelled species such as *V. mercenaria* (Carriker, 1951).

The American "Tingle", *Urosalpinx cinerea*, when feeding on *Modiolus demissus*, drills holes indiscriminately through any part of the shell of its prey (Carriker, 1951); in *Modiolus* the succulent gonad invades the lobes of the mantle and can easily be fed upon regardless of the site of the bore-hole. When feeding on oysters, *Urosalpinx* greatly prefers young, thin-shelled specimens which require relatively little drilling. These, of course, provide a relatively small meal, which has to be followed soon by another. The natural economy of effort of *Urosalpinx* therefore makes it a particularly serious pest of commercial oyster beds. The radula of

Urosalpinx bears three teeth in each transverse row, of which the scythe-shaped lateral teeth diverge widely when the radula is protruding. These lateral teeth will serve to grasp and tear off flesh after the hole has been drilled through the shell, and the tingle is feeding on the flesh within.

The precise method of drilling by *Urosalpinx* and related genera has been a matter for considerable debate. It is now understood that the bore-hole is made partly by mechanical and partly by chemical means. Firstly, the proboscis is everted, and the surface of the shell of the potential victim is explored by the rim of the mouth, which lies at the tip of the proboscis. A site for drilling having been chosen, the radula is first used to clear away encrusting Polyzoa, etc., and to begin the drilling action. At a temperature of 25°C as many as sixty strokes of the radula have been observed per minute. After a few minutes the animal ceases to drill with the radula, and glides forwards over the site of drilling. An accessory boring organ (ABO) is then everted through a pore in the anterior part of the sole of the foot, and is pressed tightly against the site of drilling and is held there for some minutes. This apparently softens the shell in preparation for the next phase of drilling. The processes of drilling and of softening are repeated alternately, each phase lasting up to 20 minutes. The odontophore can be rotated so that the axis of drilling can be systematically varied, and the resulting perforation is neat and cylindrical.

The nature of the softening agent applied to the drill-hole by the glandular tissues of the accessory boring organ is not yet known. It might be an acid which actively dissolves the calcium salts of the shell, but the presence of such an acid has not yet been finally demonstrated. Alternatively it might be an enzyme (calcase) which facilitates absorbtion of a part of the substance of the shell and which loosens the prisms of which the hard prismatic layer of the shell is constructed. The radula alone is apparently not very effective in drilling through the prismatic layer. It may take up to 2 days for a *Urosalpinx* to drill through the shell of an oyster which is $1\frac{1}{2}$ in. in diameter. In contrast, a full week has to be spent in perforating the shell of a 4-in. oyster.

Having completed the drilling, the proboscis is inserted and the radula is used to rasp the flesh within. As the radula is withdrawn the lateral teeth are drawn together and grasp and tear off fragments of flesh. Only the softer tissues can be fed upon, for the muscles which are employed in withdrawing the radula are not sufficiently strong to tear portions from the powerful adductor muscle of the prey. Having made a perforation, *Urosalpinx* may feed continuously for as long as 20 hours, which is striking testimony to the efficacy of the digestive enzymes of this animal (Carriker, 1943, 1955) (see Fig. 23). Carriker (1961) extirpated the accessory boring organ from experimental animals and found that no boring activity occurred until after this organ had begun to regenerate. Chew (1960) found that the oyster drill (*Ocinebra japonica*) is able to feed

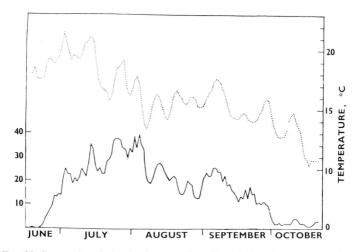

FIG. 23. Seasonal variation in the rate of feeding of *Urosalpinx cinerea*. Continuous line: the number of oyster spat destroyed by 39 specimens of *U. cinerea*; dotted line: variations in water temperature. In both cases the graphs have been smoothed by plotting averages of groups of three adjacent records. Subsequent to completion of the breeding season of the snail the rate of destruction of oyster spat is related to the ambient temperature. (Originally published in Cole, 1942, *J. mar. biol. Ass. U.K.* **25**, 501, fig. 10.)

on Manila Clams (*Venerupis japonica*), Bay Mussels (*Mytilus edulis*), Olympia Oysters (*Ostrea lurida*), and Pacific Oysters (*Crassostrea gigas*) under experimental conditions. The results of his experiments were not entirely consistent, but the overall results indicate that *O. japonica* is least inclined to attack the Pacific Oyster, and little inclined to attack Bay Mussels. There was a strong tendency for the drills to continue to attack the species on which it first fed in the course of the experiments (Table 2).

TABLE 2. *Food preferences of* Ocinebra japonica.
Data extracted from more extensive figures in Chew (1960)

Prey species	Number of specimens which fed consistently on the same prey species	Total number of specimens eaten during the course of all the experiments
Manila Clam	22	123
Olympia Oyster	22	117
Bay Mussel	8	69
Pacific Oyster	1	20

Purpura (= *Nucella*) is a predacious carnivore which is found on rocky shores, and which feeds on either *Mytilus* or on cirripedes. When feeding on the latter, *Purpura* smothers the barnacle with its foot, and thrusts its proboscis between the scuta and terga and takes its meal without having had to drill through the barnacle's shell. In Malayan waters *Thais echinata* feeds similarly on the cirripede *Tetraclita porosa*. When feeding on *Mytilus*, *P. lapillus* everts its proboscis and the radula is used to cut through the thin shell valves of the mussel. As with the perforations of the shell of *Modiolus* by *Urosalpinx*, and for the same reasons, the perforation of the shell of *Mytilus* by *Purpura* may be made at any point on the shell except in a zone very close to the shell margin. *Purpura* will feed first on the succulent gonad tissues which fill the lobes of the mantle (Pelseneer, 1924). An accessory boring organ comparable with that found in *Urosalpinx* has been found in *Purpura* and also in *Ocenebra*.

The Olividae are neogastropods which have adopted an active burrowing, carnivorous habit on clean sandy shores, and they search out and feed upon small molluscs, as do the Naticidae among the Mesogastropoda. The Olividae, however, do not drill holes through the shells of their prey.

Nassarius reticulatus is another predacious carnivore which hunts on sandy shores or in the sandy bottoms of rock pools. During the breeding season of *Lunatia* (= *Natica*) *hero*, large numbers of *N. trivittatus* congregate on the egg "collars" of *L. hero* and feed voraciously on the eggs in these collars (Clarke, 1956).

In the Toxoglossa the radular teeth are not arranged in rows attached to a basal supporting membrane, but are secreted individually and are then accumulated in a special caecum or quiver. *Philbertia*, in the Turridae, is such an example. One of the teeth is moved from the quiver and is passed up to the distal end of the proboscis, where it is firmly grasped by a sphincter muscle. Poison is secreted and accumulated in a very long poison duct which communicates with the lumen of the proboscis. Although the animals have not been seen to feed, nor have the remains of food been found in the gut, it is presumed that these snails are predacious carnivores which sting and kill their prey before consuming it. One related species has no radular teeth and it is difficult to envisage how it may feed (Smith, 1964). There are obvious similarities between the Turridae, e.g. *Philbertia* described above, and the Conidae which present a climax in the evolution of predacious carnivores among the Gastropoda.

The very large genus *Conus* is widely distributed in the tropical Pacific and Indian oceans, and all members of the genus are believed to be predacious carnivores. There may be as many as 400 good species of *Conus*, and as many as 20 may coexist on the same coral reef or in the same general area. Where only one species of *Conus* occurs in a given area its diet may tend to be more generalised than elsewhere, and *Conus* spp. are more stenophagous in regions where many species are present. Markedly ste-

nophagous habits have developed in many species. The animals are typically nocturnal, resting under stones or burying themselves in the substratum by day and crawling actively by night. Since each species normally
has well-defined food preferences and each occupies its own ecological
niche, there is little or no interspecific competition. As in the Turridae
described above, the radular teeth are elongated and lance-shaped, and
lie in a quiver at the base of the proboscis. The poison is secreted and is
stored in a poison duct, which may be many times as long as the shell
of the animal and which opens into the lumen of the proboscis. The distal
end of the poison duct is blind and comprises a large and muscular bulb,
the contraction of which doubtless supplies the force necessary to project
the poison when the animal strikes at its prey.

The majority of species of *Conus* are vermivorous, and most of these
feed chiefly on polychaet worms either in the family Nereidae or in the
family Eunicidae. Vermivorous species evidently do not use the radular
tooth to grip their prey, while several specimens of the same species of
Conus may congregate to feed on the same prey specimen. *C. lividus*
feeds principally on the family Terebellidae in Ceylonese waters, but in
Hawaiian waters the same species of *Conus* subsists chiefly on the enterop-
neust, *Ptychodera flava*. Vermivorous species and specimens are more
abundant than molluscivorous or piscivorous species and specimens.

Several species of *Conus* prey upon other gastropods. Thus *C. pennaceus*
probably feeds chiefly on the opisthobranch *Haminoea crocata*, while
C. textile is even known to feed on other species of *Conus*. In those species
of *Conus* which feed on other gastropods the radular tooth is used as a
poison dart. A single tooth is slid into position at the tip of the proboscis, the lumen of the proboscis is filled with poison which passes into the
hollow radular tooth. When *Conus* attacks a gastropod, the tooth, followed by a cloud of poison, is shot at the prey, the propulsive force being
provided by the muscular poison bulb. The snail which has been shot
immediately falls inert, the *Conus* approaches and places its expanded
mouth against the orifice of the shell of its prey, and within a short time
the flesh of the prey has been entirely removed and an empty shell is all
that remains. It is presumed that the tissues of the prey, especially the
columella muscle, are caused to break down by the action of the poison
which has been injected or by digestive juices which may be regurgitated
from the mouth of the *Conus* over its victim.

Piscivorous species are fewer, and have smaller populations, than
molluscivorous or vermivorous species. *C. striatus* and *C. catus* are known
to feed on fish, and this is probably true also for *C. halitropus*. *C. striatus*
normally buries itself in a sandy substratum from which only the siphon
protrudes. The animal becomes active at night, when their normal prey
may rest upon or close to the bottom. In an aquarium *C. striatus* becomes
active as soon as a fish is introduced into the aquarium, and the same re-

sponse is shown if water is introduced from another aquarium in which fish have been swimming. It seems probable that the presence of the fish is detected by a chemo-receptor, and that this is situated in the osphradium. In the presence of a fish *C. striatus* moves about with the proboscis everted, apparently "tracking" the fish. At this time the lumen of the proboscis is filled with poison and the radular tooth lies in readiness at the distal end of the proboscis. *C. striatus* strikes when the tip of its proboscis touches the side of the fish, the radular tooth being stabbed into the fish like a harpoon, the bulbous base of the tooth being gripped firmly by the sphincter muscle at the tip of the proboscis. Poison is injected into the fish via the cavity in the radular tooth, and the fish may struggle violently, but not for long (see Figs. 24 and 25). The venom is apparently a neuro-toxin. When the fish has ceased to struggle the *Conus* protrudes and dilates its mouth, which may expand from a few millimetres to as much as 2·0 cm in diameter. The whole fish is engulfed, tail first. This is an extreme deviation from the aboriginal molluscan habit of feeding upon minute particles.

The fish swallowed may be as long as the shell of the predator, and it cannot be passed far down the alimentary canal of the snail due to mechanical limitations imposed by the size of the fish and by the size and shape of the shell of the predator. Extracellular enzymes are presumably passed forwards in the gut, and the fish is digested while still within the oesophagus. In all cases a radular tooth is used only once. If the strike is successful the tooth is swallowed whole with the prey. If the strike is unsuccessful the tooth is discarded, and a new one is manoeuvred into position ready for the next attack.

Once having established the capacity to sting and kill prey, and to feed on large masses of flesh, *Conus* has evidently undergone a very successful adaptive radiation with consequent narrowing of the range of food taken by any one species, and with a great increase in the numbers of species in the genus. Due to stenophagous habits, many species coexist without producing any ill effects through interspecific competition. Each species is well fitted to its own particular ecological niche (see Fig. 26).

It is appropriate to add here that some species of *Conus* are capable of delivering a very painful sting, while in some species the sting is lethal to humans. *C. geographus* is the most outstanding example of the latter category, while *C. textile* and *C. tulipa* are apparently also potential killers of man. Other species may yet be added to this short list, and it is wise to take all possible precautions when handling living specimens of *any* species of *Conus*.

C. imperialis is the only species to feed on the polychaet *Eurythoe complanata*, the long setae of which penetrate human skin and cause a most unpleasant burning sensation. The gut of *C. imperialis* may be packed with these setae, without any damage being done to the gut wall. Chro-

FIG. 24. *Conus striatus* attacking, killing, and swallowing a fish. The long slender proboscis is extended with a radular tooth in position at the tip. The fish is stabbed in the side and is poisoned, the radular tooth serving as a harpoon. The fish is finally swallowed head first (see page 71). (Originally published in Kohn, 1956, *Proc. nat. Acad. Sci. Wash.* **42**, 169, figs. 3—7.)

matography has shown the presence in *C. striatus* venom of the quater-
nary ammonium compounds: *N*-methylpyridinium, homarine, and gam-
ma-butyrobetaine; unidentified indole derivatives were also present
(Kohn, 1955, 1955–6, 1956, 1958, 1959, 1959a, 1963; Kohn, Saunders
and Wiener, 1960). Endean and Rudkin (1963) found that extracts from
different parts of the poison duct may have different effects on experimental animals, which suggests that different types of gland cells are
present. This provides a means whereby the stinging apparatus can be
adapted for feeding on different kinds of prey.

Fänge (1960) found the salivary glands of *Neptunea antiqua* to contain
tetramine, but the use to which this poison is put by *Neptunea* is not known.

Kohn (1961) reviews the literature on chemo-reception in gastropods
and shows that scavengers such as *Nassarius*, *Cyclope*, and *Buccinum*,
and predacious carnivores such as *Urosalpinx* and *Conus* are able to
respond to chemical emanations from food at a distance. Even herbivores,
such as *Littorina* are able to move in the right direction to reach suitable
fucoid algae.

FIG. 25. *Conus textile*. The tip, and the base of a radular tooth. The tooth
is perforated by a central canal along which poison passes into the prey.

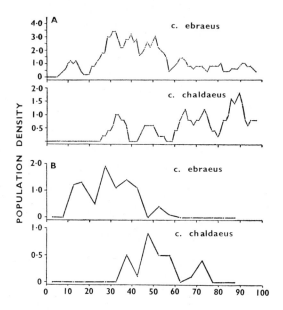

FIG. 26. The distribution of two closely related species of *Conus*, *C. ebraeus* and *C. chaldaeus*, across two coral reefs. A, Uliga Island, Majuro Atoll, in the Marshall Group. B, Milolii, Kauai, Hawaii. Whilst there is considerable overlap in the distributions of the two species on the surfaces of these two reefs, in each case *C. ebraeus* predominates over the landward half of the reef and *C. chaldaeus* predominates over the seaward half. Both species of *Conus* eat polychaets, but they differ markedly in their food preferences and there is little interspecific competition. Thus in the Hawaiian Islands *C. ebraeus* feeds principally on *Perinereis helleri* and *C. chaldaeus* feeds mainly on *Platynereis dumerilii*, though both feed to a small extent on the eunicid *Palola siciliensis*. (Originally published in Kohn and Orians, 1962, *Systematic Zoology*, **11**, 120, fig. 1.)

Cutters of algae

Not all of the Neogastropoda are carnivorous. Some members of the Columbellidae, e.g. *Columbella mercatoria*, feed exclusively on brown algae (Marcus and Marcus, 1962).

Opisthobranchia

Collectors of organic deposits

One of the most primitive members of the Opisthobranchia is *Actaeon tornatilis*, which possesses a number of features otherwise characteristic of prosobranchs, such as an operculum, a large, forwardly directed mantle cavity, and a streptoneurous nervous system. *A. tornatilis* occurs buried in clean sand near low-water mark of spring tides, in burrows the walls

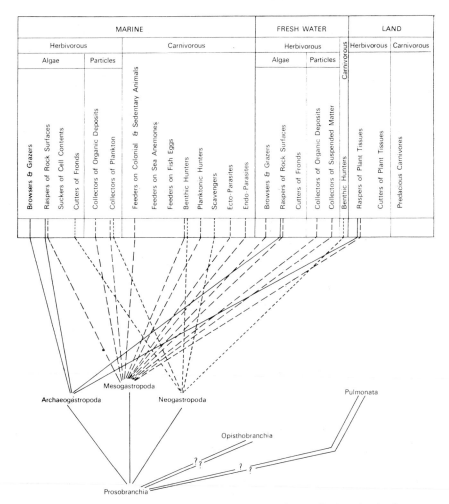

FIG. 27. Diagrammatic representation of adaptive radiation in feeding habits in the Prosobranchia.

of which are consolidated somewhat by mucoid secretions from the pedal gland. The radula is very small and insignificant, and the precise nature of the food-collecting mechanism is not yet known. The gut contents consist chiefly of sand grains (Fretter and Graham, 1954). It seems possible that the animal feeds on micro-organisms in the surface layers of the sand or in the layer of sea water lying immediately above the surface of the sand (see Fig. 127, p. 378).

Browsers and grazers on algae

Akera bullata grazes on green algae such as *Ulva*, and also feeds on comminuted material in marine deposits (Morton and Holme, 1955). As was discovered for the prosobranch *Hydrobia*, in all probability *Akera* also depends greatly on the flora of micro-organisms in the soil for its food supply. *A. bullata* can swim for short periods by beating the parapodia which are normally folded over the dorsal surface. The parapodia beat about fifty-five times per minute, the downward stroke providing

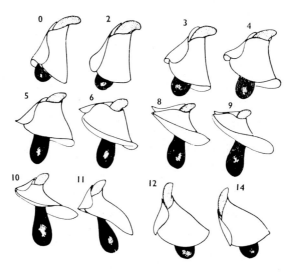

Fig. 28. *Akera bullata*, swimming. The outline drawings are taken from a ciné film record, the associated numbers indicating time intervals in sixteenths of a second. (Originally published in Morton and Holme, 1955, *J. mar. biol. Ass. U.K.* **34**, 105, fig. 1.)

the propulsive force. Swimming occurs in spring, apparently in association with the breeding season, while the capacity to swim enables *Akera* to search for suitable feeding grounds (Morton and Holme, 1955) (see Fig. 28).

The marine slug *Onchidella celtica*, now known to be an opisthobranch and not a pulmonate, occurs in small numbers near high-water mark on rocky shores; it feeds on young algae, encrusting diatoms and on foraminiferans such as *Gromia* and *Polystomella*. Once again, micro-organisms are probably an important source of food and are carried into the mouth on the debris which may thus serve principally as a vehicle for ingestion of micro-organisms. The food material is collected by the radula, which works against a chitinous jaw on the anterior wall of the oral tube (Fretter, 1943). In Singapore waters *Onchidella* spp. are dominant on the upper half of sheltered rocky shores. When the tide is in the *Onchidella* cannot be seen as they are all hidden in rock crevices or under stones, but as soon as they are exposed by the receding tide they emerge and crawl rapidly over the comparatively bare rocks, leaving long faecal cords behind them. In these waters *Onchidella* is particularly abundant on sheltered beaches where the rocks tend to be covered with a thin, slippery layer of organic debris which is doubtless rich in micro-organisms and on which *Onchidella* feeds.

Suckers of cell contents

A more advanced herbivorous feeding mechanism is by slitting open algal cells and sucking out the contents of the cells. This is of general occurrence in the Sacoglossa (= Elysiomorpha of Pelseneer). Here the radula is "monoseriate", i.e. there is only one median tooth in each "transverse row" so that these form a single longitudinal series of teeth. The leading tooth of the series is used to slit open the algal cell and old, discarded teeth are stored in an ascus sac below the mouth. *Elysia viridis* lives and feeds on *Codium* and, having slit open a cell wall, it sucks up the cell contents by expansions and contractions of the buccal mass. The facial region is firmly sealed to the surface of the weed by adhesive secretions from the buccal and pedal glands. Generally these sacoglossans are extremely stenophagous; thus *Actaeonia cocksi* and *Limapontia capitata* feed on species of *Cladophora*, while *L. depressa* feeds on species of *Vaucheria* (Gascoigne, 1956). Similar feeding mechanisms have been described for *Hermaea* and for *Caliphylla* (Fretter, 1940) and *Tridachia* is also believed to feed in this way (Yonge and Nicholas, 1940).

Cutters of algae

Aplysia feeds in an entirely different way, on *Ulva, Polysiphonia, Zostera*, etc. Here the weed is grasped firmly by the anterior end of the foot and the free end of the weed is manipulated so as to enter the mouth. The weed is drawn upwards by the radula until a piece about 2 cm long has been ingested. The jaws then tighten on the weed and the 2 cm strip is torn off by further radular action. Thus *Aplysia* feeds on comparatively large pieces of seaweed, which is a significant deviation from the primitive molluscan type of food. This has been rendered possible by the very efficient trituration of the food in the gizzard—not in the buccal mass— which exposes the cell contents for digestion. A cellulase is present, but it seems to be rather weak in its action (Howells, 1942).

Freshwater herbivores

The Succineidae are apparently unique among the Opisthobranchia in having invaded freshwater systems. *Succinea putris*, for example, lives in damp situations on the sides of lakes where it feeds by rasping green plant tissues with its radula. Secretions of the supra-pedal gland serve to lubricate the substratum, and some of this secretion is mixed with the food as it is swallowed. The Succineidae have been regarded as pulmonates until recently, but Rigby (1965) has brought forward evidence indicating that they are better placed among the Opisthobranchia.

Collectors of plankton

The thecosomatous pteropods are a group of opisthobranchs which have adopted a planktonic mode of life and which feed as herbivores on minute particles which are suspended in the sea water. In *Limacina retroversa* the ciliary fields on the wings beat outwards and do not aid in the feeding process. Food material consists of dinoflagellates and suitably shaped diatoms which are collected by ciliary mechanisms in the mantle cavity in spite of the fact that the ctenidium has been lost. A strong ciliary current enters the mantle cavity on the left and emerges on the right. Food particles are collected in a mucoid cord in mucus secreted by the

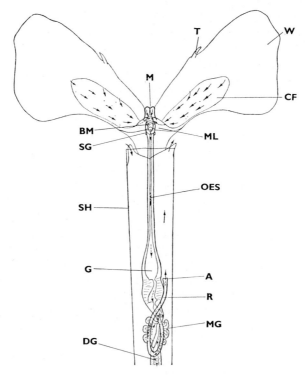

FIG. 29. *Creseis acicula*, showing ciliary feeding currents on the bases of the wings. For interpretation of lettering, see pp. 93–4. (Reproduced with permission, from Yonge, 1926, *J. Linn. Soc. Lond.* **36**, 420, fig. 2.)

pallial gland, and this mucoid cord is collected by the radula on the right side of the head. Note the parallel evolution with regard to *Crepidula* and related prosobranchs. In the oesophageal gizzard the food particles are crushed by four lateral gizzard teeth. *L. retroversa* frequently swims upwards to the surface, and then allows itself to sink passively with the wings held erect and so offering minimal resistance (Morton, 1954).

In related genera the feeding organs are a pair of flat ciliated lobes which are derived from the foot and which are innervated from the pedal ganglion. These ciliated lobes are continuous with the wing-like lobes which flap rhythmically to produce the swimming action. The ciliary currents on these ciliated feeding fields are directed orally and cause small particles to pass towards the mouth, to be ingested there (see Fig. 29). Jaws and radula are present in *Cavolinia* and in *Creseis*; they are present but only as vestiges in *Cymbulia*, and they are absent in *Gleba*. Arranging these four genera in this particular sequence provides a series showing increasing efficiency of the ciliary feeding fields, which interfere progressively less with the motion of the wings. The alimentary canal of the Thecosomata includes a muscular gizzard which is lined with plates, and which would seem to be useless in an animal which feeds on minute particles. The stomach possesses what appears to be a style sac; a crystalline style is present in *Creseis*, and is said to be present in the remainder of the Thecosomata. Assuming that the possession of a muscular triturating gizzard implies that the Theocosomata had a carnivorous ancestry, it would be necessary to assume further that such a hypothetical carnivorous ancestor had a free extracellular protease in its gut, and that it would therefore have lost all trace of any crystalline style which might have been present in a still earlier, herbivorous stage. On this premiss it would have to be concluded that the style-like structure in the stomach of *Creseis* is a secondary structure, analogous in structure, position, and function with the crystalline style of primitively herbivorous prosobranchs. This view has been put forward and the conclusion drawn that the Thecosomata have secondarily assumed a herbivorous diet (Yonge, 1926). It seems improbable, however, that a structure so similar to the crystalline style should have evolved in response to a change from carnivorous to herbivorous habits. Perhaps it would be better to accept a more economical hypothesis—always a sound scientific procedure—and argue that the lineage has always been herbivorous, that an ancestral form had fed on macroscopic algae or algal fragments and had acquired a muscular gizzard to crush them, as in the case of *Aplysia*. In that event the orginal crystalline style need never have been lost, and the only change to be postulated is a return to feeding on small particles, which may have occurred at the time when the planktonic mode of life was adopted.

Feeders on colonial animal growths and on sedentary animals

Grazing on animal substrates such as sponges is characteristic of many of the dorids or "sea lemons". Not infrequently the dorid has a cryptic coloration and closely resembles the species of sponge on which it feeds. Thus the pale yellowish *Jorunna tomentosa* feeds on sponges such as *Halichondria panicea*, making visible trails as it immerses itself by feeding on the sponge. *Jorunna* has no jaws. When feeding, the buccal mass is erected and the outer lips of the mouth are pressed downwards against the substrate of sponge. The odontophore is short, broad, and deeply bilobed by a median longitudinal groove. The radula on its upper surface is similarly shaped, and carries long transverse rows of similar, recurved radular teeth. There are three successive cylindrical lips to the mouth and these are relaxed in turn to allow the odontophore to protrude and press the radula against the substrate. The radula is then moved forwards and upwards, all the exposed rows of teeth scraping on the substrate. Fragments of sponge are torn off indiscriminately, and are carried into the mouth on the tips of the teeth. The whole of each rasping action takes less than 20 seconds and is quickly repeated (Millott, 1938). The brick-red dorid *Rostanga rufescens* is often found on the similarly colour-ed sponge, *Microciona atrasanguinea*, on which it feeds.

In contrast to the above, *Adalaria proxima* feeds mainly on the polyzoan *Electra pilosa*, but will also feed on *Membranipora membranacea*, *Flustrella hispida*, or on *Alcyonidium polyoum*, if *Electra* is scarce. When feeding, the radula is used in short bursts of three to eight strokes and then there is a short pause before renewal of action. After the exoskeleton of the polyzoan has been broken by the radula the soft parts are *sucked out* by the action of a buccal pump, and finally the dorid scrapes out any remaining flesh by using the radula (Thompson, 1958).

Pleurobranchus membranaceus feeds voraciously on compound and on solitary ascidians. When feeding on a solitary ascidian a hole is cut at any point in the tunic of the ascidian and the long proboscis is inserted to feed on the tissues within. *P. membranaceus* can swim freely, lying upside down with the mantle hanging down, by undulations of the left and right epipo-dial lobes. These lobes beat alternately and the animal rolls 45° on either side of the vertical plane (Thompson and Slinn, 1959) (see Fig. 30).

Onchidoris fusca feeds on the barnacles *Balanus balanoides* and *B. porca-tus*. Many of the smaller sea slugs feed upon a number of gymnoblast and or calyptoblast hydroids, sometimes having a strong preference for one particular species. A few are stenophagous, e.g. *Cuthona amoena*, which feeds solely on *Halecium* spp. Other nudibranchs are more catholic, e.g. *Doto coronata*, which has been known to feed on twenty different spe-cies of hydroid (Thompson, 1964). Useful lists of foods of various species of nudibranchs have been given by Graham (1937), by Miller (1961), and by Thompson (1964). An eolid exudes a copious supply of mucus over its prey before it begins to feed, the source of the mucus being the pedal gland, the gland cells at the anterior end of the foot, and also the glands in the vestibule. This secretion will lubricate the anterior part of the gut when the food is swallowed, but in addition it is possible that the mucus in some way inactivates the nematocysts of the prey so that the eolid is not stung while feeding (Graham, 1937).

Some of the larger sea slugs feed on sea anemones. *Eolidina alderi*, for example, feeds on the anemone *Heliactis bellis*. In *Eolidina* there is a pair

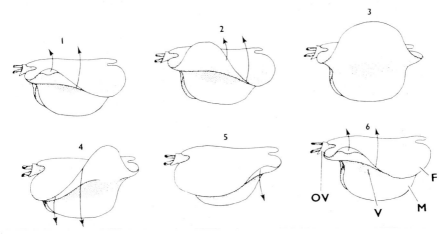

FIG. 30. The mechanism of swimming in *Pleurobranchus plumula*, seen from the right side; the gill is not figured. The animal turns over on to its back and swims by rhythmic alternating undulations of the left and the right epipodial lobes of the foot. Due to the lack of synchrony of beating of the epipodial lobes the animal rolls heavily to about 45° to either side of the vertical plane. For interpretation of lettering, see pp. 93–4. (Originally published in Thompson and Slinn, 1959, *J. mar. biol. Ass. U.K.* **38**, 516, fig. 7.)

of chitinous jaws with sharp cutting edges. These jaws hang freely in the buccal cavity and are used to cut small pieces of flesh off the anemone. The radula is then used to rake these fragments into the oesophagus. The radula is modified to serve as a conveyor belt. Apparently numerous teeth are secreted individually but subsequently fuse together in groups, each group forming a comb-like plate the free edge of which has a number of denticles (Graham, 1937). Similarly *Aeolidia papillosa* feeds on anemones such as *Actinia equina* and *Metridium senile*. Stehouwer (1952) subjected *Aeolidia papillosa* to choice tests under controlled conditions in aquaria. A number of *A. papillosa* were placed in an aquarium into which water entered at the same speed via two tubes from two contrasted sources. When the water from one of these sources contained anemones and the other source contained no anemones, the slugs tended to move towards and into the tube leading to the vessel containing the anemones. Analysis of the results showed that *A. papillosa* preferred *M. senile* to the three remaining species of anemone that were tested, namely *Diadumene cincta*, *Actinothoe anguicoma*, and *Tealia felina*. In further experiments the anemones were tested against one another in pairs to determine their relative attractiveness to the slug. Once again, *M. senile* was found to be most attractive, and no clear preferences were shown among the three remaining species.

Calma glaucoides is a nudibranch which has adopted the remarkable habit of feeding on the eggs and embryos of various shore fishes such as *Cottus*, *Lepadogaster*, *Liparis*, *Blennius*, and *Gobius*. These eggs are laid in sheets on the undersides of rocks or in rock crevices, and are only available at certain seasons. The adaptations of *Calma* to this feeding habit are particularly interesting. The mouth is modified to form a hood which fits closely over the fish egg, so ensuring that there will be no wastage of the contents when the egg membrane is punctured. The five anteriormost radular teeth are uniseriate and resemble those of other eolids. They are presumably used by the young *Calma* in orthodox eolid feeding mechanisms prior to assumption of the adult habit of feeding on fish eggs. Posterior to these five teeth the structure of the radula changes in adaptation to the mode of feeding and the diet of the adult animal, the radular membrane being developed into a firm, gently curved, cylindrical rod which bears closely set uniseriate teeth, resembling the blade of a fret-saw. In the adult stage *Calma* only uses the posterior part of the radula to slit open the egg membrane, and it then sucks the contents of the egg into the oesophagus. The structure of the whole radula therefore serves as a record of the feeding methods of the individual throughout its life, and likewise recapitulates its recent ancestry among orthodox eolids.

The contents of the eggs are highly digestible and, in consequence, the gut of *Calma* has become blind, there being no anus and no intestine. Suitable eggs and embryos can only be obtained at certain seasons, so provisions had to be made to enable *Calma* to live long enough, while feeding in this specialised way, to breed and to supply a generation for the following year, for the animal is an annual. This was achieved by feeding to excess when fish eggs are available and storing yolk and embryos in the stomach for long periods. The food material is stored in a specially large gastric sac and is slowly digested as required. Due to its specialised diet the adult *Calma* has no opportunity to swallow nematoblasts of coelenterates, as do orthodox eolids, and it is interesting to note that accordingly the cerata of *Calma* do not contain cnidosacs (Evans, 1922; Rowett, 1946).

Benthic hunters

Several opisthobranchs plough through the surface layers of sandy shores, or sub-littoral sandy deposits, and prey on comparatively large animals

such as bivalve molluscs, which they swallow whole. *Philine aperta* is such
a predacious carnivore. *Philine* has a reduced internal shell, its white
fleshy body is subdivided into four lobes, and it might not be recognised
at first sight as a gastropod. This species is typically found sub-littorally,
in sand, or in muddy sand. It feeds on a variety of organisms including na-
viculate diatoms, unicellular algae, bottom-living foraminiferans, carcases
of planktonic organisms that have died and sunk to the bottom, and even
larger animals such as bivalve molluscs (Fretter, 1938). Each transverse
row on the radula consists of only two teeth which are triangular and have

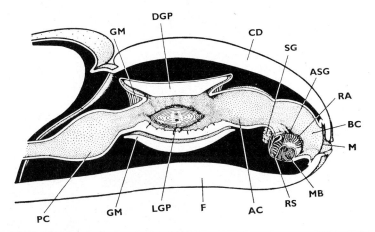

Fig. 31. *Philine aperta*. Sagittal section through the anterior part of the
alimentary canal. The mouth (M) leads into the buccal cavity (BC) in the
floor of which lies the radula (RA). The salivary gland (SG) opens into
the buccal cavity laterally (ASG). The anterior crop (AC) passes back into
a large and muscular gizzard the walls of which are provided with three
calcareous plates, the dorsal plate (DGP) being shown in longitudinal
section. The left gizzard plate (LGP) is seen lying on the left wall of the
gizzard. The gizzard passes back into a posterior crop (PC). For interpretation of other lettering, see pp. 93–4. (Originally published in Fretter, 1939,
Trans. roy. Soc. Edinb. **59**, 601, fig. 1.)

hooked tips. When the radula is protruded these teeth diverge, and when
the radula is withdrawn again the teeth converge and grasp selected objects
in the sand. The form of the teeth is admirably suited to the function of
grasping and ingesting objects (Brown, 1933).

The food is swallowed undamaged. It is passed to a muscular gizzard
armed with three powerful crushing plates, which serve to liberate the body
juices of the prey and to expose its flesh to the action of the digestive enzymes. Here we have a significant departure from the ancestral habit of
swallowing only finely divided food particles, and it has been made possible through the development of a new organ, the gizzard, and the development also of an extracellular protease (see Figs. 31 and 32).

82

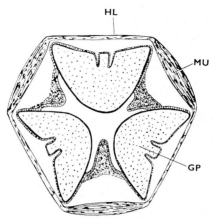

FIG. 32. *Philine aperta.* Transverse section through the gizzard, showing three large gizzard plates (GP) which serve to crush the prey. Powerful sheets of muscle fibres (MU), linked by "horny" layers (HL) provide the crushing force. (Originally published in Brown, 1934, *Trans. roy. Soc. Edinb.* **58**, 189, fig. 18.)

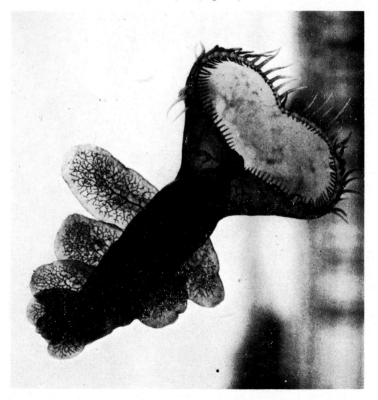

FIG. 32a. *Melibe leonina.* Photograph of a healthy specimen living in an aquarium. (This photograph was provided by Dr. Anne Hurst, Department of Zoology, University of Reading.)

Scaphander lignarius similarly feeds on bivalves and on gastropods, and whole shells of the snail *Turritella communis* have been found in its gizzard (Fretter, 1938). *Haminea* is another genus which feeds by swallowing bivalves whole and crushing them in its gizzard (Graham, 1953).

The eolid *Melibe* (see Fig. 32a) has elaborated a truly remarkable feeding procedure by casting a cephalic veil, like a net, and trapping small crustaceans under the veil. The head of this sea slug bears a highly mobile, circular cephalic veil of which the tentaculate margin is contractile. The mouth lies on a short snout which is situated centrally under the veil. When *Melibe* encounters actively moving copepods, or other small crustaceans, it moves forwards and casts its veil expertly over the prey it has selected. The margin of the veil then contracts, drawing the trapped animal close to the snout. *Melibe* has no jaws, and the animal is swallowed whole. It is a most voracious feeder, and the limbs of crustaceans more than an inch long have been found in its stomach (Elliott, 1902). The food is passed to a gizzard, where it is triturated (Agersborg, 1923).

Planktonic hunters

The nudibranch *Fiona pinnata* feeds upon the siphonophores *Velella* and *Porpita* (Thompson, 1964).

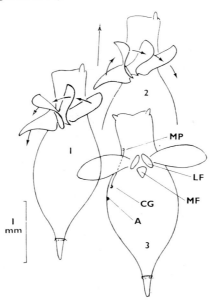

FIG. 33. *Clione limacina*. Outline diagrams showing successive positions of the wings while swimming. The downward stroke (1) and the upward stroke (2) are illustrated for the left wing only. The animal is also shown in ventral view (3) in which both wings are shown. For interpretation of the lettering, see pp. 93–4. (Originally published in Morton, 1958, *J. mar. biol. Ass. U.K.* **37**, 288, fig. 1.)

The gymnosomatous pteropods are fast-swimming predacious carnivores which prey on other planktonic organisms, particularly upon thecosomatous pteropods (see p. 89) which they have no difficulty in finding in spite of having greatly reduced eyes. *Clione limacina* is normally found in

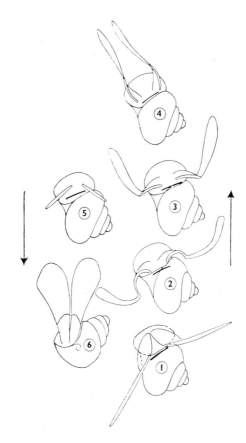

Fig. 34. *Limacina retroversa*, swimming. 1, 2, 3, and 4 show successive positions of the wings while the animal is swimming upwards; 5 and 6 are two different views of the animal while sinking with the wings held motionless and vertical. (Originally published in Morton, 1954, *J. mar. biol. Ass. U.K.* **33**, 299, fig. 1.)

company with larger numbers of *Limacina retroversa* (Fig. 34), upon which it presumably feeds. *C. limacina* swims gracefully either vertically or horizontally, sculling with the short wings which are set transversely near the anterior end of the body. The wings beat synchronously (see Fig. 33). *Clione* has no jaws but on each side of the mouth there is a hook-sac which contains about fifteen chitinous blades, and also three stalked adhesive organs known as "cephaloconi". The hooks and the cephaloconi are presumably used to grasp and to manipulate the prey. There is a mid-ventral radula which will serve to cut up and ingest the prey once this has been seized (Morton, 1958).

Ecto-parasites

Although the Pyramidellidae have certain features which are typically found in the Prosobranchia, it has been shown that this family should be reclassified and placed in the Opisthobranchia (Fretter, 1949). Little is known of the biology of the majority of species in this family, but those which have been studied have been shown to be ecto-parasites on a restricted number of host species. *Turbonilla elegantissima* lives in the laminarian zone on the sea shore, in silt under boulders or in rock crevices. It is found only in proximity to either *Audouinia tentaculata* or to *Amphitrite gracilis*, and it feeds by sucking body fluids from either of these worms. When the tide is out the worm may be retracted, and the snail will be found in the silt, unassociated with the worm. When the tide rises and the worm spreads its tentacles to feed, the snail approaches and attaches itself to the worm by a sucker at the tip of its long, slender proboscis. This may be attached to the tentacles of either species of worm, or to the gills of *Amphitrite*. In these places the body wall of the worm is sufficiently thin to be penetrated in order to suck the body fluids. There is no radula. The long proboscis is protruded by blood pressure and becomes attached to the host by its terminal sucker-like orifice. The jaws have become modified to form a tubular stylet which surrounds the combined duct from the two salivary glands. The stylet punctures the body wall of the host worm at the same time that the sucker becomes attached. Presumably secretions from the salivary glands are injected into the puncture, and may serve as an anti-coagulin. The body fluids are then sucked up by a buccal pump and passed to the stomach for digestion (Fretter, 1951).

In some species of Pyramidellidae the proboscis is transparent, and the action of the buccal pump lying within the proboscis can be clearly seen when the snail is feeding (see Fig. 35). *Odostomia unidentata* and *O. lukisii* feed upon the worm *Pomatoceros triqueter* and attach the proboscis to one of the branchial filaments of the worm. Other species of this genus may feed upon bivalve molluscs, e.g. *O. scalaris*, which stands on the shell of *Mytilus edulis* close to the ventral margin of the shell and attacks the margin of the mantle cautiously when the shell valves of the bivalve are parted. A number of these snails may be seen feeding side by side. *Chrysallidea spiralis* feeds on *Sabellaria* spp. In this case the proboscis is thrust into the mouth of the worm, and it is not possible to see whether the snail is feeding on the contents of the gut of the worm or upon body fluids sucked from the lining of the gut (Fretter, 1949). Cole and Hancock (1955) reported the occurrence of *O. eulimoides* and of *C. obtusa* feeding on *Ostrea edulis*. These parasites feed on the mantle edge, which retreats from the edge of the shell in response to the constant irritation. In due course the edge of the shell may become distorted at the point of attack and the shell may become incapable of closing properly at this point. Sand and silt may then enter, and the oyster may die as a result of the parasitic attack.

This survey of feeding mechanisms in the Opisthobranchia shows that there is less variety in feeding methods in this sub-class than was noted in the Prosobranchia. The adaptive radiation of the Opisthobranchia has been less spectacular, and there has been a notable failure to gain a foothold on dry land. Only the Succineidae among the Opisthobranchia have successfully invaded fresh waters.

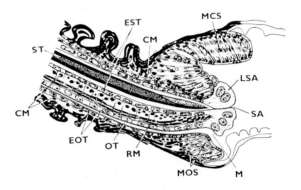

Fig. 35. *Odostomia plicata*. Longitudinal section through the mouth and the anterior part of the alimentary canal. The mouth (M) leads into the oral tube (OT). The stylet (ST) lies in a stylet tube which opens by a separate aperture (SA) above the mouth. The two apertures lie within a circular sucker provided with muscles (MCS, MOS), by means of which the proboscis is attached to the prey. For interpretation of other lettering, see pp. 93–4. (Originally published in Fretter and Graham, 1949, *J. mar. biol. Ass. U.K.* **28**, 502, fig. 4.)

We regard the Opisthobranchia as having been derived from the Prosobranchia by a process of detorsion, which has carried the mantle cavity backwards along the right side of the body and in advanced cases has led to the loss of this cavity and its contained organs. The process has also been accompanied by reduction and, in extreme cases, loss of the shell. The Prosobranchia are typically bi-sexual, whereas the Opisthobranchia are typically hermaphrodite. On these grounds we rightly regard the Opisthobranchia as having arisen more recently, and from a more highly advanced stock than did the Prosobranchia. For these reasons it is not surprising to find in the Opisthobranchia a paucity of the more primitive feeding mechanisms such as browsing and grazing on algae, rasping of algae off rocks, collecting organic deposits, etc. The Opisthobranchia probably arose from one—or more?—prosobranch stocks which had already begun to develop more specialised feeding habits.

Physiological limitations have undoubtedly played a major role in preventing opisthobranchs from invading fresh waters in abundance; feeding mechanisms may also have played some part in maintaining this limit to their distribution, for animals which have already adopted specialised feeding mechanisms, and which are relatively stenophagous, will be limited in their distribution by that of their basic food supply. The invasion of estuaries and river systems would be easier for the Prosobranchia with comparatively primitive feeding mechanisms than for the Opisthobranchia with more specialised and restrictive feeding methods. This view is supported by the fact that among the Prosobranchia apparently only one predacious carnivore, the buccinid *Antenome*, occurs in fresh waters.

It is possible that the Opisthobranchia are polyphyletic, and that the term "opisthobranch" primarily represents a structural and functional stratum in the evolution of gastropods. In the Heteropoda the sequence *Atlanta, Carinaria, Pterotrachea* points suggestively towards the Opisthobranchia. *Atlanta* has a well-developed coiled shell and an operculum, and is characteristically prosobranch; *Carinaria* has no operculum and its shell is uncoiled and of reduced size; *Pterotrachea* has no shell and no operculum, and is opisthobranchiate in form (Yonge, 1942). The Pyramidellidae used to be classified in the Prosobranchia and possess a number of prosobranch features, but now that they are better known they are recognised as being assignable to the Opisthobranchia. This suggests that a number of prosobranch lineages may have evolved towards the opisthobranch "stratum" and that two or more of these may have "broken through" and become recognised as true members of the Opisthobranchia. On these grounds the sub-class may well be polyphyletic. This would account for difficulty in agreeing upon a natural classification within the sub-class.

Pelseneer recognised two subdivisions, namely the Tectibranchia and the Nudibranchia. In the more recent classification of Thiele there are a number of striking contrasts:

1. The thecosomatous and gymnosomatous pteropods have been withdrawn from two divisions of the Tectibranchia (Bullomorpha and Aplysiomorpha respectively) to form a separate order Pteropoda.
2. The Elysiomorpha are withdrawn from the Nudibranchia to form a separate order Sacoglossa.
3. The Pleurobranchomorpha are withdrawn from the Tectibranchia and are associated with the dorids and nudibranchs in the order Acoela.

FIG. 36. Diagrammatic representation of adaptive radiation in feeding habits in the Opisthobranchia.

The classification of Thiele has therefore raised the number of orders from two to four, and offers a considerably different assessment of the systematic relationships of many of the constituent families. These clashes of opinion become understandable if the Opisthobranchia are not a natural assemblage, but are polyphyletic—in which case they will defeat all our attempts to force them into a neat, interrelated scheme. The classification of Wilbur and Yonge (1964) has increased the number of orders to eight

TABLE 3. *Schemes of classification of the sub-class Opisthobranchia*

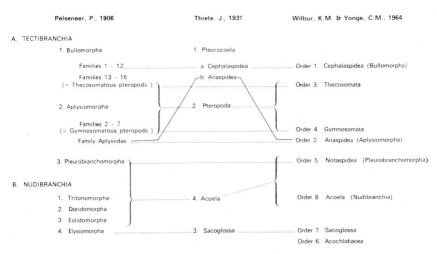

The Pyramidellidae and the Succineidae, recently transferred to the Opisthobranchia from the Prosobranchia and from the Pulmonata respectively, should presumably be assigned to the Bullomorpha (= Cephalaspidea).

As indicated by the numerals, the schemes of classification of Thiele and of Wilbur & Yonge have been slightly re-arranged so as to bring out the features of similarity of the three schemes. The most significant development is the abandoning of any attempt to group the orders into higher taxonomic categories.

The nomenclature adopted by Wilbur & Yonge is not entirely satisfactory; Cephalaspidea and Anaspidea are too easily confused with the Cephalaspida and Anaspida (categories of fossil fish), while the Acoela of Wilbur & Yonge differ in content from the Acoela of Thiele.

and offers no comments on the possible relationships between these orders.

Pulmonata

The Pulmonata have most effectively colonised dry land, and have spread from the surface of the land upwards into an arboral habitat, e.g. *Amphidromus* in Malaysia; into fresh waters, e.g. *Lymnaea, Planorbis, Ancylus*; into brackish waters such as mangrove swamps, e.g. *Ellobium*; and on the sea shore, e.g. *Otina* and *Leucopepla* in British seas, *Trimusculus* and *Siphonaria* in tropical seas. In many of the aquatic pulmonates respiration is still aerial, but the marine snail *Trimusculus* is amphibious and can breathe either air or water, while *Siphonaria* is wholly adapted to the aquatic habitat, the mantle cavity is filled with sea water, and a secondary gill has developed in place of the ctenidia.

Typically slugs and snails in the Pulmonata feed voraciously on plant material, commonly preferring the more succulent growths. Some slugs feed on bulbs, tubers, or roots, e.g. *Limax* spp., while *Arion subfuscus* feeds on fungi (Graham, 1955). In certain cases a powerful cellulase is known to be present, and this permits complete digestion of plant tissues— it being assumed that carbohydrases, lipases, and proteases are still being retained. Armed with this digestive tool, slugs and snails are able to cut and ingest comparatively large pieces of leaf, and so feed with rapidity. Growth rates and rates of reproduction may be high, and especially in the tropics. The Giant African Snail, *Achatina*, is particularly formidable as a feeder on vegetable material, and careful steps are being taken to limit further spread of this major economic pest. The snail *Partula*, which occurs in many Pacific islands, apparently subsists exclusively on the mycelia of fungi which grow on decaying plant materials, and they do not appear ever to feed on angiosperm tissues (Crampton, 1925).

Many slugs and snails will readily feed on flesh, even upon that of the same species. Some, however, have become entirely carnivorous. This is generally true of the Oleacinacea and the Streptaxacea (Graham, 1955). In the Oleacinacea the slug *Testacella* spends much of its time in the soil, and apparently feeds principally on earthworms and on small slugs. In the Streptaxacea the genera *Haplotrema*, *Edentulina*, *Streptaxis*, and *Gonaxis* are well-known carnivorous examples. *E. affinis* provides a very effective natural control of *Achatina fulica* (Burch, 1960). In 1950 about 300 specimens of *Gonaxis kibweziensis* were liberated experimentally on the island of Agiguan, off Tinian in Micronesia, in an attempt to control the population of *Achatina* there. The *Gonaxis* fed voraciously on the eggs and young specimens of *Achatina*, they increased very rapidly in numbers, and effected a very considerable reduction in the numbers of *Achatina*. *Achatina* laid egg clutches of 100–400 eggs, and these were often totally destroyed by *Gonaxis* (Anon.).

The freshwater snail *Lymnaea stagnalis* feeds on aquatic weeds, detritus, and on carrion, and regularly consumes sand with its food. This sand is essential for the proper functioning of the gizzard in the process of trituration of the food material. When deprived of the opportunity to collect sand, the sand content of the gizzard gradually diminished and when no sand was left in the gizzard the animals ceased to feed (Carriker, 1946).

Hubendick (1957) has described the feeding process in *L. stagnalis*. The animal slowly creeps forwards, swinging its head from side to side and regularly applying its mouth to the substratum to rasp at leaves, etc. Each rasping action was found to comprise a sequence of stages (see Fig. 37). The buccal bulb and the odontophore are rotated so that the antero-dorsal end of the odontophoral "cartilage" is lowered, pressed through the mouth, and applied to the substratum with the radula stretched tightly over what is now the ventral end of the odontophore. At first the exposed end of the radula only touches the substratum laterally on each side; later the convexity of the radular membrane is reduced, causing the lateral parts of the radula to converge somewhat, scraping the

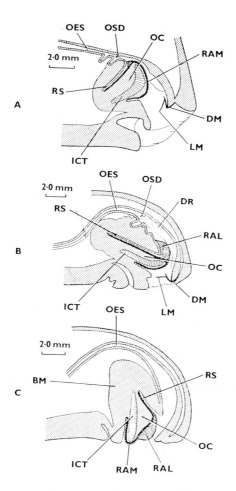

FIG. 37. *Lymnaea stagnalis*. Diagrammatic vertical sections through the head representing A, the resting stage; B, the first stage in the feeding cycle; C, the last stage in the feeding cycle of movements. The buccal mass (BM) as a whole, and also the odontophoral cartilage (OC) in addition, are rotated so that the antero-dorsal end of the cartilage is lowered and pressed through the mouth in order to apply the radula to the substratum. First of all the lateral parts of the radula (RAL) engage with the substratum, then the median part of the radula (RAM) makes contact with the substratum, and then the odontophore begins to rotate back towards its resting position, so pushing the median part of the radula forwards over the substratum. For interpretation of other lettering, see pp. 93–4. (Redrawn in simplified form from the original figures in Hubendick, 1957, *Ark. Zool.* **10**, figs. 2, 3, and 5.)

substratum towards the median line. Then the median part of the radula makes contact with the substratum, the odontophore begins to rotate backwards to its resting position and this causes the median part of the radula to be pushed forwards over the surface of the substratum. Food is taken into the mouth at this stage and the mouth begins to close posteriorly. Finally, the buccal bulb, the odontophore, and the radula are brought back to the resting position. Three stages in this process are illustrated (Fig. 37) and the difference between this process and the feeding process of a rhipidoglossan or of a taenioglossan is emphasised (see p. 45).

Various pulmonates occupy marine habitats and live on the sea shore. The tiny *Leucopepla* occurs high on the beach, near high-tide mark, and can be found on muddy deposits under stones at this level; it does not suffer long submergence in sea water. The pulmonate limpet *Trimusculus* occurs lower down on rocky shores, on the under sides of rocks or on the roofs of small caves and crevices. *Trimusculus* feeds on fine encrusting vegetable material, scraping this off the rock surface by protruding its odontophore and rasping the rock with its radula (Yonge, 1958). In contrast, the genus *Siphonaria* occurs on upward facing rock surfaces at or below mean sea level on tropical shores. Observation of *S. atra* in Singapore waters showed that this species is inactive, and is firmly clamped down to the rock for long periods when it is completely submerged, and it only begins to move about and feed when it begins to be exposed by the falling tide. It then feeds on encrusting algae, detritus, diatoms, etc., returning after such foraging trips to its "home" which is recognisable as a conspicuous scar on comparatively soft rocks. These foraging trips are of short duration, for the exposed rocks soon become hot and dry due to sun and wind, and the animals must return to their "homes" before this happens. In this habit of feeding rapidly for a very short time, only at the time of exposure by the falling tide, *Siphonaria* differs markedly from the prosobranch limpets which otherwise it closely resembles.

S. thersites has a shell of reduced size, into which it cannot withdraw, and an extensive area of the margin of the mantle is permanently exposed. The exposed area of mantle bears an abundance of glands which may be repugnatorial and so protect the animal against predators—it has no other form of protection as it is unable to clamp its reduced shell tightly against the rock surface. This species is considerably more mobile than is typical of limpets; it probably avoids the effects of insolation and desiccation in the summer by hiding in rock crevices, from which it emerges to feed on *Fucus*, in which it excavates rounded pits as it feeds. *S. thersites* is thus adapted to a specialised and restricted type of habitat and is able to survive on beaches which are swept by storms and exposed to desiccation by sun and wind. Perhaps for these reasons it has been able to extend its range as far north as Alaska, which is far beyond the range of other species of marine pulmonates (Yonge, 1960).

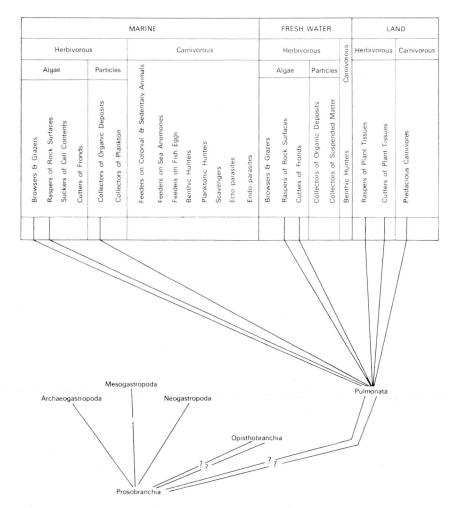

Fig. 38. Diagrammatic representation of adaptive radiation in feeding habits in the Pulmonata.

Key to the Lettering on the Figures

A	Anus.	CG	Common genital aperture.
AC	Anterior crop.	CM	Circular muscles.
ASG	Aperture of duct of salivary gland.	CT	Cephalic tentacle.
		DG	Digestive gland.
BC	Buccal cavity.	DGP	Dorsal gizzard plate.
BM	Buccal mass.	DM	Dorsal mandible.
C	Ctenidium.	DR	Dorso-lateral ridge in oesophagus.
CD	Cephalic disc.		
CF	Ciliated field.	E	Exhalant water current.

EN	Endostyle.	MP	Male genital aperture.
EOT	Epithelium of oral tube.	MS	Mucoid string containing food
ES	Exhalant siphon.		particles.
EST	Epithelium of stylet tube.	MU	Muscle sheets operating the
EYE	Eye.		gizzard.
F	Foot.	O	Osphradium.
FG	Food groove.	OC	Odontophoral cartilage.
FP	Food pouch.	OES	Oesophagus.
G	Gizzard.	OP	Operculum.
GM	Circular muscle band connect-	OSD	Orifice of salivary duct.
	ing gizzard plates.	OT	Oral tube.
GP	Gizzard plate.	OV	Oral veil.
GR	Groove along the proboscis.	PC	Posterior crop.
H	Head.	PR	Proboscis.
HL	"Horny" layer connecting	R	Rectum.
	muscle sheets.	RA	Radula.
I	Inhalant water current.	RAL	Left lateral part of the radula.
ICT	Infra-cartilage tensor mus-	RAM	Median part of the radula.
	cles.	RM	Retractor muscles of the lips
L	Lip of mouth of zooid.		of the stylet aperture.
LF	Lateral lobe of the foot.	RS	Radular sac.
LGP	Left gizzard plate.	SA	Stylet aperture.
LM	Left lateral mandible.	SG	Salivary gland.
LSA	Lip of stylet aperture.	SH	Shell.
M	Mouth.	SI	Siphon.
MA	Mantle.	ST	Stylet.
MB	Muscles of buccal mass.	T	Tentacle.
MC	Mantle cavity.	V	Visceral mass.
MCS	Muscles arching sucker.	W	Wing.
MF	Median lobe of foot.	X	Region where food collects
ML	Middle lobe of foot.		on the propodium, below the
MOS	Muscles flattening sucker.		proboscis.

Reference List

AGERBSORG, H. P. (1923) The morphology of the nudibranchiate mollusc *Melibe* (syn. *Chioraea*) *leonina* (Gould), *Quart. J. micr. Sci.* **67**, 507–592.

ANDREWS, E. B. (1965) The functional anatomy of the gut of the prosobranch gastropod *Pomacea canaliculata* and of some other pilids, *Proc. zool. Soc. Lond.* **145**, 19–36.

ANKEL, W. E. (1937) Wie bohrt *Natica*?, *Biol. Zbl.* **57**, 75–82.

ANKEL, W. E. (1938) Erwerb und Aufnahme der Nahrung bei den Gastropoden, *Verh. dtsch. zool. Ges. Leipzig* **40**, 223–295.

ANON. (1955) Biological control of *Achatina* by *Gonaxis;* the biological control of the giant african snail, *Pacif. Sci. Ass. Inf. Bull.* **7**, No. 6.

ANSELL, A. D. (1960) Observations on predation of *Venus striatula* (da Costa) by *Natica alderi* (Forbes), *Proc. malac. Soc. Lond.* **34**, 157–64.

BAER, J. G. (1952) *Ecology of Animal Parasites*, Univ. of Illinois Press, Urbana.

BAKKER, K. (1959) Feeding habits and zonation in some inter-tidal snails, *Arch. néerl. Zool.* **13**, 230–57.

BERRILL, N. J. (1930) The natural history of *Bulla hydatis* Linn. *J. mar. biol. Ass. U.K.* **17**, 567–71.

BROWN, A. C. (1961) Physiological-ecological studies on two sandy-beach gastropoda from South Africa; *Bullia digitalis* Menschen and *Bullia laevissima* (Gmelin), *Z. Morph. Ökol. Tiere* **49**, 629–57.

BROWN, A. C. (1961a) Chemo-reception in the sandy-beach snail *Bullia*, *S. Afr. J. Lab. clin. Med.* **7**, 160.

BROWN, H. H. (1934) A study of a tectibranch gastropod mollusc, *Philine aperta* (L.), *Trans. roy. Soc. Edinb.* **58**, 179–210.

BURCH, J. B. (1960) Some snails and slugs of quarantine significance to the United States, *Sterkiana*, **2**, 13–53.

CARRIKER, M. R. (1943) On the structure and function of the proboscis in the common oyster drill *Urosalpinx cinerea* Say., *J. Morph.* **73**, 441–91.

CARRIKER, M. R. (1946) Observations on the functioning of the alimentary system of the snail *Lymnaea stagnalis appressa* Say., *Biol. Bull. Woods Hole*, **91**, 88–111.

CARRIKER, M. R. (1951) Observations on the penetration of tightly closing bivalves by *Busycon* and other predators, *Ecology* **32**, 73–83.

CARRIKER, M. R. (1955) Critical review of biology and control of oyster drills *Urosalpinx* and *Eupleura*, *Fishery Prod. Rep. Fish. Wildl. Serv. U.S.*, No. 148, 1–150.

CARRIKER, M. R. (1958) Additional information on the mechanical–chemical nature of drilling by the gastropods *Urosalpinx* and *Eupleura*, *A.S.B. Bull.* **5**, (1).

CARRIKER, M. R. (1961) Comparative functional morphology of boring mechanisms in gastropods, *American Zoologist* **1**, 263–6.

CHEW, K. K. (1960) Study of food preference and rate of feeding of Japanese oyster drill *Ocinebra japonica* (Dunker), *Fishery Prod. Rep. Fish. Wildl. Serv. U.S.*, No. 365, 1–27.

CLARKE, A. H. (1956) Natural biological control of a *Mya* predator, *Nautilus* **70**, 37–38.

COLE, H. A. (1942) The American whelk-tingle, *Urosalpinx cinerea* Say. on British shores. *J. mar. biol. Ass. U.K.* **25**, 477–508.

COLE, H. A. and HANCOCK, D. A. (1955) *Odostomia* as a pest of oysters and mussels, *J. mar. biol. Ass. U.K.* **34**, 25–31.

COOK, P. M. (1949) A ciliary feeding mechanism in *Viviparus viviparus* (L.), *Proc. malac. Soc. Lond.* **27**, 265–271.

CRAMPTON, H. E. (1925) Studies on the variation, distribution, and evolution of the genus *Partula*: the species of the Mariana Islands Guam and Saipan, *Publ. Carneg. Inst.*, No. 228 A, 1–116.

ELLIOT, SIR C. (1902) On some nudibranchs from Zanzibar, *Proc. zool. Soc. Lond.* 62–72.

ENDEAN, R. and RUDKIN, C. (1963) Studies on the venoms of some Conidae, *Toxicon* **1**, 49–64.

EVANS, T. J. (1922) *Calma glaucoides*: a study in adaptation, *Quart. J. micr. Sci.* **66**, 439–55.

FÄNGE, R. (1960) The salivary gland of *Neptunea antiqua*, *Ann. N.Y. Acad. Sci.* **90**, 689–694.

FORREST, J. E. (1953) Symposium on the form and function in the molluscs: on the feeding habits and the morphology and mode of functioning of the alimentary canal in some littoral dorid nudibranchiate molluscs, *Proc. Linn. Soc. Lond.* **164**, 225–35.

FRETTER, V. (1939) The structure and function of the alimentary canal of some tectibranch molluscs, with a note on excretion, *Trans. roy. Soc. Edinb.* **59**, 599–646.

FRETTER, V. (1940) On the structure of the gut of the ascoglossan nudibranchs, *Proc. zool. Soc. Lond.* **110**, 185–98.

FRETTER, V. (1943) Studies in the functional morphology and embryology of *Onchidella celtica* (Forbes and Hanley) and their bearing on its relationships, *J. mar. biol. Ass. U.K.* **25**, 685–720.

FRETTER, V. (1948) The structure and life history of some minute prosobranchs of rock pools: *Skeneopsis planorbis* (Fabricius), *Omalogyra atomus* (Philippi), *Rissoella diaphana* (Alder), and *Rissoella opalina* (Jeffreys), *J. mar. biol. Ass. U.K.* **27**, 597–632.

FRETTER, V. (1949) The structure and mode of life of the Pyramidellidae, parasitic opisthobranchs, *J. mar. biol. Ass. U.K.* **28**, 493–532.

FRETTER, V. (1951) *Turbonilla elegantissima* (Montagu), a parasitic opisthobranch, *J. mar. biol. Ass. U.K.* **30**, 37–47.

FRETTER, V. (1951a) Observations on the life history and functional morphology of *Cerithiopsis tubercularis* (Montagu) and *Triphora perversa* (L.), *J. mar. biol. Ass. U.K.* **29**, 576–86.

FRETTER, V. (1951b) Some observations on the British cypraeids. *Proc. malac. Soc. Lond.* **29**, 14–20.

FRETTER, V. (1955) Observations on *Balcis devians* (Monterosato) and *Balcis alba* (da Costa), *Proc. malac. Soc. Lond.* **31**, 137–143.

FRETTER, V. and GRAHAM, A. (1949) The structure and mode of life of the Pyramidellidae, parasitic opisthobranchs, *J. mar. biol. Ass. U.K.* **28**, 493–532.

FRETTER, V. and GRAHAM, A. (1954) Observations on the opisthobranch mollusc *Actaeon tornatilis* (L.), *J. mar. biol. Ass. U.K.* **33**, 565–85.

GASCOIGNE, T. (1956) Feeding and reproduction in the Limapontiidae, *Trans. roy. Soc. Edinb.* **63**, 129–51.

GRAHAM, A. (1931) On the structure and function of the alimentary canal of the limpet, *Trans. roy. Soc. Edinb.* **57**, 287–308.

GRAHAM, A. (1937) The structure and function of the alimentary canal of aeolid molluscs, with a discussion on their nematocysts, *Trans. roy. Soc. Edinb.* **59**, 267–307.

GRAHAM, A. (1938) On a ciliary process of food-collecting in the gastropod *Turritella communis* Risso, *Proc. zool. Soc. Lond.* **108**, 453–63.

GRAHAM, A. (1953) Form and function in the molluscs, *Proc. Linn. Soc. Lond.* **164**, 213–17.

GRAHAM, A. (1955) Molluscan diets, *Proc. malac. Soc. Lond.* **31**, 144–57.

GRAHAM, A. (1964) The functional anatomy of the buccal mass of the limpet (*Patella vulgata*), *Proc. zool. Soc. Lond.* **143**, 301–29.

GRAHAM, A. (1965) The buccal mass of ianthinid prosobranchs, *Proc. malac. Soc. Lond.* **36**, 323–38.

GRAHAM, A. and FRETTER, V. (1947) The life history of *Patina pellucida* (L.), *J. mar. biol. Ass. U.K.* **26**, 590–601.

GRAHAM, A. and FRETTER, V. (1949) Feeding and reproduction in the Pyramidellidae, *Nature, Lond.* **163**, 361.

HABE, T. (1952) Parasitic gastropods found in echinoderms from Japan, *Publ. Seto. mar. biol. Lab.* **11** (2), 73–85.

HIRASE, S. (1932) The adaptive modifications of the gastropod *Stilifer celebensis* Kuk., a parasite on the starfish *Certonardoa semiregularis* (Müll. and Trosch), *Proc. malac. Soc. Lond.* **20**, 73–76.

HOWELLS, H. H. (1942) The structure and function of the alimentary canal of *Aplysia punctata*, *Quart. J. micr. Sci.* **83**, 357–97.

HUBENDICK, B. (1957) The eating function in *Lymnaea stagnalis* (L.), *Ark. Zool.* **10**, 511–21.

JUTTING, W. S. S. VAN BENTHEM (1956) Systematic studies on the non-marine mollusca of the Indo-Australian archipelago. V. Critical revision of the Javanese freshwater gastropods, *Treubia* **23**, 259–477.

KOHN, A. J. (1955) Studies on food and feeding of the cone shells, genus *Conus*, *Ann. Rept. Amer. Malacol. Union. Bull.* **22**.

KOHN, A. J. (1955–6) Feeding in *Conus striatus* and *C. catus*, *Abstr. Proc. Hawaii Acad. Sci. 31st Ann. Meeting*.

Kohn, A. J. (1956) Piscivorous gastropods of the genus *Conus*, *Proc. nat. Acad. Sci. Wash.* **42**, 168–71.

Kohn, A. J. (1958) Recent cases of human injury due to venomous marine snails of the genus *Conus*, *Hawaii Med. J.* **17**, 528–32.

Kohn, A. J. (1959) Ecological notes on *Conus* (Mollusca: Gastropoda) in the Trincomalee region of Ceylon, *Ann. Mag. nat. Hist.* **13**, 309–20.

Kohn, A. J. (1959a) The ecology of *Conus* in Hawaii, *Ecol. Monogr.* **29**, 47–90.

Kohn, A. J. (1961) Chemo-reception in gastropod molluscs, *Amer. Zool.* **1**, 291–308.

Kohn, A. J. (1963) Venomous marine snails of the genus *Conus*, in *Venomous and Poisonous Animals and Noxious Plants of the Pacific Area*, Pergamon Press, Oxford.

Kohn, A. J. and Orians, G. H. (1962) Ecological data in the classification of closely related species, *Systematic Zoology*, **11**, 120–7.

Kohn, A. J., Saunders, P. R. and Wiener, S. (1960) Preliminary studies on the venom of the marine snail *Conus*, *Ann. N.Y. Acad. Sci.* **90**, 706–25.

Lebour, M. V. (1932) Cowries, *Natural History*, *J. Amer. Mus. Nat. Hist.* **32**, 188–94.

Magalhaes, H. (1948) An ecological study of the genus *Busycon* at Beaufort, North Carolina, *Ecol. Monogr.* **18**, 379–409.

Marcus, E. and Marcus, E. (1962) Studies on the Columbellidae, *Bol. Fac. Filos. Ciena S. Paulo (Zool.)* **24**, 335–84.

Miller, M. C. (1961) Distribution and food of the nudibranchiate molluscs of the south of the Isle of Man, *J. Anim. Ecol.* **30**, 95–116.

Millott, N. (1938) On the morphology of the alimentary canal, process of feeding, and physiology of digestion of the nudibranch mollusc *Jorunna tomentosa*, *Phil. Trans.* B, **228**, 173–217.

Morton, J. E. (1950) Feeding mechanisms in the Vermetidae (order Mesogastropoda), *Nature, Lond.* **165**, 923–4.

Morton, J. E. (1951) The ecology and digestive system of the Struthiolariidae (Gastropoda), *Quart. J. micr. Sci.* **92**, 1–25.

Morton, J. E. (1951a) The structure and adaptations of the New Zealand Vermetidae. Part I. The genus *Serpulorbis*. Part II. the genera *Stephopoma* and *Pyxipoma*. Part III. *Novastoa lamellosa* and its affinities, *Trans. roy. Soc. N.Z.* **79**, 1–51.

Morton, J. E. (1954) The biology of *Limacina retroversa*, *J. mar. biol. Ass. U.K.* **33**, 297–312.

Morton, J. E. (1958) Observations on the gymnosomatous pteropod *Clione limacina* (Phipps), *J. mar. biol. Ass. U.K.* **37**, 287–297.

Morton, J. E. and Holme, N. A. (1955) The occurrence at Plymouth of the opisthobranch *Akera bullata*, with notes on its habits and relationships, *J. mar. biol. Ass. U.K.* **34**, 110–112.

Newell, R. (1962) Behavioural aspects of the ecology of *Peringia* (= *Hydrobia*) *ulvae* (Pennant) (Gastropoda, Prosobranchia), *Proc. zool. Soc. Lond.* **137**, 49–75.

Newell, R. (1965) The role of detritus in the nutrition of two marine deposit feeders, the prosobranch *Hydrobia ulvae* and the bivalve *Macoma balthica*, *Proc. zool. Soc. Lond.* **144**, 25–45.

Orton, J. H. (1914) On ciliary mechanisms in brachiopods and some polychaets, with a comparison of the ciliary mechanisms on the gills of molluscs, protochordata, brachiopods, and cryptocephalous polychaets, with an account of the endostyle of *Crepidula* and its allies, *J. mar. biol. Ass. U.K.* **10**, 283–311.

Pelseneer, P. (1924) Comment mangent divers gastropodes aquatiques. I. Gastropodes marins carnivores, *Natica* et *Purpura*, *Ann. Soc. zool. Belg.* **55**, 31–45.

Rigby, J. E. (1965) *Succinea putris*: a terrestrial opisthobranch mollusc, *Proc. zool. Soc. Lond.* **144**, 445–86.

ROBERTSON, R. (1961) The feeding of *Strombus* and related herbivorous marine gastropods, *Notul. nat. Acad. Philad.*, No. 343, 1–9.

ROWETT, H. G. Q. (1946) A comparison of the feeding mechanisms of *Calma glaucoides* and *Nebaliopsis typica*, *J. mar. biol. Ass. U.K.* **26**, 352–7.

SHARMAN, M. (1956) Note on *Capulus ungaricus* (L.), *J. mar. biol. Ass. U.K.* **35**, 445–50.

SIMPSON, A. C. (1960) Some economic aspects of the fisheries for molluscs in England and Wales, *Proc. malac. Soc. Lond.* **34**, 144–56.

SMITH, E. H. (1964) Ph. D. thesis, University of Glasgow.

STEHOUWER, H. (1952) The preference of the slug *Aeolidia papillosa* (L.) for the sea anemone *Metridium senile* (L.), *Arch. néerl. Zool.* **10**, 161–170.

THOMPSON, T. E. (1958) The natural history, embryology, larval biology, and post-larval development of *Adalaria proxima* (Alder and Hancock) Gastropoda, Opisthobranchia), *Phil. Trans.* B, **242**, 1–58.

THOMPSON, T. E. (1964) Grazing and the life cycles of British nudibranchs, Brit. Ecol. Soc. Symp., No. 4, *Grazing in Terrestrial and Marine Environments*, ed. D. J. Crisp, Blackwell, Oxford.

THOMPSON, T. E. and SLINN, D. J. (1959) On the biology of the opisthobranch *Pleurobranchus membranaceus*, *J. mar. biol. Ass. U.K.* **38**, 507–24.

TURNER, H. J. (1953) The drilling mechanism of the Naticidae, *Ecology* **34**, 222–3.

TURNER, R. D. (1959) Notes on the feeding of *Melongena coronata*, *Nautilus* **73**, 11–13.

WELLS, H. W. (1958) Predation of pelecypods and gastropods by *Fasciolaria hunteri* (Perry), *Bull. mar. Sci. Gulf Caribb.* **8**, 152–166.

WELLS, H. W. (1958a) Feeding habits of *Murex fulvescens*, *Ecology* **39**, 556–8.

WERNER, B. (1951) Über die Bedeutung der Wasserstromerzeugung und Wasserstromfiltration für die Nahrungsaufnahme der ortsgebundenen Meeresschnecke *Crepidula fornicata* L. (Gastropoda, Prosobranchia), *Zool. Anz.* **146**, 97–113.

WERNER, B. (1953) Über den Nahrungserwerb der Calyptraeidae (Gastropoda, Prosobranchia); Morphologie, Histologie und Function der am Nahrungserwerb beteiligten Organe, *Helgoländ. wiss Meeresunters.* **4**, 260–315.

WHITAKER, M. B. (1951) On the homologies of the oesophageal glands of *Theodoxus fluviatilis* (L.), *Proc. malac. Soc. Lond.* **29**, 21–34.

WILBUR, K. M. and YONGE, C. M. (1964) *Physiology of Mollusca*, vol. 1, Academic Press, New York and London.

WILSON, D. P. and Wilson, M. A. (1956) A contribution to the biology of *Ianthina janthina* (L.), *J. mar. biol. Ass. U.K.* **35**, 291–305.

YONGE, C. M. (1926) Ciliary feeding mechanisms in the thecosomatous pteropods, *J. Linn. Soc. (Zool.)* **36**, 417–29.

YONGE, C. M. (1932) Notes on feeding and digestion in *Pterocera* and *Vermetus*, with a discussion on the occurrence of the crystalline style in the gastropods, *Sci. Rep. Gt. Barrier Reef Exped.* **1**, 259–81.

YONGE, C. M. (1937) The biology of *Aporrhais pes-pelecani* and *A. serresiana*, *J. mar. biol. Ass. U.K.* **21**, 687–704.

YONGE, C. M. (1938) The prosobranchs of Lake Tanganyika, *Nature, Lond.* **142**, 464–8.

YONGE, C. M. (1938a) Evolution of ciliary feeding in the Prosobranchia, with an account of feeding in *Capulus ungaricus*, *J. mar. biol. Ass. U.K.* **22**, 453–468.

YONGE, C. M. (1942) Ciliary currents in the mantle cavity of the Atlantidae, *Quart. J. micr. Sci.* **83**, 197–203.

YONGE, C. M. (1953) Observations on *Hipponyx antiquatus* (Linnaeus), *Proc. Calif. Acad. Sci.* **28**, 1–24.

YONGE, C. M. (1958) Observations in life on the pulmonate limpet *Trimusculus* (*Gadinia*) *reticulatus* (Sowerby), *Proc. malac. Soc. Lond.* **33**, 31–37.

YONGE, C. M. (1960) Further observations on *Hipponyx antiquatus* with notes on North Pacific pulmonate limpets, *Proc. Calif. Acad. Sci.*, 4th. Ser., **31**, 111–19.

YONGE, C. M. and ILES, E. J. (1939) On the mantle cavity, pedal gland, and evolution of mucus feeding in the Vermetidae, *Ann. Mag. nat. Hist.*, Ser. 11, **3**, 536–56.

YONGE, C. M. and NICHOLAS, H. M. (1940) Structure and function of the gut, and symbiosis with zooxanthellae in *Tridachia crispata* (Oerst.), *Papers from the Tortugas Laboratory* **32**, 287–301.

CHAPTER 3

FEEDING METHODS AND EVOLUTION IN THE BIVALVIA

Synopsis

The Bivalvia evolved in the process of colonisation of deposits of sand or sandy mud in the Pre-Cambrian epoch, these animals feeding directly upon organic matter in these unstable substrates. Adaptations to suit the new mode of life included enclosure of the entire body by paired, lateral folds of the mantle which secreted lateral shell valves articulating mid-dorsally; other adaptations were the loss of the head, the buccal mass, the radula, and all cephalic sense organs. Sensory functions were taken over by the margin of the mantle. These early bivalves were proto-branchiate in form and collected particulate food directly from the substratum by ciliary collecting organs which may have been closely comparable with the palp proboscides of modern protobranchs. Among recent bivalves we can recognise three main feeding types: the Protobranchia, the filter-feeding bivalves (generally known as lamellibranchs), and the carnivorous Verticordiidae and Septibranchia.

The Protobranchia represent a continuation to the present day of the ancestral protobranchiate stock. With the exception of the Solenomyidae, the feeding mechanisms of which are not yet known, the remainder of the Protobranchia feed primarily by means of the ciliated palp proboscides upon organic matter in the substratum. The ctenidia principally subserve respiration and are only slightly concerned in the collection of food particles. The ciliary feeding mechanisms of various protobranchs are described in detail.

Septibranchs have evolved from a filter-feeding parentage in adaptation to life in deep-water sedimentary deposits where there is little suspended particulate food. Initially the ctenidium was simplified and all available digestible material was accepted, regardless of size. Then the ctenidia were adapted to form a muscular pumping septum by which carcases of small crustaceans could be sucked into the mantle cavity and consumed. Concurrently the stomach was simplified, and adapted as a muscular, crushing gizzard, thereby completing the requirements for a new, scavenging or carnivorous mode of life.

The great bulk of the Bivalvia, with filibranch, pseudolamellibranch, or eulamellibranch ctenidia, constitute the Polysyringia (no other name is available for these bivalves—although the term "Lamellibranchia" is suitably descriptive it is undesirable since it has long been used as the name of the whole class). The Polysyringia arose directly from a proto-branchiate ancestry by development of the ciliary food-collecting mechanisms of the ctenidia and the association between the ctenidia and the labial palps, thereby initiating the collection of food particles suspended in the super-natant water and liberating the Polysyringia from dependence on sedimentary deposits. This development initiated an extensive phase of adaptive radiation in the exploitation of the new source of food material; various lineages became adapted to life on rocky shores, on coral reefs, in sedimentary deposits, or in fresh waters, etc. At the same time there was improvement in efficiency of the ctenidia as food-collecting organs resulting in parallel evolution by a number of independent lineages through the functional strata which we term "filibranch", "pseudolamellibranch", and "eulamellibranch" respectively.

The adaptive radiation of the Polysyringia is expressed chiefly in relation to the mode of life and the habitat and, unlike that of the Gastropoda, is little influenced by the mode of feeding.

There is much diversity in the nature of the sorting mechanisms on the frontal surfaces of the ctenidial filaments of polysyringian bivalves, by means of which particulate material is sorted and potential food material is selected and passed to the labial palps. These sorting mechanisms involve interaction between ciliary activity, the degree of contraction or relaxation of muscle fibres, and the amount of secretion by mucous glands. These mechanisms are described in detail for many different examples. In general the function of the ctenidia is to reject large particles and large quantities of small particles, and only to accept small quantities of very small particles. This object is achieved in many different ways in different phylogenies, which indicates the occurrence of an explosive radiation in ctenidial structure and function among the filter-feeding bivalves (Polysyringia). The degree of efficiency of filtration of suspended particles by the ctenidium seems to be related to the diameters of the particles. Changes have been reported in the filtration rate in response to the addition of eggs or sperm of the same species, or of the toxic *Gymnodinium veneficum*, to the water and this suggests the existence of micro-sense organs. No such sense organs have yet been identified, and in the main the selection of food particles appears to be purely mechanical.

The labial palps of filter-feeding bivalves tend towards uniformity in structure and in function. Interaction of ciliated tracts on the opposed, grooved surfaces of the palps, and degree of contraction of sub-epithelial muscle fibres, provides a very efficient mechanism for the grading of particles. Only the finest particles pass up the grooves into the lateral oral groove and thence to the mouth. The ciliary sorting mechanism of the palps of *Barnea parva* are described as a typical example. In unusual examples living in oceanic water which is completely free from unwanted silt the labial palps may be greatly reduced and play little part in regulating the admission of food into the mouth, e.g. *Xylophaga dorsalis*. In some (and possibly in all) monomyarians the lips of the mouth are hypertrophied in an unusual way, e.g. in *Pecten maximus*; achievement of the monomyarian form has resulted in constriction of the volume of the anterior part of the mantle cavity, in consequence of which the water currents emerging from the oral grooves anteriorly are liable to be unusually strong. The great development of the lip apparatus in *P. maximus* and in other monomyarians probably provides the necessary protection to the mucus-bound food cord from detachment by these water currents.

The byssus gland was probably originally an organ of the post-larva which served to anchor it to grains of sand in the surface layer of the substratum at the time of metamorphosis. Extension of the activity of the byssus gland into adult life, by a process of paedomorphosis, enabled many lineages of bivalves to colonise hard, rocky substrates and to become members of the attached epifauna. Byssal attachment has in many cases had a profound effect on the growth form of the individual, with the achievement of the anisomyarian, or of the monomyarian form. In the monomyarian form the single, posterior adductor muscle is nearly central in position, and there is a secondary approach to radial symmetry—which characterises sedentary and attached forms. The detailed consequences of byssal attachment on growth form will be fully discussed in Chapter 4. Other bivalve lineages have colonised the rocky shore not by byssal attachment, but by cementation, e.g. *Ostrea*, and *Chama*; yet others have adopted the habit of boring into the substratum, e.g. *Hiatella*, *Lithophaga*, *Zirfaea*, etc. Various members of the Adesmacea, e.g. *Martesia*, *Xylophaga*, *Teredo*, and *Bankia* have adopted the habit of boring into wood and the last two examples, which are ship-worms, have acquired the capacity to digest the fragments of wood which are passed through the alimentary canal. In the Pholadidae, and in various other families, the stomach bears an appendix which serves to store large sand grains which have entered the stomach and which might otherwise damage the interior of the gut. In *Xylophaga* and in the ship-worms this appendix has been enlarged to form a wood-storing caecum in which fragments of wood derived from the wood-boring activity are stored prior to digestion. The appendix in the Adesmacea was therefore of pre-adaptive value with regard to the evolution of the wood-boring mode of life in the ship-worms.

Deposits of gravel, sand, sandy mud, and mud were originally colonised by the protobranchiate ancestors of modern bivalves, which were dependent on the organic matter they could collect from the substratum itself. These sedimentary deposits have subsequently been colonised on a number of occasions by septibranchs and by polysyringians, both of which draw their sustenance from the supernatant water. Certain of the latter, however, have resumed the practice of obtaining particulate organic matter from the substratum, using the suctorial powers of the long, slender inhalant siphon for the purpose e.g. members of the families Tellinidae and Semelidae. In all other cases the polysyringian members of the infauna are only dependent on the substratum for protection against wave action or for concealment from predators. The bivalves of the infauna may be surface dwelling or deep-burrowing forms, and their mode of life is imprinted on their growth form in ways which will be described in detail in Chapter 4.

A number of bivalve lineages have successfully invaded estuaries and freshwater systems by suitable adaptations to their excretory and reproductive processes.

We may presume that the class Bivalvia came into being in the process of colonisation of sub-littoral deposits of sand or muddy sand by some Pre-Cambrian molluscan stock capable of maintaining itself partly buried in an unstable, shifting substratum, and of feeding directly upon organic matter contained in the surface layers of the soil. Adaptations to suit this habitat and this mode of life included the realignment of the mantle/shell to provide the necessary *lateral* protection from abrasion and contamination while moving through the substratum, and also the loss of the head and all cephalic structures such as tentacles, eyes, buccal mass, and radula.

Lateral compression caused the mantle to become developed as two lateral lobes joined mid-dorsally by the mantle isthmus (Owen, Trueman, and Yonge, 1953). The mantle lobes secrete the two shell valves, which join with the resilient ligament in the mid-dorsal line, and the inner layer of the ligament is secreted by the mantle isthmus. Growth is effected by marginal accretions to the shell valves, so both the rate of growth and the growth form are determined by the mantle/shell. The mantle lobes are attached peripherally to their respective shell valves along the pallial line by a series of pallial muscle fibres by means of which the margins of the mantle may be retracted when necessary. Due to lateral compression of the mantle/shell the left and right shell valves came to lie very close to each other dorsally, anterior, and posterior to the mantle isthmus and the overlying ligament. Cross-fusion of the pallial muscle fibres at these two points of approximation of the shell valves led to the development of the anterior and posterior adductor muscles which act in opposition to the opening thrust exerted by the elastic ligament. The two shell valves enclosed the entire body including the head (if this was then still present), and this may have assisted in the reduction and loss of the head and associated structures. The only tissues now in direct contact with the environment were the margins of the mantle lobes, and these assumed the necessary sensory functions, with the development of marginal tentacles, and, in some cases, pallial eyes.

According to Dollo's law once an organ has been lost it can never be regained; the loss of the cephalic sense organs was made good in part by the acceptance of similar functions by the margin of the mantle, but the loss of the radula must have imposed severe restrictions on the evolutionary potential of the newly emerging bivalve lineage. Nevertheless, *evolution by loss* may remove some restricting factor and so permit the establishment of some new relationship between the organism and its environment or entry into some entirely new environment, thereby creating new possibilities for exploitation in adaptive radiation. Thus the biological model which we now characterise as "bivalve" evolved to meet the special requirements of life as members of the infauna of marine deposits, and at this formative stage in their history the early bivalves must have all been feeders on organic particles which were mainly collected by ciliary means directly from the substratum.

Once the biological model had evolved to meet these special requirements it was inevitable that it would be applied in due course, with modifications, to a variety of different circumstances, i.e. a phase of adaptive radiation would commence. If the model were highly efficient, and if there were little interference from competitors for food, or from predators, then the ensuing radiation would probably be dramatic and far-reaching. This was the case for the Bivalvia, where the primary phases of adaptive radiation have been concerned with important changes in the feeding mechanism.

Ancestral bivalves were protobranchiate, with respiratory ctenidia, and food-collecting palp proboscides. This biological model enjoyed limited evolutionary success and only three families survive as the sub-class Protobranchia. In contrast a sub-class of filter-feeding bivalves arose from the same ancestry by adaptation of the ctenidia to filtering food particles from the respiratory water current and passing these via the palps to the mouth. This sub-class Lamellibranchia includes all bivalves in which the ctenidia are lamellar, with three evolutionary stages, or "functional strata"—the Filibranchia, the Pseudolamellibranchia and the Eulamellibranchia (Fig. 39). A number of phylogenies probably evolved in parallel through these strata. Fortunately examples of the earlier stages survive, enabling us to reconstruct the probable evolutionary sequence. The sub-class Lamellibranchia, exploiting the efficiency of filter-feeding, underwent a major adaptive radiation producing some 72 recent families. Some of these families invaded situations which could not be exploited solely by filter-feeding. Thus the Verticordiidae and the septibranch families Cuspidariidae and Poromyidae colonised deep-water sediments where they acquired scavenging or carnivorous habits by divergent feeding methods supported by parallel adaptations of the mouth, the oesophagus and the stomach. In the Verticordiidae small animals are trapped in sticky threads secreted by pallial tentacles, whereas in septibranchs convulsive movements of a muscular septum derived from the cotenidium trap small animals by sucking them into the mantle cavity.

THE BIOLOGY OF THE MOLLUSCA

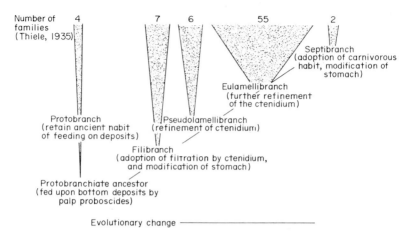

FIG. 39. The acquisition of new feeding mechanisms and exploitation of new sources of food had a profound effect on evolution of the class Bivalvia. This is one single facet of bivalve evolution.

Protobranchia

Feeding methods in the Nuculidae and the Nuculanidae remain essentially the same as those postulated above for the protobranchiate ancestral stock. A long palp proboscis is attached to the base of each of the outer labial palps; this proboscis can be protruded into the substratum in order to collect small particles and pass them along a ciliated groove on the ventral surface of the proboscis to the base of the outer labial palp. The opposed surfaces of the inner and outer palps are corrugated with close parallel ridges and grooves which are covered with a ciliated epithelium. Here the particles collected by the palp proboscides are subjected to ciliary sorting mechanisms. The cilia on different parts of the folds beat in different directions and so exercise a selective sorting of the particles collected by the palp proboscides. Fine particles are carried oralwards over the crests of the folds, or alternatively dorsalwards in the depths of the grooves, towards the lateral oral groove lying between the bases of the palps, and then along the lateral oral groove towards the mouth. In contrast, excess material is carried ventralwards along the distal sides of the folds to the free ventral border of the palp, whence it is transferred to the ciliary cleansing currents on the inner face of the mantle (see Figs. 40 and 41). Stasek (1965) described nine distinct ciliated tracts on the ridged, opposed surfaces of the palps of *Yoldia ensifera* and showed that the ciliary sorting mechanism is basically comparable with that described elsewhere in this chapter for polysyringian bivalves.

There can be no doubt that in the Nuculidae and in the Nuculanidae food is collected partly from the substratum by the palp proboscides and partly from particles in suspension, by the ctenidia. Orton (1912) observed the collection of potential food material from the ctenidia by the palp appendages in *Nucula*, and Stasek (1961) showed that in *Acila castrensis*, another nuculid, there was an effective food-collecting association between the ctenidia and the labial palps. In the Nuculanidae, Atkins (1937)

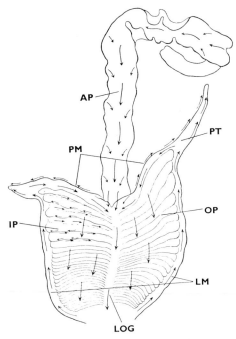

FIG. 40. *Nuculana minuta*, the inner and outer labial palps parted to show the ciliary sorting mechanisms on their opposed, folded surfaces. Potential food material is collected from the substratum, and perhaps also from the ctenidium to a certain extent, by the palp appendage (AP) and is passed towards the lateral oral groove (LOG) and thence to the mouth. Ciliary rejection currents are seen passing backwards along the free ventral margins of the inner palp (IP) and outer palp (OP). For interpretation of other lettering, see p. 142. (Originally published in Atkins, 1937, *Quart. J. micr. Sci.* **79**, 203, fig. 8.)

found that in *Nuculana minuta* some of the particles collected on the ctenidia are transferred to the labial palps. In *Y. ensifera* the ctenidia collect and grade particles by size. Fine particles are manipulated by the fine frontal cilia and tend to be passed to the labial palps at a point close to the proximal oral groove where they will be subjected to minimal sorting by the palps. Larger particles are handled by the coarse frontal cilia and tend to be passed to the palps at a level more distant from the proximal oral groove, where they will be subjected to far more rigorous sorting by the palps.

There is a ciliated distal oral groove between the anterior end of the ctenidium and the posterior end of the proximal oral groove, and potential
food material is passed forwards along this groove (Stasek, 1965). We do
not yet know the extent to which these protobranchs derive their food
from suspended particles collected by the ctenidia as compared with deposited particles collected by the palp proboscides; food may be collected pre-
ponderantly from the substratum.

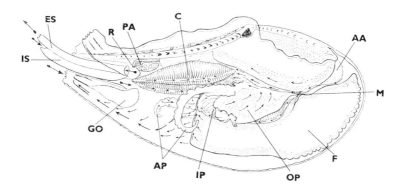

Fig. 41. *Nuculana minuta*, seen from the right side after removal of the
right shell valve and with the right mantle lobe folded back. Note the large
size of the labial palps (IP, OP) as compared with the ctenidium (C), and the
elongated palp appendages of the left and right sides (AP). The taxodont
nature of the hinge is indicated. Other organs displayed include the anterior
adductor muscle (AA), the posterior adductor muscle (PA), the rectum
(R), the exhalant siphon (ES), and the inhalant siphon (IS). For interpretation of other lettering, see p. 142. (Originally published in Atkins, 1937,
Quart. J. micr. Sci. **79**, 187, fig. 1.)

Nucula normally feeds just below the surface of the sand in which it lives.
In *Yoldia* the posterior third of the body may be above the surface of
the mud and the palp proboscides collect food particles from the surface
of the substratum (Atkins, 1937; Yonge, 1939). In spite of the rigorous
sorting mechanism on the opposed surfaces of the labial palps, a certain
amount of sand is accepted and passed into the stomach, where it doubtless plays an essential part in the trituration of diatoms, etc., due to muscular contractions of the stomach wall. Judging by the abundance of specimens of various species of Nuculidae and Nuculanidae in suitable marine
deposits, this mode of feeding is highly successful.

In the remaining family of the Protobranchia, the Solenomyidae, the
ctenidia are relatively enormous, cilia on their frontal surfaces convey particles towards the median ventral line and thence forwards on to the base
of the foot. The labial palps are very small, do not possess a palp proboscis, and do not bear any ridges and grooves on their opposed surfaces. The

palps collect material brought forward from the frontal surfaces of the ctenidia, and pass this material to the mouth, where it is ingested. What is extraordinary about the Solenomyidae is the small size of the stomach and the digestive diverticula. The stomach is only the faintest dilation on the course of the short and slender gut, the tubules of the digestive diverticula are scarcely any more conspicuous, and it is hard to envisage how the animal can obtain the necessary nourishment via the alimentary canal.

In the Solenomyidae the ctenidia are definitely concerned in the feeding process, but the finer details of this process remain undetermined (Yonge, 1939). With the lack of the usual elaborate ciliary sorting mechanisms on the opposed surfaces of the labial palps, how is it possible that all unwanted, indigestible material is prevented from entering the mouth? Is it possible that the food material is so readily disrupted that it does not need to be subjected to the process of trituration which normally occurs in the stomach of a bivalve? Is it possible that the food material is already soluble, and does not need to be subjected to digestive processes? Only in such extreme circumstances would it seem possible for an animal of the size of *Solemya parkinsoni* to obtain the necessary nourishment solely through its diminutive gut. It has been suggested that preliminary digestion of food material may occur in the infrabranchial chamber (Owen, 1961).

Septibranchia

In the Septibranchia feeding is by drawing small crustaceans, carcases of these, or pieces of organic debris, into the mantle cavity in a sudden inflow of water. The mantle cavity is divided by a longitudinal muscular septum into infra-septal and supra-septal chambers which communicate by small pores, or branchial sieves in the septum (see Fig. 42). The septum is considered to be derived from the left and right ctenidia due to fusion of these, invasion of muscle fibres, and the co-ordination of the whole into a muscular platform which can be suddenly raised, thereby sucking a volume of water into the infra-septal chamber, and possibly drawing in at the same time a substantial object suitable for consumption. The perforations in the septum are few in number, amounting to four, or five small slit-shaped pores in the various species of *Cuspidaria*. In the genus *Poromya* the septum is delicate, and is perforated by an anterior and a posterior pair of small branchial sieves which usually possess between five and eight filaments, according to the species. These filaments are regarded as the remnants of the ctenidia (Yonge, 1928).

It had been suggested that the Septibranchia arose directly from a protobranchiate ancestry, by adoption of a muscular pumping action of the ctenidia (Purchon, 1962, and the 1st edition of this book). However, an important review of the family Verticordiidae by Allen and Turner (1974)

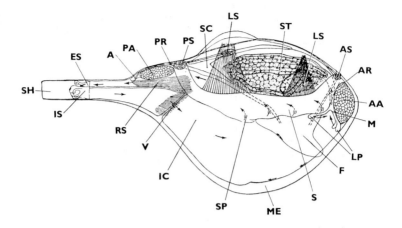

FIG. 42. *Cuspidaria rostrata*, seen from the right side after removal of the right shell valve and right mantle lobe. Note the absence of a ctenidium and the presence, in lieu, of a large and muscular septum (S) perforated by a series of small pores (SP). The septum is operated by four stout muscle bundles, anterior (AS), posterior (PS), and lateral (LS). The labial palps (LP) are simplified and reduced in size, while the mouth (M) is relatively large. The siphons lie within a sheath (SH) and are withdrawn by retractor muscles (RS). A valve (V) guards the entry from the inhalant siphon into the infraseptal mantle cavity. For interpretation of other lettering, see p. 142. (Originally published in Yonge, 1928, *Phil. Trans.* B, **216**, 224, fig. 1.)

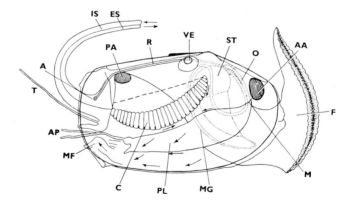

FIG. 43. *Malletia obtusata*, viewed from the right side, by transparency. The ctenidium (C) is shown in its ventralmost position. The broken line indicates the position of the dorsal surface of the ctenidium when pulled upwards in respiratory movements. For interpretation of other lettering, see p. 142. (Originally published in Yonge, 1939, *Phil. Trans.* B, **230**, 99, fig. 16.)

shows that these eulamellibranchs are adapted to life in abyssal sediments in a manner approaching that of the septibranchs *Cuspidaria* and *Poromya*. In the absence of sufficient suspended minute particles the ctenidium has become simplified and lacks ciliary sorting mechanisms; it has reverted to its original respiratory function, but its cilia convey any material entering the mantle cavity towards the large, funnel shaped mouth. These bivalves can now accept small particles and also carcases of small crustaceans as food material. This scavenging, or carnivorous, habit has necessitated modifications of the oesophagus and stomach on the same lines as in the septibranch *Cuspidaria* (Purchon, 1956). The structure and mode of life of the Verticordiidae clearly shows a stage in the probable course of evolution of septibranchs from a near-verticordiid eulamellibranch ancestry. We should now regard septibranchs as highly specialised eulamellibranchs, and not as a separate sub-class Septibranchia.

Lamellibranchia ($\equiv$ Polysyringia)

It has been stated above that all the Bivalvia *must* at some time have passed through a protobranchiate stage in which the ctenidia served wholly or primarily for respiration, and when feeding was by collection of small particles of organic matter directly from the substratum by ciliary mechanisms on protrusible palps, or palp proboscides (Yonge, 1939). While discussing the feeding mechanisms of the Protobranchia, it was indicated that the palp proboscides might, to a small extent, collect particles directly from the ctenidia and accept these as food material. Some such association between the ctenidia and the labial palps, whereby the ciliary *cleansing* mechanisms of the ctenidia chanced to accumulate what became a *potential source of food*, must have supplied the impetus which led to the evolution of the ciliary filter-feeding mechanisms which occur in the Polysyringia (bivalves with lamellar ctenidia).

There is an interesting parallelism between the origin of filter-feeding mechanisms in the "higher" bivalves, and that which occurred in *Crepidula*, for example, among the taenioglossan Gastropoda. The Gastropoda evolved primarily as colonists of the rocky shore, but have radiated widely from this origin. Some lineages of the radiation had embarked upon deposit-feeding, and the adoption of filter-feeding mechanisms merely provided a slightly different kind of "deposit" to be picked up by the radula. Consequently, the development of filter-feeding in the Gastropoda made a comparatively small contribution to the scope of the radiation of the class as a whole.

In the Bivalvia, the situation was quite different. We judge the class Bivalvia to have evolved primarily as colonists of littoral and sub-littoral deposits, feeding directly upon these deposits. The evolution of filter-feeding mechanisms in the Bivalvia rendered them independent of the substratum as a direct source of food, and laid the basis for a very extensive adaptive radiation. This radiation of filter-feeding bivalves mainly concerned the colonisation of a very wide variety of habitats—excluding only the colonisation of land. Apart from diversity in details in the ciliary feeding mechanisms, the Polysyringia in general exhibit some monotony of feeding methods, and the adaptive radiation itself finds its principal expression in the diversity of form of the animal. This in turn depends partly on the mode of life, but mainly on the nature of the habitat which is being colonised.

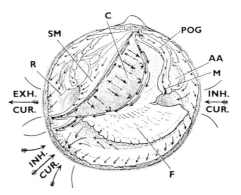

FIG. 44. Ciliary currents in the mantle cavity of *Glycymeris glycymeris*. On all four faces of the ctenidium (C) there are antagonistic ciliary currents. Coarse frontal cilia beat downwards into the posteriorly directed rejection current at the ventral margin of each demibranch. Fine frontal cilia beat dorsalwards into the oralward feeding currents at the dorsal margin of each face of the ctenidium. The latter pass food material into the oralward current in the proximal oral groove (POG). For interpretation of other lettering, see p. 142. (Originally published in Atkins, 1937, *Quart. J. micr. Sci.* **79**, 218, fig. 11.)

Once the material removed from the frontal surfaces of the ctenidial filaments began to assume importance as a source of food, evolutionary processes would intervene to amplify the ciliary filter-feeding mechanism (see Fig. 44). Increase in the efficiency of the mechanism could be effected by increase in the surface area of the ctenidia. This may have been effected by hypertrophy of the mid-region of the filaments of both demibranchs, or of only the inner demibranch in certain instances (Ansell, 1962; Mortimer, 1963). Increase in surface area has also been effected by developing the originally flat surfaces of the ctenidial lamellae into a series of shallow folds, as in *Petricola*, *Anodonta*, or *Aspatharia* or into a series of very deep folds, as in *Pecten*, *Pinna*, *Tridacna*, *Egeria*, or *Brechites*. Increase in the efficiency of the mechanism has also been effected by union of adjacent filaments by transverse strands of tissue, so that the ctenidium is converted into one single complex rectangular meshwork, and also by elaboration of the muscular and ciliary mechanisms operating on the ctenidial filaments.

Special attention has been paid to the degree of tissue fusion between adjacent filaments, and three different conditions have been recognised; in the *filibranch* condition the filaments may be united to adjacent filaments by interlocking groups of cilia known as ciliated discs. This is regarded as the most primitive condition among bivalves with lamellar ctenidia. In the *pseudolamellibranch* condition there are various degrees of tissue fusion of a minor nature between adjacent filaments (there is tissue union along the tops of the ascending limbs of the filaments in the Mytilidae, and certain species of *Anomia*, *Arca*, etc.); in some cases there is also organic union along the free ventral borders of the filaments, e.g. in the Vulsellidae; finally, there may be a regular series of tissue junctions between adjacent filaments up and down the length of the filaments, as in the Ostreidae, Pinnidae and Limidae (Ridewood, 1903; Purchon, 1960), but the total number of these tissue junctions between adjacent filaments is small. In the *eulamellibranch* condition the filaments are far more intimately bound together by innumerable tissue junctions, so that the whole ctenidium is welded into a compact rectangular meshwork in which, however, the vertically running filaments remain easily distinguishable.

It seems clear that these three conditions of the ctenidium—filibranch, pseudolamellibranch, and eulamellibranch—are progressive stages in the structural and functional development of the ctenidium as an organ serving for the filtration of particles of food from the inhalant water current. The pseudolamellibranch condition is intermediate between the other two stages, and is itself divisible into various intermediate degrees of fusion between the filaments. While these three terms remain valuable for descriptive purposes, it is becoming increasingly recognised that they are not relevant to a phylogenetic classification.

The emergence of ctenidial filter-feeding in this lineage (Polysyringia) led to the elaboration of the ctenidium from a simple series of slender parallel filaments into a complex rectangular meshwork; it may be presumed that development on these lines led to increase in the efficiency of the filter-feeding mechanism, but this remains to be verified experimentally. The success of this novel feeding mechanism led to the adaptive radiation which was to be expected, the main feature in this adaptive radiation being the application of the new model to colonisation of a wide variety of habitats previously unavailable to the earlier bivalves.

At the same time the efficiency of the ctenidium was amplified by elaboration of the ciliary mechanisms on the frontal surfaces of the filaments; by the development of special devices for selection and rejection which will be discussed below, and perhaps by the development of micro-sense organs capable of making an appraisal of the nature of material borne in the inhalant stream. It is probable that much parallel and/or convergent evolution took place in the refinement of the ctenidium for filter-feeding in the various lineages of bivalves that were developing during this process.

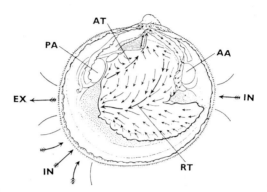

FIG. 45. Ciliary cleansing currents on the surface of the foot and visceral mass of *Glycymeris glycymeris*, seen from the right side after removal of the right shell valve, right mantle lobe, and most of the right ctenidium. Waste material collected from these surfaces is bound with mucus and transferred to the surface of the mantle lobes, whence it is ejected from the infra-branchial chamber as pseudofaeces by sudden clapping of the shell valves. For interpretation of lettering, see p. 142. (Originally published in Atkins, 1937, *Quart. J. micr. Sci.* **79**, 235, fig. 17.)

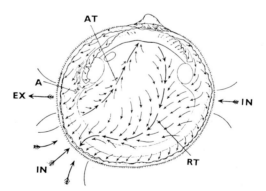

FIG. 46. Ciliary cleansing currents on the inner surface of the mantle of *Glycymeris glycymeris*. One main ciliated pathway (AT) accompanies the dorsal margin of the ascending lamella of the outer demibranch, and this may perhaps constitute a feeding current. The other major ciliated pathway (RT) is a rejection tract. Heavy arrows indicate the positions of two inhalant and one exhalant water streams; the presence of an anterior inhalant water stream is atypical, and may be primitive. For interpretation of other lettering, see p. 142. (Originally published in Atkins, 1937, *Quart. J. micr. Sci.* **79**, 237, fig. 19.)

Similarities, and differences, in the details of ciliary patterns on the ctenidial filaments may not necessarily be reliable indicators of phylogenetic relationship between families or orders.

Ciliary cleansing currents on the surface of the visceral mass and of the foot, and on the inner face of the mantle, ensure that the organs in the mantle cavity do not become occluded by deposits of sediment (see Figs. 45 and 46).

The adoption of the filter-feeding habit also led to elaboration of the ciliary sorting mechanisms on the inner surface of the stomach, thus permitting continuation within the stomach of the process of grading and

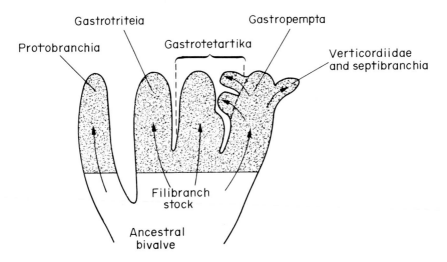

FIG. 47. Diagrammatic representation of possible lines of evolution within the Bivalvia, based on consideration of feeding mechanisms and on the structural organisation of the stomach. The filibranch stock adopted the practice of feeding by filtration of small particles from the inhalant water current by ciliary and other devices on the ctenidia and labial palps; the filter-feeding bivalves were very successful as indicated by the generation of orders, families and genera, as compared with the deposit-feeding Protobranchia. Three main lineages are envisaged within the filter-feeding bivalves, though only two of these may be monophyletic. The condition in the Verticordiidae and in the septibranchs is a secondary specialisation.

selection of food particles. A survey of the internal architecture of the stomachs of a wide variety of bivalves reveals much diversity and suggests that in the adaptive radiation of the Polysyringia there may be three principal lineages (see Fig. 47) (Purchon, 1960, 1962).

Structure and Function of the Ctenidia

The cilia on the frontal surface of the filaments are sometimes arranged in two, and sometimes in three distinct longitudinal bands, one of these consisting of comparatively long cilia, and the other band(s) consisting of shorter cilia (see Fig. 48). These frontal cilia beat parallel with the

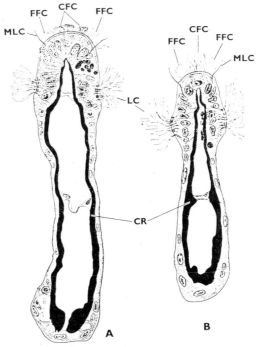

FIG. 48. Transverse sections through ctenidial filaments of A, *Glycymeris glycymeris*, and B, *Arca tetragona*. The lateral cilia (LC) are responsible for creating the flow of water through the ctenidium; the frontal cilia are divided into three parallel, longitudinal tracts, one of coarse cilia (CFC) on the crest of the frontal surface, and two of fine cilia (FFC) on either side. The coarse and the fine frontal cilia beat in opposite directions, and particles of suitably small size are thereby isolated and passed in due course towards the mouth for ingestion. Latero-frontal cilia (MLC) filter particles from the water current passing between the ctenidial filaments, and pass these particles on to the frontal surfaces of the filaments. For interpretation of other lettering, see p. 142. (Originally published in Atkins, 1937, *Quart. J. micr. Sci.* **79**, 224, fig. 14.)

long axis of the filament and drive small particles along the filaments, either upwards or downwards; in some cases delicate investigation may show that adjacent particles are travelling in different directions, some travelling dorsalwards under the influence of the tract(s) of short frontal cilia, while others are travelling ventralwards under the influence of the tract of long frontal cilia. By these means particles may be driven to one or more of five points on the ctenidium: the free ventral margins of the inner and the outer demibranchs, where there may or may not be a marginal groove for reception of the mucus-bound material; the ctenidial axis, dorsally, between the bases of the two demibranchs; the dorsal borders of the ascending limbs of the filaments, where these approach —and often fuse with—the mantle, externally, or the visceral mass, internally. At any of these five points there may exist an oralward ciliary current that will convey the mucus-bound food cord forwards towards the mouth. In some cases, however, posteriorward rejection currents may be encountered at the free margin of the demibranch.

Figure 49 indicates the variety which may be encountered in the development and disposition of the ctenidial lamellae, and the ciliary currents

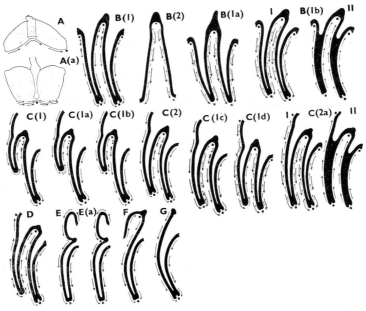

Fig. 49. Ciliary sorting mechanisms on the frontal surfaces of the ctenidia of a wide variety of bivalves. In each case the inner demibranch lies on the right side of the figure. A, Protobranchia. B(1), Mytilidae, Pinnidae. B(1a), Arcidae, Anomiidae. B(1b), most Pseudolamellibranchia: I, ordinary filaments, II, principal filaments. B(2), *Heteranomia*. C(1), many Eulamellibranchia. C(1a), *Pholadidea loscombiana*. C(1b), *Venus fasciata*. (C1c), *Barnea candida*. C(1d), *Cultellus pellucidus*. C(2), *Lutraria lutraria*. C(2a), *Solen* and *Ensis*: I, ordinary filaments; II, principal filaments. D, Unionidae and *Beguina semiorbiculata*. E, Tellinidae, Semelidae, Anatinacea. E(a), *Tellina crassa, Scrobicularia plana*. F, *Lasaea rubra*. G, Lucinidae, Montacutidae, Teredinidae. ● represents an oralward feeding current; ✕ represents material destined to be rejected. (Originally published in Atkins, 1937b, *Quart. J. micr. Sci.* vol. **79**, 416, 417, fig. 18.)

on their frontal surfaces, as seen in diagrammatic transverse sections. The primitive condition of the ctenidium as seen in transverse secton is probably the W-shaped figure shown in B(1a) for the Arcidae and Anomiidae. Derived conditions are seen in B(2), F and G in which portions of the ctenidium have been deleted, and in E and E(a) in which the outer demibranch has been turned dorsalwards. It will be noted that in B(1a) opposing ciliary currents are shown on all four frontal surfaces, and this is considered to be the primitive condition for ciliary currents. In all other cases one or more lamellae bear ciliary currents which pass in one direction only, and it is suggested that these are derived conditions which arose by deletion of one or other of the original ciliary tracts. It appears easier to account for this diversity of ciliation as the result of *evolution by loss* than by postulating that new ciliary currents were independently created in certain positions in various bivalve groups. Thus the condition shown in C(1c) occurs in *Barnea candida* and in *Petricola pholadiformis*, two unrelated forms; in this case *parallel evolution by loss* is considered more probable than the creation of opposing ciliary currents on the same lamella through convergence. Again, the condition shown in D was originally found only in the Unionidae, but has since been discovered in *Beguina semiorbiculata*, which lives embedded in massive sub-littoral coral masses (Purchon, unpublished work). This should be interpreted as chance similarity caused by loss of the same ciliary currents in the two cases, and not as signs of common ancestry, or of convergent evolution.

In many, but not all cases in which the ctenidia are deeply plicate, the filament which lies centrally in the bottom of the trough of the plica differs structurally from the remainder, and is known as a principal filament; the ciliary currents on the principal filament and on some ordinary filaments lying adjacent to it may differ from those on the remainder of the filaments of the plicae. Thus in most of the pseudolamellibranchiate forms (B(1b)), and in the Solenidae (C(2a)) the ciliary currents on the principal filaments and a few adjacent filaments are directed upwards only, whilst ciliary currents pass both upwards and downwards on the remainder of the ordinary filaments.

If the inhalant water stream contains relatively few and small particles the ctenidia may be well expanded, and the plicae will be held widely open, thus exposing the principal filaments in the troughs of the plicae. At the same time the tracts of coarse frontal cilia, which beat ventralwards, are inactive. Under these circumstances the majority of the particles falling on the face of the ctenidium will be conveyed dorsalwards, under the influence of the tracts of fine frontal cilia. Mucus-bound particles reaching the ctenidial axis, or the dorsal borders of the ascending lamellae, will then be carried forwards in safety to the palps and eventually to the mouth.

If, on the other hand, the inhalant water stream carries a comparatively heavy load of suspended matter, the ctenidia are liable to contract, thus accentuating the folding of the surface due to plication, and obscuring the principal filaments which lie far below the general surface, in the troughs of the plicae. At the same time the tracts of coarse frontal cilia become stimulated to activity, there is considerable secretion of mucus, and mucus-bound material is driven ventralwards to the free ventral borders of the ctenidia. Under such circumstances the activity of the tracts of fine frontal cilia will be negated by the adjacent and more vigorous coarse cilia, so that little material passes dorsalwards up the filaments. Much of the material which reaches the free ventral borders of the ctenidia is likely to fall off the ctenidia, to be shaken off by muscular movements of the ctenidia, or to be dragged away through contact with ciliary cleansing currents on the surface of the visceral mass, or the inner surface of the mantle.

Even under such extreme circumstances, however, it may be possible for the animal to feed. The heaviest load of sediment will be removed from the inhalant water stream by the ctenidial filaments which lie close to the apices of the plicae, and the water which penetrates further, to be filtered by the filaments in the depths of the troughs, adjacent to the principal filaments, will be far cleaner. This water coming into contact with the principal filament and other filaments lying nearby will still contain a small amount of suspended particles of small size only, and such particles will be conveyed dorsalwards in the troughs between the plicae. Such particulate material can then be carried forwards along the dorsal borders of the lamellae towards the palps and the mouth.

It will be seen that the principal function of the ciliary mechanisms on such a ctenidium is to divide the material collected by the filaments into two fractions. Large particles, and also large quantities of fine particles, will travel ventralwards and will tend largely to become discarded; small particles, in small quantities, will tend to travel dorsalwards and to be accepted. The interaction between muscular activity, ciliary activity, and secretory activity by mucus glands to effect this discriminatory process should be emphasised. It will be noticed that in some other examples there is division of suspended matter into two separate routes, dorsalwards for acceptance, and ventralwards for rejection or for "less certain acceptance", but that the process does not apply to all four faces of the ctenidium. Thus the process is operative on three frontal surfaces of the ctenidia of *Cultellus* (C(1d)), and on one frontal surface only on the ctenidia of *Barnea candida, and Petricola pholadiformis* (C(1c)).

In other types of ctenidial organisation the ciliary currents are in one direction only on each frontal surface, except for very small areas close to the ctenidial axis, where there may be dorsalward currents passing particles into the oralward food current at the axis. Nowhere on these gills are there adjacent currents beating in opposite directions on the same filament. Even here the same principle may apply since the oralward ciliary current at the ctenidial axis is a well-protected route, while the oralward currents along the free margins of the demibranchs are sometimes only weakly developed, and may be subject to interference by the cleansing mechanisms on the ciliated surfaces of the visceral mass and the mantle (see Fig. 50).

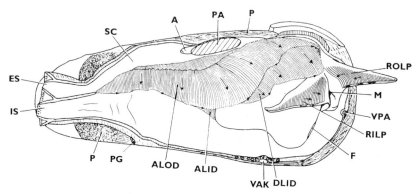

Fig. 50. Organs in the mantle cavity of *Martesia striata*, seen from the right side after removal of the right shell valve, the right mantle lobe, and the right side of the siphonal process. The outer labial palp has been turned forwards to expose the ridged, opposed surfaces of the two palps. Note that the foot is very much reduced in this wood-boring pholad. For interpretation of the lettering, see p. 142. (Originally published in Purchon, 1956, *Proc. zool. Soc. Lond.* **126**, 250, fig. 3.)

It has been suggested that in some bivalves the meshwork of the ostia, or of the interlocking lateral cilia is too coarse to trap the finest particles, and that the efficiency of the ctenidia is increased by the continuous secretion of a sheet or net of mucus which is capable of trapping particles as small as 1–2 μ (MacGinitie, 1941; Jorgensen, 1949; Jorgensen and Goldberg, 1953). Doubt has been cast on the possibility of such a mucus net being continuously secreted and passed over the surface of the ctenidium, for the presence of such a net would presumably hinder the ciliary sorting mechanisms of the frontal cilia (Ballantine and Morton, 1956). I have myself seen a thin mucus sheet being detached from one side of a ctenidium, by contraction of the filaments and the ejection of water backwards through the ostia. This action was interpreted by me as a method of ridding the ctenidium of unwanted material on the frontal surfaces, for the mucus sheet was immediately entangled in the ciliary cleansing currents of the visceral mass and removed from the ctenidium. Further evidence would be desirable before agreeing that the ctenidium can be aided in collecting very small particles by the continuous passage of a fine mucoid net over the frontal surface.

Tammes and Dral (1955) investigated the capacity of *Mytilus edulis* for straining suspensions of particles of various sizes, and concluded that the degree of efficiency of the ctenidium for retention of particles was related to the diameter of the particles in the suspension. Thus *Lycopodium* spores (diameter 30–40 μ) were strained out with great efficiency, whereas carbon black particles (diameter 10–300 mμ) were hardly retained at all. The retention of particles of intermediate size, e.g. blood corpuscles (diameter 7–8 μ) was very variable, which was possibly due to variation in the spacing of the ctenidial filaments. Tammes and Dral made direct observations on minute specimens of *M. edulis* through the transparent shell valves, and observed that the latero-frontal cilia were sometimes active and were sometimes at rest. When fed with minute carmine particles these particles were sometimes seen to adhere to a single latero-frontal cilium, causing this cilium to decrease its rate of beating. The particle was then wiped off on to the row of frontal cilia "by a peculiar movement". This observation does not support McGinitie's theory of feeding by use of a continuous mucous sheet.

Mucus is secreted, of course, to bind the particles accepted into mucoid cords which are passed forwards towards the mouth. Mucus may be secreted copiously if excessive turbidity is encountered, to rid the ctenidia of unwanted material.

In some cases the marginal grooves at the free ventral borders of the demibranchs are comparatively deep, and may be held widely open or tightly closed according to the suitability or unsuitability of the inhalant water stream (see Fig. 51). It is not yet known whether the groove opens widely in response to the presence of particular types of food particles or whether it closes in response to generally unsuitable conditions; nor do we yet know the nature and the location of the sense organs involved. When the marginal groove is tightly closed there may still be a superficial oralward current along the margin, but any material travelling along this route is only precariously secured and is unlikely to reach its destination. If the ctenidium is plicate, the terminations of the plications may project as a regular series of lobes at the free margin of the demibranch; in such cases, e.g. *Pinna* or *Brechites*, even when these lobes are drawn over the marginal groove, small particles approaching the groove along the principal filaments in the troughs of the plicae can enter the marginal food groove, as is indicated in the figure. Under conditions of heavy turbidity the ctenidium will be contracted, the marginal groove closed, and the ventralward ciliary currents over the apical parts of the plicae will serve to reject particles, for all particles collected here will be unable to enter the depths of the marginal groove. Such particles as are collected in the ciliary currents in the troughs between the plicae will enter the marginal groove, and so the animal can continue to feed. In the cases cited, which live on tropical shores of muddy sand, minor wave action may render the water highly turbid, and the possession of ctenidia of the type described is clearly of adaptive value.

In some cases sorting of particles occurs in close proximity to the marginal grooves, small quantities of minute particles being admitted into the depths of the groove where their transmission forwards towards the mouth is secure against interference, while larger particles are only able to make

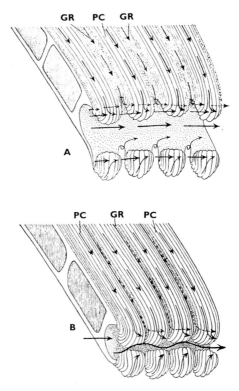

FIG. 51. The ventral margin of a demibranch of *Pinna fragilis*, showing plication and the marginal food groove. A, the ctenidial musculature is relaxed, the plicae are widely expanded so that the plical grooves (GR) are exposed, and the marginal food groove is open. B, contraction of the ctenidial musculature causes reduction of the size of the plicae so that the plical crests (PC) are drawn close together, and the principal filaments in the plical grooves are obscured; the marginal groove is tightly closed. (Originally published in Atkins, 1937a, *Quart. J. micr. Sci.* **79**, 348, fig. 2.)

a precarious forward progress along the rims of the groove. Thus in *Musculus marmoratus*, which has filibranch ctenidia, fans of relatively stiff and long cilia arch over the marginal groove, and are only active intermittently (see Fig. 52). Fine particles are able to pass round the sides of the tips of the filaments, along tracts of fine frontal cilia, and slip under these guarding fans of long cilia, and so enter the marginal groove;

other material passes medially along the filaments to their tips under the direction of a tract of coarse frontal cilia and, further progress being barred by the fans of guarding cilia, is unable to enter the safety of the marginal groove. In another case, e.g. *Entovalva*, the fans of guarding cilia occur on one side of the marginal groove only, and presumably serve to hinder the admission of particles into the marginal groove from one of the two lamellae.

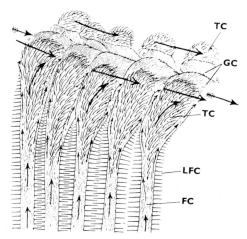

FIG. 52. *Musculus marmoratus*. The marginal groove at the free edge of a demibranch is protected by fans of guarding cilia (GC). Small arrows indicate the routes taken by fine particles which are able to pass into the depths of the marginal groove, under the guarding cilia. Coarser particles are retained on the frontal surfaces of the ctenidial filaments and pass forwards precariously over the tips of the filaments until they eventually fall on to the cleansing tracts of the visceral mass and of the mantle. Latero-frontal cilia (LFC) filter the potential food material from the inhalant water stream; Frontal cilia (FC) direct particles towards the free edge of the demibranch; coarse terminal cilia (TC) direct coarse, unsuitable particles external to the fans of guarding cilia. (Originally published in Atkins, 1937a, *Quart. J. micr. Sci.* **79**, 354, fig. 4.)

Examination of the contents of the stomachs of many different kinds of bivalves often reveals the presence of coarse particles, such as sand grains, which one would have expected to have been excluded by the rigorous sorting mechanisms which occur on the ctenidia, and often also on the labial palps. It seems probable that in many cases it is advantageous for small quantities of fine sand to be admitted to the stomach, where it may aid in the trituration of food particles. It is suggested, therefore, that the function of the sorting mechanisms on the ctenidia is to limit the total quantity of material to be ingested and to ensure that a high proportion of this consists of very fine particles which may include much organic matter.

In three families, the Lucinidae, Montacutidae, and Teredinidae, and also in *Xylophaga* (Purchon, 1941) the ctenidium has been reduced to one demibranch only, as shown in Fig. 49. G. This seems to reinforce the impression that in some cases the efficiency of the ctenidia is so great that the animal is faced with the problem of preventing an excess of food material being passed into the gut. Stasek (1962) reports a unique condition in the ctenidia of the giant clam *Tridacna derasa*. The ctenidia are plicate and on all lamellae the ctenidial filaments are interrupted at regular intervals by ridges which run transversely across the plicae and parallel with the marginal food groove. Ciliary currents following these ridges abstract potential food material from the frontal surfaces of the filaments and direct it transversely across the plica into the adjacent groove between two plicae, whence it passes directly to the marginal food groove. When the ctenidia are in their normal disposition the ridges on successive plicae are arranged in lines parallel with the marginal food groove; the number of these "plical nodes", as they are named by Stasek, increases with the age and size of the individual, a specimen of 115 mm shell length having as many as twelve series of plical nodes. The purpose of this device is evidently to ensure that all potential food material collected on the surfaces of the lamellae is secured and passed to the marginal food groove. This may seem unusually unselective and voracious for a bivalve, but the water over a coral reef may be very deficient in nutriment and, moreover, *T. derasa* is the largest existing bivalve and will require much food in order to maintain its growth rate.

When *Lasaea rubra* was fed upon uni-algal cultures, it was found that filtration ceased when the toxic *Gymnodinium veneficum* was offered. Other uni-algal cultures were filtered at different rates according to the species. Differences in the sizes of the algal cells having been considered and excluded, it appears that there must be some sense organ which is sensitive to differences in the effluvia from these various algae. It is possible that minute chemo-receptors are sited within the siphons, at the osphradium, or even on the ctenidial filaments themselves; it is presumed that stimulation of such sense organs may cause the ctenidia to contract or to expand. Filtration rates for three species of uni-algal cultures, selected from many which were studied, were in the following order: *Chlorella stigmatophora* < < *Phaeodactylum tricornutum* < *Isochrysis galbana*, the highest rate of filtration being for the last-named species (Ballantine and Morton, 1956).

In various other bivalve species it has been noted above that under certain circumstances the ctenidial filaments may be held widely apart, or somewhat contracted; where there is a deep marginal groove at the free ventral border of the demibranch this groove may sometimes be held widely open or it may be tightly closed. Differences of this sort will have a profound effect on the functioning of the ctenidium as a feeding organ, but the underlying mechanisms are as yet unknown. It is probable that the ciliary mechanisms themselves, on the ctenidia and on the labial palps are solely concerned with the grading of particles by size, taking no account of the chemical composition of the particles or their suitability for digestion. Microscopic chemo-receptors may detect the presence of toxic, or of acceptable organisms; impulses issuing from such sense organs may effect changes in tonus of the muscle fibres in the ctenidia, and the labial palps and so bring about changes in the disposition of the parts, and also the rates of filtration of the inhalant water current. If this were true, then the interaction of micro-sense organs, muscle fibres, and ciliary action would bring about some degree of qualitative selection of the material passed to the mouth for ingestion.

Nelson and Allison (1940) have shown that in *Ostrea* (*Gryphaea*) *virginica* the ctenidium is capable of responding to the presence of the hormone diantlin which emanates from spermatozoa of the same species, when sperm is present in sufficient quantity in the inhalant water current. It is possible that expansion of the plicae, and of the marginal food grooves, in some cases is similarly due to perception of chemical substances emanating from certain types of food particles borne in the inhalant stream. Experimental work in this field is greatly needed.

Jorgensen emphasises the extreme sensitivity of some bivalves to mechanical interference and to the very slightest abnormality that may arise in aquaria. Meaningful observations, he contends, can only be obtained from studies on undisturbed animals living under ideal conditions. Jorgensen considers that it may be possible to reconcile the views of McGinitie with those of other workers on the following lines—certain bivalves may feed when the water is very clean by use of a mucous net secreted continuously over the ctenidia; when the water is less clean the same bivalves no longer form a mucous net, and suspended particles are collected and sorted by ciliary mechanisms on the ctenidia. However, the ciliary currents beat *posteriorly* at the free ventral margins of the ctenidia in the Arcidae and Anomiidae, and bivalves in these families therefore cannot feed by means of a mucous net on the lines suggested by McGinitie (Jorgensen, 1955).

The rate of pumping of water may best be investigated by determining the rate of clearance of suspended organisms or particles, the bivalves under investigation being fully expanded at the time and not subjected to any mechanical disturbance. Such experiments were carried out by Jorgensen on *Mytilus edulis*, using the diatom *Asterionella japonica* (colonies up to 200 μ in diameter); colloidal graphite—Aquadag A and Prodag (the dominant particle size of freshly prepared suspensions being 1–2 μ in diameter); haemocyanin from the lobster *Homarus americanus* and haemoglobin in solution from the dogfish *Acanthias squalus*. The concentrations of the last three suspensions were determined by spectrophotometer readings at 6100 Å, 5000 Å, and 4100 Å respectively. When particles were present in low concentration the rate of clearance of the suspension was independent of the concentration. In general no pseudofaeces were formed in these experiments and all particulate material cleared from suspension must have been ingested. It was found that *M. edulis* was able to retain particles as small as 1–2 μ in diameter, and the average rate of water transport was found to be approximately 1·5 litres per hour at 14 °C. The presence of haemocyanin, or of haemoglobin, had no effect on the porosity of the ctenidia or on the rate of uptake of particles of graphite. Haemocyanin was only taken up extremely slowly, while the rate of clearance of haemoglobin was too low to be determined (Jorgensen, 1960).

In *M. californianus* at any given temperature the rate of transport of water was found to be higher in mussels from high latitudes than in mussels from lower latitudes (Rao, 1953). Different rates of transport were found in specimens from different tidal levels; the highest rates were found in specimens from sub-littoral sites, and the lowest rates were found in specimens taken from a high intertidal level (Segal, Rao, and James, 1953). Jorgensen (1960) was unable to obtain comparable results when investigating *M. edulis* taken from different tidal levels, but his samples from a large mid-tidal pool could not be regarded as typical of the inter-tidal zone since they were presumably always submerged and never exposed to rhythmic exposure to air by tidal changes.

Although, as indicated above, some bivalves have been shown capable of retaining particles as small as 1–2 μ in diameter, conflicting results have been obtained by other investigators who found only incomplete retention of organisms of 60 μ diameter, e.g. in *Ostrea virginica*. This might be due to great expansion of the interfilamentar spaces, as may occur when a female is spawning, when eggs are passed between the ctenidial filaments into the infrabranchial cavity.

Structure and Function of the Labial Palps

The labial palps of the filter-feeding bivalves vary considerably in size, but tend towards uniformity in structure and function. Typically there are two palps, one inner and one outer, on each side of the mouth; they are triangular in shape and approximately equal in size, and are attached antero-dorsally to the visceral mass by a broad base. Prolongations of the inner and outer palps pass forwards on each side as low ridges on the visceral mass and meet anteriorly to form the posterior and anterior lips of the mouth, respectively (see Fig. 53). A ciliated groove, the lateral oral groove, lies between the bases of the pair of palps and in this groove the cilia beat forwards towards the mouth. This groove is continued forwards as a proximal oral groove which passes forwards between the two above-mentioned ridges and leads directly into the mouth. The opposed surfaces of the two palps are covered with close set ridges and grooves which usually pass at right angles, or obliquely, outwards from the lateral oral groove towards the free ventral border of each palp. These ridges and grooves are covered with a ciliated epithelium in which the cilia are arranged in as many as eight distinct tracts (Allen, 1958), the cilia in each tract beating in a particular direction in relation to the ridges and grooves on the surface of the palp.

At the anterior end of the ctenidium the inner demibranch is often deeper than the outer, and this inner demibranch alone may project forwards and lie between the inner and outer labial palps. The ciliary currents in

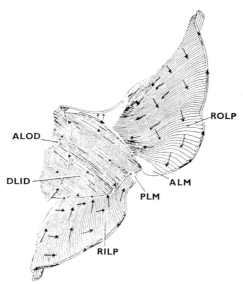

FIG. 53. Ciliary sorting mechanisms on the labial palps of *Petricola pholadiformis*. The outer labial palp (ROLP) has been turned forwards to expose the anterior end of the ascending lamella of the outer demibranch (ALOD) and the descending lamella of the inner demibranch (DLID), and the ridged, ciliary sorting surfaces of both palps. For interpretation of other lettering, see p. 142. (Originally published in Purchon, 1955a, *J. mar. biol. Ass. U.K.* **34**, 262, fig. 5.)

the ctenidial axis often deliver material to the base of the labial palps, into the lateral oral groove, where such material commonly passes forwards directly to the mouth, to be ingested without having been subjected to any ciliary sorting mechanisms on the faces of the two palps.

Material which is brought forward in the marginal groove of the outer demibranch may often be transferred to the descending lamella of the inner demibranch; material brought forwards in the marginal groove or on the anterior filaments of the inner demibranch will be presented to the ciliary sorting mechanisms on the folded opposed surfaces of the two palps, where it is subjected to rigorous sorting processes.

While there are variations in detail in the various bivalves in which the palps have been closely studied, the disposition of the ciliary tracts on the folds of the labial palps of *Barnea parva* may be taken as characteristic (Purchon, 1955) (see Fig. 54). In life each fold overlaps the adjacent fold in front so that the deep groove between the folds, which is clearly displayed in the figure, is not easily visible in the living animal. When particulate material is placed on the folded surface of the labial palp it is rapidly subjected to a process of ciliary sorting which often causes the material to be divided into two fractions quite quickly and dramatically. One fraction,

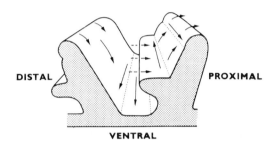

FIG. 54. Ciliary sorting currents on the opposed, ridged surfaces of the labial palps of *Barnea parva*. A diagrammatic representation of two folds parted more widely than would ever occur under natural conditions. The oralward side is on the right of the figure. Oralward ciliary currents pass over the crests of the folds, down the proximal slopes of the folds, and up the middle part of the distal slopes; a re-sorting ciliary current passes dorsalwards along the shelf on the distal slope; a rejection current in the bottom of the groove between two folds discards large and heavy particles; another rejection current on the distal side of the crests of the folds disposes of lighter particles if these are present in excessive quantity. (Originally published in Purchon, 1955, *Proc. zool. Soc. Lond.* **124,** 876, fig. 9.)

appearing to consist mainly of larger particles, tends to be thrown forwards and ventralwards under the combined action of the oralward and the ventralward ciliary currents which occur at the crest of each fold. In this way the surface of the palp is cleared of the greater part of the material added, in only a few seconds. This material is passed to the ventral margin of the palp, where it is passed posteriorly along the margin and is finally transferred to the ciliary cleansing currents on the mantle or on the visceral mass.

A second fraction, consisting mainly of smaller particles, slips into the upper half of each groove and travels dorsalwards along the groove under the influence of a tract of cilia lying half-way up the distal surface of the fold, which beats towards the lateral oral groove. This material becomes bound into cords by secretion of mucus, and these mucous cords tend to be thrown over the crests of the folds forwards into the next groove, but often travelling a little ventralwards at the same time. Muscular rippling of the folds may occur, allowing large particles, and strips of mucus-bound cords to fall into the lowest part of the grooves, and such material is then carried ventralwards in the bed of the groove, to be discharged as unwanted material. As the folds approach the lateral oral groove they usually gradually become finer, and the selective process seems to become more critical in these regions; only small numbers of the very finest particles reach the lateral oral groove and are passed forwards and are accepted into the mouth.

In this sorting process we find interaction between the ciliary tracts and the tonus of the subcutaneous muscle fibres; the latter will influence the degree of overlapping of the folds and the degree of exposure of the rejection currents in the bed of the grooves, and thereby determine the degree of selective action by the labial palps.

In some cases, e.g. in the Solenidae (Graham, 1931) the ciliary tracts on the folds of the labial palps are even more diversified; this may signify that in these occupants of inter-tidal sands the sorting mechanisms of the palps need to be even more highly organised in order to exercise adequate control over the material passed to the mouth. In contrast, the labial palps are greatly reduced in *Xylophaga* and in those members of the Teredinidae which occupy baulks of timber in the open sea; here the sea water is completely free from fine grains of sand, there seems to be no need for rigorous sorting mechanisms on the labial palps, and the palps are accordingly much reduced.

Gilmour (1964) has drawn attention to an interesting functional problem that probably concerns all monomyarian bivalves, and he shows how this problem has been overcome in different, but comparable ways in two species of *Lima* and in *Pecten maximus*. Mucous food strings passed forwards in the proximal oral groove towards the mouth are accompanied by oralward water currents generated by ciliary activity in the groove. At the mouth the food material is swallowed and the water is deflected into the anterior part of the infra-branchial cavity. Evolution leading to the adoption of the monomyarian condition involves reduction of the anterior end of the body and forward displacement of the foot; Gilmour considers that this constriction of the free space at the anterior end of the mantle cavity will lead to an increase in the velocity of water currents emerging anteriorly from the oral grooves and this could cause dislodgement of the mucous food cords. This difficulty has been overcome in *L. excavata*, *L. hians*, and *P. maximus* by the development and arching over of the lips to enclose the proximal oral grooves. The method of enclosure differs in each case and we have here an interesting example of parallel evolution. Complete fusion of the lips over the proximal oral groove would prevent the entry of the ciliary water current which ordinarily accompanies the mucous food cord to the mouth, and this would perhaps be disadvantageous. For this reason, doubtless, complete fusion of the lips over the groove has not occurred in any of these examples. In *L. excavata* and in *P. maximus* the anterior and posterior lips have been modified with the formation of lobes which interlock over the proximal oral groove, but which do not fuse with each other. Water can escape from the proximal oral grooves into the infra-branchial cavity via the interstices between these interdigitating folds. In *L. hians*, on the other hand, extensive fusion has occurred between the anterior and posterior lips, providing a permanent roof over the proximal oral grooves, but there is a series of perforations along the line of fusion and water can escape from the lip apparatus via these pores. (see Figs. 73 and 74, pp. 191-2).

Yonge (1926) showed that in *Ostrea edulis* the upper lip of the mouth is enlarged to form a hood which encloses the thickened lower lip, thereby giving comparable protection to mucous food cords in the proximal oral grooves.

Bivalves which lie on one side on the surface of the substratum may require to clap the shell valves vigorously from time to time in order to expel foreign matter which has gained entry to the infra-branchial cavity. Enclosure of the proximal oral grooves in the manner described above will also protect the food train from disturbance by such intermittent violent water movements. It will be of considerable interest to learn the condition of the proximal oral groove in other monomyarians such as *Spondylus*, as well as in other species of *Lima*, *Pecten*, and *Ostrea*. The proximal oral groove of *Tridacna* spp. is not specialised in spite of the monomyarian condition and constriction of the anterior end of the body, and in spite of the habit of clapping the shell valves occasionally to expel water and foreign matter from the mantle cavity.

Colonisation of Rocky Exposures

The early protobranchiate ancestral bivalves were confined to feeding on organic matter in marine deposits, and presumably to living as members of the infauna of such deposits. This remained true for their modern descendants in the Protobranchia. The evolution of efficient ctenidial filter-feeding mechanisms, as described in detail above, enabled the Polysyringia to become independent of marine deposits as a direct source of food, to undergo an adaptive radiation, and to invade many environments hitherto unexploited by the Bivalvia.

The establishment of bivalves on rocky shores, on coral reefs, and on buttress roots at the seaward fringes of mangrove swamps, etc., required some form of permanent attachment to the substratum. In this way there would not be unnecessary expenditure of energy in hanging on to the rock when this is submerged, whilst when the tide ebbed the foot could be withdrawn and the shell valves closed to prevent water loss through desiccation.

Such attachment has been provided in various ways, with profound effects on the morphology and mode of life of the various lineages concerned (these effects will be discussed in detail elsewhere).

First and foremost, attachment has been by use of the byssus gland, a pedal gland which was probably originally an organ of the post-larva and which originally served at the time of metamorphosis to anchor the post-larva to a suitable substratum. Extension of the activity of the byssus gland into adult life may be regarded as an example of paedomorphosis, and the widespread occurrence of this (in eighteen super-families) points to the importance of this development in the radiation of the Bivalvia (Yonge, 1962) (see Fig. 76, p. 202).

In the temperate Atlantic byssally attached forms on rocky coasts include the well-known *Mytilus edulis*, the small *Arca lactea* which occurs in crevices and under rocks near low-water mark, some members of the Pectinidae such as *Chlamys distorta* which occurs sub-littorally, and various members of the Anomiidae. In the last-named family the byssus is in the form of a massive sub-central calcified pillar, e.g. *Anomia ephippium*. Special mention should be made of the minute *Lasaea rubra*, which is one of the commonest of bivalves on British rocky shores, but which is often overlooked on account of its small size. It occurs over a very wide vertical range on the shore, attaching itself by one or two byssus threads in small crevices, e.g. in the empty shells of dead barnacles, etc. The species extends upwards into the upper limits of distribution of the barnacle *Chthamalus stellatus* and in fronds of *Pygmaea pumila*. At these high levels on the shore the animal may only be wetted by the tide for about one hour in each tidal cycle, and the shells may be quite dry at other times. *Lasaea* has become adapted to survive under these extreme conditions by the ability to react quickly to the smallest splashes of water which may wet it, and to filter these with rapidity (Ballantine and Morton, 1956; Morton, Boney and Corner, 1957). It has been shown that feeding and digestive processes in *Lasaea* show a regular periodicity which is linked to that of the tidal cycle (Morton, 1956).

On tropical rocky shores species of *Malleus* may be byssally attached in inter-tidal crevices, e.g. *M. regula*, while the delicate, flattened *Pedum spondyloideum* and the more robust *Beguina semiorbiculata* occur embedded and byssally attached in crevices in massive sub-littoral coral growths. Species of *Tridacna* are byssally attached, *T. crocea* burrowing downwards into soft inter-tidal coral boulders, while *T. squamosa* lies sublittorally, attached by a massive byssus and embedded between coral colonies (see Figs. 68 and 69 on pp. 176-7). In the Tridacnidae feeding has been augmented by the farming of zoo-xanthellae in the mantle tissues. The adoption and development of this auxiliary feeding method has brought about extensive forward migration of the mantle tissues surrounding the siphons, thereby exposing the maximum area of flesh to sunlight and making the most of the photosynthetic powers of the zoo-xanthellae. "Hyaline organs" may enable light to penetrate more deeply into the mantle tissues (see Fig. 55). The zoo-xanthellae are subsequently digested and provide a substantial part of the food of the clam. The consequences of this symbiotic association upon the morphology of the clam will be discussed in Chapter 4. Species of *Pteria* may form dense colonies on the buttress roots on the seaward fringes of mangrove swamps, where they are firmly attached by a mass of byssus threads. Utilisation of the post-larval byssus gland as an organ of attachment of the adult animal has thus facilitated a widespread colonisation of various hard substrates by bivalves.

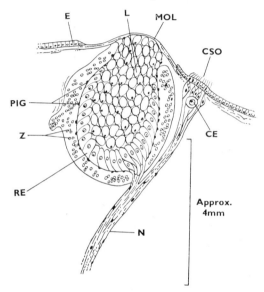

Fig. 55. Diagrammatic reconstruction of the pallial eye and the associated ciliary sense organ (CSO) of *Tridacna maxima*. Note the multicellular lens (L), the membrane overlying the lens (MOL), the retinal layer (RE), the common nerve serving both the eye and the ciliary sense organ (N), and the zoo-xanthellae (Z) lying in the perioptic space. For interpretation of other lettering, see p. 142. (Redrawn from the original illustration in Stasek, 1966, *Occ. Pap. Calif. Acad. Sci.* **58**, 7, fig. 5.)

In a number of phylogenies one shell valve has become cemented to the substratum by a fluid secretion from the periostracal groove, after a very brief post-larval phase of byssal attachment in some cases. In *Ostrea edulis* at metamorphosis a single drop of adhesive fluid is exuded from the byssus gland and this serves to cement the lowermost shell valve to the substratum (Cole, 1938). In the tropics *O. cucullata* may form an extremely dense zone inter-tidally on exposed rocky shores, while the mangrove oyster *O. parasitica* may form dense incrustations on the buttress roots of mangrove plants. On exposed rocky shores the solitary

Chama and *Spondylus* lie compactly against the rock to which they are cemented, and are rendered inconspicuous by a covering growth of epiphytic weeds. More colourful and highly ornamented specimens of thorny oyster (*Spondylus*) occur sub-littorally amongst the coral. Various other marine genera, e.g. *Cleidothearus*, are cemented to the substratum, and cementation to the substratum also occurs in the Etheridae, in tropical freshwater systems (Yonge, 1962a). There has been appreciable convergent evolution as regards cementation to the substratum in the Bivalvia, with consequent increase in the degree of colonisation of hard substrates by bivalves (see Fig. 76, p. 202).

Colonisation of the rocky shore, and of coral reefs, has also been effected by the adoption of the habit of rock-boring. *T. crocea*, and various members of the Pholadidae such as *Zirfaea crispata*, and also the unrelated *Hiatella arctica*, all bore into rocks by mechanical, rocking action of their shell valves. In contrast the date mussel, *Lithophaga*, bores only into soft calcareous rocks and coral boulders by chemical means, contributing in the process to the erosion of the rocks on the foreshore.

Thus by the contrasted methods of attachment by byssus threads, or by cementation, and by drilling into the hard rocky substratum, a considerable variety of bivalve lineages have invaded and established themselves on rocky exposures on the sea shore and in freshwater systems. Adaptations to new habitats and to new modes of life as indicated above have called forth—or have been dependent upon—morphological changes which will be examined in detail in the next chapter.

Colonisation of Wood

Mention has already been made of the attachment of bivalves to the buttress roots of mangrove plants by byssal attachment (*Pteria*) and by cementation (*Ostrea*). Attention is now drawn to the invasion of wood surfaces by bivalves employing mechanical methods of burrowing. This is likely to have evolved in mangrove swamps and tropical estuaries. In mangrove areas much wood is available for infestation, in fallen logs and dead stumps which are invariably present, and in old aerial roots of *Avicennia* which protrude upwards from the surface of the soil at the foot of the tree, on more exposed beaches. Ship-worms of undetermined genus have been found burrowing downwards and reaching appreciable sizes in the aerial roots of mangrove trees a little above mean sea level on shores in the neighbourhood of Singapore. Tropical rain forests are subjected to sudden and torrential rainfall over wide catchment areas, and forest rivers may rise 50 ft in only a few hours. Huge forest trees may be undermined and torn from the soil in such spates, and carried down to the sea to be attacked and slowly demolished by wood-boring bivalves.

In the Pholadidae the different genera bore into a variety of substrates including hard rocks, softer shales, clays, and one genus, *Martesia*, bores into wood. *Martesia* enters the wood at right angles to the surface and bores an oval oblong burrow of approximately the same size as its shell, across the grain of the wood. It only penetrates about an inch into the wood and does not do a great deal of damage individually. It is probable that *Martesia* does not digest the wood through which it bores, but it benefits from its boring action in the same way as do the rock-borers in the same family, i.e. it obtains a foothold, and protection from predators. In a similar way the remarkable genus *Xylophaga* bores into wood in the open sea, particularly at great depths, occurring in a variety of water-logged material such as *Pandanus* fruits, and quite small twigs (Knudsen, 1961). The pear-shaped burrows of *Xylophaga* lie at right angles to the grain of the wood, the animal lies at the innermost end of the burrow and communicates with the sea water by a siphonal process which is commonly surrounded by tightly packed faecal masses. These masses are the mucus-bound remains of minute fragments of wood which were produced by the action of boring into the wood, and which have been passed through the alimentary canal. There is some evidence that *Xylophaga* may be able to digest part of these wood fragments (Purchon, 1941), but further experimental work is required in this matter. It has been shown that *Teredo* is able to digest the wood fragments that are swallowed (Harington, 1921; Potts, 1923; Dore and Miller, 1923; Boynton and Miller, 1927).

In many members of the Pholadidae the stomach bears a small appendix, the function of which seems to be the temporary storage of small fragments of rock or sand. Such fragments enter the stomach as the result of the action of boring into the rock, and are thought to be liberated periodically into the stomach where they may aid in the trituration of the stomach contents (Purchon, 1955). This appendix is clearly homologous with the far larger caecum which is found in the wood-boring *Xylophaga* and in the Teredinidae where it reaches still greater size (Purchon, 1955, 1960). In these wood-borers the caecum serves a similar function, namely the storage of the products of the boring action, and the liberation of these tiny fragments of wood into the stomach in small quantities.

In the Teredinidae, and perhaps also in *Xylophaga*, these fragments of wood can be digested and are a major source of food. It is clear, therefore, that the occurrence of an appendix on the stomach in the rock-boring Pholadidae, and the functions which it serve in this family, proved to be pre-adaptive towards the emergence of the wood-boring family Teredinidae and to the peculiar genus *Xylophaga*. In these animals a new source of timber may be heavily colonised by ship-worm larvae, which bore into the wood as rapidly as possible and consume the wood fragments which are produced by the boring mechanism. In this way a store of food is set aside in the caecum (= appendix) for future digestion. In the ship-worms the anatomy of the animal is grossly distorted in order to accommodate this food reserve, and the mass of living tissues which comprise the body are appreciably less than would appear at first sight. With a substantial reserve of food for future use, the ship-worm can then mature and enter a protracted phase of heavy reproduction. The great success of this particular variant of the basic bivalve model is witnessed in the abundance of species and also of specimens of ship-worms, especially in tropical seas.

In the Teredinidae the greatest importance has been placed on rapid boring to acquire a sufficient reserve of food to ensure that breeding can occur. Growth of the shell valves—the orthodox symbol of growth in bivalves—is subordinated to growth of the soft parts to accommodate the increasing store of food in the caecum and to enable the animal to bore ever deeper into the wood and yet maintain respiratory contact with the sea water outside. The protective functions normally provided by the shell are provided instead by the walls of the burrow, which are normally lined internally with a smooth calcareous deposit. The shell valves, being freed from their primary function of protection, have been converted to all intents and purposes into jaws.

The shell valves are relatively small and light, and can be easily moved. The ligament has been reduced to a small point near the dorsal surface, the hinge teeth have been lost, and the dorsal margin of the shell at the hinge line is no longer straight in the antero-posterior axis, but is slightly curved laterally. The shell valves have acquired a small knob for articulation on the ventral border, and the two adductor muscles have come to contract alternately. Consequently the shell valves no longer open and close with the hinge line as a fulcrum, but rock about a vertical axis. When the anterior adductor is contracted the anterior ends of the shell valves are approximated. The posterior adductors are large in cross-section and when these contract the anterior ends of the shell valves are powerfully parted, this being the effective boring action when the animal is lying undisturbed in its burrow. The muscular and suctorial foot emerges through a wide pedal gape in the shell valves anteriorly, and grips the anterior end of the burrow. The outer surface of the shell valves is covered with fine rows of denticulations anteriorly, and when these ends of the shell valves are parted under contraction of the posterior adductor, the denticulations are rasped against the wood. This rasping action removes small particles of wood from the interior of the burrow, and these particles are passed into the mouth. From this account it becomes clear that the shell valves now function primarily as jaws, once it has been established that the wood fragments are indeed a major source of food.

The main factors in the emergence of this entirely new feeding mechanism have been the pre-adaptive occurrence of an appendix to the stomach, the habit of passing the waste products of the boring mechanism through the alimentary canal, and finally the adaptation of the digestive diverticula with the production of a powerful cellulase. A new kind of substrate has been subjected to feeding attack, from the chemical point of view the food is a novelty, yet the Teredinidae continue to follow the normal bivalve procedure of feeding on comparatively small particles. (In point of fact the wood fragments are somewhat larger than the minute particles collected by the typical filter-feeding bivalve.) Dr. Ruth Turner (1966) has performed an invaluable service in reviewing present knowledge of the Teredinidae, including the anatomy of the soft parts, functional morphology, and physiology, and an illustrated catalogue of all known species in fourteen genera.

Although a powerful cellulase has only been demonstrated in a restricted number of molluscan types, it is probable that cellulases of lesser activity are much more widely distributed in the phylum than has yet been demonstrated. It may reasonably be assumed that the diet of the earliest molluscs consisted largely of living plant material or of organic debris of plant origin. One would therefore expect the array of digestive enzymes possessed by such early molluscs to have included at least a weak cellulase. Stone and Morton (1958) review the literature on the occurrence of cellulases in the Mollusca and, using substrates less refractory than filter paper and sawdust, have demonstrated the occurrence of cellulase activity in extracts of the digestive diverticula of *Scrobicularia plana*, *Cardium edule*, *Mya arenaria*, and *Anodonta cygnea*, and in extracts of the crystalline styles of the first three of these. They obtained positive results when using soluble laminarin, cotton cellulose which had been swollen in phosphoric acid, and sodium carboxymethylcellulose as substrates, but the digestive action was most effective with soluble laminarin. The capacity to digest cellulose and related substances may therefore occur at a very low level in a wide variety of bivalves, and only in specialised lineages such as the Teredinidae has this digestive capacity been developed to an outstanding degree.

Colonisation of Deposits of Gravel, Sand, Sandy Mud, and Mud

These mobile substrates have been subjected to a number of independent waves of colonisation by bivalve molluscs. One of these substrates is deemed to have been the original home of the earliest protobranchiate bivalves which fed upon organic matter lying in the surface layers of the substratum. With the emergence of the carnivorous, or scavenging, mode of life, these deposits were subjected to a second colonisation by the Septibranchia. A third colonisation was effected by the filter-feeding bivalves or Polysyringia. Since these three groups of bivalves are in general dependent on different sources of food, there should be no competition between them for survival so far as food is concerned*. Some other reason should be sought for the tendency for exclusion of septibranchs from the littoral mobile substrates which are so extensively colonised by eulamellibranchs.

In the filter-feeding bivalves the post-larva is aided when establishing itself in the substratum by possession of a byssus gland. This gland is embedded in the hind part of the foot, and serves to secrete a small number of threads with adhesive tips by means of which the post-larva is securely anchored to grains of sand or other small objects in the sand. The byssus gland is essentially an organ of the post-larva, and although traces of the gland may persist into adult life, the gland is normally inactive in the adults of bivalves inhabiting substrates of sand, sandy mud, or of mud. Exceptions to this generalisation are found in the genera *Pinna* and *Atrina* which live partly buried in the substratum, anchored in each case by a luxuriant growth of byssus threads. It was suggested above that the colonisation of rocky surfaces by filter-feeding bivalves was brought about by a process of paedomorphosis (Yonge, 1962) whereby the byssus gland of the post-larva was maintained as an actively functioning organ of attachment during adult life. We may surmise that *Pinna* and *Atrina* were originally colonists of rocky surfaces and retained their byssus gland into adult life as a result of paedomorphosis, doubtless acquiring a mytiliform appearance at this stage due to byssal attachment; we may further surmise that these genera later became adapted to life on unstable shores where they were liable to be partly buried by drifts of sand, and that they finally became wholly adapted to living three-quarters buried in substrates of sand, mud or of mud (see Fig. 56). If this were true this would constitute a very minor re-invasion of unstable substrates by animals which were previously colonists of rocky surfaces.

The filter-feeding bivalves are in most cases only dependent on the substratum for a foothold, for protection against instability or dislodgement by wave action, and for concealment from some of their enemies; they are characteristically dependent on the super-natant water for their supplies of food and of oxygen, maintaining contact with the overlying water by means of two orifices—one inhalant and the other exhalant. Many such bivalves lie very close to the surface of the sand and they possess comparatively short siphons, e.g. *Cardium edule*, the common cockle of european shores, in which the inhalant and exhalant siphons form two short separate tubes which are only united at their bases. Generally bivalves occupying a superficial position in the sand are comparatively small, but bivalves of greater size are known in this position, such as

* As is mentioned below, the Tellinidae and the Semelidae feed upon organic matter in the sedimentary deposits in which they live.

Cyprina islandica which occurs sporadically on British shores, and *Geloina ceylanica* which may occur in some abundance in the surface layers of the mud in mangrove swamps in the Far East.

In contrast to these relatively immobile bivalves, members of the Solenidae are deep-burrowers which are capable of active movement within the more or less permanent vertical burrows which they occupy.

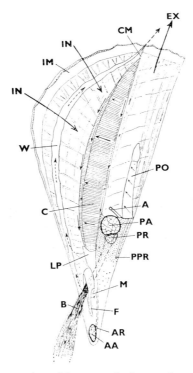

FIG. 56. *Pinna carnea*, view of the organs in the mantle cavity after removal of the right shell valve and the right mantle lobe. The animal is wedged vertically in the substratum with a considerable portion of the posterior end of the shell exposed to the risk of damage by abrasion by rolling fragments of coral, etc. The shell is anchored firmly by a substantial mass of byssus threads (B). The mantle, and the ctenidium (C) retreat extensively from the posterior shell margin when the shell has been damaged, the pallial organ (PO) is dilated by blood withdrawn from elsewhere, and it serves to brush away fragments of shell prior to repair to the shell. Note the ciliated waste canal (W) by which pseudofaeces are removed vertically from the mantle cavity. For interpretation of other lettering, see p. 142. (Originally published in Yonge, 1953, *Phil. Trans.* B, **237**, 350, fig. 9.)

Ensis siliqua, for example, lies vertically in the surface layers of sand near low-water mark of spring tides and its shell may even protrude slightly above the surface of the sand when the animal is undisturbed.

In accordance with this superficial feeding position, the inhalant and exhalant siphons are short and separate tubes. The animal possesses a large, muscular foot which is well adapted for burrowing, and on the slightest disturbance the animal plunges to safety at the bottom of its burrow.

Some bivalves lie more deeply embedded in the substratum, maintaining their contact with the water by means of a single stout siphonal process which is formed by the union of the inhalant and exhalant siphons. The siphonal process is relatively stiff and inflexible, and is generally protected against abrasion by a stout sheath of periostracum. These types are generally immobile, their only movement being to push slowly further downwards into the soil as they grow. Examples of such bivalves include *Mya* and *Lutraria* which may be found intertidally in stiff muds. Similar conditions occur in *Barnea candida* which 'may colonise stiff clay or peat outcrops on the foreshore.

Radically different conditions are found in the Lucinacea which have colonised a wide variety of substrates ranging from muddy gravel and coarse sand to very fine muds. Habitats colonised by members of the Lucinacea do not generally support a rich infauna and it is thought that in these environments the food supply is so low that all available particulate material would have to be accepted by any bivalve member of the infauna if it were to survive (Allen, 1958). The Lucinacea appear to comprise an independent lineage of polysyringian bivalves which have specialised in the colonisation of such impoverished habitats. Within the Lucinacea specialisation seems to have proceeded furthest in the family Lucinidae. The shell lies fairly deeply buried in the substratum, and the animal maintains communication with the overlying water by means of a vertical channel in the substratum which lies on the *anterior* side of the shell. This open channel in the substratum is established, and is maintained by action of the very long, slender, cylindrical foot, which can be protruded upwards through the sand or mud as much as six times the diameter of the shell or even more. Here we have a novel feeding mechanism in which the foot and the anterior adductor muscle play important roles. Examples of this remarkable feeding mechanism, which is described in detail on p. 158, are found in *Lucina*, *Thyasira*, *Phacoides*, etc. (Allen, 1958).

It is curious that some filter-feeding bivalves have adopted the habit of feeding upon organic matter deposited in the surface layers of the substrates in which they lie buried, "browsing" upon the surface of the mud or muddy sand with the tip of the inhalant siphon. In these cases the inhalant and exhalant siphons are long, separate, and highly mobile tubes, and the inhalant siphon is protruded for some distance and may explore the surface of the substratum actively, sucking up food material from its surface. The food material thus collected is particulate, and is presumably subjected to the normal processes of ciliary selection and rejection on the ctenidia and the labial palps. This condition is found in the Tellinacea, especially in the families Tellinidae and Semelidae. The success of this return to the ancestral habit of feeding upon organic matter deposited in the substratum is indicated by the great diversity of habitats which have been colonised by members of these families, and the high densities reached in populations of some species, e.g. species of *Abra* in sub-littoral muds; *Tellina tenuis*, over 4000 per square metre, in sub-littoral sands; *Macoma balthica*, 3071 per square metre in inter-tidal sandy mud or stiff mud; *Scrobicularia plana*, up to 1000 per square metre intertidally in soft estuarine muds (Yonge, 1949). It must be emphasised that although the Tellinidae and Semelidae feed upon deposited material and can be compared with modern protobranchs and with their remote common ancestor in this respect, they do so by an entirely new technique. The long, slender, and highly mobile inhalant siphon collects material indiscriminately after the manner of a vacuum cleaner; this material accumulates in the infra-branchial chamber in large quantities that would completely clog the ctenidia of a "typical" filter-feeding bivalve, and large quantities of material are rejected by the ciliary sorting mechanisms of the ctenidia and palps, to be voided as pseudofaeces via the inhalant siphon.

Special adaptations to this method of "suction-feeding" include the great length and mobility of the inhalant siphon, its habit of groping about on the surface of the substratum and the insensitivity of its tip. Another feature of interest is the presence of a waste canal near the postero-ventral margin of each mantle lobe, in which pseudofaeces can gradually accumulate without interfering with the inflow of water through the inhalant canal. With the accumulation of large quantities of the substratum within the mantle cavity as a result of the deposit-feeding habit, it is inevitable that some of the material entering the stomach will be unsuitable for digestion. In spite of rigorous sorting mechanisms on the ctenidia and on the labial palps, large particles such as sand grains will occasionally enter the stomach, and this is particularly true for deposit feeders such as *Scrobicularia*, *Macoma*, etc. It is interesting to note that in the Tellinacea, doubtless correlated with the adoption of the deposit-feeding habit, the stomach has acquired a postero-dorsal caecum or appendix, and that sand grains and similar indigestible fragments tend to accumulate in this large and distensible sac (Yonge, 1949). The role played by this sac in the processes of digestion will be discussed elsewhere.

The ability of *Scrobicularia* and of *Macoma* to feed upon organic matter in the substratum has undoubtedly contributed to the success of these genera in colonising estuaries where the turbidity of the water may be very high. In the upper reaches of the Bristol Channel and in the estuary of the river Severn as much as 30 grams of sediment may be suspended per litre of water, due to the strength of the water currents (Purchon, 1937). Filter-feeding organisms such as bivalves would find great difficulty in contending with such large quantities of sediment, the immobilisation of which would require an excessive secretion of mucus. Bivalves such as *Scrobicularia* and *Macoma* may be able to satisfy their respiratory needs after a fairly short period of siphoning, and may be able to select appropriate moments for siphoning when the turbidity is temporarily at a low level. They are not wholly dependent on siphoning for feeding, they probably feed a certain amount after they have been uncovered by the ebbing tide, and so can penetrate the higher reaches of estuaries undisturbed by the high levels of turbidity that would prevent survival of normal filter-feeding bivalves.

It seems probable that organic debris in the substratum is not itself directly available for digestion by bivalves such as *Scrobicularia* and *Macoma*. These bivalves are probably feeding upon the flora of micro-organisms that naturally develops in sediments of fine grade. Thus Newell (1965) found that faeces freshly collected from *M. balthica* were high in carbon and low in nitrogen content, whilst older faecal material had developed a relatively high nitrogen content as the result of the activity of micro-organisms developing therein (see Fig. 127, p. 378).

The invasion of freshwater systems by bivalves must have been exclusively via estuaries and river mouths, the main problems being the adaptation of the excretory organs to combat the in-flow of water from the external medium due to endosmosis, and also adaptation of the processes of reproduction to suit the entirely different environmental circumstances. The very wide distribution of some freshwater bivalve genera, e.g. *Pisidium*, and the extreme specialisations in the developmental processes, e.g. in the Mutelidae, indicate the great antiquity of the invasion of freshwater systems in these lineages. In contrast the byssally attached *Dreissena polymorpha* of western Europe may be a comparatively recent addition to the freshwater fauna. The Unionacea and the Sphaeriacea are exclusively freshwater groups which inhabit the surface layers of the muddy beds of rivers and lakes, and which feed by orthodox filter-feeding methods, e.g. *Anodonta cygnea* and *Sphaerium corneum* respectively.

We may conclude this chapter by observing that the emergence of ctenidial filter-feeding in one lineage of the Bivalvia set the stage for a major phase of adaptive radiation, liberating the animal from dependence on the substratum for a source of food. With the aid of the post-larval byssus gland, the activity of which was extended into adult life in many lineages by a process of paedomorphosis, the potential of this adaptive radiation was fully exploited, and in consequence the Polysyringia, or "filter-feeding" bivalves, became established in almost every conceivable aquatic habitat.

Key to the Lettering on the Figures

A	Anus.	ME	Mantle edge.
AA	Anterior adductor muscle.	MF	Mantle fold.
ALID	Ascending lamella of inner demibranch.	MG	Mid-gut.
		MLC	Micro-latero-frontal cilia.
ALM	Anterior lip of mouth.	N	Common nerve.
ALOD	Ascending lamella of outer demibranch.	MOL	Membrane overlying the lens.
		OP	Outer palp.
AP	Appendage of palp.	P	Periostracum.
AR	Anterior retractor muscle of the foot.	PA	Posterior adductor muscle.
		PC	Plical crests.
AS	Anterior septal muscle.	PG	Periostracal groove.
AT	Acceptance tract.	PIG	Pigment granules.
B	Byssus threads.	PL	Palp lamella.
BC	Basal layer of darkly staining cells in hyaline organ.	PLM	Posterior lip of mouth.
		PM	Posterior margin of palp.
C	Ctenidium.	PO	Pallial organ.
CE	Spherical cell associated with nerve-cell bodies of the ciliary sense organ.	POG	Proximal oral groove.
		PPR	Posterior pallial retractor muscle.
		PR	Posterior retractor muscle of the foot.
CFC	Coarse frontal cilia.		
CM	Ctenidial membrane.	PS	Posterior septal muscle.
CR	Chitinous rods supporting the ctenidial filaments.	PT	Tongue-like process of outer palp.
CSO	Ciliary sense organ.	R	Rectum.
CW	Capsular wall around the hyaline organ.	RE	Retina.
		RILP	Right inner labial palp.
DLID	Descending lamella of the inner demibranch.	ROLP	Right outer labial palp.
		RS	Retractor muscle of siphons.
E	Epithelium of the mantle.	RT	Rejection tract.
ES	Exhalant siphon.	S	Septum.
EX	Exhalant water current.	SC	Supra-branchial, and supra-septal, mantle cavity.
F	Foot.		
FC	Frontal cilia.	SH	Sheath surrounding the siphons.
FFC	Fine frontal cilia.		
GC	Guarding cilia.	SM	Suspensory membrane of the ctenidium.
GO	Glandular organ.		
GR	Grooves between crests of plicae of ctenidium.	SP	Septal pore.
		ST	Stomach.
IC	Infra-septal mantle cavity.	T	Tentacle.
IM	Inner fold of mantle edge.	TC	Terminal frontal cilia.
IN	Inhalant water current.	V	Valve at junction of inhalant siphon with infra-septal cavity.
IP	Inner palp.		
IS	Inhalant siphon.		
L	Multicellular lens.		
LC	Lateral cilia.	VAK	Ventral articular knob.
LFC	Latero-frontal cilia.	VE	Ventricle.
LM	Lower margin of palp.	VPA	Vestigial pedal aperture.
LOG	Lateral oral groove.	W	Waste canal.
LP	Labial palps.	Z	Zoo-xanthellae lying in the circum-optic space.
LS	Lateral septal muscles.		
M	Mouth.		

Reference List

ALLEN, J. A. (1958) On the basic form and adaptations to habitat in the Lucinacea (Eulamellibranchia), *Phil. Trans. B,* **241,** 421–84.

ALLEN, J. A. and TURNER, J. F. (1974) On the functional morphology of the family Verticordiidae (*Bivalvia*) with descriptions of new species from the abyssal Atlantic. *Phil. Trans. B,* **268,** 401–536.

ANSELL, A. D. (1962) The functional morphology of the larva, and the post-larval development of *Venus striatula* (da Costa), *J. mar. biol. Ass. U.K.* **42,** 419–43.

ATKINS, D. (1937) On the ciliary mechanisms and interrelationships of lamellibranchs. Part I. New observations on sorting mechanisms, *Quart. J. micr. Sci.* **79,** 181–308.

ATKINS, D. (1937a) On the ciliary mechanisms and interrelationships of lamellibranchs. Part II. Sorting devices on the gills, *Quart. J. micr. Sci.* **79,** 339–73.

ATKINS, D. (1937b) On the ciliary mechanisms and interrelationships of lamellibranchs. Part III. Types of lamellibranch gills and their food currents, *Quart. J. micr. Sci.* **79,** 375–421.

BALLANTINE, D. and MORTON, J. E. (1956) Filtering, feeding, and digestion in the lamellibranch *Lasaea rubra, J. mar. biol. Ass. U.K.* **35,** 241–274.

BOYNTON, L. C. and MILLER, R. C. (1927) The occurrence of a cellulase in the shipworm, *J. Biol. Chem.* **35,** 613–8.

COLE, H. A. (1938) The fate of the larval organs in the metamorphosis of *Ostrea edulis, J. mar. biol. Ass. U.K.* **22,** 469–484.

DAVENPORT, D. (1955) Specificity and behaviour in symbiosis, *Quart. Rev. Biol.* **30,** 29–46.

DORE, W. H. and MILLER, R. C. (1923) The digestion of wood by *Teredo navalis, Univ. Calif. Publ. Zool.* **22,** 383–400.

GILMOUR, T. H. J. (1964) The structure, ciliation and function of the lip-apparatus of *Lima* and *Pecten* (Lamellibranchia), *J. mar. biol. Ass. U.K.* **44,** 485–98.

GRAHAM, A. (1931) On the morphology, feeding mechanisms, and digestion of *Ensis siliqua* (Schumacher), *Trans. roy. Soc. Edinb.* **56,** 725–51.

HARINGTON, C. R. (1921) A note on the physiology of the shipworm (*Teredo norvegica*), *Biochem. J.* **15,** 737–41.

HUNTER, W. R. (1949) The structure and behaviour of *Hiatella gallicana* (Lamarck) and *H. arctica* (L.) with special reference to the boring habit, *Proc. roy. Soc. Edinb.* **63,** 271–89.

JORGENSEN, C. B. (1949) The rate of feeding by *Mytilus* in different kinds of suspension, *J. mar. biol. Ass. U.K.* **28,** 333–44.

JORGENSEN, C. B. (1955) Quantitative aspects of filter feeding in invertebrates, *Biol. Rev.* **30,** 391–454.

JORGENSEN, C. B. (1960) Efficiency of particle retention and rate of water transport in undisturbed lamellibranchs, *J. Cons. Int. Explor. Mer.* **26,** 94–116.

JORGENSEN, C. B. and GOLDBERG, E. D. (1953) Particle filtration in some ascidians and lamellibranchs, *Biol. Bull. Woods Hole* **105,** 477–89.

KNUDSEN, J. (1961) The bathyal and abyssal *Xylophaga* (Pholadidae, Bivalvia), *Galathea Rep.* **5,** 163–209.

MACGINITIE, G. E. (1941) On the method of feeding of four pelecypods, *Biol. Bull., Woods Hole* **80,** 18–25.

MORTIMER, J. E. (1963) A comparative study of post-larval feeding mechanisms in the bivalvia, Ph. D. thesis, University of Glasgow.

MORTON, J. E. (1956) The tidal rhythm and action of the digestive system of the lamellibranch *Lasaea rubra, J. mar. biol. Ass. U.K.* **35,** 563–86.

MORTON, J. E., BONEY, A. D. and CORNER, E. D. S. (1957) The adaptations of *Lasaea rubra* (Montagu), a small inter-tidal lamellibranch, *J. mar. biol. Ass. U.K.* **36,** 383–405

NELSON, T. C. and ALLISON, J. B. (1940) On the nature and action of diantlin, a new hormone-like substance carried by the spermatozoa of the oyster, *J. exp. Zool.* **85**, 299–338.

NEWELL, R. (1965) The role of detritus in the nutrition of two marine deposit feeders, the prosobranch *Hydrobia ulvae* and the bivalve *Macoma balthica, Proc. zool. Soc. Lond.* **144**, 25–45.

OLDFIELD, E. (1955) Observations of the anatomy and mode of life of *Lasaea rubra* (Montagu) and *Turtonia minuta* (Fabricius), *Proc. malac. Soc. Lond.* **31**, 226–49.

ORTON, J. H. (1912) Mode of feeding of *Crepidula*, and in gastropods and lamellibranchs, *J. mar. biol. Ass. U.K.* **9**, 444–78.

OWEN, G. (1961) A note on the habits and nutrition of *Solemya parkinsoni* (Protobranchia: Bivalvia), *Quart. J. micr. Sci.* **102**, 15–21.

OWEN, G., TRUEMAN, E. R. and YONGE, C. M. (1953) The ligament in the lamellibranchia, *Nature, Lond.* **171**, 73.

PELSENEER, P. (1906) *A Treatise on Zoology*, ed. E. R. Lankester, V, *Mollusca*, A. & C. Black, London.

POTTS, F. (1923) The structure and function of the liver of *Teredo*, the shipworm, *Proc. Camb. Phil. Soc. biol. Sci.* **1**, 1–17.

PURCHON, R. D. (1937) Studies on the biology of the Bristol Channel. 2. An ecological study of the beach and dock at Portishead. *Proc. Bristol Nat. Soc.*, 4 ser., **8**, 311–29.

PURCHON, R. D. (1941) On the biology and relationships of the lamellibranch *Xylophaga dorsalis* (Turton), *J. mar. biol. Ass. U.K.* **25**, 1–39.

PURCHON, R. D. (1955) The structure and function of the British Pholadidae (rock-boring lamellibranchia), *Proc. zool. Soc. Lond.* **124**, 859–911.

PURCHON, R. D. (1955a) The functional morphology of the rock-boring lamellibranch *Petricola pholadiformis* Lamarck, *J. mar. biol. Ass. U.K.* **34**, 257–78.

PURCHON, R. D. (1956) A note on the biology of *Martesia striata* L. (Lamellibranchia), *Proc. zool. Soc. Lond.* **126**, 245–58.

PURCHON, R. D. (1960) Phylogeny in the Lamellibranchia, *Proc. Cent. and Bicent. Congr. Biol. Singapore*, 1958, pp. 69–82.

PURCHON, R. D. (1962) Phylogenetic classification of the Bivalvia, with special reference to the Septibranchia, *Proc. malac. Soc. Lond.* **35**, 71–80.

RAO, K. P. (1953) Rate of water propulsion in *Mytilus californianus* as a function of latitude, *Biol. Bull. Woods Hole* **104**, 171–81.

RIDEWOOD, W. G. (1903) On the structure of the gills of the Lamellibranchia, *Phil. Trans.* B, **195**, 147–284.

SEGAL, E., RAO, K. P. and JAMES, T. W. (1953) Rate of activity as a function of intertidal height within populations of some littoral molluscs, *Nature, Lond.* **172**, 1108–9.

STASEK, C. R. (1961) The ciliation and function of the labial palps of *Acila castrensis* (Protobranchia; Nuculidae), *Proc. zool. Soc. Lond.* **137**, 511–38.

STASEK, C. R. (1962) The form, growth and evolution of the Tridacnidae (Giant Clams), *Arch. de Zool. exp. gen.* **101**, 1–40.

STASEK, C. R. (1965) Feeding and particle-sorting in *Yoldia ensifera* (Bivalvia; Protobranchia) with notes on other nuculanids, *Malacologia* **2**, 349–66.

STASEK, C. R. (1966) The eye of the giant clam (*Tridacna maxima*), *Occ. Pap. Calif. Acad. Sci.* **58**, 1–9.

STONE, B. A. and MORTON, J. E. (1958) The distribution of cellulases and related enzymes in the Mollusca, *Proc. malac. Soc. Lond.* **33**, 127–41.

TAMMES, P. M. L. and DRAL, A. D. G. (1955) Observations on the straining of suspensions by mussels, *Arch. néerl. Zool.* **11**, 87–112.

THIELE, J. (1935) *Handbuch der systematischen weichtierkunde*, Bd. 2.

TURNER, R. D. (1966) *A Survey and Illustrated Catalogue of the Teredinidae*, Museum of Comparative Zoology, Harvard University, Cambridge, Mass.

YONGE, C. M. (1926) Structure and physiology of the organs of feeding and digestion in *Ostrea edulis*, *J. mar. biol. Ass. U.K.* **14**, 295–386.

YONGE, C. M. (1928) Structure and function of the organs of feeding and digestion in the septibranchs *Cuspidaria* and *Poromya*, *Phil. Trans.* B, **216**, 221–63.

YONGE, C. M. (1936) Mode of life, feeding, digestion and symbiosis with zooxanthellae in the Tridacnidae, *Sci. Rep. Gr. Barrier Reef Exped.* **1**, 283–321.

YONGE, C. M. (1939) The protobranchiate mollusca; a functional interpretation of their structure and evolution, *Phil. Trans.* B, **230**, 79–147.

YONGE, C. M. (1949) On the structure and adaptations of the Tellinacea, deposit-feeding Eulamellibranchia, *Phil. Trans.* B, **234**, 29–76.

YONGE, C. M. (1953) Form and habit in *Pinna carnea* Gmelin, *Phil. Trans.* B, **237**, 335–74.

YONGE, C. M. (1957) Mantle fusion in the Lamellibranchia, *Pubbl. Staz. zool. Napoli* **29**, 151–71.

YONGE, C. M. (1962) On the primitive significance of the byssus in the Bivalvia and its effects in evolution, *J. mar. biol. Ass. U.K.* **42**, 113–25.

YONGE, C. M. (1962a) On *Etheria elliptica* Lam. and the course of evolution, including assumption of monomyarianism, in the family Etheriidae (Bivalvia: Unionacea), *Phil. Trans.* B, **244**, 423–58.

CHAPTER 4

ADAPTIVE RADIATION IN THE POLYSYRINGIAN BIVALVES

Synopsis

Diversity of form in the filter-feeding bivalves (=Polysyringia) is correlated with the diversity of habitats which have been colonised and the variety of modes of life which have been adopted. It is suggested that the growth form of the unspecialised, ancestral stock was probably comparable in many ways with that of the modern genus *Glycymeris*, i.e. approximately circular in outline, equivalve, equilateral, isomyarian, without pallial fusions and without a pallial sinus. Specialised types can be compared with such a "typical" example and the mode of development of the specialisations can be deduced by consideration of the mechanical stresses arising from the habitat and mode of life. The mode of growth of a shell valve is described in terms of radial, transverse, and tangential growth factors.

Adaptive radiation may be considered as deviations from the "typical" form described above, it being emphasised that adaptive radiation concerns all organ systems and all life processes, not merely the gross appearance of the shell.

In the infauna of sedimentary deposits a rounded, tumid shape is suitable for relatively motionless surface forms such as *Cardium*, but actively moving members of the infauna tend to be more streamlined, e.g. the elliptical *Abra* and *Tellina*. Deep-burrowing members of the infauna require long siphons to maintain communication with the supernatant water, and in consequence the shells of these tend to be inequilateral, with a deep pallial sinus, e.g. *Mya*. The Lucinacea are modified in various ways enabling them to live in regions too deficient in food material for most other animals, and relatively large particles are permitted to enter the ducts to the digestive diverticula of these animals. Members of the Tellinidae and Semelidae are unusual in that the siphons are separate and very mobile, being formed only from the inner lobe of the mantle margin; in these families the animals feed by sucking up particulate matter from the substratum by means of the inhalant siphon, e.g. *Abra*, *Macoma*, and *Scrobicularia*. In deep-burrowing forms such as *Mya* the two siphons are fused to form a single, rather stiff siphonal process which is derived from all three lobes of the mantle margin; the siphonal process of *Mya* is protected from abrasion by a tough sheath of periostracum. Where the siphons are very long, or very substantial, it may not be possible for them to be extended by blood pressure. In *Mya* there is a very extensive mid-ventral fusion of the left and right mantle lobes, and the pedal gape is small; the pedal gape and the siphonal apertures can be closed by muscular contractions, and then contraction of the adductor muscles forces the siphonal process to extend by hydraulic pressure of the sea water in the mantle cavity. The Solenidae are adapted for feeding at the surface of the substratum and for rapid retreat to greater depths when alarmed; the shell is produced posteriorly and, being an almost straight cylinder, can easily slide up and down a semi-permanent burrow in the sand. The elongated foot is adapted as a powerful digging organ and is aided in digging by jets of water which loosen the sand below the animal, e.g. *Ensis.*

In many members of the infauna the byssus gland retains its original status as an organ of the post-larva, serving only to anchor the animal to sand grains in the substratum at the time of metamorphosis and until the young individual is able to fend for itself, e.g. *Mercenaria.* In many lineages of bivalves a process of paedomorphosis has extended the period of activity of the byssus gland into adult life, thereby enabling such bivalves to attach themselves to rocks, etc. In some cases such byssiferous forms have returned to live in or on sedimentary deposits, e.g. *Modiolus* and *Malleus* which live byssally attached at the surface, and *Pinna* which lies partly buried in the sediment. The adaptations of *Pinna* and of *Brechites* —the "watering-can shell"—to life as members of the infauna are described in detail.

The earliest bivalves, having no means of attachment in adult life, were unable to colonise the rocky shore. In certain lineages paedomorphosis extended the activity of the byssus gland into adult life, thereby enabling these bivalves to colonise rocky surfaces. *Lasaea* and *Turtonia* are two minute byssally attached occupants of the rocky shore and their small size may be due to neoteny; these unrelated genera exhibit convergent evolution, and their adaptations to life on the rocky shore are described in detail.

Byssal attachment tends to cause deviation from the isomyarian (*Arca*) to the anisomyarian (*Mytilus*) and eventually to the monomyarian condition (*Pecten, Chlamys*) by progressive diminution of the anterior end of the body including the anterior adductor muscle. In the extreme condition the posterior adductor is sub-central in position and the ctenidium, draped around the adductor, filters water drawn from about 220° of arc—thus this sedentary animal has acquired a close approximation to radial symmetry. A similar approach to radial symmetry has been obtained independently by parallel evolution in other byssally attached forms, e.g. *Lima, Pinctada*, and *Anomia*. It is paradoxical that the ultimate achievement in several of these byssally attached lineages has been a free-living mode of life (*Placuna, Pecten, Hippopus*) and even an actively swimming habit (*Lima, Chlamys, Amussium*). In the genus *Lima* some species are byssally attached, some build an extensive nest of byssus threads, while others possess no byssus in the adult, and can swim.

In the family Tridacnidae byssal attachment, associated with the farming of symbiotic zoo-xanthellae, has also led to the monomyarian condition, with extensive changes in the relations between the viscera and the mantle/shell. Symbiosis has enabled *Tridacna gigas* to exceed the size limits otherwise imposed on the Bivalvia by virtue of their feeding habits.

In some occupants of the rocky shore attachment is not by byssus threads but by cementation of the lowermost shell valve to the substratum. In this respect there has been parallel evolution in several unrelated genera, e.g. *Ostrea, Spondylus, Hinnites, Chama*, etc.

Several unrelated lineages of bivalves have adopted the habit of boring into rocks. In some this habit has had little effect on growth form, e.g. *Tridacna crocea, Petricola pholadiformis*, and *Hiatella arctica*, all of which burrow mechanically into comparatively soft rocks. In contrast *Lithophaga* bores chemically only into calcareous rocks or corals, the chemical probably being an acid secreted by glandular thickenings of the anterior margin of the mantle; the shell of *Lithophaga* is adapted for boring by development of a thick, protective coat of periostracum and by assumption of a cylindrical form which is suitable for moving up and down, and for rotating within, its cylindrical burrow. The byssus is retained in *Lithophaga* and movement in the burrow is effected by the byssal retractor muscles. The family Pholadidae exhibit extreme adaptation to the rock-boring habit, including alternate action of the adductor muscles which rock the shell valves about a vertical axis passing through the umbo; reduction of the ligament and loss of the hinge teeth permit this new axis of movement. The anterior end of the shell is emarginated for protrusion of the suctorial foot, and the pedal muscles are inserted on to a blade-shaped apophysis which projects downwards from the umbonal region of the shell. The mantle tissues and the siphonal process are protected by the development of up to six accessory shell plates. In some pholads, when boring is no longer possible, there is a marked change in growth form and the anterior pedal gape is closed by the rapid formation of a callum, an extension of the shell valve which lacks growth lines and abrasive denticulations, e.g. *Pholadidea loscombiana*.

One member of the Pholadidae, *Martesia striata*, has specialised as a wood-borer and this is true also of *Xylophaga* and of the closely related family Teredinidae which have the same major adaptations to boring as do the Pholadidae. In the pholads some of the products of rock-boring are swallowed and passed into the stomach, where these small fragments of rock are isolated in an appendix to the stomach. In *Xylophaga* and in the shipworms the stomach bears a wood-storing caecum which is obviously homologous with the appendix of the Pholadidae, and the fine fragments of wood derived from wood-boring are swallowed and accumulated in the wood-storing caecum. Due to the evolution of a cellulase in the digestive system of the ship-worms the by-products of boring activity have become an important item of food, and it is clear that the possession of an appendix to the stomach in the ancestral adesmacean was pre-adaptive towards the wood-boring habit in the Teredinidae. In the Teredinidae the visceral mass has been greatly extended to accommodate the huge wood-storing caecum and this has been achieved by extension of the mantle cavity far beyond the limits imposed by the size of the shell valves; the ctenidia and most of the visceral mass lie in a fleshy tubular extension of the mantle which lies between the siphons posteriorly and the shell valves anteriorly. In the ship-worms growth is primarily seen as increase in length of the worm-like soft parts, and growth of the valves is far less noticeable. The shell valves serve as jaws, supplying fine particles of wood as a source of food and the rearrangement of the viscera in the ship-worms could be interpreted as providing a secondary cephalisation.

Various lineages of bivalves have given rise to free-living members of the epifauna lying on the surface of a number of different kinds of substratum; these include *Pecten*, *Pandora*, and *Placuna*, which habitually rest on the same side and consequently have become markedly inequivalve. Another member of the epifauna is the heavily built *Hippopus*. All these exhibit adaptations to the free-living mode of life as compared with related attached forms. Some members of the free-living epifauna are able to swim, e.g. *Amussium*, and certain species of *Chlamys* and *Lima*.

Various members of the Erycinacea exhibit different degrees of overgrowth of the shell valves by the mantle, and reduction of the shell valves, e.g. *Scintilla*, *Devonia*, and *Phlyctaenachlamys*: again, various members of this family exhibit commensalism, with echinoids (*Montacuta*), with an ophiuroid (*Mysella*), with a holothurian (*Devonia*), and with a stomatopod (*Phlyctaenachlamys*). The significance of either of these trends is not yet known. Parasitism could be the final outcome of commensalism, and an example is found in *Entovalva* which occurs in the gut of a holothurian.

Functional hermaphroditism occurs in *Pecten maximus*, and in the Anomalodesmata, while protandric hermaphroditism with sex reversal is found in some species of *Ostrea* and of *Teredo*. Incubation of developing embryos in the mantle cavity is found in some species of *Ostrea* and of *Teredo*, and in *Lasaea*, *Montacuta*, *Anodonta*, *Sphaerium*, etc. This is discussed in detail in another chapter.

Several lineages of bivalves have invaded freshwater systems, the most ancient of these being cosmopolitan, e.g. the Sphaeriidae and Unionidae; *Dreissena* is the only freshwater genus still to possess a free-swimming veliger larva; the Sphaeriidae incubate the young to a late stage of development, while in the Unionidae and Mutelidae the larvae are ecto-parasitic on fish for a time. Various typically marine families of bivalves have contributed single genera or species to freshwater faunas, e.g. *Tanysiphon rivalis* (Solenacea), *Martesia rivicola* (Pholadidae), *Teredo minima* (Teredinidae), *Egeria radiata* (Donacidae), etc., these species are relatively restricted in their distribution.

———————————————

The Bivalvia exhibit a truly remarkable diversity of form which may be correlated, on the one hand, with the wide variety of habitats which they have invaded, and, on the other hand, with the variety of modes of life which they have adopted. There are in all some seventy-five families of living bivalves, and by reviewing this vast assemblage it is possible to see that some are relatively unspecialised in form and could be designated as "typical" bivalves, whereas others exhibit various degrees of specialisation which can often be directly correlated with their special habitats or modes of life. In many cases it is possible to offer a plausible account of the way in which specialisations may have arisen, through descent with modification, to meet the requirements of the present habitat and mode of life. This can be done by direct comparison of these specialised types with some selected unspecialised or "typical" bivalve. In other words one can "explain" the specialised bivalve by assuming that it arose from a less specialised or unspecialised ancestry.

In this process of explanation we are seeking to satisfy our own rather urgent need for orderliness in nature by deriving specialised anatomical forms from some hypothetical ancestral "norm"; we are reducing present diversity of form to this norm, or highest common factor; we are trying to substitute order for disorder. We must recognise two important limitations in this process of explaining, the first of which is that we are not necessarily revealing any phylogenetic relationships between various forms when we arrange them in a series and show how one can be'derived from another as a result of certain mechanistic considerations. The series may be purely of our own making; the units in the series may have evolved independently on parallel lines in response to comparable influences, but to different degrees. Above all, we must avoid selecting a certain form as "typical" and then allowing it to become designated as ancestral to other, more specialised forms. The other limitation is set by the very nature of the "explanation", which is normally of a mechanistic nature, in relation to the exploitation of some functional advantage. It is implied in the explanation that change in form brings some functional advantage, but there is no explicit statement in the explanation as to the precise origin of this change, nor any hint as to the means of transmission of the change to the next generation. In some cases the explanation involves mechanical stresses resulting from the mode of life which might influence the growth form of the individual during its life time. Even if this were the case, and it might well be so, the explanation does not attempt to deal with the means of transmission of change from generation to generation or the means of accumulation of change of form. That being so, we should not look for, or think we have found, any specifically Darwinian or Lamarckian undertones in our subject-matter.

Elementary Diversities of Form

We have suggested elsewhere that the bivalve model originally emerged as a colonist of marine deposits of sand, sandy mud, or mud, the body being laterally compressed and enclosed within a pair of shell valves which lie on left and right sides of the body. The basic form of the Bivalvia is bilateral symmetry, and we should consider the shell valves typically to be mirror images of each other, i.e. to be *equivalve*, for there is no reason to expect unequal pressures to be exerted upon the left and right shell valves when the animal moves through the substratum. In many bivalves each shell valve is *equilateral*, i.e. the valve can be divided into equal anterior and posterior moieties by a line dropped from the umbo at right angles to a line joining the centres of the two adductor muscle scars; these moieties are approximately mirror images; the adductor scars are comparable in area, i.e. the animal is *isomyarian*; the course of the pallial line is approximately parallel with the margin of the shell. This condition may be regarded as "typical" and is found in *Glycymeris* in which each shell valve is almost circular in outline.

The mantle cavity is divided into infra-branchial and supra-branchial chambers by the two ctenidia which are commonly in contact with each other posteriorly in the median line, and with the lobes of the mantle laterally. Posteriorly the ctenidia reach almost to the edge of the mantle and form a horizontal, or W-shaped, partition which separates the exhalant area above from a region of inhalant water currents below. The left and right lobes of the mantle may remain separate, or there may be one or more fusions between them, thereby constituting an exhalant aperture, and perhaps also an inhalant aperture, each with a continuous oval rim. These exhalant and inhalant apertures may remain close to the general level of the margin of the mantle or they may become raised on the extremities of shorter or longer tubes, the exhalant and inhalant siphons. These siphons may remain separate, when they are usually very flexible, or they may become fused together to form a single more rigid siphonal process divided internally by a longitudinal partition which separates the exhalant and inhalant water currents. The mantle edge is mobile, being extended by blood pressure and being retracted by a regular series of pallial muscle fibres which are inserted basally on to the inner side of the shell along the pallial line.

Where the exhalant and inhalant apertures are borne at the tips of two siphons or on the extremity of a siphonal process, these marginal pallial muscle fibres are extended and co-ordinated to form well-defined siphonal retractor muscles which serve to withdraw the siphons when the animal is disturbed. The siphonal retractor muscles are inserted on to the inner surface of the shell at the pallial line, but in order to increase the length of the retractor muscle, and to make room for accommodation of the retracted siphons within the shell valves, the posterior part of the pallial line retreats from the primitive position close to the shell edge, thus forming a *pallial sinus*. Where the siphons are very long, and can be completely withdrawn between the shell valves when necessary, the pallial sinus is particularly deep. In some cases the siphonal process can be only partly withdrawn in between the shell valves. These variations in form of the mantle, and the degree of development of the pallial muscle fibres at the sites of the exhalant and inhalant water streams, are related to the habitats and the modes of life of the various types of bivalves. The primitive ancestral condition was presumably that in which no mantle fusions had occurred, and no siphonal retractor muscles had yet been elaborated.

Mode of Growth

The shell and the mantle are interdependent, the shell being produced by the secretory activity of the mantle, and the mantle being supported by the shell and attached to it at the pallial line and at the site of the adductor muscles. It is now customary to regard the two inseparably as the "mantle/shell" (Yonge, 1953a). While the substance of the shell is increased over its whole inner surface by secretion of shell substance—nacre—from the whole outer surface of the mantle, increments to the area of the shell are made at its rim, by secretions at the margin of the mantle. Here, the inner face of the outer fold secretes the periostracum, which forms an outer layer to the shell, while the outer border of the outer fold (Yonge, 1957) of the mantle margin secretes a layer of prismatic material immediately below the periostracum (see Fig. 57). While increments are being made to the margin of the shell valves in this manner, the margin of the mantle is extended beyond the margin of the shell by blood pressure.

Since the shell supports the mantle, it is clear that the disposition of the edge of the shell will influence its own extension by addition of fresh shell material secreted by the mantle. Any local injury to the margin of the shell will obviously affect the orientation of the adjacent mantle lobes and so influence the deposition of new shell material. If the animal indulges in any activities which affect the margin of the shell, e.g. by abrasion against the substratum, these activities will presumably also influence the effective rate of growth of the shell at the locus of such abrasion. This implies that during the lifetime of a bivalve, its shell form may be influenced to some extent by its own vital activities; reference will be made to this from time to time.

Increments to the margin of the shell are not equal at all points on the periphery of the shell, growth being least along the hinge line close to the umbo, this possibly being due in part to mechanical contact with the corresponding part of the opposite shell valve on the slightest closure of the shell valves, as indicated above. Passing away from the hinge line anteriorly, and posteriorly, such contacts are of decreasing significance, and the

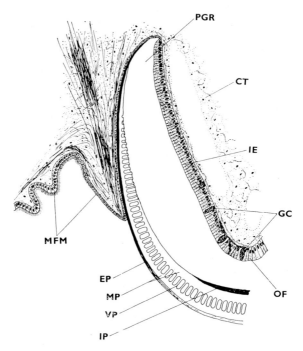

FIG. 57. Transverse section through the margin of the mantle of *Mytilus edulis*, illustrating the mode of formation of the periostracum. The periostracal groove (PGR) lies between the outer fold of the mantle (OF) and the middle fold of the mantle margin (MFM). The periostracum is seen to comprise four layers, an external layer (EP) of constant thickness; a middle layer (MP) which contains a central series of vacuoles (VP); and, finally, a thin internal layer (IP). For interpretation of other lettering, see p. 201–4. (Originally published in Beedham, 1958, *Quart. J. micr. Sci.*, **99**, 343, Fig. 1.)

increments of fresh shell material increase in extent until a point is reached where the increments are maximal. Where the shell margin is approximately circular, as in *Glycymeris*, marginal increments are least in the region of the hinge line, and are greatest at a single zone mid-ventrally where the shell valves gape most widely; in between these extremes, marginal increments are graded in amount, with the result that as the shell grows it maintains a more or less circular periphery (Yonge, 1952) (see Fig. 58 and 59). Since increments at the hinge line are minimal, the umbones of the left and right shell valves remain very close together, and the animal remains relatively slender in transverse section. In the Arcidae, on the other hand, secretory activity at the hinge line is not so largely suppressed, and new shell material secreted here gradually forces the umbones apart with the production of a far more tumid form, e.g. *Arca tetragona* or *Anadara granosa*.

The growth of a valve of the bivalve shell may be dependent on one, two, or sometimes three factors. The first of these is a radial growth factor and

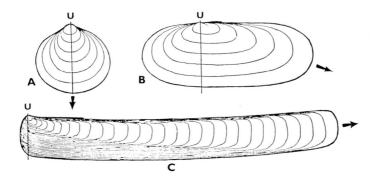

FIG. 58. Three shell valves of contrasted growth form, seen from the left side. A, *Glycymeris* sp. B, *Siliqua patula*. C, *Ensis arcuatus*. The vertical line passing through the umbo (U) is the median vertical axis of the shell. The heavy arrows indicate the direction of greatest marginal increments to the shell during growth. (Originally published in Yonge, 1952, *Univ. Calif. Publ. Zool.* **55**, 430, fig. 6.)

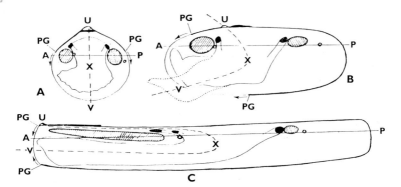

FIG. 59. Diagrams showing the effect of the form of the shell upon the shape and disposition of the soft parts within. A, *Glycymeris* sp. B, *Siliqua patula*. C, *Ensis arcuatus*. The only internal structures depicted are the adductor muscles (hatched), the pedal retractor muscles (black in section at the point of insertion into the shell), the outline of the foot, and the positions of the mouth and anus. The line A——P represents the antero-posterior axis of the body; The line U— —X— —V passes through the umbo (U), the base of the foot (X), and the functional mid-ventral point (V). This second line indicates the division between the anterior and the posterior parts of the body. (Originally published in Yonge, 1952, *Univ. Calif. Publ. Zool.* **55**, 432, fig. 7.)

where this factor alone is active the shell valve so produced is flat, e.g. the upper valve of *Pandora inaequivalvis* or of *Pecten maximus*. In the majority of cases, however, there is a second growth factor which operates transversely to the radial component, so causing the shell valve to be curved in transverse section in the form of a plane logarithmic spiral. This is true for the lower valve of *P. maximus*, and for both valves of a large number of examples. This may remain true even if the outline of the shell is not circular, but is oval e.g. *Tellina*, and even if the shell is inequilateral, e.g. *Anodonta*, as long as the umbo is not tilted forwards or backwards. The ratio of these two factors will determine the value of the spiral angle and so the degree of convexity of the shell valves.

In other cases a third growth factor is operative; this acts tangentially to the margin of the shell causing the demarcation line* (Yonge, 1955) of the shell to lie in a turbinate spiral instead of a plane spiral. This condition may be recognised by examining the umbones, which are tilted slightly forwards as in *Glossus* or backwards as in *Nucula* (Owen, 1953, 1953a).

The "Typical" Growth Form

As a basis for contemplation of the structures and growth forms of various specialised types of bivalves, we may now itemise those characters considered to be primitive, or "normal", and which may still be possessed by unspecialised types. These are: shell equivalve, equilateral, isomyarian, rounded in outline, pallial line approximately parallel with the edge of the shell, pallial sinus absent, median dorsal mantle isthmus, and separate left and right mantle lobes, exhalant and inhalant areas isolated solely by the transverse partition formed by the hind ends of the ctenidia, siphons not present, foot wedge-shaped, without a functional byssus organ in the adult. The genus *Glycymeris* may serve to illustrate these generalised features, but it must be emphasised once more that any comparisons drawn between this generalised condition and other selected specialised types do not imply that *Glycymeris* or any comparable modern bivalve is ancestral to any more specialised type.

Adaptive Radiation

Adaptive radiation does not solely concern diversity in form of the shell and such external features as can be discerned by casual inspection of the whole animal. On the contrary, it concerns variations in any organ system,

---

* The demarcation line passes through three points: 1, the umbo; 2, the point of maximal width of the shell; 3, that part of the free ventral margin of the valve which exhibits the greatest concavity. When a line joining these three points deviates from a line drawn through the umbo perpendicular to the long axis of the body, the growth form of the shell is a turbinate spiral, e.g. *Glossus humanus*.

or in any vital process, by means of which the possessor is better adapted to survive in its own special circumstances. Attention will therefore be drawn at appropriate points to the roles played by the various organ systems in the adaptive radiation of bivalves.

The Infauna of Gravels, Sands, and Muds

The rounded form exemplified by *Glycymeris* is suited to a relatively inactive bivalve, and is also found in many other inactive members of the infauna which may be highly specialised from certain other points of view. As examples we may take the large and inactive genera *Glossus* (Owen, 1953), and *Geloina*, which lie embedded and more or less motionless in the surface layers of soft or stiff muds respectively, and *Cyprina* which colonises firmer sand or sandy mud. In other cases a slow movement through the substratum may be necessary to enable the animal to re-embed itself if it has been partly exposed by wave action, for example; a more streamlined shape would facilitate such movement. The rounded-triangular *Egeria* which occurs in the sandy beds of certain West African rivers is able to re-embed itself in the sand fairly quickly if it is partly exposed, but probably does not move much after it has adopted a safe feeding position. The ability to re-embed in the sand quickly is an important adaptation to the risk of being exposed by local variations in the current near the bed of the river. Many rounded-triangular forms which occur littorally or in shallow water, e.g. *Cardium*, *Mactra*, or *Spisula*, are doubtless comparable. Even greater burrowing activity may be displayed by some bivalves which have acquired an oval, or elliptical shell outline, e.g. *Abra*, and many species of *Tellina*. In spite of this change in shape of the shell by elongation in the antero-posterior axis, the shell may still remain equilateral in form. The change in shape is effected by a change in the gradient of secretory activity along the margin of the mantle; instead of there being a single zone of maximal secretion in a mid-ventral position, there are two such zones, one at the anterior and one at the posterior end of the shell.

In many members of the infauna, however, the shell has lost its primitive equilateral form, the posterior end of the shell being moderately or greatly produced; this may be explained by the occurrence of only one region of maximal secretory activity by the shell valves, this region being sited at the posterior border of the shell. The reason for this modification is not far to seek; with the anterior end of the shell most deeply embedded in the substratum, the highest importance is centred on the posterior end of the shell by which the animal maintains contact with the life-giving supernatant water. The basic functions of feeding, respiration, and emission of gametes in the breeding season are all effected via the inhalant and exhalant apertures, which are situated posteriorly. It is not surprising that a growth differential should develop in members of the infauna, with relative aggrandisement of the posterior part of the shell. As compared with the "typical" form the shape of the shell is adapted to suit the stresses and requirements of the environment and these deformations can be expressed as distortions of *radial* coordinates. The organs of the body are not simply compressed to fit the confines of the now modified shell valves, but are themselves variously adapted to suit the mechanical functions imposed by the mode of life which has been adopted. In contrast to the shell valves, the form of the internal organs can only be interpreted on the basis of distortion of *rectangular* coordinates (Yonge, 1952).

Burrowing in a substratum not only affects the form of the shell but also the nature of the inhalant and exhalant apertures, and various other internal organs, particularly the foot and the muscular system. The order Lucinacea is specially interesting in this connection. The Lucinacea comprise three families which are progressively specialised for living in a variety of substrates, ranging from gravel and coarse sand to fine mud, which apparently have as a common feature an impoverished infauna and a deficiency of food material. Members of the Lucinacea are evidently adapted to survive under conditions which seem to be too harsh for the majority of bivalves, and of the three constituent families the Lucinidea is the most highly specialised. In the Lucinacea the posterior inhalant and exhalant orifices are not borne on siphonal extensions (except for the exhalant siphon of the family Lucinidae) and little water is drawn in through the posterior inhalant siphon. The foot is very long and vermiform, and is protruded in front of the anterior adductor and upwards to the surface of the substratum; in some cases the foot can extend up to ten times the length of the shell. When the foot is so extended, a ring of particles from the substratum becomes embedded in a layer of mucus which is secreted at the tip of the foot. This ring of mucus-bound particles is carried by ciliary action to the base of the foot. More rings are similarly formed and are added on distally to the first one at the base of the foot, and all these rings become welded together into a semi-permanent mucus-bound chimney which extends from the anterior end of the shell to the surface of the substratum. The foot is then withdrawn and this anterior inhalant tube then provides the main channel for inhalation of the respiratory and feeding water current. Accessory pallial gills have developed close to the opening of this anterior inhalant tube.

Food particles brought in via this anterior water current are sorted to a small extent on a novel ciliary sorting field which has developed on the surface of the anterior adductor muscle which is here developed to form a flattened plate lying below the mouth. Particulate material is apparently scarce in the environments colonised by members of the Lucinacea for sorting of these particles is only slight; after they have been admitted to the stomach little further sorting occurs, and relatively large particles are accepted into the ducts of the digestive diverticula (Allen, 1958).

As mentioned above, fusion of the left and right mantle lobes may occur at one, two, or more points posteriorly, providing the exhalant aperture and perhaps also the inhalant aperture with a continuous oval rim formed by the margin of the mantle. This fusion may concern only the innermost muscular fold of the mantle margin, e.g. in the Tellinacea, in *Petricola*, or in *Tridacna*; this type A of Yonge (1948, 1957a). Alternatively, and most commonly, the fusion involves both the inner muscular and the middle sensory folds of the mantle margin, e.g. in the widely distributed genera *Cardium* and *Venus*; this is type B of Yonge. In a third condition which also occurs widely the fusion also includes the inner face of the outer fold, which is responsible for secretion of the periostracum and the siphonal process so formed is therefore covered with a layer of periostracum secreted by this area. This condition is found in *Mactra* and in the deep-burrowing *Mya*; this is type C of Yonge.

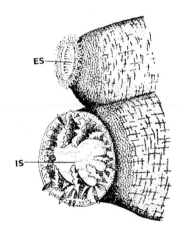

FIG. 60. The tip of the siphonal process of *Zirfaea crispata*, showing the exhalant aperture (ES) and the inhalant aperture (IS). The pinnate straining tentacles which arch across the orifice of the inhalant tube will serve to prevent entry of coarse particles of sand, etc., which might otherwise damage the organs in the mantle cavity. Such a device is important for species which occur in wave-swept areas where turbulent water is likely to contain much coarse suspended matter. (Originally published in Purchon, 1955a, *Proc. zool. Soc. Lond.* **124**, 861, fig. 1.)

Where there is danger of entry of sand grains, etc., into the mantle cavity via the siphons, the siphonal orifices are commonly guarded by well-developed incurving tentacles which serve as strainers (Purchon 1955) (see Fig. 60).

In the well-known freshwater genus *Anodonta*, which moves slowly through the surface layers of mud on the beds of rivers and lakes, there is only one pallial suture, and the exhalant and inhalant areas are not developed into siphons; this is adequate since the posterior margin of the shell lies at the surface of the substratum. In the cockle, *Cardium*, which may live in dense colonies in inter-tidal sands, the shell lies a short distance below the surface of the sand—there are two pallial sutures and two short separate siphons. These siphons comprise the inner and also the middle folds of the mantle margin, and bear numerous scattered short sensory tentacles. These short siphons suffice to maintain contact with the water above; they require the development of retractor muscles, but as these are comparatively short the pallial line is not affected and so there is no pallial sinus.

The Veneridae also have specialised as shallow burrowers and are a very successful group which colonise a variety of soft substrates and which have representatives in all seas. Members of the family show considerable variety in the length and degree of fusion of the two siphons according to the habitat and to the posture that is adopted (Ansell, 1961). In *Venerupis pullastra*, which typically occurs in shallow pockets of sand or gravel under flat stones, almost any posture is adopted in nature; this may be correlated with the habitat chosen. Quayle (1949) showed that when *V. pullastra* buries itself the digging action involves obtaining anchorage by the extended tip and heel of the foot followed by contraction of the valves to eject a powerful downward jet of water anteriorly. This jet loosens the sand and enables the animal to enter the sand slightly on contraction of the anterior pedal retractor muscles. In this digging action Quayle suggests that the flattened lunule of the shell acts as a pressure plate and so aids in the burying process.

Trueman, Brand, and Davis (1966) have studied the dynamics of burrowing into the substratum by *Tellina tenuis*, *Donax vittatus*, *Macoma balthica* and *Cardium edule*, and concluded that the digging cycle appears to be basically similar in all burrowing bivalves which have been studied to date. Adduction of the shell valves raises the pressure of blood in the pedal haemocoele, so enabling the foot to penetrate the substratum. When, as the result of terminal dilation, the foot has obtained a firm anchorage in the substratum, contraction of the anterior and posterior pedal retractor muscles draws the shell downwards into the soil. The contractions of these muscles are not synchronised, and so the shell valves rock to and fro in each digging cycle. As mentioned above, downward jets of water from the mantle cavity loosen the soil below and aid in the burrowing action.

As a general rule burrowing bivalves lie in the soil with their median plane vertical, but this is not invariably the case. Just as some surface-dwelling forms lie on one side, e.g. *Pandora*, so also do some burrowing forms such as *Cochlodesma*.

In this last genus there are two separate siphons, each of which constructs a mucus-lined tube in the substratum and then withdraws from it. The inhalant siphon and tube extend upwards, but the exhalant tube extends horizontally and terminates within the substratum (Allen, 1958).

Deeper burrowing necessitates the development of longer siphons, as in the Tellinacea in which the two long separate siphons are formed by fusion of the inner fold of the mantle margin only (Yonge, 1949) (see Fig. 61). These long siphons can be retracted by extensive retractor muscles

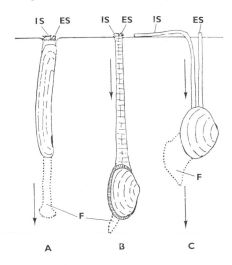

FIG. 61. Normal positions of various members of the bivalve infauna: A, Solenidae. B, *Mya*. C, *Scrobicularia*. The extent of the foot when fully protruded is indicated by dotted lines in each case. Arrows indicate movements which occur when the animal is alarmed. (Originally published in Yonge, 1949, *Phil. Trans.* B, **234**, 58, fig. 21.)

and the pallial line is deeply embayed to form a pallial sinus which accommodates the siphons when these are retracted. In the Tellinidae and Semelidae these siphons are extremely mobile, and the inhalant siphon gropes over the surface of the substratum sucking up potential food material from the bottom deposit, e.g. *Abra*, *Macoma*, and *Scrobicularia*. *Macoma* makes U-shaped movements through the substratum, the first limb of the U being photopositive and is formed just after exposure by the receding tide, while the second limb of the "U" formed later on is photonegative. This behaviour enables the animal to sample different areas of substratum and never completely exhaust the store of deposited organic matter which principally lies on the surface, while maintaining approximately the same station on the shore. When the tide is in *Macoma* ceases to feed on deposits and commences normal filter-feeding (Brafield and Newell, 1961).

It is important for a deposit-feeder like *Scrobicularia* that the siphons should not diminish in diameter as they extend in length as would be the case if their extension and retraction were due to antagonistic action of longitudinal and circular muscle fibres. Chapman and Newell (1956) show that there are no circular muscles, but collagen fibres instead; as the siphons extend the siphonal walls become thinner and the diameter of the lumen of the siphon is not diminished, and there is no reduction in the rate of inflow of water.

In deep-burrowing forms such as *Mya* and *Lutraria* there is a single rather rigid siphonal process which bears separate inhalant and exhalant apertures at its extremity; this siphonal process is surrounded by a sheath of periostracum, since it comprises the inner and middle folds of the mantle margin together with the inner face of the outer fold. This periostracal coat provides valuable protection against abrasion and so is an important contribution to the deep-burrowing habit. In such cases the pallial sinus is very deep, but, nevertheless, the siphonal process is too large to be wholly withdrawn between the shell valves.

The adductor muscles can be contracted either alternately or independently since the posterior ends of the shell valves of *Mya* can be diverged in order to accommodate the base of the siphonal process when this is contracted. Reference will be made to this when considering the rock-boring members of the Pholadidae. The siphonal process of *Mya* is extended by a series of stages in each of which the siphonal aperture is closed and more water is injected into the lumina of the process from the mantle cavity by adduction; further water is then drawn into the mantle cavity through the opened siphons when the shell valves are parted. This use of sea water in the mantle cavity for hydraulic purposes is made possible by the very extensive ventral fusion of left and right mantle lobes with consequent reduction of the pedal aperture, and also by the valvular nature of the pedal aperture itself (Chapman and Newell, 1956).

The opening moment of the ligament of *Mya* is insufficient to cause the valves to part against the resistance of the substratum. By closing all orifices to the mantle cavity, i.e. the pedal gape and the two siphonal orifices, and then presumably by contraction of the siphonal process, hydraulic pressure in the mantle cavity can be raised sufficiently to force the valves apart against the pressure of the surrounding soil (Trueman, 1954). *Mya* normally occurs in rather stiff muds which will not collapse against the shell valves when these are adducted, and the need to force the valves by hydraulic pressure against the soil would seem to serve mainly to enlarge the chamber slightly and so allow for growth of the shell. Both the form and the habits of *Mya* are thus suited to the circumstances in which it lives.

When considering the growth form of an animal in relation to its habitat and mode of life, the question may often arise as to which is cause and which is effect—whether the growth form of an animal permits the adoption of a particular mode of life or whether the mode of life that was adopted brought about a reshaping of the growth form of the animal. The answer will probably generally lie somewhere between the two extremes—the form of the animal and its mode of life having mutated together. In the case of *Mya* it seems probable that the incorporation of the inner face of the outer fold in the structure of the siphonal process thereby protected the siphonal process against abrasion from movements in the substratum; in this way the siphonal process became better fitted to the function which it had to perform; we may perhaps conclude that the form adopted by the siphonal process enabled the animal to exploit more fully the deep-burrowing mode of life with consequential further development of the siphonal process, siphonal retractor muscles, and pallial sinus. Thus form and function here are inextricably interrelated.

Among the infauna an extreme condition is found in razor shells of the genus *Ensis* which are adapted in form for feeding while lying at the surface of the substratum, and for rapid retreat to a greater depth in the sand when alarmed. Here a very small and more or less uniform increment is added to the whole ventral free margin of the shell, which remains straight, while a relatively large increment is added to the whole of the posterior free margin (Owen, 1952); the shell maintains the form of an approximately straight cylinder which is not quite circular, but slightly elliptical in transverse section. Such a shell is able to slide easily up and down the semi-permanent burrow in the sand. Since the animal comes up to the surface of the sand for filter-feeding, it has very short siphons; the small development of siphonal retractor muscles only requires the slightest embayment of the pallial line. The main means of protection against predators is by very rapid descent of the burrow when disturbed. This is effected by the remarkably long foot which can be protruded downwards to a distance equivalent to the length of the shell. When a razor shell is burying itself there is a regular sequence of movements of the foot. When the foot is being protruded by blood pressure the tip is pointed, so facilitating penetration of the soil; when fully extended in the soil the tip of the foot is then swollen by blood pressure, thereby creating a firm anchorage; contraction of the pedal retractor muscles then causes the cylindrical shell to shoot downwards into the soil. The strength of the digging process can be appreciated by grasping the posterior end of a specimen which is trying to bury itself. Drew (1907) described the burrowing action of *Ensis directus* Con., a species which is abundant along the east coast of the United States. He observed that just before the retraction of the foot, the siphons and the ventral pallial aperture are closed so that the sudden withdrawal of the foot results in the emission of a powerful jet of sea water between the anterior end of the mantle and the sides of the foot. This downward jet of water will loosen the soil below the animal and so facilitate the burrowing action. In my own experience of *E. ensis*, on the other hand, disturbance of the animal at the surface of the sand resulted in a jet of water shooting upwards into the air for a foot or more. It is possible that in addition some water was forced downwards into the sand below the foot, as in *E. directus*. At the pedal gape the inner fold of the mantle margin is well developed and forms a valve which wipes sand off the sides of the foot when the latter is withdrawn (Yonge, 1959). The genus *Solen* is comparable in many respects, but favours more muddy conditions.

Another important adaptation for survival which has been noted in the Solenidae is the capacity to shed a terminal ring of flesh from the tip of the siphonal process. If a browsing fish were to bite the protruding tip of the siphonal process of a razor shell, the capacity for autotomy might well save the life of the shellfish. This capacity for autotomy has been carried to an extreme in the case of the little *Solen delesserti* which is often found exposed on the surface of the sand at low water mark among fronds of the weed *Enhalus* in the Far East. Here the siphonal process is up to an inch long and consists of a series of rings which make it resemble the stem of a Horsetail (*Equisetum*). I have myself seen such an individual shed a series of rings of siphonal tissue in rapid succession when it was disturbed. The successive rings dropped off as though sliced off by an invisible knife. The Solenidae are the only bivalves which exhibit autotomy. *S. delesserti* can also perform a swimming escape reaction by expulsion of a series of jets of water from the inhalant siphon. This may be advantageous when the animal is exposed at the surface by wave action, or after autotomy if dragged out of the sand by a predator. On the other hand Drew (1907) noted that *Ensis directus*, which typically occurs in sandy mud, moved posterior end foremost when swimming. While the animal was swimming the foot was active, and water was expelled in a powerful jet anteriorly between the mantle collar and the sides of the foot.

In many burrowing bivalves there is the problem of the disposal of sand grains, etc., which enter the mantle cavity in spite of the occurrence of incurved straining tentacles at the inhalant orifice. Such unwanted material may have to be transported vertically against gravity by ciliary currents, and in some cases the efficiency of these ciliary cleansing tracts has been increased by the development of a longitudinal flap of mantle tissue which subtends the ciliated waste canal and protects the train of mucus-bound waste material from being dislodged by movements of the organs in the mantle cavity. Kellogg (1915) recorded such a waste canal in members of the Mactridae and Tellinidae, while Yonge (1952) records a slightly different type of waste canal in *Siliqua* in the Solenidae.

Byssus-bearing Forms

After a week or more in the plankton the veliger larva acquires a foot and approaches the substratum. It may now settle and crawl about by means of its foot, searching for a suitable spot on which to settle and metamorphose. It may return temporarily to swim about in the plankton, and in the case of *Mercenaria mercenaria* this searching behaviour may continue for several days. A weak byssal attachment is then formed and for some weeks the young *Mercenaria* may vary between crawling and byssal attachment (Carriker, 1956). This byssus gland seems to have evolved as an organ of the post-larval stage, which originally served for anchorage of the animal until metamorphosis was completed and the animal was securely buried in the substratum (Yonge, 1962). By a process of paedomorphosis the active phase of this organ was projected into adult life in certain lineages, thereby enabling the possessors to attach themselves permanently to a firm substratum by the secretion of byssus threads. Certain of these paedomorphic forms occur as infauna or as epifauna of sedimentary bottoms. It is probable that the byssogenous habit developed during an earlier phase when the forms in question were colonists of firm rocky substrates, and that the invasion of sandy, or muddy shores is a more recent event. Thus *Modiolus*, the horse-mussel, may be found on the surface of muddy shores, attached by byssus threads to small pebbles or gravel in the substrate. *Malleus albus* is commonly found lying on the surface of muddy sand near low-water mark on tropical shores; it may either lie free or it may be attached by a small byssus to stones or pieces of shell in the substratum. The form of the shell of this animal seems well adapted to preventing it from sinking into the substratum. The hinge line is developed into two conspicuous anterior and posterior projections, while the margin of the main body of the shell is developed into a series of deep saddle-shaped flares, those of the left valve fitting exactly into those of the right valve. More often than not the shell is not anchored by the byssus, and it would seem that the T-shaped shell, with deep flares on both sides of the stem of the T, may slide over the surface of the sand under the influence of wave action, but it is very unlikely to sink into the soil and become smothered.

The ear-shell *Pinna*, which abounds near low-water mark in muddy sands on tropical shores, lies three-quarters buried and is securely anchored by a luxuriant mass of byssus threads attached to bits of shell, or small stones, lying deep in the soil. The markedly anisomyarian structure is the consequence of byssal attachment. The exposed portion of the shell is commonly thin and flexible, this part of the shell not being strengthened by any internal layer of nacreous shell material. It is suggested that in an exposed habitat, with surf waves rushing over the reef flat, the posterior end of the shell of *Pinna* is likely to become badly damaged by moving coral boulders and logs, or by attacks by predacious fish, etc. *Pinna* is adapted to repair such damage to the shell in various related ways, and so to be able to survive under these conditions (see Figs. 62 and 63). When the shell is badly damaged the mantle lobes and the ctenidia are strongly contracted, this being possible due to the lack of marginal pallial attachments, the development of

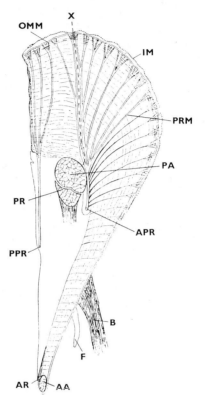

Fig. 62. *Pinna carnea*. View of the animal from the left side after removal of the left shell valve. Note the powerful anterior pallial retractor muscle (APR) and the smaller posterior pallial retractor muscle (PPR) both of which serve to retract that part of the mantle which lies above the posterior retractor muscle (PA). X— — — — signifies the boundary between the infra-branchial and supra-branchial chambers. For interpretation of other lettering, see pp. 201–4. (Originally published in Yonge, 1953, *Phil. Trans.* B, **237**, 348, fig. 7.)

extensive pallial retractor muscles, and the lack of hinge teeth. The blood withdrawn from the mantle and ctenidium becomes available to engorge a unique finger-like pallial organ which lies above the rectum, in the exhalant chamber. When engorged this pallial organ can be moved about in order to sweep away such fragments of shell and other debris as obstruct the mantle cavity. The whole of the mantle behind the posterior adductor

muscle is free from the shell, there being no pallial line; the outer epithelium of the mantle bears numerous mucous glands, which is most unusual,
and these will presumably aid the movements of the mantle by providing
a mucoid lubricant. Contraction of the mantle lobes is effected by two

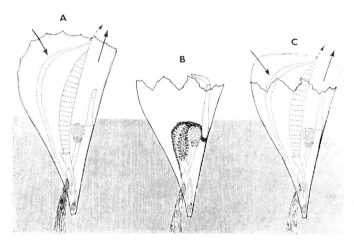

Fig. 63. *Pinna carnea*. Diagrammatic representation of the process of
repairing of the shell. A, the normal, undamaged individual, one half to two-
thirds buried in the substratum (stippled). B, the shell has suffered major
damage by a wave-rolled log or boulder; the mantle and the ctenidia are
deeply contracted and the blood withdrawn from these tissues has been
used to distend the pallial organ, which moves and sweeps away the debris.
C, the mantle and the ctenidia expand to their original dimensions and the
shell is repaired by rapid secretions of the outer surface of the mantle.
(Originally published in Yonge, 1953, *Phil. Trans.* B, **237**, 363, fig. 15.)

branched pallial retractor muscles which are inserted into the shell in front
of the posterior adductor muscle. Once the fragments of broken shell have
been swept away by the movements of the pallial organ, the retracted
mantle lobes and the ctenidium can be extended once more by blood pressure, and the posterior ends of the shell valves can be rapidly recreated by
fresh secretion of prismatic material by the mantle lobes. Extensive damage
to the shell valves can therefore be repaired fairly quickly (Yonge, 1953). In
more sheltered waters, of course, damage to the shell would be less likely
to occur and the shell valves would naturally gradually become sufficiently
robust to resist such damage. In the sheltered waters around Singapore,
for example, the fragile shells of *Pinna hanleyi* might well be so broken and
so repaired, and this species was only found on the most sheltered beaches.
P. atropurpurea is more robustly built, it only occasionally showed signs of
damage and repair, and it was found on both sheltered and on slightly exposed beaches. The related ear-shell *Atrina* is much more powerfully built

and did not show signs of damage; this genus was usually found on the most exposed of local beaches.

The posterior byssal retractor muscle is particularly well developed and its contraction will force the shell downwards against its byssal anchorage and so grind the shell more deeply into the soil. This enables the animal to bury itself at about the same rate as it grows, and causes some erosion of the anterior end of the shell. If partly uncovered by storm waves, *Pinna* can also re-bury itself more rapidly by forcing a jet of water downwards from the anterior apex of the shell, so loosening the substratum, the posterior end of the infra-branchial chamber being closed by apposition of the left and right mantle lobes at the time (Grave, 1909). Contraction of the byssal retractor muscle will then draw the animal downwards relatively easily into the loosened soil. Two parallel flaps of mantle tissue enclose a ciliated waste canal which passes sand grains, etc., directly upwards, so clearing the mantle cavity of any soil which may enter as a result of burying action. These pseudofaeces are voided *continuously* and are carried away by the powerful adjacent exhalant stream. This waste canal is necessary to combat the danger of the animal becoming silted up; without this adjunct the mode of life of *Pinna* could never have been adopted.

An Enigmatic Case

Too little is known of the fascinating genus *Brechites* to offer any certain explanation as to the way in which it acquired its remarkable growth form, which is probably the most bizarre in the whole class. The shell is a slightly irregular tube tapering a little towards its posterior extremity, which protrudes very slightly above the surface of the soil, and which protects the siphonal process within. At the opposite end, which lies deepest in the soil, the anterior end of the shell is widest and is closed by a transverse disc of shelly material perforated by many downward-projecting tubules. Near this end of the shell the dorsal surface is marked externally by the two post-larval shell valves, which are incorporated immovably into the substance of the shelly tube. A saddle-shaped area lying below these umbones is differentiated from the remainder of the tube by the occurrence of irregular concentric lines of growth, and examination of the interior of the shell shows that the periphery of this saddle-shaped area is marked by an extremely broad pallial line, which is surrounded by a line along which periostracum is inserted into the inner side of the tube.

Nothing is yet known of the young stages of the animal. All shells found are of the adult form and are very approximately of comparable size. Since a tubular shell cannot increase in diameter except by extension at one end, it seems probable that there are two phases in the growth of the animal. In the first phase the shell consists of a small saddle-shaped area enclosing the visceral mass and principal organs, and there may also be an independent adventitious shelly tube which surrounds the siphonal process, comparable to that lining the burrow of a ship-worm. At this stage there is probably no anterior perforated shelly plate, and the animal may perhaps possess a fairly well-developed foot. When the animal has grown to maturity, it is thought that a second stage is reached in which the adventitious shelly tube becomes fused to the saddle-shaped shell valves, the latter now becoming wholly immobile, and the anterior end of the tube becomes closed by the secretion of the transverse perforated plate. The adult animal possesses only a few strands of anterior adductor

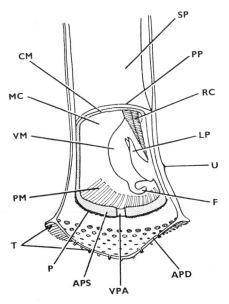

Fig. 64. *Brechites penis*. Vertical section through the anterior end of the "watering can shell", a rare tropical bivalve which burrows vertically in stiff muddy sand. An anterior pallial septum (APS), operated by powerful pallial muscles (PM), acts as a muscular pump. Water is drawn from the substratum through the pores in the anterior face of the tubular shell (APD), and enters the mantle cavity (MC) through the much reduced pedal orifice (VPA). By this means the animal is able to bury itself more deeply in the substratum when necessary. For interpretation of other lettering, see pp. 201–4. (Originally published in Purchon, 1960, *Proc. malac. Soc. Lond.* **34**, 20, fig. 1.)

muscle, and no posterior adductor muscle, and it is possible that the loss of the adductor muscles is consequent upon completion of the shelly tube. It seems that only two living specimens have ever been studied (Purchon, 1956; 1960) and the young stages have never been seen so that this suggested growth sequence is mainly conjectural.

In its adult form *Brechites* shows interesting adaptations to its habitat and mode of life. Associated with the immobility of the shell "valves", one adductor muscle has disappeared entirely and the other has been reduced to three or four small bundles of fibres. The left and right lobes of the mantle have fused very extensively in the mid-ventral line, this fusion having also extended forwards to form an anterior limit to the extent of the mantle cavity. The anterior part of the mantle lobes thus forms a transverse partition, parallel with the transverse perforated shell plate, there being a small pedal orifice which is the sole remnant of the ventral gape between the two mantle lobes. This anterior pallial partition is highly muscular, due to the very great development of the pallial musculature, and it serves as a muscular water pump which draws water from the substratum, through the perforated shelly plate, via the valvular pedal aperture and into the mantle cavity (see Fig. 64). The purpose of this water pump is probably to embed the shell more deeply in the substratum if it is partly exposed by wave action. It is also possible that fine particles of organic matter may be brought into the mantle cavity as a result of the pumping action, and this may be ingested and provide an auxiliary source of food.

Colonists of Rocky Surfaces

Byssal Attachment

It has already been said that the colonisation of rocky surfaces by bivalves was facilitated by a process of paedomorphosis whereby the byssus gland of the post-larva became projected into adult life as a permanent organ of attachment. The minute *Lasaea rubra* can readily be envisaged as such a paedomorphic form, but it is not known whether its small size is primitive or whether it is a secondary adaptation to the habitat it has adopted in the splash zone. Due to its minute size and to the temporary nature of its byssal attachment, this has not led to relative reduction in the organs at the anterior end of the body and to the anisomyarian condition. *Lasaea* is gregarious and occurs in large numbers high on the sea shore in tufts of lichen or among barnacle shells where it may only be wetted for short periods by splashes. It is adapted to this habitat in various ways including the incubation of larvae in the suprabranchial chamber, the suppression of the planktonic veliger stage, the possession of an *anterior* inhalant siphon (Oldfield, 1955), and the development of rhythmic digestive processes (Morton, 1956).

Turtonia minuta is another very tiny bivalve which resembles *Lasaea* in various ways but is apparently unrelated, so these similarities are due to convergent evolution. *Turtonia* is unique among bivalves in laying few large eggs in a gelatinous capsule which is secured to the substratum (Oldfield, 1955).

A characteristic example of byssal attachment in a primitive state is found in species of *Arca*, e.g. the small but numerous *A. lactea* which

occurs in small groups, byssally attached in rock crevices and on the undersides of rocks near low-water mark. Here the animal hangs by its byssus
in a sheltered situation and there is little if any modification to the form
of the shell; the shell is equilateral and approximately isomyarian. This
condition can be seen in the larger, but much more rare *A. tetragona*.

In species of *Mytilus*, on the other hand, the mechanical effects of byssal
attachment have wrought a considerable change in the form of the shell.
The animal occurs on exposed rocky surfaces where wave impact may
exert a considerable drag on the retractor and protractor muscles of the
byssus apparatus. The shells often occur in dense colonies and under
these conditions the evolutionary trend has been towards diminution
of the anterior end of the animal and enlargement of the posterior end
which houses the all-important inhalant and exhalant areas (see Fig. 65).

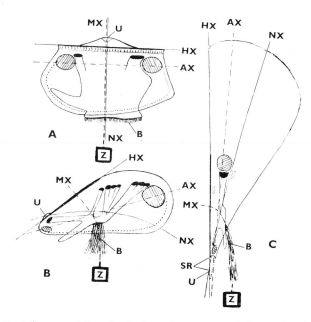

Fig. 65. Influence of byssal attachment upon growth form. A, *Arca*. B,
Mytilus. C, *Pinna*. The axes of the mantle/shell are indicated by continuous
lines: HX——hinge axis; NX——normal axis. The axes of the body are
represented by broken lines: AX — — — antero-posterior axis; MX— — —
median axis. Each example is shown attached by byssus threads (B) to a fixed
point in the substratum (Z). *Arca* is attached to rocks in sheltered crevices,
where it does not experience drag from wave impact, and byssal attachment has
not interfered with growth form. In *Mytilus* and in *Pinna* byssal attachment
has resulted in diminution of the anterior part of the mantle/shell and of the
contained body. For interpretation of other lettering, see pp. 201–4. (Originally published in Yonge, 1953, *Phil. Trans.* B, **237**, 365, fig. 16.)

In this way the relatively inactive anterior end is tucked down close to the substratum, while the posterior—food-collecting—end is reared upwards into a position where it will sample clean water and where it will not be fouled by waste matter from adjacent animals or by sediment from the sea water. The mechanical consequences of byssal attachment are therefore reduction of the anterior end of the body and especially the anterior adductor muscle, and the assumption of the anisomyarian condition. This process could be termed "mytilisation".

This evolutionary trend has been carried considerably further in the Pectinidae in which the anterior adductor has been completely lost, the enlarged posterior adductor muscle has taken up an approximately central position, and the shell valves have secondarily acquired an almost circular outline. The small byssus emerges via a notch close to the hinge line, this notch frequently bearing a number of ribs or scales which presumably grip the byssus when the shell valves are tightly closed, so reducing the strain on the rather weak pedal musclature. In the byssally attached *Pedalion*, *Pinctada*, and *Chlamys*, which lie on the right shell valve, the right anterior pedal retractor muscle atrophies so that the mouth is not occluded when the pedal muscles contract (Yonge, 1955). The "ventral" gape between the two shell valves now exceeds three-quarters of the circumference of the shell, the ctenidium has been draped around the posterior adductor muscle and filters water through some 220° of arc, and a close approach has been made towards a secondary radial symmetry (see Figs. 73 and 74, pp. 191–2). An extraordinary feature of this pectinid lineage is that a monomyarian form and radial symmetry having been achieved as a result of byssal attachment, in many cases the byssus apparatus has then degenerated with the production of unattached, free-living, and even swimming forms. The development of a pallial curtain was necessary in all attached forms in which the left and right mantle lobes remain separate, in order to permit control of inhalant water currents by the formation of temporary local inhalant or exhalant "apertures". Attention will be paid to the swimming forms under the sub-heading "The free-living epifauna", but it is of interest to note here that the development of these pallial curtains in attached forms was pre-adaptive towards the development of the swimming habit (Yonge, 1936).

The monomyarian condition with secondary approach to radial symmetry has been acquired independently in other lineages as well as in the Pectinidae, e.g. in the pearl oyster *Pinctada margaritifera*, in *Pedalion* (= *Isognomon*), and in the saddle oyster *Anomia*.

In the Limidae, another monomyarian family, some species are permanently attached to the under sides of boulders by a few byssus threads, e.g. *Lima lima*; some lie embedded in a "nest" composed of innumerable interlaced byssus threads, e.g. *L. hians*, while some, such as *L. fragilis* are free living and can even swim—*L. hians* also can swim when disturbed from its nest (see Fig. 66). In the genus *Lima* the foot has been rotated through 180° to point away from the mouth, and this presumably prevents the mouth from being occluded when the pedal retractor muscles are contracted. As a result of this realignment of the foot, when *Lima* crawls it does so with the hinge end foremost (Yonge, 1936, 1955).

When *L. hians* is burrowing into a substratum of loose stones and gravel it claps the shell valves vigorously and darts forwards into a crevice,

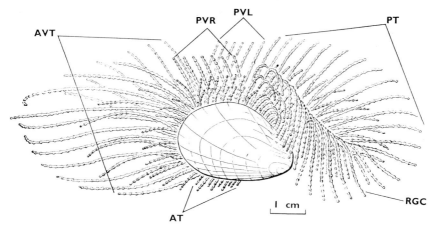

FIG. 66. *Lima hians* swims with the shell valves held vertically and with the ventral margins of the valves foremost. Water is expelled from the posterior end of the mantle cavity by vigorous contractions of the posterior adductor muscle. Swimming is aided by "rowing" action by the tentacles which arise from the middle fold of the mantle margin. Each tentacle is held rigid during its effective stroke, but is relaxed during the recovery movement, the movement of the tentacles being comparable with those of ciliary movement. (Originally published in Gilmour, 1963, *Proc. malac. Soc. Lond.* **35**, 81, fig. 1.)

tucking its antero-ventral tentacles between the shell valves and pulling the shell valves as close together as possible. The adductor muscle is then relaxed, the shell valves are parted by the opening thrust exerted by the ligament, and this pushes the stones apart, so widening the crevice. When the animal is sufficiently embedded it commences to consolidate its nest by secreting byssus threads. One thread is attached to a stone, the foot is then rotated through 180° and a second thread is attached at its tip to another stone and at its base to the base of the first byssus thread. The animal then disengages itself from the two byssus threads which remain binding the two stones together. Further activity on these lines gradually consolidates the stones and gravel to form a compact refuge within which the bivalve periodically changes its position so that its exhalant water currents, passing in various directions, can prevent the nest from being fouled by sediment. Where a fast bottom current flows over clean shell gravel, *L. hians* may live in dense communities (Gilmour, 1963).

The condition in the saddle oysters is worthy of special attention. Here the byssus does not consist of numerous flexible threads, but of a single calcified column which appears to perforate the lower (right) shell valve. The byssus does not lie in the plane of the primary bilateral symmetry of the animal as is customary, but lies at right angles to this and is attached vertical to the substratum. To permit this arrangement the lower shell valve is very deeply emarginated, and the position of byssal attachment is approximately central. The shell is markedly inequivalve, the lower valve being fitted very closely to the substratum and the margin of the concave upper valve fitting very closely to this. The axis of attachment is vertical to the substratum, sub-central in position, and the centre of the animal's body is occupied by the posterior adductor muscle and the retractor muscle of the byssus apparatus. Adduction and byssal retraction are thus coaxial, and an extremely efficient "limpet" form has been achieved. With this central byssal attachment the approach to a secondary radial symmetry is even closer than that achieved by the Pectinidae. It is remarkable that the byssally attached Anomiidae have also given rise to free-living and to mobile forms—which will be discussed with other members of the free-living epifauna (see Fig. 76, p. 202).

The monomyarian form is also found in the Tridacnidae, where the form of the animal has been influenced not only by byssal attachment but also by the occurrence of a symbiotic association with zoo-xanthellae. These zoo-xanthellae are invariably carried within phagocytes and are confined to the spaces of the vascular system, being "farmed" by exposure to sunlight in the superficial mantle tissues. The zoo-xanthellae contain large accumulations of starch, and they do not possess any cellulose walls so they can be digested easily by the phagocyte. The zoo-xanthellae are eventually transported to the region of the digestive diverticula prior to digestion, and vast numbers of phagocytes containing zoo-xanthellae in all stages of digestion are to be found here. In contrast the digestive diverticula are reduced in numbers (Yonge, 1936a).

In the Tridacnidae there has been a considerable rearrangement as compared with the "typical" bivalve growth form; thus the hinge lies immediately below the mouth and immediately anterior to the foot. Again, the widest gape of the shell valves lies furthest from the substratum, and immediately above the pericardium (see Fig. 67). These are most unusual dispositions and there has been much discussion as to the manner in which they have been brought about. Yonge (1936; 1953) stated that the point of byssal attachment marked the ventral surface, and concluded that there had been a rotation of the mantle/shell about a fixed visceral mass—the umbo, for example, rotating by about 180° of arc. The reason brought forward by Yonge to account for this process of rotation was the occurrence of zoo-xanthellae in the mantle tissues in the vicinity of the siphons and the advantage conferred on the bivalve by this symbiotic association. Growth of the zoo-xanthellae provides the clam with a valuable source of food material, and any change which tends to increase the area of mantle tissues infected with zoo-xanthellae and exposed to sunlight

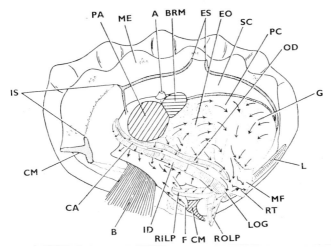

FIG. 67. Ciliary cleansing and feeding currents, and organs in the mantle cavity of *Tridacna crocea*, seen from the right side after removal of the right shell valve and the right mantle lobe. The spatial relations between the viscera and the mantle/shell are unique in the family Tridacnidae; the umbo lies antero-ventral to the mouth, while the greatest gape between the shell valves lies above the pericardium. The region of byssal attachment (B) may be regarded as a fixed point. The farming of symbiotic zoo-xanthellae in the mantle tissues surrounding the exhalant aperture (ES) led to hypertrophy of this region; forward migration and expansion of the tissues in which algae are farmed resulted in forward migration of the umbo and hinge to the maximum possible extent. For interpretation of lettering, see pp. 201–4. (Originally published in Purchon, 1955b, *Proc. malac. Soc. Lond.* **31,** 97, fig. 1.)

would be advantageous. Under these circumstances the umbo gradually migrated forwards and the siphonal area spread upwards and forwards in the wake of the umbo.

The concept of rotation of the mantle/shell employed by Yonge suffices to explain the unexpected relationships of parts mentioned above, but Yonge did not attempt to explain the mechanism effecting the rotation. Stasek (1962) studied the growth process as exemplified by a series of young stages and found that rotation of the mantle/shell upon the visceral mass did not, in fact, occur. Stasek found that as growth proceeds there is a gradual change in the angle which the hinge line subtends to the substratum, due to hypertrophy of the postero-ventral part of the shell. Thus as the animal grows the disposition of the shell is gradually changed; that part of the shell which began in a mid-ventral position remains there, but more rapid growth of the postero-ventral margin of the shell causes this to spread up and over the dorsal surface of the body. Thus Stasek carried the explanation a stage further, providing some details of the process by which the growth form of the adult *Tridacna* (see Figs. 68, 69) is attained. The growth form of the Tridacnidae is the outcome of the interaction of byssal attachment and symbiosis with zoo-xanthellae.

Stasek (1965) reports that in *Tridacna maxima* the exhalant siphon can be manipulated so that a jet of water can be directed at any point on the mantle which is irritated mechanically. This would serve, on the one hand, to dislodge foreign bodies such as coral rubble which might be thrown over the clam by wave action, and, on the other hand, might drive away a fish attempting to browse on the exposed and otherwise defenceless mantle tissues. Stasek noted that *T. maxima* responded to the movement of objects in the vicinity, e.g. fish, even if they did not cast a shadow on

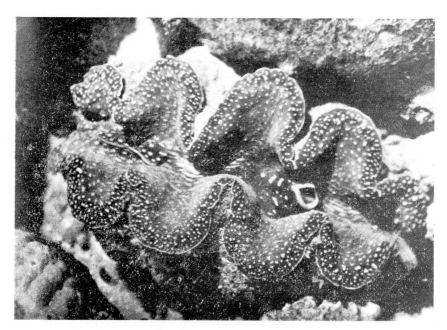

FIG. 68. *Tridacna derasa*. View from above of a fully expanded specimen which is just covered with water. The long slit-shaped inhalant aperture is seen closed on the left, while the conical exhalant siphon on the right is seen opened. (Reproduced by permission of the Trustees, British Museum, (Natural History) from Yonge, 1936a, *Sci. Rept. Gt. Barrier Reef Exped. Brit. Mus.*, vol. 1, plate I, fig. 2.)

the clam. Stasek (1966) subsequently showed that the "hyaline organs" of *T. maxima* are eyes (see Fig. 55, p. 132).

The evolution of the monomyarian condition in various lineages of bivalves involved a diminution of the anterior end of the body and great reduction of the free space in the anterior part of the mantle cavity. Gilmour (1964) concludes that this will inevitably cause an increase in the velocity of water currents emerging from the proximal oral grooves, and that this could interfere with the passage of mucous food strings towards the mouth. Enclosure of the proximal oral grooves would protect the food stream from such interference, and the complex lip apparatus of the Pectinidae and Limidae probably evolved for this reason. It is postulated that the lips of

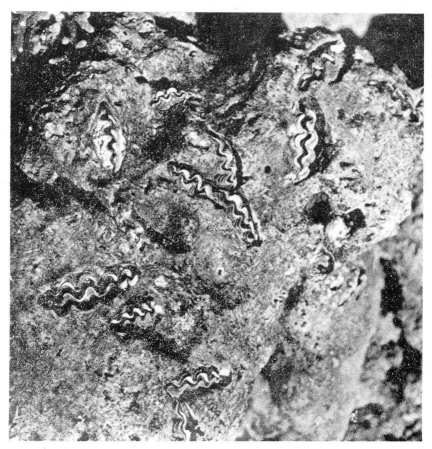

FIG. 69. *Tridacna crocea.* View of a coral boulder into which several specimens of the clam have burrowed; the boulder is uncovered by the tide and the mantles of the clams are withdrawn. (Reproduced by permission of the Trustees, British Museum (Natural History) from Yonge, 1936a, *Sci. Rept. Gt. Barrier Reef Exped. Brit. Mus.*, vol. 1, plate II, fig. 3.)

other monomyarian bivalves are also hypertrophied for the same reason (Gilmour, 1964).

There is presumably some theoretical limit to the size which can be attained by a bivalve which feeds exclusively by filtration of the supernatant water. Such a limit does not apply in the case of the Tridacnidae, the food supplies of which have been vastly augmented by the adoption of symbiosis, and *T. gigas*, the Giant Clam, is the largest bivalve to have evolved in the history of the world. This species lies wedged between coral masses, as also does the considerably smaller *T. squamosa*. The sharp semicircular flanges which ornament the outer surface of the shell of the latter are the relics of the shell margin at earlier stages of growth, the semicircular flares of the shell edge serving to provide a broad base over which the mantle tissues can expand to allow maximal photosynthesis by the zoo-xanthellae. The persistence of traces of these flanges on the outer surface of the shell, even close to the umbo, and the position adopted by these species wedged between coral masses, is proof that they do not bore into the substratum; nor does the related genus *Hippopus*, which lies unattached upon the surface of the reef flat. Rock-boring, therefore, has presumably played no part in the moulding of the growth form of the Tridacnidae. However, one species does bore into soft intertidal rocks, e.g. beach limestone, or coral boulders, namely *T. crocea*. Since the specimen grows as it burrows, it becomes imprisoned in its own burrow. If, through splitting of a boulder, *T. crocea* became detached and fully exposed on the beach, it would be unable to recommence to bore at a new point on the rock surface. The general form of *T. crocea* is comparable to that of other species in the genus, the only way in which the habit of rock-boring has affected its appearance being the absence of flanges on the outer surface of the shell. Abrasion of the shell against the wall of its burrow makes the outer surface of the shell comparatively smooth, especially near the umbo.

Cementation

Attachment to a rock surface by cementation of one shell valve has occurred independently in at least four different lineages, of which the oyster is best known due to its commercial importance. The lowermost valve, which is the left in *Ostrea* but the right in *Spondylus* and in *Hinnites*, is cemented to the substratum by a secretion from the foot at the time of metamorphosis: thereafter the lowermost valve grows for some time in the closest proximity to the substratum. These genera are inequivalve and monomyarian. In *Ostrea* the hinge is edentulous, protection against opening of the shell by a predator being by the very powerful and long sustained action of the adductor muscle, aided by the comparatively extensive ligament. In contrast closure of the shell valves of *Spondylus*, which is equally powerful, is effected by the adductor muscle, fine interlocking denticulations around the free edges of the two valves, and a pair of very powerfully built hinge "teeth" which cannot be separated without breaking; the ligament is a stout internal resilium. *Chama*, on the other hand, is an isomyarian, the adductor scars being very well defined; the hinge teeth are irregular in form and bear fine serrations which interlock with those on the opposite hinge tooth. Both *Chama* and *Spondylus* commonly occur attached securely to the sides of littoral boulders, where they are usually well hidden by fine growths of algae which extend equally over the rock and over the shell. Attachment by cementation also occurs in the freshwater family Aetheriidae, in which striking changes in growth form are encountered.

In *Hinnites*—a sub-genus of *Pecten*—cementation occurs at a relatively late stage in development. The larva usually settles in a small crevice, being byssally attached at first, and lying upon its right shell valve. After a period of byssal attachment the right mantle lobe is protruded and shell material is secreted in such a way that it cements the rim of the right shell valve to the substratum. Thereafter the shell may become irregular in form, conforming to the shape of the underlying rock. If, as is usual, the hinge line lies in a crevice and movement of the upper valve becomes restricted as the animal grows, then the hinge line gradually migrates ventralwards, carrying the axis of opening of the shell valves gradually out of the crevice. In this way as much as 20 per cent of the dorso-ventral length of the shell may come to lie dorsal to the hinge line. If shell movement is not restricted in this way by the confines of a crevice, then the hinge line does not migrate ventralwards (Yonge, 1951b). It is striking that in the family Pectinidae byssal attachment has given way to cementation in *Hinnites*, and at least twice to adult freedom (*Amussium* and certain species of *Pecten* and *Chlamys*) (see Fig. 76, p. 202).

Rock-boring

This mode of life may have originated in two contrasted ways; by further specialisation of deep burrowers into soft substrates, and alternatively by development of the "nestling" habit of forms which are byssally attached to the surface of soft rocks (Yonge, 1955). The extent to which the rock-boring mode of life affects the growth form is naturally dependent on the manner in which rock-boring is carried out. It has already been said that rock-boring has had relatively little influence on the form of *T. crocea*. The same may be said of *Petricola pholadiformis*, which only penetrates comparatively soft matrices and which might be described as a burrower rather than as a borer. *Hiatella arctica* is a true rock-borer which occurs only in comparatively soft rocks. In *Hiatella* the two lobes of the mantle are fused in the median ventral line except for the two siphonal orifices and for a very small pedal orifice anteriorly. The young animal first enters the rock with its ventral gape lowermost, but it soon turns and lies with its anterior end at the bottom of the excavation. Closure of the siphonal aperture, followed by contraction of the siphonal process, causes the latter to bulge and grip the walls of the burrow firmly. This use of the sea water in the siphonal process as a hydro-skeleton provides the animal with a fixed point, or fulcrum, on which the shell valves can move; the adductor muscles can be contracted independently causing the shell valves to bear against the walls of the burrow and thereby to deepen the burrow (Hunter, 1949).

Examination of *Botula*, a sub-genus of *Lithophaga* in the family Mytilidae (Thiele, 1935), indicates the possible sequence of events in the evolution of the capacity of *Lithophaga* to bore by chemical attack on calcareous rocks. The starting point is considered to have been a byssally attached mytiliform shell which nestled in crevices in soft inter-tidal rocks and which tended to "crouch" into the bottom of the crevice when disturbed. Where the rock was sufficiently soft, and the "crouching" action was repeated sufficiently frequently, mechanical boring would naturally ensue. But deep burrowing could not be developed by a mytiliform bivalve, and it is necessary to predicate a reversal in the earlier phase of diminution of the anterior end of the shell due to byssal attachment. This has occurred in *Botula*, which has secondarily resumed an isomyarian form and a more or less cylindrical shell which permits deep penetration into the rock.

Rock-boring by *Botula* is solely by abrasion of the soft rocks by the shell valves, especially at the umbonal regions which become deeply worn. Relaxation of the adductor muscles causes the shell valves to part under the opening thrust of the elongated ligament; contraction of the anterior byssal retractor muscles then pulls the animal deeper into the burrow and the shell valves bear against the rock walls of the burrow. The elongated, cylindrical form of the shell permits movement of the animal up and down the burrow in this way. The burrow is modelled on the shape of the shell, there being a well-defined "dorsal" ridge which lies between the umbones, and a lesser "ventral" ridge which lies opposite the ventral gape of the shell. *Botula* cannot rotate in its burrow but always stays in the same plane. The fused inner folds of the mantle margin lying between the dorsal edges of the shell valves anterior and posterior to the ligament are liable to become covered by fragments of rock which are the product of rock boring. These are eliminated by local secretion of mucus and by ciliary waste currents.

In *Lithophaga* (*sensu stricto*) the story has been carried a stage further: the animal occurs only in calcareous rocks, or in coral growths, into which it bores solely by chemical means. The mantle tissues lying between the dorsal edges of the shell, both anterior and posterior to the ligament, are hypertrophied and contain special glandular patches. The anterior glandular areas are particularly well developed in *Lithophaga*. These glandular thickenings can be protruded and pressed against the inner wall of the burrow, and it is thought that these glands have developed an acid character. Apposition of these glandular patches to the wall of the burrow has actually been observed in *L. plumula*, and the direct action of their acid secretions may cause erosion and thereby deepen and widen the working end of the burrow (Yonge, 1955).

In addition to the presumed secretion of acid by the thickened part of the edge of the mantle, the shell is modified in two respects towards the rock-boring mode of life. Firstly, the shell is nearly circular in transverse section, i.e. the transverse component of shell growth is unusually high (Owen, 1952). This enables the animal to fit easily into a cylindrical shaft, along which it can move comparatively freely, rotating on its long axis if necessary. In contrast a shell which is flattened from side to side, as is typical, could not enjoy the same degree of freedom within a burrow. Secondly, the outer coat of periostracum is unusually thick, and this is thought to protect the shell against the risk of abrasion by friction against the walls of the burrow, followed by the danger of etching by any free acid secretions within the burrow. Finally, it should be added that the animal retains its byssus attachment and by means of this it is able to move up and down the burrow according to need. Posteriorly, the margins of the mantle are elongated to form a siphon-like extension which lines the outer part of the burrow while the animal is undisturbed. This outer part of the burrow is constricted by concentric laminae of a hard calcareous deposit thus converting the cross-section of the posterior part of the burrow to the characteristic "hour-glass" shape. This calcareous secretion is undoubtedly laid down here by the posterior extensions of the mantle margin. It seems probable that the source of the material is the head of the burrow where it may not be dissolved by an acid secretion, but where it may be digested by enzymic action and absorbed by the anterior margin of the mantle. This requires further investigation.

Rocellaria (= *Gastrochaena*) occurs near low-water mark on coral reefs in tropical waters, burrowing only into the soft reef limestone or into coral boulders, as does *Lithophaga*, but in this case boring is by mechanical means which is surprising, for the shell is comparatively thin and fragile. The shell lies in a pear-shaped chamber in the coral, its presence often being revealed by a delicate white calcareous "chimney" which protects the siphonal process. The coral rock is comparatively soft, and is apparently abraded by the periodical opening and closing of the shell of *Rocellaria*, this movement being effected in the normal way, with the dorsal hinge line as the fulcrum. Inspection of the animal shows that *Rocellaria* is modified for rock boring in three main ways. The antero-ventral border of the shell is very deeply emarginated, providing a long and wide permanent pedal gape through which the foot can protrude to grip the end of the burrow; the foot, instead of being wedge-shaped with a pointed tip, is more cylindrical in form, is directed anteriorly, and has a broad flat suctorial termination; the retractor muscles of the foot are inserted on to a small elevation of the inner surface of the shell which lies close to the hinge line. This elevation, or boss, on to which the retractor pedis muscle is attached, may effect a slight change in the direction of action of the muscle which is advantageous to the now forwardly directed foot. The suctorial—or sticky—disc-shaped tip of the foot is permanently exposed, due to emargination of the shell, and is able to grip the end of the burrow, sometimes being aided in this by secretion of a single byssus thread. Opening and closing of the shell then wears away the interior of the burrow, broadening and deepening it. It seems probable that this action will inevitably abrade the antero-ventral border of the shell, thereby slowing down the rate of growth of the shell in this region and thereby contributing to the maintenance of a broad anterior pedal gape. Thus the form of the shell may be partly dependent on the immediate consequences of boring activity.

In the family Myidae one genus, *Platyodon*, appears to have evolved from the habit of burrowing in rather stiff substrates, with the production of the capacity to drill into soft rocks. There is a long siphonal process of type C, near the tip of which the periostracum is thickened to form four pointed scales. These scales may serve by abrasion to smooth the inner wall of the upper part of the burrow, and they give protection by meeting centrally over the apex of the siphonal process when its tip is contracted and withdrawn. Rock boring is by mechanical abrasion of the shell valves against the walls of the burrow. The shell is wedged in position by engorgement of the mantle in the mid-ventral line, which forces the shell valves apart; contraction of the posterior adductor muscle then causes the anterior part of each shell valve to be scraped against the walls of the burrow. *Platyodon* cannot rotate in its burrow, and the shape of the excavation is modelled on that of the shell, there being two longitudinal ridges on the walls of the burrow, much as in *Botula* (Yonge, 1948, 1951).

We must now turn to the family Pholadidae, the members of which are all very highly specialised for the purpose of rock or wood boring. Perhaps the most striking modification is the capacity to rock the shell valves about a vertical axis passing through the umbo instead of about a horizontal axis passing along the hinge line. This is achieved by three or perhaps four associated changes. Firstly, the ligament is reduced in extent until it constitutes no obstruction to the new direction of motion; secondly, the hinge teeth have been necessarily obliterated; thirdly, the anterior and posterior adductor muscles alternate in action so that in one extreme the anterior ends of the shells are approximated while in the other extreme they are widely divergent; fourthly, in certain genera a boss has been developed on the ventral border of each shell valve to provide a point of articulation at the ventral end of the new axis of movement. The first three of these modifications may have been associated with an earlier, deep-burrowing mode of life for we see somewhat similar conditions in *Mya* and *Platyodon* which can contract the adductors independently in order to accommodate the base of the siphonal process between the posterior ends of the shell valves.

The Pholadidae are the only bivalves which have developed a number of accessory shell plates—up to six—some of which protect the mantle tissues along the mid-dorsal line while one, the callum, is developed in some genera to close the large pedal gape after the animal has ceased to bore into the substratum. The antero-dorsal margin of the shell is reflected outwards and the anterior adductor has migrated upwards on to this reflection of the shell where it occupies a position well suited to oppose the action of the posterior adductor muscle (see Figs. 70 and 71). In *Zirfaea*

Fɪɢ. 70. *Zirfaea pilsbryi*. Ventral view of a pair of empty shell valves, showing the wide pedal gape at the anterior end, and the blade-like apophysis within; the external surface of the anterior part of the shall valves bears rows of sharp serrations by means of which the animal bores into the substratum. (Originally published in Turner, 1954, *Johnsonia*, **3**, 62, plate 34.)

crispata the anterior adductor muscle is in two parts, which contract alternately. The anterior of these parts, which may be termed the accessory anterior adductor, contracts at the same time as a ventral adductor muscle which has developed at the posterior end of the pedal gape. The synchronous contraction of these two muscles draws the anterior ends of the valves together. The posterior moeity of the anterior adductor muscle has an entirely different function; contraction of this muscle is accompanied by relaxation of the ventral adductor muscle and as a result of this the valves spread apart ventrally and so bear against the sides of the burrow. Contraction of the posterior adductor causes the anterior ends of the valves to part and so scrape against the anterior walls of the burrow. A powerful jet of water can be forced out of the mantle cavity anteriorly, ventral to the foot, and this serves to flush the working end of the excavation (Turner, 1954, 1955).

As in *Rocellaria*, in most of the Pholadidae the shell is deeply emarginated to provide a permanent pedal gape, and the powerful forwardly directed foot bears a terminal suctorial—or sticky—disc which grips the anterior end of the burrow. In the Pholadidae, however, the pedal gape is more

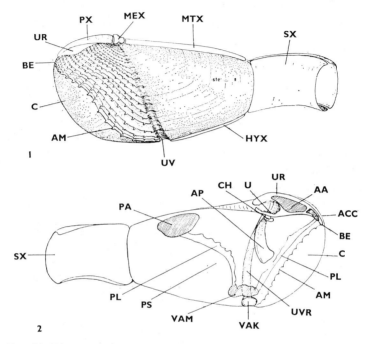

FIG. 71. Diagrammatic representation of the shell in the Pholadidae. 1, Exterior, showing the positions of the accessory shell plates. 2, Interior, showing the apophysis (AP), the umbo (U) which serves as a dorsal point of articulation, the ventral articular knob (VAK) and the insertional scars of the anterior and posterior adductor muscles (AA, ACC, PA). For interpretation of other lettering, see pp. 201–4. (Originally published in Turner, 1954, *Johnsonia*, **3**, 11, plate 6.)

anterior in position than in *Rocellaria*. In both cases the shape of the pedal gape seems closely related to the axis of movement of the shell valves, and indicates the regions of the shell margins which bear against the wall of the burrow. Thus it seems that in the Pholadidae, also, the shape of the shell is influenced and partly maintained by abrasion of the shell margin against the walls of the burrow—this abrasion will inevitably slow down the rate at which this zone of the shell grows by marginal accretions. Comparison of *Barnea candida* and *Petricola pholadiformis* which are confined to fairly soft substrates such as peat and clay, and the shells of which both lack a pedal gape, with *B. parva* which bores into harder rocks and the shell of which possesses a pedal gape, suggests that the mechanical action of boring may itself influence the rate of growth of the anterior part of the shell margin. This view is further strengthened by consideration of *Pholadidea loscombiana*, the shell of which may be found in either of two conditions. In young specimens which are actively boring, the shell possesses a conspicuous pedal gape and is not unlike that of *B. parva*. In older specimens which have ceased to bore this pedal gape has been obliterated by the development of the callum, a thin dome-shaped layer of shell which lacks lines of growth and has presumably been secreted fairly rapidly. This anterior extension of the shell is underlaid by a comparable extension of the mantle which fuses with its partner to reduce the pedal orifice to a small anterior slit. At the same time the foot becomes reduced in size. Thus on cessation of boring the shell is quickly restored to a complete oval outline, which seems to support the contention that the form of the shell was originally dictated to some extent by the mechanics of the rock-boring mode of life.

Another most interesting modification towards the rock-boring habit is the development of the apophysis, a long, slender down-growth from the umbonal region of the shell which projects into the substance of the visceral mass, and which is the site of insertion of the pedal retractor muscles. This is evidently important in adjusting the direction of the pedal retractor muscle fibres to suit the more anteriorly directed foot, and the stresses upon this during boring action. The functional considerations are doubtless comparable to those which evoked the formation of a small boss at the point of insertion of the retractor muscle in *Rocellaria*, and the two cases may be seen as examples of parallel evolution.

The Pholadidae are very slightly anisomyarian, the posterior adductor being slightly larger in transverse section (or in area of adductor scar) than the anterior adductor muscle. This is because the effective boring action is provided by contraction of the posterior adductor muscle, which causes the anterior ends of the shell to diverge and so scrape against the walls of the burrow near its innermost, blind end. The anterior adductor muscle, or a part of this, provides a less vigorous recovery movement. In some cases, e.g. *Pholas dactylus*, the antero-dorsal border of the shell has been raised upwards and reflected over the umbonal and anterior part of the shell, and the insertion of the anterior adductor muscle has been carried upwards to this superficial position. This places the muscle in a direct line with the reduced ligament and the posterior adductor, and may confer some slight mechanical advantage in operating the shell valves about the new axis of movement. This change in position of the anterior adductor muscle removes it from the place of protection it originally enjoyed within the confines of the shell, and this may be the reason for the development of one or two accessory shell plates which are secreted by a median lobe of the mantle margin which overlies the anterior adductor in some genera. In some cases a similar accessory plate lies behind the umbo, overlying the region of the pericardium and rectum, which would otherwise be exposed when the posterior ends of the shell valves diverge.

The long siphonal process, e.g. in *Zirfaea crispata*, is of type C (Yonge, 1948) and is protected by a sheath of periostracum against abrasion on the walls of the burrow. This may be regarded as a parallel evolution with burrowing forms such as *Mya* and *Lutraria*, for incorporation of the inner surface of the outer mantle fold in the substance of the siphonal process is an important precursor to both burrowing and rock-boring, providing a protective coat of value in both modes of life.

The results of the field studies of Evans (1966) on *Penitella penita* suggest that during the juvenile period the energy expended in rock-boring prevents the accumulation of the food reserves necessary for the development of the gonads and maturation of the gametes. In relatively soft rocks, boring is easy and there is little hindrance to the growth of the shell, which is of light weight and is long relative to its depth. In harder rocks growth is hindered by the limited dimensions of the burrow, and the shell is both heavier and shorter in relation to its depth. The ease of burrowing in soft rocks results in the early acquisition of sexual maturity and the early production of the next generation of rock-borers. This makes for maximal erosion by rock-borers at those parts of the coast which are least capable of withstanding attack. In contrast, those individuals which bore into harder rocks experience a delay in the onset of sexual maturity, and benefit by enjoying a longer life. Six specimens from the moderately hard rocks of South Jetty were successfully re-planted in artificial burrows in the softer rocks at Fossil Point, where in the following year they grew 1·7 times as rapidly as comparable specimens native to the area. The higher growth rate exhibited by the transplanted specimens may be attributable to the acquisition of greater vigour in the initial period spent boring in the harder rocks at South Jetty. Alternatively, the higher growth rate might be attributable to a genetical difference between specimens of *P. penita* from the two areas. There are drawbacks to the acceptance of either theory (Table 4).

Wood-boring

The transition from rock-boring to wood-boring is a small one, and it is not surprising to find that certain members of the Pholadidae are not found in such hard substrates as rock, but only in wood, e.g. *Martesia striata* which has a worldwide distribution in tropical waters. This species burrows into wood in just the same way as other pholads burrow into rocks, and it lies in a short pear-shaped burrow which lies at right angles to the surface of the wood. *Martesia* is just a pholad which has specialised in boring into wood; it has not exploited all the possibilities of this specialised mode of life since the requisite powerful cellulase has apparently not evolved in this genus (Purchon, 1956).

TABLE 4. *Summary of Results of Field Investigations on* Penitella penita, the most abundant and most widely distributed rock-boring bivalve on the Pacific Coast of America (Data extracted from Evans, 1966)

Locality	Relative hardness of rock	Approximate age at metamorphosis to adult stage	Appearance of shells of uniform depth	Comments
Fossil Point	1·0	$2\frac{1}{2}$ years	Small in size, light in weight, relatively long	Sexual maturity is reached at a relatively early age
South Jetty	1·7	$4\frac{1}{2}$ years	Intermediate in all respects	
Cape Blanco	3·8	18 years	Largest and heaviest, but relatively short	The growth rate and sexual maturity are retarded by 6–7 times

The genus *Xylophaga* is generally placed in the family Pholadidae, although unusual features in its reproductive system—which will be discussed elsewhere—have been advanced for placing it in a separate family (Purchon, 1941). It is not known for certain whether *Xylophaga* can digest wood, but this seems a distinct possibility since it possesses a stomach caecum in which wood fragments are stored, as also do the Teredinidae, or ship-worms, some of which definitely can digest wood fragments. It is interesting to consider the origins of these wood-digesting bivalves. The shells of the Teredinidae are basically comparable in structure and in function to those of the Pholadidae described above, and there is no doubt that the two families sprang from a common ancestry; there is the same reduction of the ligament, loss of hinge teeth, migration of the anterior adductor muscle, development of a ventral articular boss on the shell margin, alternating action of the adductor muscles, and development of an apophysis to carry the insertion of the pedal retractor muscles. *Xylophaga* is generally similar, but lacks an apophysis.

The main difference between the Teredinidae and the Pholadidae is the vast development of the stomach caecum for the purpose of storing minute fragments of wood; in order to accommodate this organ the visceral mass has been extended beyond the confines of the shell, enclosed in a worm-like extension of the mantle which passes backwards from the shell valves, along the long and meandering tunnel through the wood to the point where this communicates with the exterior. Here there is a pair of typical siphons. The posterior part of the tunnel is lined by a smooth tubular coat of calcium salts, and a pair of calcareous pallets has been evolved at the posterior end of the mantle, and lateral to the bases of the siphons, to close the opening in the wood when the siphons are withdrawn. The ctenidia have been displaced to the posterior part of the mantle cavity, mostly posterior to the visceral mass, and in order to maintain communication with the ctenidia, the heart is turned upside down and is very greatly elongated. In *Xylophaga* the stomach caecum is not so greatly enlarged, none of the organs are displaced from the normal position, and all are accommodated within the shell valves.

The stomach caecum of the Teredinidae and of *Xylophaga* is basically comparable to the appendix of the stomach which is found in most

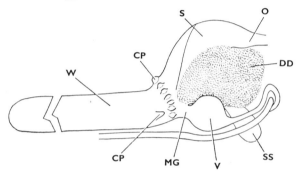

Fig. 72. *Psiloteredo amboinensis*, view of the gut from the right-hand side, showing the stomach (S) with its huge wood-storing caecum or appendix (W), digestive diverticula (DD), style sac (SS), and mid-gut (MG). A swelling at the proximal end of the mid-gut contains a swelling of the major intestinal typhlosole and this serves as a valve (V) which controls the passage of the contents of the stomach into the mid-gut. For interpretation of other lettering, see pp. 201–4. (Originally published in Purchon, 1960a, *Proc. zool. Soc. Lond.* **135**, 475, fig. 18.)

members of the Pholadidae, and to the postero-dorsal stomach caecum of the Tellinacea, both in structure, position and in function. In all cases small fragments derived from the substratum, which happen to enter the mantle cavity and become swallowed, are passed from the stomach into this caecum or appendix to be stored there temporarily (see Fig. 72). In the Tellinacea it is sand grains which find their way to the caecum, while in the Pholadidae it is small fragments of rock which have been produced during the rock-boring action. There seems little doubt that the postero-dorsal caecum of the Tellinacea, the appendix of the Pholadidae, and the stomach caecum of *Xylophaga* and of the Teredinidae are homologous.

It seems probable that early in the history of one lineage of bivalves a caecum evolved and served as a temporary storage organ for relatively hard, rough, and indigestible fragments which were inadvertently admitted into the stomach, these fragments being passed back into the stomach in small quantities later on, perhaps to aid in trituration of stomach contents. This lineage subsequently specialised on two lines, one giving rise to the Tellinacea which occupy a variety of unstable substrates, and the other giving rise to the Adesmacea (Pholadidae plus Teredinidae) which specialised in boring into hard substrates such as rock or wood. The caecum continued to function as originally in both lineages, i.e. to store particles inadvertently ingested together with food material. But in the wood borers this material consisted of fine fragments of wood and was capable of being digested if only a cellulase became available. In *Martesia* this did not occur; in *Xylophaga* it may have occurred—for what other reason was the caecum so greatly enlarged as compared with the appendix of the Pholadidae? In the Teredinidae a cellulase was evolved, one particular portion of the digestive diverticula being modified by enlargement of its ducts to aid in the digestion of these fragments of wood. With the possibility of digesting these wood fragments, increase in size of the appendix was of great functional significance—it became possible for the animal to lay up a store of food which could then be digested in leisure after the external supply of timber had long been exhausted. We may conclude that the emergence of the postero-dorsal caecum (= stomach caecum, or appendix) was an important pre-adaptive step towards the evolution of the ship-worms.

To conclude this survey of rock and wood boring it is interesting to note that this habit has been adopted by as many as nineteen genera belonging to seven different lineages of bivalves. This emphasises the importance of habitat and mode of life in the adaptive radiation of bivalves. Where the same mode of life has developed in many different lineages it is inevitable that there will have been instances of parallel or of convergent evolution. Even the skeletal parts—the shell valves and their hinge teeth—prove to be highly plastic, so it is not surprising that it is not yet possible to recognise the main phylogenetic relationships of the many orders of filter-feeding bivalves (see Fig. 76, p. 202).

The Free-living Epifauna

We now have to consider a variety of bivalve genera which lie free and unattached upon the surface of the substratum. These are found on a variety of substrates and have evolved independently from a number of distinct lineages.

It is remarkable that within the Pectinacea the extreme condition of monomyarianism having been imprinted on the body form as the outcome of byssal attachment, some members have abandoned byssal attachment and have become free living, while others have gone even further and have acquired the capacity to swim. During the long phase of byssal attachment the sub-central posterior adductor muscle developed in two parts; one of these consisted of unstriated "catch" muscle fibres which served for sustained closure of the shell valves when the animal is disturbed, while the other part consisted of striated "quick" muscle fibres which served to provide a rapid series of powerful contractions to eject unwanted material from the mantle cavity. At the same time the innermost muscular fold of each mantle margin became developed to form a pallial curtain which served to control inhalant and exhalant water currents. These features of the soft parts were pre-adaptive towards the free-living mode of life which was ultimately achieved by certain members of the order.

Certain species in the Pectinidae have specialised as members of the epifauna living upon certain kinds of substratum. Thus *Pecten maximus* and *Chlamys distorta* require a hard bottom in deep water, while *C. varia* requires a hard bottom in shallow water. In contrast *C. septemradiata* lives upon deep water muds. Some other species are more catholic, e.g. *C. opercularis* (Allen, 1953). In *C. opercularis* in European waters, and in *P. irradians* in American waters, vigorous contractions by the "quick" muscle cause a series of jets of water to be ejected from the mantle cavity. The ligament is in the form of a highly elastic internal resilium which is well adapted to cause opening of the shell immediately after each contraction of the quick muscle; a fresh supply of water enters the mantle cavity on each occasion, ready to be expelled on the next contraction of the muscle. The pallial curtains form a valve which allows free entry of water over the whole extent of the shell gape, but which restricts the outward movement of water to two areas anterior and posterior to the hinge line. Thus a small inflow of water over the whole area of the gape is converted to two powerful jets directed dorsalwards at either end of the hinge, and the animal is projected ventral margin foremost in a jerky swimming action. When alarmed, however, *Chlamys* makes an escape reaction in which the pallial curtains are drawn backwards and there is no obstacle to the ejection of water along the whole length of the wide pedal gape; this causes the animal to shoot rapidly away in the opposite direction with the hinge foremost (Yonge, 1936). These animals rest on the right valve on sandy bottoms and can use the swimming action either to make a limited move to escape an enemy such as a starfish (Rees, 1957), or it may make more extensive swimming movements.

The evolution of the swimming habit has required no further change in the structure of the shell or of the soft parts, except for some elongation and obliquity of alignment of the quick muscle which permits wider opening of the shell valves, entry of a larger volume of water, and a greater movement of the animal when the water is expelled (Yonge, 1936). This component of the adaptive radiation of bivalves has thus had no major morphological consequences; had this lineage died out many millions of years ago it is unlikely that palaeontologists could have deduced any capacity to swim from the fossil remains, but on the contrary they might have envisaged continuance of byssal attachment (see Fig. 73 and 74).

Amussium pleuronectes, which is probably the best swimmer among the bivalves, always rests on the right shell valve (Atkins, 1937) which is white on its outer side, whereas the upper shell valve is a rich chestnut

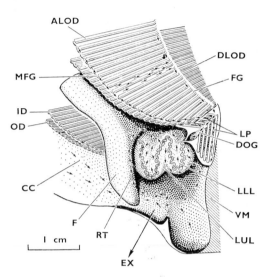

FIG. 73. *Pecten maximus.* Detailed structure of the anterior end of the ctenidium, the labial palps and the lip-apparatus, viewed from the right side. The anterior and posterior lips of the mouth are very deeply folded and interlocked, so providing protection for the mucus-bound food train entering the mouth. This is an important adaptation associated with the need for occasional vigorous clapping of the shell valves to rid the mantle cavity of detritus. For interpretation of lettering, see pp. 201–4. (Originally published in Gilmour, 1964, *J. mar. biol. Ass. U.K.* **44**, 490, fig. 6.)

brown colour; this is a striking example of adaptive coloration among the bivalves. This species is unusual in that the lower valve bears a regular series of radial ridges on its inner surface, the functional significance of these ridges being unknown.

As has been mentioned previously (p. 164), certain members of the Solenidae are able to swim for short distances as an escape reaction. *Solen delesserti* has been observed to swim with its anterior end foremost by jets of water expelled from the inhalant siphon, whereas *Ensis directus* has been observed to swim in the opposite direction as the result of jets of water expelled at the anterior end, around the base of the foot, as the result of repeated pedal movements.

The habit of resting on the substratum with the same valve always lowermost may bring about an inequivalve condition, the upper valve being less deeply convex, or even being flat. In *Pandora* the right valve is uppermost and is either flat or even slightly concave. *P. inaequivalvis* lies on a bottom of muddy sand in shallow water, while *P. pinna*, the other British species, favours depths exceeding 40 fathoms (Allen, 1954). In *Pecten maximus*, too, the upper valve is quite flat. In these two cases

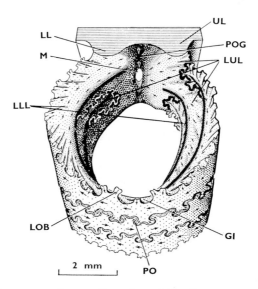

FIG. 74. Stereogram of a small section of the lip apparatus of *Pecten maximus*, showing the deep interdigitation of the lobes of the anterior and posterior lips of the mouth and ciliary cleansing currents thereon. For interpretation of lettering, see pp. 201–4. (Originally published in Gilmour, 1964, *J. mar. biol. Ass. U.K.* **44**, 491, fig. 7.)

growth of the upper, flat shell valve is solely radial in form, the transverse component of growth of this valve having been suppressed in some way (Owen, 1952). *Aloidis gibba*, in the Myacea, is inequivalve, the left valve being smaller and less convex than the right valve; when the valves are closed the edges of the left valve fit within the contour of the larger valve and this may give added protection against potential predators. One might expect *Aloidis* to lie upon the surface of the substratum, but this is not so; *Aloidis* lies buried in muddy gravel (Yonge, 1946).

The close approach to radial symmetry by *P. maximus*, which rests on the bottom, is worthy of mention. This pectinid is unable to swim like the related *C. opercularis*, but is capable of making righting movements if it is turned upside down by wave action. This is done by manipulating the pallial curtains so that on sudden closure of the shell valves a powerful stream of water is ejected downwards along the free margin of the shell, thus causing the shell to be raised and turned over with the straight hinge line as a fulcrum (Yonge, 1936).

Some, but not all, species of *Lima* can swim, e.g. *L. hians* and *L. fragilis*. Here the animal swims with the sagittal plane held vertically, whereas the valves of *Chlamys* are disposed horizontally while swimming. *L. lima* is byssally attached under boulders and has a heavily built shell which can enclose all the organs of the body. In contrast *L. fragilis*, which can swim, has a small fragile shell which cannot be closed and the extensive and beautifully coloured pallial tentacles are always fully exposed.

Gilmour (1963) has described the swimming mechanism for *L. hians* (see Fig. 66, p. 173). The margin of the mantle bears numerous long, hollow, sticky tentacles. At the base of each tentacle there is a transverse septum containing a sphincter muscle and with a central pore, and when this pore is closed the tentacle becomes an independent hydrostatic unit and can perform independent elongating and bending movements. Prior to and during swimming all the tentacles of the middle fold of the mantle margin are fully extended, so increasing the surface area and decreasing the sinking rate. The anterior and the postero-ventral series of tentacles execute rowing movements by alternate contractions of their anterior and posterior longitudinal muscle fibres. The tentacles are rigid during the effective beat, and are relaxed during the recovery movement, being comparable in this respect to cilia, but the tentacles of *Lima* differ from cilia in being able to reverse and to beat in the opposite direction. The tentacles are also advantageous in serving a repugnatorial function, which protects the animal from attack by fish such as *Pleuronectes*, and by crabs.

If it is surprising that the Pectinidae should give rise to free living and even to swimming forms, it is even more remarkable that the lineage of the Anomiidae should also have produced free-living and even motile forms. For the term "anomiid" immediately evokes the concept of a flat lower shell valve which is very deeply emarginated to accommodate a powerfully built sub-central, calcified byssal pillar. Yet this condition is lacking in both *Placuna* and *Enigmonia*.

E. rosea occurs in tropical mangrove swamps, being found on the leaves, trunks, or buttress roots of the mangrove plants. The lower, right shell valve is either flat or is slightly concave to fit the curvature of the part of the plant on which it lies, and it is deeply emarginated in the normal anomiid fashion. *E. rosea* is motile at least in its early stages, up to a length of 1 in. or more; the foot can be protruded through the byssal notch and by means of this the animal can glide slowly over the surface of the plant. There may possibly be a temporary non-calcified byssal attachment during part of the time when the animal is exposed by the ebbing tide. It is not known whether older animals retain the potentiality for movement, or whether they become permanently attached at a later stage in their life history. The largest specimens are found on the trunks of the mangrove trees, their growth form being a large oval outline oriented vertically on the trunk and being slightly concave in transverse section to fit the curvature of the trunk. Specimens found on the buttress roots were also oriented with their long axes parallel with the axis of the buttress root, but seemed to be somewhat stunted in form and both shell valves were very deeply convex to fit closely round the comparatively slender buttress roots. Since these specimens were wrapped round about two-thirds of the circumference of the root, it is unlikely that they could move about even if they were not permanently attached by some form of byssus. In this species it is obvious that a protracted phase of mobility enables the animal to explore a far larger area than could be explored by a post-larva at metamorphosis, and so select a large smooth area which is unlikely to impose a severe limit on growth. This is particularly important for a bivalve with a delicate growth form which is closely applied to the substratum.

Another, and far larger, anomiid is the windowpane shell, *Placuna*, which lies unattached on the surface of sheltered beaches of muddy sand, below mean sea level. One of the species, *P. placenta*, is almost perfectly circular in outline, and up to 4 or 5 in. in diameter. The shell is slightly inequivalve, one valve being flat and the other only very slightly convex, so that the animal is extremely thin in relation to its size. In my own experience, the flat shell is lowermost, but it has been stated elsewhere that the slightly convex valve is lowermost, thus raising the shell gape slightly above the substratum, so reducing the danger of entry of unwanted mud, etc., from the substratum. Another species, *P. sella*, has a shell which differs in being irregularly saddle-shaped, the two shell valves fitting each other very closely. In *Placuna* the monomyarian condition, with a centrally placed posterior adductor muscle, suggests that the genus has passed through an earlier formative period when it was attached to a firm substratum by a byssus. Such a byssal attachment might have been of the type characteristic of the Anomiidae, through an embayment of the lower shell valve, but if so there now remains no trace of this earlier condition in the adult *Placuna*. If there were an earlier "anomiiform" stage in the racial history of *Placuna*, then it follows that this lineage has become secondarily adapted to its present mode of life among the epifauna of tropical deposit beaches. It is not known by what growth processes the byssal notch would have been obliterated. In the absence of any byssus pillar, and if the foot itself is inactive from an early stage in ontogeny, there will be no cause for localised abrasion at the site which normally bears the byssal embayment. Under these circumstances the secretory activity of the margin of the mantle at this point would presumably be the same as in adjacent areas, and there would be normal increments to the shell at this point. If this were true, absence of the byssus pillar and inactivity of the foot would themselves be sufficient explanation for the disappearance of the byssal notch in *Placuna*. As in *Pecten maximus*, there is an approach to radial symmetry also in the free-living genus *Placuna* (see Fig. 76, p. 202).

Among the Tridacnidae, the monotypic genus *Hippopus* alone is unattached and lies freely upon the surface of the reef flat. As compared with the genus *Tridacna*, there are two major modifications in the growth form of *Hippopus*, adapting it to this free-living existence. In *Tridacna* the ventral margin of the shell valves is not straight, but decidedly concave, thereby providing an aperture through which the substantial mass of the byssus can protrude. In *Hippopus* there is no byssus and if this gape were to remain, it would provide a wide gap in the animal's defences whereby carnivorous worms, etc., could enter and destroy the *Hippopus*. There is no such danger in fact, for the edges of the two shell valves are straight and meet each other exactly along the ventral border. It seems probable that the presence of a large mass of byssus threads in *Tridacna* will necessarily cause some regular abrasion of the adjacent margins of the shell valves and so slow down the rate of accretion to the shell in this region as the animal grows, thereby automatically providing a byssal gape of appropriate dimensions. In *Hippopus* the lack of activity by the byssus gland automatically removes any local interference with the rate of growth of the ventral edges of the shell valves, so these meet each other and the byssal gape is thus obliterated. In other words it is argued that in the Tridacnidae, as also in the Anomiidae, the typical byssal notch is directly dependent in part on the mode of life of the animal, not solely on the genetical constitution.

Having no form of attachment to the substratum, *Hippopus* is liable to be rolled about by the surf which rushes over the reef flat at certain stages of tide. Since *Hippopus*, like *Tridacna*, benefits from a growth of symbiotic zoo-xanthellae in the mantle tissues which are exposed when the shell valves are widely parted, it is important that the shell should lie with these tissues uppermost and fully exposed to the tropical sun. The shell is elongated in the antero-posterior axis and approximates to being circular in cross section; it will therefore tend to roll in the transverse plane under wave action. The mantle margin does not overlap the edge of the shell, and the shell does not possess crescentic scalloped "ornamentations" as in *T. squamosa*, so there is consequently little hindrance to lateral rolling of the shell of *Hippopus* due to wave action. The risk of being rolled and left upside down is obviated in an extremely effective way by particularly heavy depositions of nacre in the interior of the umbonal part of the shell. This provides an automatic righting system, for the centre of gravity of the animal is very low and, no matter how much the shell is rocked by the waves, the shell always comes to rest with the umbones and hinge line lowermost, and with the photosynthetic mantle tissues uppermost.

So far in the examples we have discussed there has been no difficulty in apprehending the significance of the adaptive changes noted. In some cases we have been concerned with the execution of a particular mechanical activity, and the adaptation concerns the principles of mechanics: thus the interrelations of adductor muscles and the shell in the Pholadidae is a matter of levers, and of mechanical advantages. In other cases the "activity" that is being exploited is a passive one, e.g. the farming of zoo-xanthellae in the mantle tissues in the Tridacnidae and here too the advantage gained is easy to envisage. Yet there are other, and important, branches in the adaptive radiation of the filter-feeding bivalves which are not so obvious, or so well understood. There may be striking variations in form, in the reproductive processes, or in the life history, the significance of which may be far from clear. Again, several lineages of bivalves have successfully invaded freshwater systems, and we are ignorant of the sequence of physiological changes, and changes in the processes of reproduction, that were involved in these invasions. We will now give brief consideration to such cases.

Commensalism

In the Erycinacea there are two major variations in form and habit, but these appear to be unconnected. In some genera the mantle is reflected over the shell to greater or lesser extent, and the shell may be reduced in size and strength. In *Kellia laperousii* the mantle margin extends over and partly covers the edges of the shell valves and is developed to form a long *anterior* inhalant siphon and a short posterior exhalant siphon. All these mantle tissues can be withdrawn into the shell. *K. laperousii* is modified to suit its mode of life, attached by byssus threads in the shelter of deep empty burrows made by piddocks (Pholadidae) (Yonge, 1952a). Again, *Scintilla deshayesi* possesses a lightly built oval shell which is normally overlaid both anteriorly and posteriorly by dark brown reflections of the mantle, and a small number of long and mobile pallial tentacles tend to obscure the underlying bivalve form. This species occurs on the under sides of littoral boulders near low-water mark on tropical beaches. The foot is extremely mobile and the animal can move about in rather convulsive jerks, partly by vigorous movements of the foot and partly by sudden ejection of jets of water from the mantle cavity. *S. timorensis* possesses similar extensions of the mantle bearing numerous small yellowish papillae. Both of these species, but particularly the latter, tend to be social and commonly occur in small clusters. In *Devonia* the shell is similarly covered with mantle tissues, but here the shell valves are held widely apart and cannot be completely closed. In *Phlyctaenachlamys* the shell is wholly enclosed by permanent extensions of the mantle and has become secondarily internal in position and is much reduced in size (Popham, 1939). The significance of enclosure of the shell by the mantle and reduction of the shell valves is not known.

A second feature of interest in the Erycinacea is the tendency towards commensalism. While some members of the order are free living, e.g. *Kellia, Lasaea, Scintilla,* and *Galeomma,* other genera are commensal with a variety of invertebrate forms. Thus *Montacuta ferruginosa* occurs in the sand close to the anal area of the heart urchin, *Echinocardium cordatum.* Similarly *M. substriata* is normally found attached by two or three byssus threads to the anal spines of *Spatangus purpureus* (Popham, 1940). *Mysella bidentata* occurs buried in mud with the ophiuroid *Ophiocnida brachiata,*

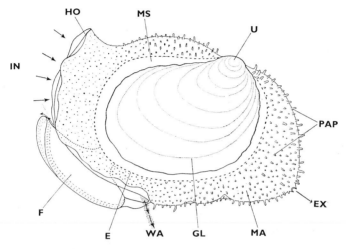

FIG. 75. *Devonia perrieri,* seen from the left side. The shell valves are partly covered by reflections of the mantle (MA) which is covered with papillae (PAP). The inhalant water current (IN) is anterior, and the exhalant water current (EX) is posterior. The foot (F) bears a broad, flat creeping sole by which the bivalve is attached to the skin of the holothurian *Leptosynapta inhaerens.* For interpretation of other lettering, see pp. 201–4. (Originally published in Popham, 1940, *J. mar. biol. Ass. U.K.* **24,** 573, fig. 19.)

but is apparently never attached to it. *Devonia perrieri* also occurs buried in mud, and is normally found crawling about on the skin of the holothurian *Leptosynapta inhaerens,* the foot of this bivalve being unusual in possessing a broad flat creeping sole (see Fig. 75). *Phlyctaenachlamys* is commensal with the stomatopod *Lysiosquilla maculata,* usually occurring attached by byssus threads to the walls of the long and deep burrow of the stomatopod (Popham, 1939). These commensal bivalves occur in a variety of habitats; they are either buried in clean sand or in mud or are attached to the smooth wall of a burrow; they are either attached to the body of the commensal partner or they lie separately but close at hand. Many, but not all, are associated with echinoderm partners. Thus there are two trends in the order Erycinacea—one towards enclosure of the shell valves by the mantle and the other towards commensal association; these trends do not seem to be correlated, nor is it clear in what ways the individual genera may benefit from either trend.

The Erycinacea are unique among filter-feeding bivalves in possessing an anterior inhalant orifice or inhalant siphon. Popham (1940) who reviewed the group (under the title Leptonacea) considered this to be a reversion to the primitive condition, but it could alternatively be regarded as the retention of a primitive state. *Montacuta substriata* and *M. ferruginosa* are not invariably found in association with *S. purpureus* and with *Echinocardium cordatum* respectively, but each may be found with a number of other species of echinoid. These bivalves can detect the presence of echinoids, presumably as a result of chemical emanations passing from the echinoid into the water. Specimens of *Montacuta* spp. which are isolated in aquaria tend to become quiescent, but when appropriate echinoids are introduced into the aquaria the bivalves soon exhibit pedal activity. Experiments in the laboratory demonstrated that both species tend to move upstream with lateral deviations, this being characteristic rheo-klino-taxism; both *M. substriata* and *M. ferruginosa* tended to aggregate in the vicinity of the echinoids due to delay in leaving the area once this had been reached—they displayed a positive chemo-ortho-kinesis. The anterior inhalant siphon is well suited to this form of behaviour, since it enables the animal to orient itself so as to move towards the source of the chemical attractant. Both species of *Montacuta* exhibited geotaxism, but this differed in sign. *M. substriata* exhibited negative geotaxism, while *M. ferruginosa*, which lives at a greater depth in the substratum with *E. cordatum*, showed a positive geotaxism (Gage, 1966, 1966a).

Commensalism is not found solely in the order Erycinacea, but is also known for *Cryptomya californica*, a member of the Myacea. This species lives commensally within the burrows usually of *Callianassa californiensis*, but sometimes also of *Upogebia*, or of the echiuroid, *Urechis caupo*. *Cryptomya* occurs at a depth of up to 50 cm, but as it draws its water supply from the burrow of its host it possesses a short siphonal process and no pallial sinus; it would be impossible to judge from its anatomy that it is such a deep burrower (Yonge, 1951) (see Fig. 76, p. 202).

Parasitism

Parasitism may perhaps be the final outcome of commensalism; an example of endo-parasitism is found in *Entovalva mirabilis* (Montacutidae) which occurs in the gut of a holothurian (Popham, 1940).

Reproductive Processes, Including Hermaphroditism

This subject will be treated more thoroughly in another chapter, but it is necessary to record here that significant changes have occurred in the reproductive system, in methods of dispersal, and in life history, and that these are relevant to a study of adaptive radiation of the filter-feeding bivalves.

There is some evidence to support the view that in various phyla the individual has a basic capacity to function both as a male and as a female, but that this fundamental totipotency is modified as a general rule to guide the individual either into male or into female condition. This guidance may be provided by the genetical constitution, or perhaps in some cases by environmental stimuli. As a result of this "guidance", we consider the normal condition of modern bivalves to be dioecious. Further, the spermatozoa and ova are normally broadcast, and fertilisation occurs at random in the surrounding water. In many cases there are striking variations from this norm. Thus the hermaphrodite condition is found in certain of the Pectinidae, e.g. *Pecten maximus*, and also in many, if not all of the Anomalodesmata, e.g. *Laternula* and *Brechites*. In these cases the ovary and the testis develop at approximately the same time, and form complementary halves of the gonad.

In many other cases the individual is a protandric hermaphrodite, and sex reversal occurs possibly more than once in the life of the individual; this is true of some but not all species of *Ostrea*, e.g. *O. virginica*, in some species of *Teredo*, and in *Xylophaga dorsalis* (Purchon, 1941). Whether hermaphroditism or sex reversal lead to self-fertilisation, or to any increase in the percentage of ova which become fertilised, is a matter for conjecture.

In some bivalves the eggs are retained in the suprabranchial chamber, are fertilised there, and the young developing embryos are incubated there for some time. This is true for many members of the Erycinacea, e.g. *Lasaea*, *Kellia*, and *Galeomma* which are free living, and for the commensal *M. ferruginosa* (Popham, 1940). This is also true of some, but not all species of *Ostrea* and also of *Teredo*. Enormous numbers of glochidia larvae are incubated in the outer demibranch of *Anodonta*, while in *Sphaerium* it is considered that the small number of larvae which are retained in the gill are actually nourished by maternal secretions. In some of the most ancient fresh water lineages the part of the ctenidium which bears the developing embryos is modified considerably to form a marsupium. Little seems to be known of the functional benefits of the development of such a marsupium. *Xylophaga* appears to be unique in possessing a receptaculum seminis, which lies in the suspensory membrane of the ctenidium, close to the orifice of the genital duct. Spermatozoa are stored here and it is presumed that this will ensure fertilisation of the ova which are produced at the end of the female phase (Purchon, 1941). In freshwater forms the free-

living larval stage is suppressed, the few young of *Sphaerium* being liberated as small adults. In the Unionidae the glochidia larvae, when liberated, become ecto-parasites upon the skin of fish for a period, thereby ensuring that the species is widely distributed through the freshwater system
in question. In the Mutelidae also the larvae are ectoparasitic on fish, but
here the larvae are extremely modified in form (Fryer, 1961).

There may be a very broad correlation between these various manifestations and the exigencies of the environments in question. Oysters are
gregarious, and favour certain specialised settling grounds, often in estuaries or near the mouths of rivers; the liberation of larvae at an advanced
stage of development with the capacity to settle after a relatively short
planktonic life may enable large numbers of larvae to settle on the parental breeding grounds. Liberation of unfertilised eggs would involve a far
longer planktonic larval phase, wider distribution, and failure of the majority of larvae to concentrate and settle in close proximity to their parental
grounds. Clearly incubation of larvae is advantageous in the case of the
oyster, thereby maintaining the formation of densely colonised oyster
grounds and the prospect of a high degree of fertilisation in future years.
The same may be said of ship-worms; if the larvae are able to settle upon
wood almost immediately after they have been liberated from the parental
gill chamber, there will be the maximal colonisation of any stray piece of
timber which becomes infected with ship-worms. This in turn will ensure
maximal fertilisation and a large output of larvae. It is not surprising that
the most ancient of freshwater bivalves should incubate their larvae. Suppression of a free-swimming larval phase is advantageous in river systems
where otherwise the larvae would be swept downstream.

The Invasion of Freshwater Systems

At least four major lineages and a number of minor lineages of bivalves
have invaded freshwater systems. The most ancient of these are cosmopolitan in distribution, their origin as freshwater forms, and their dispersal
through the freshwater systems of the whole world having preceded the
subdivision of the land mass into separate continents. Thus *Pisidium* and
Sphaerium in the family Sphaeriidae are found in fresh waters in all continents. Great antiquity is also revealed by the development of special
adaptations, for example in the family Mutelidae the peculiar "haustorius"
larva is an ecto-parasite of fish (Fryer, 1961), and the less extensively
modified glochidia larvae of the family Unionidae are also ecto-parasites
of fish; the development of a portion of the ctenidium as a special organ,
for the incubation of larvae, the marsupium, is another indication of the
great antiquity of certain freshwater groups. It may be presumed that the
Unionacea were among the first to colonise fresh waters and the large
number of genera in this order (110) indicates the extent of this branch

of bivalve adaptive radiation, as well as the importance of isolation and of lack of competition in adaptive radiation (See Fig. 76, p. 202).

In contrast the less highly modified *Dreissena polymorpha* (Dreissenacea) is the only freshwater bivalve which is known to have a free-swimming veliger larva; *Egeria radiata* (Donacidae) is confined to relatively short lengths of certain West African rivers, immediately above the main region of salt water penetration, and may possibly be dependent on increased salinity for breeding purposes (Purchon, 1964); These examples may be regarded as comparatively recent additions to the fresh water fauna.

Other genera to have become established in fresh water systems include several members of the Mytilidae (*Brachyodontes* sp., *Limnoperna* sp., *Sinomytilus* sp.); several members of the Solenacea (*Tanysiphon rivalis* and *Novaculina gangetica* from India and the Far East, and *Glaucomya chinensis*); Three genera in the Cardiidae—*Monodacna*, *Didacna*, and *Adacna*—occur in the Caspian Sea; one pholad, *Martesia rivicola*, and one ship-worm, *Teredo minima*, are also reported from fresh waters.

Nothing is known of the sequence of events in the physiological adaptation of these various lineages, which enabled them first to invade fresh waters and gradually to adapt themselves wholly to this new environment.

Key to the Lettering on the Figures

A—P	Antero-posterior axis.	CM	Cut edge of mantle.
A	Anus.	CP	Ciliated pockets at entrance
AA	Anterior adductor muscle.		to the wood-storing caecum.
ACC	Accessory anterior adductor	CT	Connective tissue.
	muscle.	DD	Digestive diverticula.
ALOD	Ascending lamella of outer	DLOD	Descending lamella of outer
	demibranch.		demibranch.
AM	Anterior margin of shell.	DOG	Distal oral groove.
AP	Apophysis.	E	Edge of shell.
APD	Anterior perforate disc of	EO	Excretory organ.
	shell.	EP	External layer of periostra-
APR	Anterior pallial retractor mus-		cum.
	cle.	ES	Exhalant siphon.
APS	Anterior pallial septum.	EX	Exhalant water current.
AR	Anterior retractor muscle of	F	Foot.
	foot.	FG	Food groove at the upper
AT	Anterior tentacles.		margin of the ascending la-
AVT	Antero-ventral tentacles.		mella of the outer demibranch.
A—X	Antero-posterior axis.	G	Gonad.
B	Byssus threads.	GC	Gland cells.
BE	Beak.	GI	Groove on the inner surface
BRM	Retractor muscle of byssus		of lobe of lower lip.
	apparatus.	GL	Growth lines on shell.
C	Callum.	HO	Hood.
CA	Ctenidial axis.	H—X	Hinge axis.
CC	Ciliary cleansing current.	HYX	Hypoplax.
CH	Chondrophore.	ID	Inner demibranch.

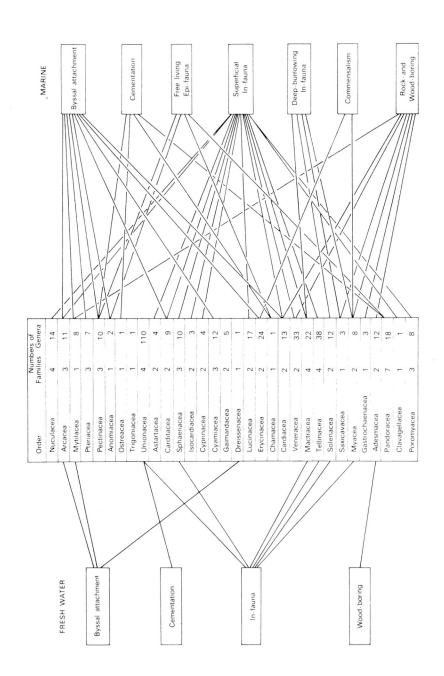

FIG. 76. Modes of life and habitats of bivalves. The classification and numbers of constituent families and genera are taken from Thiele (1935), the term "order" having been substituted for the "stirps" of Thiele. It must be appreciated that classification must inevitably remain in a state of flux and that in various important respects opinions have changed since the publication of the monograph of Thiele. Various authoritative works have recommended changes in the composition of certain orders, while many of Thiele's genera have been subdivided; thus we now admit more than one genus in the Ostreacea. The data shown graphically here are confined to examples quoted in the chapter, for any attempt to deal with all the orders would render the figure too complicated to be useful. Attention is drawn to the following major features: (1) Convergent, or parallel evolution, whereby a number of independent orders have given rise to animals with comparable modes of life, e.g. parallelism in the case of byssal attachment; e.g. convergence in the case of rock boring. (2) The majority of marine orders comprise comparatively few genera, the Tellinacea (38), Veneracea (33), Erycinacea (24), and Mactracea (22) being exceptional; here the main phase of adaptive radiation must have been very ancient, and led to the establishment of the great variety of marine orders. (3) In several orders more than one mode of life is to be found, e.g. the Cardiacea includes examples in the superficial infauna (*Cardium*), others which are byssally attached (*Tridacna*), rock boring (*T. crocea*), and even free living (*Hippopus*); this indicates the occurrence of a second phase of adaptive radiation. (4) The enormous number of freshwater genera (110) in the order Unionacea; this underlines the evolutionary importance of entry into a new habitat, previously unexploited by the biological model which we know as "bivalve".

IE	Inner epithelium of the outer mantle fold.	MP	Middle layer of periostracum.
IM	Inner fold of mantle edge.	MS	Mantle, reflected over the shell.
IN	Inhalant water current.		
IP	Internal layer of periostracum.	MTX	Metaplax.
		M−X	Median axis.
IS	Inhalant siphon.	N−X	Normal axis.
L	Ligament.	O	Oesophagus.
LL	Lower lip.	OD	Outer demibranch.
LLL	Lobe of lower lip.	OF	Outer fold of mantle margin.
LOB	Lobule.	OMM	Outer and middle folds of mantle margin.
LOG	Lateral oral groove.		
LP	Labial palps.	P	Periostracum.
LUL	Lobe of upper lip.	PA	Posterior adductor muscle.
M	Mouth.	PAP	Papillae.
MA	Mantle.	PC	Pericardium.
MC	Mantle cavity.	PG−PG	Limits of pedal gape.
ME	Mantle edge.	PGR	Periostracal groove.
MEX	Mesoplax.	PL	Pallial line.
MF	Part of the mantle, folded outwards.	PM	Pallial muscles operating the anterior pallial septum.
MFG	Marginal food groove.	PO	Pore.
MFM	Middle fold of mantle margin.	POG	Proximal oral groove.
MG	Opening of the mid-gut into the stomach.	PP	Line of insertion of periostracum on to inner surface of shell.

PPR	Posterior pallial retractor muscle.	SX	Siphonoplax.
PR	Posterior retractor muscle of foot.	T	Hollow, tubular prolongations of the shell.
PRM	Pallial retractor muscles.	U	Umbo.
PS	Pallial sinus.	UL	Upper lip.
PT	Posterior tentacles.	UR	Umbonal reflection.
PVL	Postero-ventral tentacles of left side.	UV	Umbonal-ventral sulcus.
		UVR	Umbonal-ventral ridge.
PVR	Postero-ventral tentacles of right side.	U—X—V	Line separating anterior and posterior halves of the body.
PX	Protoplax.	V	Valve at the proximal end of the mid-gut.
RC	Right ctenidium.	VAK	Ventral articular knob.
RGC	Ring of gland cells.	VAM	Ventral adductor muscle.
RILP	Right inner labial palp.	VM	Visceral mass.
ROLP	Right outer labial palp.	VP	Vacuolated layer of periostracum.
RT	Rejection tract.		
S	Stomach.	VPA	Vestigial pedal aperture.
SC	Supra-branchial chamber.	W	Wood-storing caecum.
SP	Siphonal process.	WA	Waste material.
SR	Shell, eroded at anterior end.	Z	Substratum.
SS	Style sac.		

Reference List

ALLEN, J. A. (1953) Observations on the epifauna of the deep-water muds of the Clyde Sea area, with special reference to *Chlamys septemradiata* (Muller), *J. Anim. ecol.* **22**, 240–60.

ALLEN, J. A. (1954) On the structure and adaptations of *Pandora inaequivalvis* and *P. pinna*, *Quart. J. micr. Sci.* **95**, 473–82.

ALLEN, J. A. (1958) On the basic form and adaptations to habitat in the Lucinacea (Eulamellibranchia), *Phil. Trans.* B, **241**, 421–84.

ALLEN, J. A. (1958a) Observations on *Cochlodesma praetenue* (Pulteney) Eulamellibranchia, *J. mar. biol. Ass. U.K.* **37**, 97–112.

ANSELL, A. D. (1961) The functional morphology of the British species of Veneracea (Eulamellibranchia), *J. mar. biol. Ass. U.K.* **41**, 489–515.

ATKINS, D. (1937) On the ciliary mechanisms and interrelationships of Lamellibranchs. Part V. Note on the gills of *Amussium pleuronectes*, *Quart. J. micr. Sci.* **80**, 321–9.

BALLANTINE, D. and MORTON, J.E. (1956) Filtering, feeding and digestion in the lamellibranch *Lasaea rubra*, *J. mar. biol. Ass. U.K.* **35**, 241–74.

BEEDHAM, G. E. (1958) Observations on the non-calcareous component of the shell of the lamellibranchia, *Quart J. micr. Sci.* **99**, 341–57.

BRAFIELD, A. E. and NEWELL, G. E. (1961) The behaviour of *Macoma balthica* (L.), *J. mar. biol. Ass. U.K.* **41**, 81–87.

CARRIKER, M. R. (1956) Biology and propagation of young hard clams, *Mercenaria mercenaria*, *J. Elisha Mitchell sci. Soc.* **72**, 57–60.

CHAPMAN, G. and NEWELL, G. E. (1956) The role of the body fluid in the movement of soft-bodied invertebrates. II. The extension of the siphons of *Mya arenaria* L. and *Scrobicularia plana* (da Costa), *Proc. roy. Soc.* B, **145**, 564–80.

DREW, G. A. (1907) The habits and movements of the razor shell clam, *Ensis directus* Con., *Biol. Bull. Woods Hole* **12**, 127–40.

EVANS, J. W. (1966) The ecology of the rock-boring clam *Penitella penita* (Conrad,1837) Ph.D. thesis, University of Oregon.

FRYER, G. (1961) The developmental history of *Mutela bourguignati* (Ancey) Bourguignat (Mollusca, Bivalvia), *Phil. Trans.* B, **144**, 259–98.

GAGE, J. (1966) Observations on the bivalves *Montacuta substriata* and *M. ferruginosa* "commensals" with spatangoids, *J. mar. biol. Ass. U.K.* **46**, 49–70.

GAGE, J. (1966a) Experiments with the behaviour of the bivalves *Montacuta substriata* and *M. ferruginosa* "commensals" with spatangoids, *J. mar. biol. Ass. U.K.* **46**, 71–88.

GILMOUR, T. H. J. (1963) A note on the tentacles of *Lima hians* (Gmelin), *Proc. malac. Soc. Lond.*, **35**, 81–85.

GILMOUR, T. H. J. (1964) The structure, ciliation and function of the lip-apparatus of *Lima* and *Pecten* (Lamellibranchia), *J. mar. biol. Ass. U.K.* **44**, 485–98.

GRAVE, B. H. (1909) Anatomy and physiology of the wing-shell *Atrina rigida*, *Bull. U.S. Bur. Fish.* **29**, 409–439.

HUNTER, W. R. (1949) The structure and behaviour of *Hiatella gallicana* (Lamarck) and *H. arctica* (L.) with special reference to the boring habit, *Proc. roy. Soc. Edinb.* **63**, 271–89.

KELLOGG, J. L. (1915) Ciliary mechanisms of lamellibranchs with descriptions of anatomy, *J. Morph.* **26**, 625–701.

MORTON, J. E. (1956) The tidal rhythm and action of the digestive system of the lamellibranch *Lasaea rubra*, *J. mar. biol. Ass. U.K.* **35**, 563–8.

OLDFIELD, E. (1955) Observations on the anatomy and mode of life of *Lasaea rubra* (Montagu) and *Turtonia minuta* (Fabricius), *Proc. malac. Soc. Lond.* **31**, 226–49.

OWEN, G. (1952) Shell form in the Lamellibranchia, *Nature, Lond.* **170**, 148.

OWEN, G. (1953) On the biology of *Glossus humanus* (L.) (*Isocardia cor* Lam.), *J. mar. biol. Ass. U.K.* **32**, 85–106.

OWEN, G. (1953a) The shell in the Lamellibranchia, *Quart. J. micr. Sci.* **94**, 57–70.

POPHAM, M. L. (1939) On *Phlyctaenachlamys lysiosquillina* gen. and sp. nov., a lamellibranch commensal in the burrows of *Lysiosquilla maculata*, *Sci. Rep. Gr. Barrier Reef Exped.* **6**, 62–84.

POPHAM, M. L. (1940) The mantle cavity of some of the Erycinidae, Montacutidae and Galeommatidae, with special reference to the ciliary mechanisms, *J. mar. biol. Ass. U.K.* **24**, 549–87.

PURCHON, R. D. (1941) On the biology and relationships of the lamellibranch *Xylophaga dorsalis* (Turton), *J. mar. biol. Ass. U.K.* **25**, 1–39.

PURCHON, R. D. (1954) A note on the biology of the lamellibranch *Rocellaria* (*Gastrochaena*) *cuneiformis* Spengler, *Proc. zool. Soc. Lond.* **124**, 17–33.

PURCHON, R. D. (1955) The functional morphology of the rock-boring lamellibranch *Petricola pholadiformis* Lamarck, *J. mar. biol. Ass. U.K.* **34**, 257–78.

PURCHON, R. D. (1955a) The structure and function of the British Pholadidae (rock-boring Lamellibranchia), *Proc. zool. Soc. Lond.* **124**, 859–911.

PURCHON, R. D. (1955b) A note on the biology of *Tridacna crocea* Lam. *Proc. malac. Soc. Lond.*, **31**, 95–110.

PURCHON, R. D. (1956) A note on the biology of *Brechites penis* (L.) (Lamellibranchia), *J. Linn. Soc.* **43**, 43–54.

PURCHON, R. D. (1956a) A note on the biology of *Martesia striata* L. (Lamellibranchia), *Proc. zool. Soc. Lond.* **126**, 245–58.

PURCHON, R. D. (1960) A further note on the biology of *Brechites penis* (L.) (Lamellibranchia), *Proc. malac. Soc. Lond.* **34**, 19–23.

PURCHON, R. D. (1960a) The stomach in the Eulamellibranchia, stomach types IV and V. *Proc. zool. Soc. Lond.*, **135**, 431–89.

QUAYLE, D. B. (1949) Movements in *Venerupis (Paphia) pullastra* (Montagu), *Proc. malac. Soc. Lond.* **28**, 31–37.

REES, W. J. (1957) The Living Scallop. *The Scallop; Studies of a Shell and its Influence on Humankind*, pp. 15–32, The Shell Transport and Trading Co., London.

STASEK, C. R. (1962) The form, growth and evolution of the Tridacnidae (Giant Clams), *Arch. Zool. exp. gén.* **101**, 1–40.

STASEK, C. R. (1965) Behavioral adaptation of the giant clam *Tridacna maxima* to the presence of grazing fishes, *Veliger*, **8**, 29–35a.

STASEK, C. R. (1966) The eye of the giant clam *(Tridacna maxima)*. *Occ. Pap. Calif. Acad. Sci.*, **58**, 1–9.

THIELE, J. (1935) *Handbuch der systematischen Weichtierkunde*, Teil 3, Fischer, Jena.

TRUEMAN, E. R. (1954) Observations on the mechanism of opening of the valves of a burrowing lamellibranch, *Mya arenaria, J. exp. Biol.* **31**, 291–305.

TRUEMAN, E. R., BRAND, A. R. and DAVIS, P. (1966) The dynamics of burrowing of some common littoral bivalves, *J. exp. Biol.* **44**, 469–92.

TURNER, R. D. (1954) The family Pholadidae in the Western Atlantic and the Eastern Pacific. Part I. Pholadinae, *Johnsonia* **3**, 1–64.

TURNER, R. D. (1955) The family Pholadidae in the Western Atlantic and the Eastern Pacific. Part 2. Martesiinae, Jouannetiinae and Xylophaginae, *Johnsonia* **3**, 65–160.

YONGE, C. M. (1936) The evolution of the swimming habit in the Lamellibranchia, *Mem. Mus. Hist. nat. Belg.*, ser. 2, **3**, 78–100.

YONGE, C. M. (1936a) Mode of life, feeding, digestion and symbiosis with zooxanthellae in the Tridacnidae, *Sci. Rep. Gt. Barrier Reef. Exped.* **1**, 283–321.

YONGE, C. M. (1946) On the habits and adaptations of *Aloidis (Corbula) gibba, J. mar. biol. Ass. U.K.* **26**, 358–76.

YONGE, C. M. (1948) Formation of siphons in Lamellibranchia, *Nature, Lond.* **161**, 198.

YONGE, C. M. (1949) On the structure and adaptations of the Tellinacea, deposit-feeding Eulamellibranchia, *Phil. Trans.* B, **234**, 29–76.

YONGE, C. M. (1951) Studies on Pacific Coast Molluscs. I. On the structure and adaptations of *Cryptomya californica* (Conrad), *Univ. Calif. Publ. Zool.* **55**, 395–400.

YONGE, C. M. (1951a) II. Structure and adaptations for rock-boring in *Platyodon cancellatus* (Conrad), *Univ. Calif. Publ. Zool.* **55**, 401–8.

YONGE, C. M. (1951b) III. Observations on *Hinnites multirugosus* (Gale), *Univ. Calif. Publ. Zool.* **55**, 409–20.

YONGE, C. M. (1952) IV. Observations on *Siliqua patula* Dixon and on evolution in the Solenidae, *Univ. Calif. Publ. Zool.* **55**, 421–38.

YONGE, C. M. (1952a) VI. A note on *Kellia laperousii* (Deshayes), *Univ. Calif. Publ. Zool.* **55**, 451–4.

YONGE, C. M. (1953) Form and habit in *Pinna carnea* Gmelin, *Phil. Trans.* B, **237**, 335–74.

YONGE, C. M. (1953a) The monomyarian condition in the Lamellibranchia, *Trans. roy. Soc. Edinb.* **62**, 443–78.

YONGE, C. M. (1953b) Mantle chambers and water circulation in the Tridacnidae (Mollusca), *Proc. zool. Soc. Lond.* **123**, 551–61.

YONGE, C. M. (1955) Adaptation to rock-boring in *Botula* and *Lithophaga* (Lamellibranchia, Mytilidae) with a discussion on the evolution of the habit, *Quart. J. micr. Sci.* **96**, 383–410.

YONGE, C. M. (1957) *Enigmonia aenigmatica* Sowerby, a motile anomiid (Saddle Oyster), *Nature, Lond.* **180**, 765–6.

YONGE, C. M. (1957a) Mantle fusion in the Lamellibranchia, *Publ. Staz. zool. Napoli* **29**, 151–71.

YONGE, C. M. (1959) On the structure, biology and systematic position of *Pharus legumen* (L.), *J. mar. biol. Ass. U.K.* **38**, 277–90.

YONGE, C. M. (1962) On the primitive significance of the byssus in the bivalvia and its effects in evolution, *J. mar. biol. Ass. U.K.* **42**, 113–25.

CHAPTER 5

DIGESTION

Synopsis

Both intracellular and extracellular digestion occur in the Mollusca. Intracellular digestion requires almost continuous feeding on minute food particles, and a gut possessing large surface areas capable of phagocytosing suitably small particles and efficient ciliary sorting mechanisms; this is thought to have been the primitive method of digestion in the Mollusca. The later development of extracellular digestion permitted the isolation of secretory cells into glandular masses opening by ducts into the gut, the development of muscle fibres in the gut wall, and the manipulation of larger masses of food by muscular action. Many molluscs exhibit a combination of preliminary extracellular digestion and subsequent phagocytosis and intracellular digestion. It is probable that primitively the gut possessed only one type of "multipurpose digestive cell" which was capable of phagocytosis, intracellular digestion, absorbtion, secretion, and excretion of indigestible matter and waste products. In some cases these multipurpose cells have been retained; each cell passes through a series of distinct phases of activity and consequently there appear to be several different kinds of cell in a histological preparation. In other cases the multipurpose digestive cells have been superseded by a number of distinct cells each of which has specialised in one or perhaps two functions.

In the Mollusca each class or order has become adapted to a particular mode of life, feeding and digestion playing a large part in this adaptive radiation, and it is therefore not possible to generalise about the digestive processes of the phylum or even about individual classes. It is therefore necessary to consider each class in turn, making appropriate subdivisions where necessary.

Nothing is known for certain regarding the digestive processes of the Aplacophora or of the Monoplacophora. Comparatively little is known of digestion in the Scaphopoda; digestion may be partly extracellular, by enzymes secreted by the oesophageal pouches and by the digestive diverticula. In the Scaphopoda the thin-walled stomach is capable of contractions which force the partly digested stomach contents into the lumina of the digestive diverticula, where the products of extracellular digestion are absorbed and where small particles are phagocytosed by the digestive cells. The tips of the digestive cells are nipped off and discharged into the stomach, carrying with them spherules of indigestible matter, and this may be the source of extracellular enzymes in the lumen of the stomach.

Among the Polyplacophora many are herbivorous and are well equipped with carbohydrases, but some polyplacophorans are rapacious carnivores, and it is to be expected that their digestive processes differ to some extent. Among herbivorous polyplacophorans the acidity of the stomach reduces the viscosity of the mucus to a minimum and permits easy mixing of the enzymes and the food. Extracellular amylases are derived from the oeso-phageal glands, and an extracellular protease is supplied by the digestive diverticula; the contents of the gut are manipulated by interaction of ciliary tracts and muscle fibres in the gut wall. Phagocytes may pass through the gut wall and ingest small particles, but apart from this digestion is extracellular and the products of digestion are absorbed by cells in the digestive diverticula. Due to the lack of a powerful cellulase large quantities of algal matter pass through the gut undigested, the only material to be digested being the contents of cells which were ruptured by the rasping action of the radula, and delicate micro-organisms which probably abound in the substrates scraped by the chitons. It must be emphasised that for any molluscs which swallow much detritus the nature of the gut contents is unlikely to indicate the nature of the real food material since it is princi-pally the indigestible material which remains available for inspection.

The Bivalvia are not homogeneous and care must be taken not to make unjustifiable generalisations on the basis of the comparatively few examples which have been studied. There are striking differences among the three constituent orders of the sub-class Protobranchia. In the order Nuculidae much of the interior of the stomach is lined by a chitinous girdle, the walls of the stomach are muscular, and the stomach serves as a gizzard. There is no firm crystalline style but instead there is a viscid mixture of secretions from the style sac and from the digestive diverticula together with material ingested from the substratum. Digestion is exclusively extracellular and the products of digestion are absorbed by the gastric epithelium and the epithelium of the anterior mid-gut. In the digestive diverticula spherules are nipped off the tips of the digestive cells, and these spherules are probably the source of some of the enzymes in the stomach, enzymes in the stomach also being supplied from the epithelium of the style sac; the stomach contains a free amylase, lipase, and a protease. As compared with the Nuculidae, conditions differ in the order Nucula-nidae as regards one of the masses of digestive diverticula. Small particles are passed up the duct leading to this one mass of digestive diverticula and here only, small particles are phagocytosed and are digested intracellularly. In the family Solenomyidae nothing is known regarding digestion; the whole alimentary canal, including the digestive diverticula, is extremely small and it is doubtful whether food is digested in the alimentary canal.

The Verticordiidae and the Septibranchia have adopted a carnivorous, or scavenging, mode of life; they feed on small crustaceans, etc. The highly muscular stomach is extensively lined by a chitinous sheath, and it serves as a crushing gizzard; digestible matter is squeezed out of any carcases lying in the stomach and is passed up the two ducts to the digestive diverticula where particles are phagocytosed and digested intracellularly. The relative lack of success of this carnivorous lineage may be attributed to the failure to evolve an extracellular protease.

The Polysyringia exploited the practice of feeding on minute particles suspended in the supernatant water by means of the ctenidia. The adaptive radiation of these bivalves involved diversification of the ciliary sorting mechanisms on the ctenidia and also within the stomach. Typically, the interior of the stomach is very complex, the purpose of the ciliary sorting areas being to dispatch coarse particles via the intestinal groove under the protection of the major typhlosole into the mid-gut, and to keep the remainder in circulation while it is subjected to preliminary extracellular digestion. The crystalline style rotates and helps to stir the stomach contents, to brush particles against the ciliary sorting areas, and to liberate extracellular amylases as it dissolves. In the main ducts to the digestive diverticula cilia are restricted to a gutter and they beat outwards, towards the stomach; particles in suspension are drawn into the main ducts in a countercurrent entering above the ciliated gutter, and a hierarchical system of grading ensures that only the smallest particles reach the acini of the digestive diverticula. These particles are ingested and subjected to intracellular digestion.

Extracellular enzymes operating in the lumen of the stomach are principally derived from the crystalline style, but in some cases may be supplied by the bursting of fragmentation spherules cut off from the digestive cells in the digestive diverticula. Phagocytes may pass through the gut wall, wander through the contents of the gut ingesting particles, and then return through the wall of the gut; the activity of these phagocytes under experimental conditions should not be mis-interpreted as the action of extracellular enzymes. Investigations on a variety of bivalves have generally shown the crystalline style to contain amylases, but in some cases a lipase, a cellulase, and even an exo-peptidase have been indicated. Similarly, investigations on the digestive diverticula of various bivalves have yielded various amylases, proteases, and a lipase, while a cellulase is active in a portion of the digestive diverticula of *Teredo*. Significant differences have been found between the digestive processes of *Lima* and of *Mya*, and it is important to recognise that there is much diversification in the digestive processes of different bivalve genera. *Lasaea* is unusual in showing a digestive rhythm linked with the tidal cycle, and in showing much extracellular digestion even of proteins, the enzymes being derived from fragmentation spherules from the digestive diverticula. In ship-worms, e.g. *Teredo*, small fragments of wood are digested intracellularly by a powerful cellulase in phagocytes and in the digestive cells in part of the digestive diverticula.

It seems probable that division of the Gastropoda into Prosobranchia, Opisthobranchia and Pulmonata is an over-simplification; the Opisthobranchia and Pulmonata are polyphyletic groups closely linked with their prosobranch ancestry by early offshoots of problematic affinity. The adaptive radiation of early gastropods was largely concerned with feeding and digestion, with much parallel evolution especially between prosobranchs and opisthobranchs. It is therefore most practical in this chapter first to consider the primitive herbivorous condition and then to turn to the more specialised carnivorous state, disregarding formal classification of the Gastropoda.

The most primitive herbivorous gastropods feed on minute particles of detritus and algal material, which are first subjected to extracellular digestion by amylases supplied by the oesophageal pouches. The final phase is of phagocytosis and intracellular digestion in the digestive diverticula. In the primitive marine pulmonate, *Otina*, there is preliminary extracellular digestion by enzymes supplied by fragmentation spherules discharged from the digestive diverticula, but digestion is principally intracellular in the digestive cells of the digestive diverticula. In primitive herbivorous prosobranchs the interior of the stomach bears extensive ciliary sorting areas, and in some cases there is a protostyle which differs from a crystalline style in lacking adsorbed enzymes and in being continuous posteriorly with the faecal cord. A crystalline style has arisen from some such origin in four super-families of mesogastropods, e.g. *Aporrhais*, this being an interesting example of parallel evolution with the Bivalvia. In these style-bearing prosobranchs there are no oesophageal glands.

More advanced herbivores nibble larger pieces of plant material and may possess a weak cellulase, e.g. *Aplysia*, or a powerful cellulase, e.g. *Pterocera*, *Lymnaea*, and *Helix*. Where large pieces of leaf are swallowed grit may be an essential part of the diet, being an essential aid in the trituration of food in the muscular gizzard, e.g. *Pomacea* and *Lymnaea*. Here the stomach is more simple, and more muscular, and digestion is largely or wholly extracellular; absorbtion of the products of digestion occurs in the digestive diverticula. Generally the faeces are well consolidated, this being necessary to avoid risk of contamination of the mantle cavity. Some herbivores feed by slitting the cell walls of plants and sucking up the cell contents, e.g. *Elysia*; a muscular suction pump is required, but in other respects the gut is much simplified in such examples. The pulmonates are generally highly specialised as herbivores and are well supplied with extracellular enzymes, especially a powerful cellulase. In the Stylommatophora most of these extracellular enzymes are supplied by the salivary glands.

It is a simple transition from rasping algae from the surfaces of rocks to browsing on sponges, and it is not surprising that the latter habit should

have arisen in various primitive prosobranchs, e.g. *Diodora* and opisthobranchs, e.g. *Archidoris* and *Jorunna*. In *Diodora* digestion is largely
extracellular by enzymes derived from the oesophageal pouches and the
digestive diverticula; the indigestible sponge spicules are compacted with
mucus into a faecal rod which passes backwards into the mid-gut. In
Archidoris there is a preliminary phase of extracellular digestion by enzymes from the salivary glands and digestive diverticula, and a subsequent
phase of intracellular digestion in the cells of the digestive diverticula.
Jorunna differs markedly in that there are no salivary glands, and no extracellular digestion; in *Jorunna* all digestion is intracellular in the digestive
cells and phagocytes of the digestive diverticula. This difference between
two dorids is quite remarkable, and calls for further investigations on
related species and genera. Various mesogastropods, e.g. *Cerithiopsis*,
also feed on sponges, and here digestion is at least partly extracellular.

The eolids are carnivorous, and most feed on hydroid coelenterates.
Digestion is extracellular, and the contents of the gut are moved about by
muscular contractions of the gut wall; the soluble products of digestion
are absorbed by the epithelia of the stomach and of the digestive diverticula. Nematocysts are passed into the digestive diverticula, which extend
into the cerata, and most nematocysts are digested here. In *Aeolis pilata*
only one type of nematocyst resists digestion, and such nematocysts are
passed into the cnidosacs at the tips of the cerata, to be used in defense.
The eolid *Calma*, in contrast, feeds on the highly digestible contents of
fish eggs and the stomach is modified as a food store; the gut lacks an
anus and intestine. Many prosobranchs and opisthobranchs are predacious
carnivores, e.g. *Philine*, which feeds on small molluscs and other organisms
in the sand. Such food is crushed in a powerful gizzard; extracellular
digestion by enzymes from the salivary glands commences in the gizzard,
and continues in the stomach with the addition of enzymes from the digestive diverticula. Small particles may be ingested and subjected to intracellular digestion in the cells of the digestive diverticula, but digestion is
principally extracellular. In such predacious carnivores the stomach is
not concerned with the grading of particles by ciliary sorting areas, and
the stomach is greatly simplified in comparison with that of a microphagous herbivore. The contents of the gut are manipulated by peristalsis.

In parasitic forms, e.g. the Pyramidellidae, which feed on the body fluids
of their hosts, little digestive action is required, and the gut is greatly
simplified as in the herbivorous Elysiidae which feed on the contents of
plant cells.

The adaptive radiation of gastropods has involved a wide variety of
feeding mechanisms, food preferences, and digestive processes, and it is
not possible to designate one or two "typical" examples. The most primitive condition is found in microphagous feeders in which the stomach is
complicated by the presence of ciliary sorting mechanisms and digestion

is initially extracellular and subsequently intracellular; the gut contents are moved about by ciliary action. In more specialised types, both herbivorous and carnivorous, larger food masses are ingested, the stomach is simplified by the loss of ciliary sorting mechanisms, the gut contents are manipulated by muscular contractions, and digestion is predominantly or wholly extracellular. Each lineage in the adaptive radiation of the Gastropoda has evolved in its own direction along these lines to suit its own particular needs.

All the dibranchiate cephalopods are predacious carnivores feeding on fish, crabs, or prawns, and they possess the extremely efficient digestive systems necessary to support their very active mode of life. In those few examples which have been studied thoroughly digestion is exclusively extracellular; comparatively large pieces of meat are swallowed by peristalsis, and the digestive process is rhythmic and discontinuous. The prey is bitten and is killed by poison secreted by the posterior salivary glands. The oesophagus and the muscular stomach are lined by a chitinous sheath, and the stomach serves as a gizzard. There is a preliminary period of gastric digestion by an accumulation of enzymes secreted by the "pancreas", after which the fluid mixture of digestible matter and enzyme is passed into the caecum for further digestion by an accumulation of enzymes secreted by the "liver". Particulate waste material is collected by a pleated ciliated area adjacent to the orifice of the intestine, and is rejected. The soluble products of digestion are absorbed by the epithelium of the caecal sac and by the intestinal walls. Finally, the indigestible skeletal remains, which have been isolated in the stomach, are passed directly from the stomach to the intestine, and are voided. Manipulation of the contents of the gut is effected by a series of sphincter muscles sited at appropriate points, in such a way that the delicate walls of the caecum are not exposed to abrasion by fragments of the skeletons of the prey. Feed-back from the stomach to the central nervous system restrains the voracious appetite of the animal while a meal is being digested.

Digestion in a metazoan may be either an intracellular or an extracellular process, and we find that both occur in the phylum Mollusca. The former process, which is especially well suited for the treatment of finely divided particles of food material, is generally thought to be the more primitive.

The occurrence of intracellular digestion demands the following main conditions: first, the food material should consist of large quantities of minute particles capable of being phagocytosed by the digestive cells; second, feeding should be an almost continuous process so that the presentation of food material to the ingesting surfaces should not be concentrated into a short period of time; third, the alimentary canal should include an extremely large surface area over which particles of food can be ingested by the epithelial cells; finally, there must be some circulatory mechanism in the stomach to ensure that suitable food particles are presented to the ingesting surface in sufficient numbers while at the same time protecting the ingesting surface from damage by coarse particles that may be swallowed with the food. Larger pieces of food, such as the pieces of flesh that might be swallowed by a predacious carnivore, could not be digested by such a process.

We know nothing of the emergence of the various invertebrate phyla in the Pre-Cambrian epoch, but in many cases there must have arisen at this very early stage a new, extracellular method of digestion whereby digestive enzymes were secreted by gland cells into the lumen of the gut. This would provide a preliminary phase of extracellular digestion of larger particles and masses, thereby liberating smaller particles, the digestion of which could be completed by intracellular means.

With the development of extracellular digestion, gland cells could be concentrated into glandular masses external to the general course of the alimentary canal, into which they could discharge by short ducts, thereby making the alimentary canal itself more sturdy and compact and capable where necessary of muscular manipulation of the contents. With extracellular digestion, enzymes could be poured in greater quantities upon the food mass and the mixture of food and enzymes could be churned about by muscular action. With extracellular digestion, the digestive processes could be arranged into an appropriate sequence of events with specialisation of function of the successive parts of the gut and with consequent increase in efficiency.

Co-ordinated with the development of these new possibilities, and with the creation of new types of digestive enzymes, it was possible for new methods of feeding to be explored and for novel kinds of substrates to be attacked. In particular, it allowed the elaboration of the carnivorous mode of life. Thus we find the carnivorous habit even in so lowly a group as the diploblastic Coelenterata, as well as in the more advanced Annelida, Arthropoda, and Mollusca. We must concede that the developing capacities of the digestive system must have played a prominent role in the early phases of animal evolution, i.e. in the establishment of the phyla and their constituent classes.

Primitively food materials were probably moved through the gut principally by ciliary action, and the alimentary system probably possessed only one basic type of "multipurpose digestive cell". These cells served various purposes including phagocytosis and intracellular digestion of minute particles, absorbtion, secretion, and also excretion of indigestible remains and waste products. Such multipurpose cells may have given way in some cases to a number of contrasted cell types each of which has specialised in one or two particular functions. In other cases multipurpose cells may have been retained, and these may pass through a series of phases of activity, each phase being characterised by certain cytological features. Specialisation will have followed different lines in different phylogenies. In the filter-feeding bivalves, for example, there has been increased emphasis on intracellular digestion. In many gastropods, on the other hand, emphasis has been placed on extracellular digestion, on the replacement of ciliary action by muscular movements of the walls of the gut, and on secondary simplification of the stomach.

In the adaptive radiation of the Mollusca each class or order has become adapted in general to a certain way of life, feeding and digestion having played a prominent part in this adaptation. Consequently the digestive processes differ in the various molluscan classes, and it is not possible to make a generalised statement on digestion in the Mollusca as a whole. It will be necessary to deal with each class separately. Within each class, however, further radiation has given rise to a multiplicity of different forms, each adapted to a particular habitat and mode of life, sometimes producing great diversity in feeding habits. It is not safe to assume that what is discovered of the digestive process of one animal will necessarily also be true for others in the same class. It is only possible to make broad generalisations after the same results have been obtained in innumerable investigations on different genera. This caution is particularly necessary in respect of the Gastropoda, where remarkable diversity in feeding habits is found even within the taenioglossan prosobranchs. Even the Bivalvia, which may appear rather uniform in their habit of feeding on small particles, are not a homogeneous group as regards their digestive processes.

After these introductory and precautionary remarks, we will proceed to consider each molluscan class in turn.

Aplacophora

Nothing appears to be known of the details of the digestive processes in the Aplacophora. All members appear to be specialised as carnivores, and it may be tentatively assumed that digestion is at least partly extracellular.

Monoplacophora

Living animals of this newly discovered class have not yet been studied, and nothing is yet known of their digestive processes. Examination of the contents of the gut indicates that *Neopilina* is a detritus feeder, living on radiolarians and debris collected from the substratum. It is improbable that such material could be digested solely by intracellular enzymes. The contents of the gut do not necessarily indicate the food upon which an animal depends, but may merely be the indigestible remains of digestion. If, however, *Neopilina* does feed selectively on such material, it is probable that some extracellular enzymes would be necessary for their reduction.

Scaphopoda

Dentalium entalis is a highly selective feeder on foraminifera such as *Elphidium* (= *Polystomella*), *Bulimina, Quinqueloculina*, and *Discorbis*, which are selected from the substratum and dragged back to the mouth by the captacula. They may perhaps also derive nourishment from particles of the substratum which may be swallowed with these foraminiferans, or from accumulations of such material by the ciliary cleansing currents of the mantle cavity. Another species, *D. conspicuum*, feeds primarily on such small particles and micro-organisms (Dinamani, 1964).

When the foraminifera are drawn into the buccal cavity by the radula their skeletons may be broken by pressure against the chitinous jaw which lies on the antero-dorsal side of the buccal mass. Digestive enzymes may be poured on to the food as it passes down the oesophagus, from the pair of oesophageal pouches. Although thin-walled, the stomach acts as a gizzard and digestion is considered to be largely by extracellular enzymes which are derived from the pair of digestive diverticula which open into the stomach posteriorly. Muscular contractions of the wall of the stomach inject the contents of the stomach into the lumina of the digestive diverticula. Here the products of digestion are absorbed by the cells forming the follicles of the digestive diverticula, while in addition small particles are ingested by the follicular cells and subjected to intracellular digestion.

Indigestible material which accumulates in the digestive cells is finally rejected into cytoplasmic spherules which are nipped off the distal tips of the cells and passed into the stomach. If these spherules disintegrate in the stomach it is possible that they are the source of the extracellular enzymes which operate in the lumen of the stomach (Morton, 1959). The posterior wall of the stomach bears a series of ciliated ridges and grooves and waste particles are directed by ciliary currents in the grooves into the mid-gut. Peristaltic movements of the stomach periodically squeeze waste material from the stomach into the mid-gut. The loosely formed faeces are extruded at the anus as the result of muscular movements of the foot.

Polyplacophora

Cryptochiton stelleri, a large chiton common inter-tidally on the Pacific coast of North America, feeds on green, red, and brown algae, swallowing relatively large pieces of these weeds. Meeuse and Fluegel (1958) have investigated the digestive enzymes secreted by the sugar glands of this species, and found that there is a maltase and a powerful amylase, both of which have optima at pH $5.0-5.2$. The amylase was shown to hydrolyse alpha- and beta- glucosides and beta- galactosides. Meeuse and Fluegel concluded from the results of experiments that the secretions of the sugar glands of *C. stelleri* contained pure alpha-amylase, and that this probably comprised two distinct enzymes which split $a-1:4$, and $a-1:6$ linkages respectively.

Chitons such as *Lepidochitona cinereus* and *Acanthochitona crinitus* feed on filamentous and encrusting algae as well as on debris, and they may also rasp fragments off the surfaces of larger seaweeds. Particles of such food are brought into the buccal cavity by the radula, where they are mixed with mucus from glands in the wall of the buccal cavity to form a mucus foodchain which enters the oesophagus. Ciliary currents on the walls of the oesophagus carry the food string backwards and secretions from the glands of the anterior part of the oesophagus are poured on to it. These secretions include a diastase and a glycogenase, and probably also an acid which is responsible for the acidity of the stomach (pH 4·7–5·0). This mixture of food, enzymes, and acid is passed backwards from the oesophagus into the stomach, the oesophageal orifice into the stomach bearing a sphincter muscle which prevents regurgitation of stomach contents.

Due to the acidity of the stomach contents the viscosity of the mucus is minimal, which provides optimal conditions for mixing of the enzymes with the food material. The cells of the digestive diverticula are of two types, one of which contains many spherules of lime, and these may possibly serve for adjusting the degree of acidity of the secretions from the digestive diverticula. The other cells, which are more numerous, secrete a protease which is carried down the duct of the gland and is mixed with food in the stomach. In addition to secreting a protease, these cells absorb the soluble products of digestion. The stomach wall is strengthened by a layer of connective tissue which contains circular, oblique, and longitudinal muscle fibres. The stomach bears a ventral dilatation which is not ciliated, but is lined by a thin layer of cuticle. Muscular contractions of the walls of this ventral dilatation of the stomach cause the food particles to be thoroughly mixed with the amylases from the oesophageal glands and the protease from the digestive diverticula.

Phagocytes may migrate through the wall of the stomach, invade its lumen, and ingest small particles. Such phagocytes will return through the wall of the gut and pass the products of digestion to the cells of the stomach or intestinal wall. In the main, however, digestion is extracellular. Since there is no cellulase, it is only the contents of cells which have been broken open by the teeth of the radula which can be extracted and digested. Large quantities of plant material may pass through the gut undigested. The optimum pH for the action of the diastase derived from the oesophageal gland is about pH 5·6, the optimum for the glycogenase is from pH 6·4 to 6·8; the protease derived from the digestive diverticula has two optima, one between pH 4·3 and 4·8 and the other between pH 8·0 and 8·8.

The digestive diverticula open into the stomach by two ducts, the orifices of which are protected by sphincter muscles. These orifices lie very close to the origin of the mid-gut at the posterior end of the stomach. Two ciliated bands on the roof of the stomach pass on either side of these two orifices and direct particulate material downwards and backwards from the oesophagus to the ventral dilatation of the stomach, where it is mixed with enzymes as stated above.

A valve on the course of the mid-gut divides it into a short anterior portion and a longer posterior portion. The anterior part of the mid-gut bears a pair of lateral ciliated tracts on which the cilia beat transversely and so impart a spiral motion to the contents. Muscular contractions of the wall of the stomach forces a cord of particles, bound in mucus and mixed with enzymes, backwards into the mid-gut. Part of this cord enters the midgut valve and is held there by the contraction of the circular muscle fibres. The stomach and anterior mid-gut now act as a single unit and contractions in their walls squeeze the rotating food cord and express the soluble products of digestion into a dorsal gutter in the wall of the mid-gut. This fluid passes forwards along the gutter towards the orifices of the two ducts to the digestive diverticula, and is injected into these ducts by contraction of the walls of the stomach and the anterior midgut. The soluble products of digestion are now absorbed by the more numerous secreting and absorbing cells in the tubules of the digestive diverticula.

The mid-gut valve holds, compresses, and cuts off a small section of the food-cord and this is moulded into a faecal pellet, aided in this process by the further addition of mucus and the increasing viscosity of the mucus due to a rise in the pH of the contents of the hind part of the mid-gut (pH 6·6). No further digestion or absorbtion of soluble matter occurs in the remarkably long mid-gut, but particulate material may be ingested by wandering phagocytes.

The valve in the mid-gut serves the digestive process in many important ways—it stops food from leaving the stomach and anterior mid-gut during the process of digestion; it immobilises the food-cord and thus makes possible the extraction of the fluid products of digestion when the food cord is squeezed by the muscles in the wall of the stomach and the anterior mid-gut; it breaks up the remains of the food cord into convenient lengths for conversion into faecal pellets.

Apart from the action of amoebocytes, the only site for absorption of food matter is in the tubules of the digestive diverticula, the only site where soluble matter is absorbed, and where no ingestion of particulate material occurs (Fretter, 1937).

Bivalvia

Although at first sight the Bivalvia may seem to be a fairly homogeneous class, it has recently been established that the Protobranchia comprise an independent lineage to which might even be accorded the status of a sub-class. Studies on the digestive diverticula of a variety of bivalves have revealed striking differences between the Nuculidae, on the one hand, and the Anisomyaria and Eulamellibranchia, on the other. Again, highly specialised conditions are found in the Septibranchia, and the Verticordiidae also have embarked upon a carnivorous mode of life. Lesser specialisations in feeding and digestion occur in the Teredinidae, or ship-worms, and in the Tridacnidae. For these reasons it is best to deal with the principal divisions of the Bivalvia separately.

Protobranchia

There are three families in the order (or sub-class) Protobranchia and, since they exhibit striking differences as regards feeding and digestion, it will be necessary to consider these in turn.

Nuculidae

In the Nuculidae, Yonge (1939) showed that the digestive diverticula open by three slender ducts into a stomach with muscular walls which lie comparatively free from the adjacent tissues. The stomach is composed of two parts, of which the globular upper part receives the oesophagus and the three ducts from the digestive diverticula. A part of the inner surface of the stomach wall, lying anteriorly and on the right side of the stomach, bears a ciliated epithelium and is corrugated to form a system of ridges and grooves which act as a sorting area. The remainder of the wall of the upper, globular part of the stomach is protected internally by an extensive chitinous investment. Part of this investment is homologous with the gastric shield which occurs throughout the Bivalvia, while the remainder is a special development in the Protobranchia. The lower part of the stomach is a conical structure which is lined by a ciliated columnar epithelium and which corresponds with the combined style-sac and anterior mid-gut of many of the filter-feeding bivalves (see Fig. 77).

There is no firm crystalline style in the Nuculidae, but instead there is a viscid mass which comprises secretions from the style-sac and from the digestive diverticula, mixed with food material, organic debris, and sand grains, etc., which have been garnered from the substratum by the palp proboscides. The contents of the stomach are thoroughly mixed by muscular movements of the stomach wall and by rotation imparted by the ciliated epithelium of the style-sac. Muscular contractions of the stomach wall squeeze the soluble products of digestion out of the food/enzyme mixture,

and these are absorbed by the stomach epithelium (Owen, 1956). A sphincter muscle at the distal end of the style-sac prevents the untimely escape of
material into the mid-gut, except for small quantities which are allowed to
pass out via a ciliated groove which lies on one side of the style-sac.

The internal architecture of the stomach of *Nucula* has been studied by
Purchon (1956), while Owen (1955, 1956) has made a fresh appraisal of the
digestive processes in the Nuculidae, and has shown striking differences be-

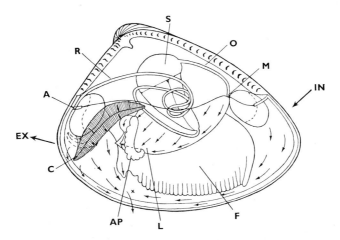

Fig. 77. *Nucula hanleyi*, seen from the right side after removal of the right
shell valve and the right mantle lobe. In addition to the organs in the
mantle cavity, the course of the alimentary canal is also shown. The stomach
(S) consists of two portions, the lower conical part containing a viscous mass,
not a well-defined crystalline style. The stomach wall is muscular, and is relatively free from attachments to the surrounding tissues; the upper globular
portion is extensively protected internally by a chitinous secretion, and
muscular contractions of the stomach wall serve to mix the ingested material with the digestive enzymes and to triturate the food material. For interpretation of other lettering, see pp. 263–4. (Originally published in Yonge,
1939, *Phil. Trans.* B, **230**, 91, fig. 9.)

tween these and the digestive processes of the Anisomyaria and the Eulamellibranchia. Owen showed that the cilia around the orifices of the ducts
from the digestive diverticula are very active and they serve to expel fluid
from the ducts into the stomach and to prevent any particles in the stomach
from entering the ducts. In contrast with the filter-feeding bivalves, the cells
forming the tubules of the digestive diverticula in the Nuculidae do not ingest particles nor do they absorb the soluble products of digestion. Digestion is exclusively extracellular, and absorbtion is solely by the epithelium
of the stomach and anterior mid-gut (Owen, 1956). There are two types
of cell in the digestive diverticula: one of these is dark-staining, and bears a
single flagellum and these cells may perhaps be secretory; the other type

of cell is taller and goes through a phase during which the distal ends of the cells are nipped off and passed down the duct to the lumen of the stomach. These fragmentation spherules may be partly excretory, and may also be the source of extracellular enzymes in the lumen of the stomach. Food reserves are stored in the form of protein spheres, fat droplets, and glycogen in the cells lining the alimentary canal.

The contents of the stomach were slightly acid, at pH 5·7–5·9, and contained an amylase, a lipase, and a weak protease. The amylase was derived from two sources—the style-sac and the digestive diverticula. The amylase from the style-sac was found to have optimal activity at pH 6·2. The amylase extracted from the digestive diverticula had two optima, one at pH 6·1–6·4, which is probably operative in the lumen of the stomach, and another at pH 4·3–4·5. This second optimum may perhaps signify the occurrence of a second enzyme which remains within the tissue. The lipase is also probably derived both from the style-sac and the digestive diverticula, but the very weak protease is provided solely by the digestive diverticula (Owen, 1956). In the classical study of the Protobranchia by Yonge (1939) it had been concluded as the result of feeding experiments using iron saccharate that absorbtion occurred in the digestive diverticula. Owen (1956), however, showed that ordinarily the tissues of *Nucula* contain appreciable amounts of iron, which could be removed by a period of starvation in filtered sea water. By this refinement Owen succeeded in showing that absorbtion of iron saccharate did *not* occur in the digestive diverticula, but only in the ciliated epithelium of the stomach and the first part of the midgut.

There is a striking series of differences in structure and in function between the digestive diverticula of the Nuculidae and those of the filter-feeding bivalves. In the Nuculidae the main ducts of the digestive diverticula have no cilia while the secondary ducts have a uniform layer of ciliated epithelium all round the duct. In these ducts the fluid only flows out wards from the diverticula into the stomach. Within the tubules themselves the dark-staining cells each bear only one single flagellum. The fluid contents of the stomach do not penetrate the digestive diverticula, and no absorbtion or ingestion of particles occurs in the diverticula. In all these respects contrasted conditions are found in the filter-feeding bivalves (see p. 232). Owen concluded from his study that the Nuculidae did not evolve from suspension feeding ancestors, that it is unlikely that the Nuculidae ever had a firm crystalline style, and that in the Nuculidae there has been a secondary loss of intracellular digestion.

It is extraordinary that in the Nuculidae the digestive diverticula should have lost the capacity to absorb as well as the capacity to ingest.

Nuculanidae

It is clear that the conditions in the Nuculanidae are partly different from those in the Nuculidae. Yonge (1939) found that relatively large particles may be swallowed by a combination of ciliary and muscular activity, and that the oesophagus is somewhat enlarged for this purpose. In general the internal architecture of the stomach is comparable to that in the Nuculidae, but one of the three masses of digestive diverticula differs from the other two and resembles the digestive diverticula of the septibranch *Cuspidaria* (see pp. 224–5). Most of the material that enters the stomach is compacted into a mucus-bound mass which is rotated by the cilia of the stylesac. The stomach is elastic and muscular, and is relatively free from the surrounding organs, as in the Nuculidae, and it acts very much like a gizzard. The rotating mucus-bound contents are crushed and compressed against the gastric shield, which causes trituration of particles and the expression of the soluble products of digestion. Yonge concluded that fine particles could escape from the revolving mass in the stomach and be subjected to selection by size on the ciliary sorting area. Here the finest particles are accepted and passed into the ducts leading to the specialised digestive diverticula, where they are ingested and subjected to intracellular digestion. Slightly coarser particles are rejected from the stomach via the ciliated rejection tract which leads down the side of the style-sac into the mid-gut.

Yonge found abundant evidence for the ingestion and intracellular digestion of minute particles by the cells of the specialised digestive diverticula. In *Yoldia* he found that the wide digestive diverticula were almost black, due to the presence of numerous fine particles of black sand in the lumina of the diverticula and also within the cells of the digestive diverticula. In the Nuculanidae the two masses of unspecialised digestive diverticula are essentially similar to the digestive diverticula of the Nuculidae in being solely secretory in function. In contrast, in the specialised mass of digestive diverticula the digestive cells are undoubtedly capable of phagocytosing relatively large particles, and of absorbing the soluble products of extracellular digestion. By virtue of the possession of two types of digestive diverticula, the Nuculanidae are capable of both intracellular and extracellular digestion, whereas in the Nuculidae digestion appears to be exclusively extracellular. As regards their digestive processes the two families appear to be evolving in different directions.

Solenomyidae

The digestive mechanisms of *Solenomya* remain a mystery. Yonge described the alimentary canal of *S. togata* as being greatly reduced and simplified; the stomach is very small and has no ciliary sorting areas, no gastric shield, and although there is a style-sac there is apparently no crystalline style. Owen (1961) found that in *S. parkinsoni* the gut was similarly extraordinarily small for so substantial an animal; the stomach was represented merely by a slight dilation associated with the point of entry of the two ducts from the greatly reduced digestive diverticula, while the mid-gut was a "slender greenish thread", less than half the diameter of the anterior aorta. The digestive diverticula consisted of a small number of distinct tubules on to the distal ends of which were inserted a number of rather stout muscle fibres. Owen found the stomach of *S. parkinsoni* to possess the normal internal features including a dorsal hood, and a gastric shield, in spite of its minute size. The style-sac, which tapered backwards towards the mid-gut, exhibited the normal ciliation (see Figs. 78 and 79).

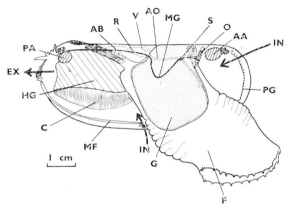

FIG. 78. *Solemya parkinsoni*, seen from the right side after removal of the right shell valve and right mantle lobe, and after further dissection to display the course of the alimentary canal. The animal lives deeply buried in intertidal sandy beaches, near low-water mark, and the enormous size of the foot (F) is doubtless associated with burrowing habit. The ctenidium (C) is suspended from the ventral border of a very large hypobranchial gland (HG). The methods of feeding and digestion in *Solemya* are not understood. The whole alimentary canal is greatly reduced; e.g. the oesophagus (O), the stomach (S), the mid-gut (MG), and the rectum (R). For interpretation of other lettering, see pp. 263–4. (Originally published in Owen, 1961, *Quart. J. micr. Sci.* **102**, 18, fig. 3.)

Specimens of *S. parkinsoni* were placed in sea water containing starch grains, titanium dioxide particles, and iron saccharate, and considerable quantities of these materials were accumulated in the mantle cavity. None, however, was passed into the gut or was found in any of the tissues. Tests showed that starch was being broken down in the mantle cavity. Very little particulate material was found in the lumen of the gut of any freshly caught specimens.

One is obliged to ask the very extraordinary question—What part is played by the reduced gut in the nutrition of this rather bulky and potentially very active animal?—but the answer is not to hand.

S. parkinsoni lives buried about 50 cm deep in sandy mud at low-water mark of spring tides; since there are no traces at the surface of the sand it presumably does not occupy a burrow—unless it makes use of a horizontal gallery of a burrowing crustacean such as *Callianassa*. Water is taken into the mantle cavity anteriorly and from beneath the surface of the substratum. As a result of muscular activity, correlated with the high degree of flexibility of the shell valves, large quantities of bottom material are taken into the mantle cavity and food material is apparently accumulated from this by the greatly enlarged ctenidia.

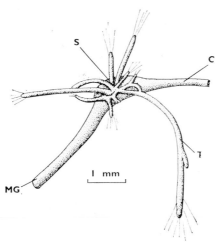

Fig. 79. *Solemya parkinsoni*, enlarged view of the anterior part of the alimentary canal, seen from the right side. The stomach (S) is only a very slight dilation as compared with the oesophagus (O) and the mid-gut (MG), while there are very few and small tubules to the digestive diverticula (T). (Originally published in Owen, 1961, *Quart. J. micr. Sci.* **102**, 19, fig. 4.)

Since the gut is so greatly reduced, it seems necessary to consider the possibility that food material is digested in the mantle cavity, and that the soluble products of such digestion are passed forward in liquid form to the mouth. This alone, it would seem, would account for the extreme reduction of the alimentary canal and its associated glands.

If digestion does occur in the mantle cavity, what is the source of the enzymes that are responsible? Owen states that there are some glands in the epithelium of the foot which are not known to occur in any other bivalves, and these might possibly be concerned. Alternatively, or additionally, can digestion in the mantle cavity be effected by wandering amoebocytes which may emerge from the epithelium of the ctenidia, or of the mantle?

Septibranchia

Cuspidaria and *Poromya* feed upon the bodies of small crustaceans, such as copepods, etc., which have been drawn into the infra-septal part of the mantle cavity and have been trapped there by the valve at the base of the inhalant siphon. In *Poromya* the food mass is actively pushed into the mouth by the small but muscular labial palps, and it is then conveyed along the oesophagus by muscular contractions, not by ciliary action. The mouth in both genera is relatively large and is not obscured by the bases of the palps. The stomach is long and cylindrical with muscular, folded walls, the inner surface of which is mainly covered by a protective chitinous sheath. There is a short oval style-sac and a crystalline style is present in spite of the carnivorous habit, but only the very tip of the style projects into the lumen of the stomach.

The digestive diverticula tend to unite with one another to form a compact structure, and the lumina of the diverticula are strikingly wide. They comprise only one kind of cell in which no traces of secretory activity were noted by Yonge (1928). The digestive diverticula of *Cuspidaria* open into the stomach by only two ducts which were thought by Yonge to be rather wide, but I myself considered them to be no wider than in most other bivalves (Purchon, 1956).

It is not known whether septibranchs depend mainly on living or on dead animals for their food. In addition to such animals, sand grains and even small pieces of slag may be found in the stomach, and these foreign bodies are doubtless useful in crushing the food. The muscular stomach serves as an efficient gizzard and the longitudinal ridges, protected by the chitinous sheath, come together and crush any organisms within the stomach. Digestible matter is squeezed out of the prey and is passed up the two ducts into the digestive diverticula where alone digestion can occur. The indigestible remains are passed from the stomach directly into the mid-gut.

Very few small particles are allowed to enter the mouth, and there is an almost complete lack of wandering phagocytes in the alimentary canal in the Septibranchia.

In feeding experiments with *Cuspidaria* blood corpuscles of fish were accepted and were passed into the digestive diverticula, where they were ingested by the cells of the diverticula. Very little iron saccharate was allowed to enter the mouth, and such that did was only absorbed in the digestive diverticula, where it was taken into large vacuoles in the digestive cells. Thus digestion in *Poromya* and in *Cuspidaria* was solely intracellular (see Fig. 80).

The persistence of the crystalline style in the Septibranchia indicates the lack of a free protease in the lumen of the gut (Yonge, 1930) and the failure to evolve an extracellular protease is doubtless the reason for the very limit-

ed success of this order of bivalves. The Septibranchia may have evolved
in deep water where there was a deficiency of minute water-borne particles, adopting the carnivorous habit instead to exploit an alternative source
of food. In consequence small particles are now prevented from entering
the mouth. When *Cuspidaria* was placed in a suspension of indian ink very
few of these particles were taken in at the mouth, and none was ingested,

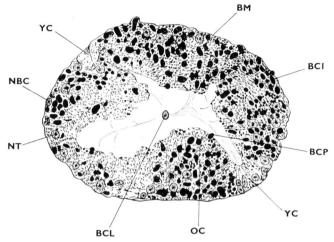

FIG. 80. *Cuspidaria cuspidata*. Transverse section through a single tubule of
the digestive diverticula. The specimen had been fixed after feeding for 10
hours on fish blood, and red blood corpuscles (BCL) can be seen in the
lumen of the tubule, in process of being ingested by the digestive cells
(BCP), and deep within the cytoplasm of the digestive cells (BCI). For interpretation of other lettering, see pp. 263–4. (Redrawn from the original which
was published in Yonge, 1928, *Phil. Trans.* B, **216**, plate 14, fig. 13.)

even in the digestive diverticula. As a result of this change in diet from small
particles to comparatively large planktonic organisms, there must have
been a change in general metabolism, for it appears that, unlike most other
bivalves, glycogen is not stored as a food reserve in the tissues in *Cuspidaria*.

Filter-feeding Bivalves (=Polysyringia)

Under this general heading we will consider those lineages of bivalves
which possess lamellar ctenidia and most of which feed by filtration of
particles suspended in the supernatant water. These bivalves possess filibranch, or pseudolamellibranch, or eulamellibranch ctenidia; there is no
generally accepted term to distinguish these bivalves collectively from the
Protobranchia and Septibranchia other than "Polysyringia".

Concurrent with the elaboration of the ctenidium for the collection of
suitable quantities of small particles from the water, there has also been

an equally important elaboration of the interior of the stomach for handling this particulate material during the early stages of digestion. In marked
contrast to the stomachs of the Protobranchia and Septibranchia, the internal architecture of the stomach of the filter-feeding bivalves is generally
very complex (Purchon, 1957, 1958, 1960) (see Fig. 81). While there is be-

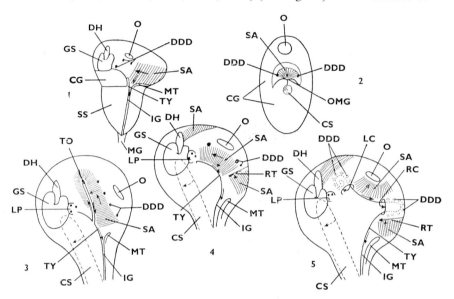

Fig. 81. Diversity in the internal architecture of the stomach in the Bivalvia. 1, Protobranchia. 2, Septibranchia. 3, 4, 5, Three contrasted conditions
found in the filter-feeding bivalves. In the stomach in the Protobranchia
the inner surface of the stomach wall is extensively protected by a lining of
chitin (CG) and ciliary sorting areas (SA) are limited in extent. In the
Septibranchia the stomach is highly muscular, the chitinous lining (CG) is
even more extensive and the ciliary sorting area (SA) is minimal in extent.
In the filter-feeding bivalves the gastric shield (GS) is relatively small,
ciliary sorting areas (SA) are highly developed, and the major typhlosole
(TY) and the intestinal groove (IG) extend far into the stomach; the crystalline style (CS) is typically very well developed. (Originally published in
Purchon, 1960, *Proc. Cent. and Bicent. Congr. Biol. Singapore*, 1958, p. 79,
fig. 7.)

wildering variety in detail in the many genera which have been studied,
the main features can be summarised as follows.

The oesophagus is slender, and particles are passed along it into the stomach by ciliary and not by muscular activity. There is a crystalline style
which is secreted in a style-sac, and which projects into the lumen of the
stomach. The style-sac may be conjoined with the anterior section of the
mid-gut and only separated from the lumen of the latter by a pair of longitudinal typhlosoles. Alternatively, the style-sac may be partly or wholly

separated from the mid-gut and will then open into the stomach by an independent orifice (see Fig. 82). This isolation of the style-sac from the mid-gut has occurred independently in many different lineages, and was apparently effected by fusion of the two longitudinal typhlosoles. The style-sac is lined by a uniform columnar ciliated epithelium, and the concerted

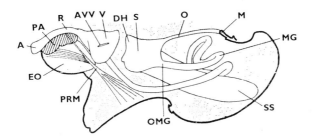

FIG. 82. The course of the alimentary canal in the rock-boring eulamelli-branch *Pholas dactylus*, the outline of the shell, mantle, ctenidia, etc., having been omitted from the figure. The relatively enormous style-sac (SS) has become completely separate from the mid-gut (MG) due to fusion of the major and minor intestinal typhlosoles, followed by separation of the two tubes so formed. This advanced condition has been achieved independently in many unrelated lineages and is without any phylogenetic significance. For interpretation of other lettering, see pp. 263–4. (Originally published in Purchon, 1955, *Proc. zool. Soc. Lond.* **124**, 887, fig. 14.)

action of the cilia causes the style to rotate on its longitudinal axis. Rotation of the crystalline style has been observed on many occasions, the direction of rotation being clockwise when viewed from the anterior end. Thus Nelson (1918) observed the maximum rate of rotation of the styles of *Anodonta* to be 11 per minute at 11·5°C, and of *Modiolus* to be 13 per minute at 25°C, Allen (1958) observed the style of *Loripes* rotating in the unopened stomach from 20 to 40 times per minute, while rotation rates up to 90 rev/min have been recorded in the larva of *Ostrea edulis* (Yonge, 1949).

Where the contents of the stomach are rather viscous, the rotation of the style may act like a capstan, mucoid threads containing food material being drawn into the stomach from the oesophagus and wound round the head of the style.

A portion of the left wall of the stomach is protected on its inner surface by a chitinous secretion, the gastric shield; this is normally held firmly in position by two saddle-shaped flares of the gastric shield which pass into deep pockets which occur on this side of the stomach. The gastric shield normally bears a small prominence, or tooth, and this may sometimes serve to hold the crystalline style so that its tip lies near the mouth of one of these pockets—which we now know as the dorsal hood.

As the crystalline style revolves in the stomach it performs a number of functions that are relevant to the digestive process (see Fig. 83). First, it may drag food material into the stomach more rapidly than it could enter under the sole impetus of the oesophageal cilia. Second, it stirs the general contents of the stomach and brushes these against certain corrugated ciliary sorting areas which may abstract unsuitable particles and reject them

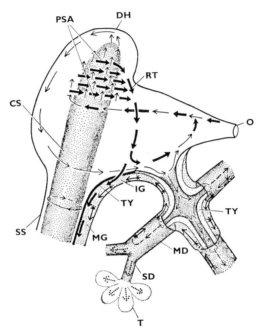

FIG. 83. Diagrammatic representation of the probable circulation of particles in the stomach and digestive diverticula of a bivalve with a gastropemptan type of stomach (ref. Fig. 81, type 5). Heavy arrows represent the course followed by coarse particles; fine arrows represent the course taken by fine particles; tailed arrows represent material that is being rejected; dotted arrows represent movement due to absorbtion of material by the digestive cells. For interpretation of lettering, see pp. 263–4. (Originally published in Owen, 1955, *Quart. J. micr. Sci.* **96**, 533, fig. 13.)

into the mid-gut. Third, it gently rubs the contents of the stomach against the gastric shield; since these contents may include sand grains, bits of organic debris, diatoms, as well as naked flagellates, etc., this rubbing may cause a certain amount of trituration of the food material, reducing the size of particles and liberating the contents of diatom frustules. Fourth, it slowly dissolves, liberating into the stomach the amylolytic enzymes which were secreted and incorporated into the substance of the crystalline style in the style-sec. These enzymes, liberated by the crystalline style, initiate a preliminary phase of extracellular digestion in the stomach. At the same time the dissolution of the style acidifies and so lowers the viscosity of the mucoid contents of the stomach.

The digestive diverticula consist of a very large number of acini, or tubules, which communicate via a system of secondary ducts into a smaller number of main ducts which in turn open into the stomach. In the Protobranchia and the Septibranchia very few main ducts open into the stomach; in the filter-feeding bivalves the number of main ducts is considerably larger, though in many cases these main ducts do not open directly into the main cavity of the stomach, but into a small number of deep embayments of the stomach wall. In the most advanced condition I have found only three such embayments which receive ducts from the digestive diverticula, namely the left pouch, the left caecum, and the right caecum. These three embayments are not universally present and in many genera alternatively—or additionally—a number of main ducts from the digestive diverticula may open directly into the main cavity of the stomach. It is generally considered that feeding and digestion are continuous processes in bivalves, new food material being added all the time, and unwanted material being as constantly eliminated by passage into the mid-gut. This continuous treatment of small particles requires special features which ensure that discarded material is kept separate from that which is still being subjected to the digestive processes. Some bivalves have a well-defined rhythm in their processes of feeding and digestion, e.g. *Lasaea* (Morton, 1956).

The elimination of unwanted material from the stomach and from the digestive diverticula is effected by a prolongation of a ciliated rejection tract from the mid-gut into the stomach; this is known as the intestinal groove. One of the two intestinal typhlosoles, which in some genera separate the lumen of the style-sac from that of the mid-gut, also penetrates the stomach and accompanies the intestinal groove on its left side to its termination. This major intestinal typhlosole forms a flap which arches over the intestinal groove and effectively separates it from the main cavity of the stomach. The two structures pass forwards across the floor of the stomach and terminate on its left anterior floor.

In some cases an extensive ciliary sorting area of ridges and grooves lies adjacent to a part of the intestinal groove. Unwanted particles become trapped in the grooves of this sorting area and are conveyed by ciliary activity along the grooves, into the intestinal groove, and thence under the protective flap of the major typhlosole to the mid-gut. Other particles encountering the sorting area may only be influenced by the cilia on the crests of the ridges, which beat transversely across the ridges, and these particles are retained in the circulating contents of the stomach.

Where numerous main ducts open from the digestive diverticula into the main cavity of the stomach, their orifices are often to be found in close association with the above-mentioned ciliary sorting area, the ridges and grooves of which penetrate the mouths of these ducts. The cilia within the ducts always beat outwards, towards the stomach and any waste material passed out from the digestive diverticula is carried immediately to the intestinal groove for elimination. In many, but not all, anisomyarian bivalves, the intestinal groove and its accompanying typhlosole are deve-

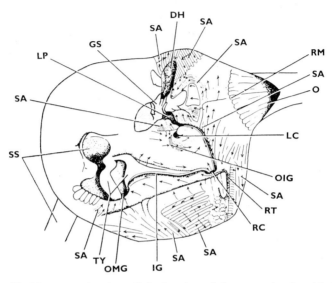

FIG. 84. *Mya arenaria*, view of the interior of the stomach after this has been opened by a mid-dorsal incision. The oesophagus (O) enters anteriorly and the style-sac (SS) and the mid-gut (OMG) leave the stomach postero-ventrally. The major typhlosole (TY), accompanied by the intestinal groove (IG), passes forwards from the mouth of the mid-gut and enters the right caecum (RC) and then crosses the floor of the stomach from right to left and terminates in the left caecum (LC). Ducts from the digestive diverticula enter only the two caeca and the left pouch (LP). The crystalline style (not shown) passes from the style-sac to the mouth of the dorsal hood (DH) and rubs against the gastric shield (GS). The stomach of *Mya* is characteristic of most filter-feeding bivalves in possessing a great elaboration of ciliary sorting mechanisms (SA), the purpose of which is to protect the entrances to the ducts of the digestive diverticula from entry of coarse particles. For interpretation of other lettering, see pp. 263–4. (Originally published in Purchon, 1960, *Proc. zool. Soc. Lond.* **135**, 470, fig. 16.)

loped into a long tongue which has this sorting area, and orifices of ducts from the digestive diverticula, on each side. This seems to emphasise the importance of this mechanism for elimination from the stomach of all unwanted material.

In about half of the eulamellibranch families the stomach possesses the three well-defined embayments which were mentioned above, and most if not all of the ducts from the digestive diverticula open into these (see

Fig. 84). In these families the major typhlosole and the intestinal groove arise within the left caecum, usually circling within this, close to the orifices of the ducts. The typhlosole and intestinal groove emerge from the left caecum, cross the anterior floor of the stomach, and enter the right caecum. The typhlosole and intestinal groove circle the interior of the right caecum and then emerge and pass back into the mouth of the mid-gut. Within each of the caeca the typhlosole usually forms a number of little tongues, each of which presses into the mouth of one of the main ducts from the digestive diverticula which open within these caeca. Some of the main ducts open into the left pouch, and some may open separately on the right side of the stomach, and these do not come into close proximity with the intestinal groove. Generally the ciliary currents in the close vicinity of the openings of all the main ducts from the digestive diverticula beat away from the diverticula and towards the stomach, and the general impression gained is that the function of these cilia is to ensure that unwanted particles cannot enter the ducts.

It is well known that these filter-feeding bivalves feed on minute particles borne in suspension, and that much digestion is intracellular, after these particles have been ingested by cells in the digestive diverticula. If the main function of the ciliary mechanisms in the stomach seems to be to hinder the entry of particles into these ducts, how do the particles effect an entry?

One suggestion that has been put forward is that the cavities of the stomach, the ducts and the tubules of the digestive diverticula are in communication and that fluid may pass in or out of the ducts as the result of changes in the pressure on the system. This could be effected by local changes in the tonus of the muscle fibres which occur in the wall of the stomach and around the individual ducts and tubules of the diverticula (Purchon, 1955). This might not be a very efficient means of ensuring circulation of food material in a large bivalve, but it does occur in the larva of *O. edulis*. There are two simple sac-like diverticula and these exhibit rhythmic alternate contractions. When one diverticulum contracts the other expands, and a quantity of the stomach contents is drawn into the latter. Contraction of each diverticulum is effected by a slender strand of muscle fibres that passes over the top of the diverticulum. The observed rate of pulsation was usually 18–21 times per minute (Millar, 1955). A similar, but less regular pulsation of digestive diverticula was observed in the brachiopod *Lingula unguis* (Chuang, 1959).

The most satisfactory answer to the problem has been found by Owen (1955) who showed that the filter-feeding bivalves differ markedly from the Nuculidae as regards the structure and function of their systems of digestive diverticula (see pp. 218-21). In the filter-feeding bivalves the system of digestive diverticula comprise large numbers of acini, or tubules, which open into a smaller number of secondary ducts, which in turn open into a still smaller number of main ducts. The main ducts, which may bear branches, possess a ciliated gutter which is subtended by two longitudinal ridges, the remainder of the circumference of the main duct being lined by an epithelium with a brush border, with no cilia. The secondary ducts, which do not branch, are lined exclusively by a brush border and there is an abrupt transition from this to the lining of digestive cells in the blind-ending tubules of the diverticula. In the main ducts the cilia in the gutter beat outwards, driving fluid out of the system of diverticula into the sto-mach, and preventing particles from entering the main ducts along the gutters (see Fig. 85). Since fluid is being driven out of the system of diverti-

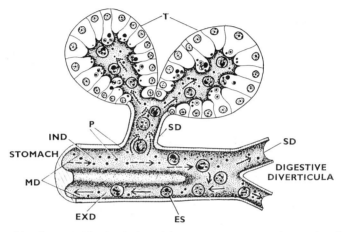

Fig. 85. The probable circulation of fluid and particles within the digestive diverticula of filter-feeding bivalves. Unbroken arrows represent outwardly directed ciliary currents; broken arrows represent an inwardly directed compensation current; dotted arrows represent movement due to absorbtion of material by the digestive cells. For interpretation of lettering, see pp. 263–4. (Originally published in Owen, 1955, *Quart. J. micr. Sci.* **96**, 528, fig. 9.)

cula in these gutters and since material in the lumen of the tubules is absorbed by the digestive cells—as will be described below—it is obvious that there must be a compensating current flowing from the stomach into the main ducts in the upper half which is lined by a brush border. Larger particles approaching the orifice of a main duct are likely to be influenced by the outward current in the ciliated gutter, and these will tend to be driven away. Smaller particles, which are more suitable for acceptance and ingestion in the tubules, are more likely to be carried into the main duct in the inwardly directed counter current. This counter current will carry a selection of the stomach contents as far as the entran-ces to the secondary ducts, but no further. Reid (1965) and Mathers (1970) have made important observations supporting this hypothesis.

Since absorption and ingestion is occurring in the tubules of the digestive diverticula, fluid must be drawn into the tubules from the secondary ducts at a corresponding rate, and this can only be done by a similar quantity of fluid being sucked into the secondary ducts from the main ducts. In this way circulation of fluid is maintained throughout the whole system of digestive diverticula. There will be a selection of particles by size at the entrances to the main ducts, the larger ones being rejected. Due to the branching of the system and the comparatively slow rate at which fluid is drawn into the secondary ducts as the result of absorption in the tubules, there will presumably be a second selection of particles by size at the entrances to the secondary ducts. Only the finest particles will pass into the secondary ducts; slightly larger particles will remain in the main ducts but will finally drop into the ciliated gutter and will be ejected into the stomach.

The selection of particles of food is the result of a hierarchical series of gradings by size which commences in the mantle cavity and is continued on the sorting areas of the stomach, at the entrances to the main ducts of the digestive diverticula, and again at the entrances to the secondary ducts. This ensures that the material which finally reaches the digestive cells is sufficiently small to be ingested intracellularly.

In many eulamellibranch bivalves the majority of the main ducts from the digestive diverticula do not open directly into the main cavity of the stomach, but into the left caecum, or the right caecum, both of which are deeply penetrated by the intestinal groove and the major typhlosole. Many of the larger particles in the contents of the stomach will be rejected at the mouth of these caeca, where the cilia beat vigorously and create powerful currents which usually sweep this area clean. That part of the major typhlosole which lies within the caecum is rolled in the form of an incomplete tube the cavity of which is continuous with that of the stomach. Small flares of the rim of the typhlosole are similarly rolled and each projects into the upper part of the mouth of a main duct (see Fig. 86). In contrast the intestinal groove, which lies below the margin of the major typhlosole, is brought into very close proximity to the openings of the ciliated gutters of the main ducts. Thus each of these caeca is in effect a pair of concentric tubes. The inner tube is formed by the major typhlosole, and this provides a route for the admission of food material to the main ducts of the digestive diverticula. The outer tube is formed by the walls of the caecum and the lower side of the flap of the major typhlosole, and this constitutes a closed route for the passage of waste material to the mid-gut (Owen, 1955).

The above account shows how in most filter-feeding bivalves the closest precautions are taken to prevent any but the finest particles from gaining access to the digestive diverticula. There are exceptions, e.g. in the Lucinacea, which generally inhabit substrates where there is apparently a deficiency of

finely divided food material. The three families in the Lucinacea have specialised progressively towards acceptance of larger particles, the most advanced
conditions being found in the Lucinidae and Thyasiridae. In these two
families the interior of the stomach is simplified by diminution and loss of
ciliary sorting areas, the digestive diverticula have few ducts with large orifices,
and there is little hindrance to the entry of comparatively large particles.
The Verticordiidae and the Septibranchia, occupying similarly impoverished
substrates, have adopted a carnivorous or scavenging habit, the interior of
the stomach being protected from damage by hypertrophy of the gastric

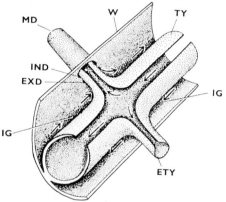

FIG. 86. Diagrammatic representation of the form of the interior of the
left and right caeca in bivalves with a gastropemptan type of stomach.
Ciliary currents are indicated by arrows. Flares of the major typhlosole
deliver a suspension of fine particles into the upper, unciliated part of the
duct to the digestive diverticulum, where these particles are accepted into
the inhalant, compensating current. For interpretation of lettering, see
pp. 263−4. (Originally published in Owen, 1955, *Quart. J. micr. Sci.* **96,**
531, fig. 11.)

shield (Allen, 1958, Allen & Turner, 1974). Thus the Lucinacea, Verticordiidae,
and Septibranchia have deviated from the main trend of evolution in the
Bivalvia in feeding upon rather larger particles. In the Teredinidae also, as will
be discussed more fully below, the ducts to the digestive diverticular are wider
than is usual, and this relates to the capacity of ship-worms to digest the
fragments of wood which are rasped off the walls of the burrow and passed
into the stomach.

Having considered the mechanics of the digestive system of the filter-
feeding bivalves, we may now study the digestive processes themselves.
Experimental studies on feeding and digestion in *Lasaea, Dreissena, Ostrea*
and *Scrobicularia* have shown the importance of considering the whole
animal in its natural environment, the possible influence of environmental
and bodily rhythms (adduction, digging, etc.) any of which may influence
feeding and digestion. Such studies (reviewed in Purchon, 1971) show that
these functions are not necessarily in a steady state. In some genera phases of

extra-cellular digestion in the stomach and intra-cellular digestion in the digestive diverticula may alternate rather than synchronise. There may be cyclical changes in the cytology of the digestive gland, in the nature of the stomach contents, and the pH of these, and troughs and peaks in enzyme activity. In any such studies it is now essential to consider the time factor and all external and internal variables.

The acid conditions in the lumen of the stomach are brought about partly by the dissolution of the head of the crystalline style and partly by fragmentation spherules derived from the digestive diverticula. This acidity reduces the viscosity of the mucus in which incoming food particles are embedded, and so prepares the stomach contents for digestion. The condition of the crystalline style differs according to the structure of the gut, being ephemeral in forms such as the oyster, in which the style-sac is conjoined with the mid-gut. Nelson (1925, 1933) stated that the oyster does not feed during the later part of the night and that at sunrise, before active feeding has begun, the crystalline style is usually thin or may even be absent. On the flood tide, when the oysters are feeding, the style is large and firm, but when the tide has ebbed and digestion is far advanced, the style is reduced to a soft and amorphous mass of jelly. Jorgensen (1955) questions the statement by Nelson that *Ostrea* may cease to filter sea water at low tide, and he suggests that adverse conditions may have developed locally at the time of low tide. Elsewhere it has been found that *O. virginica* filters water and feeds at all times of day and night, regardless of the state of the tide. Nelson claimed that in *Ostrea* the style can be completely regenerated in only 15 minutes, whereas this may take some hours in *Modiolus*, *Anodonta*, or *Lampsilis*. Specimens of *Lasaea* living at a high level on the rocky shore also show cyclical phases in the degree of development of the crystalline style, these phases being correlated with the state of the tide and the stage of digestion (Morton, 1956). In *Mya*, on the other hand, the style-sac is separated from the mid-gut and the style will remain firm even if the animal has been out of the water for many hours.

In the Tellinacea, which feed by browsing on the surface of the substratum with the inhalant siphon, the stomach contains much coarse material. Here the style is strongly formed, and it serves to triturate the stomach contents against the gastric shield (Yonge, 1949). But in *Glossus*, Owen (1953) found the style to be too delicate to be likely to perform any triturating action.

In the preliminary phase of extracellular digestion in the stomach, the crystalline style is the principal, if not the exclusive, source of digestive enzymes. An amylase appears to be generally present, having been determined in *Mya* and in *Ostrea* (Yonge, 1923, 1926a), *Ensis* (Graham, 1931), and *Martesia* (Ganapati and Nagabhushanam, 1956). The amylase in *Ostrea* had an optimum activity at pH 5·9 at 43°C, while the amylase in *Ensis* displayed optimal activity at pH 6·0 and at 35°C. Similarly, a glycogenase has been demonstrated in *Mya* and in *Ostrea* (Yonge, 1923, 1926a), and an oxidase in *Ostrea* (Yonge, 1923) and in *Ensis* (Graham, 1930). Many authorities have recorded the absence of a lipase in extracts of the crystalline style, and this must be regarded at least provisionally as the normal condition, but George (1952) claims to have obtained conclusive evidence of the extracellular digestion of fats in *Crassostrea virginica, Modiolus demissus, Atrina livida,* and *E. directus.* In all these experiments he found that the style was an important source for this extracellular lipase. Ganapati and Nagabhushanam found both a maltase and a lactase in extracts of the style of *Martesia.* A cellulase has been demonstrated in the crystalline styles of *Ostrea, Mytilus, Mactra,* and *Mya* (Lavine, 1946; Newell, 1953). Morton (1953) makes the interesting suggestion that this cellulase may be linked with the occurrence of dense populations of spiro- chaets in the crystalline styles of certain bivalves.

Coe (1948) found very small quantities of cellulase were present in the gastric juices of *Mytilus californianus, M. edulis diegensis,* and *Tivela stultorum,* but he said that the enzyme was too weak to complete the diges- tion of dinoflagellates and other phytoplankton with cellulose walls. Coe did not determine the source of this cellulase. Owen (1953), working on *Glossus,* had little doubt that extracellular lipase and protease were present in the gastric cavity in very small quantities, and thought that these might have been derived from the free phagocytes which normally abound in the lumen of the bivalve gut, or perhaps from the spherules that are dis- carded by the cells of the digestive diverticula into the stomach and which may break up there. Mansour and Zaki (1947) made the same suggestion.

The conditions in the minute bivalve *Lasaea* appear to be exceptional, and are doubtless correlated with its diminutive size and the position it occupies high up on the shore where it is only covered by the sea for short periods each day. Morton found definite evidence of extracellular di- gestion, both in the stomach and in the lumen of the digestive diverticula, of *Phaeodactylum ("Nitzschia")* the flagellate *Isochrysis* and (more slowly) the diatom *Thalassiosira,* and even dogfish blood corpuscles. The blood corpuscles may be digested in as little as 2 hours, which indicates the presence even of an extracellular protease in the lumen of the gut.

These extracellular enzymes are thought to be derived from the spherules that are nipped off the tips of the cells of the digestive diverticula and are passed down to the stomach, where presumably they are broken and so release any contained enzymes. In the case of *Lasaea* this appears to be a major contribution to the programme of digestion, and not merely a minor side effect. *Lasaea* is also remarkable in the absence of wandering phagocytes in the lumen of the gut, and it may be that these cells would be superfluous in the presence of these powerful extracellular enzymes (Ballantine and Morton, 1956; Morton, 1956).

The crystalline style is composed partly of protein and is liable to be dissolved if an extracellular protease is present in the lumen of the stomach. This may not matter, since the style is destined to dissolve slowly in the stomach, so long as the dissolution of the style occurs at the right time and is confined to the lumen of the stomach. The presence of an extracellular protease in the stomach of *Lasaea* may even be advantageous in accelerating the liberation of enzymes from the style during the short phase of active digestion in this bivalve. Reid (1964) has suggested that in the cases of *Lima* and *Mya* the style must be protected in some way from proteolytic attack, an exo-peptidase being incorporated in the substance of the style in both of these.

With the exception of *Lasaea*, wandering phagocytes generally abound in the lumen of the gut of filter-feeding bivalves, where they are able to ingest particles which are too large to be passed into the digestive diverticula, and being then able to withdraw from the lumen, through the gut wall. The extent to which they contribute to the digestive process is variable. Phagocytes were abundant in the tissues and in all parts of the gut, including its lumen, in *Ostrea edulis*. Specimens which had been starved for several weeks were then fed with a suspension of dogfish corpuscles, which were taken freely into the gut. This caused a vast increase in the number of phagocytes, which ingested blood corpuscles and digested them intracellularly with the production of a mass of fat globules. The phagocytes migrated through the gut wall and the fat was transferred to the epithelial cells. Phagocytic activity was also seen in the ducts to the digestive diverticula, but not in the tubules themselves; their purpose is mainly to engulf such larger particles as would be unacceptable to the cells of the diverticula. The phagocytes also ingested and digested *Nitzschia*, and droplets of olive oil which had been fed to the oysters (Yonge, 1926a). An abundance of phagocytes and similar digestive activity by these was also found in *Cardium*, *Mya*, and *Teredo* (Yonge, 1923, 1926). Potts (1923) found that phagocytes occurred in the specialised part of the digestive diverticula of *Teredo*, and that they both ingested and digested wood fragments in the lumina of the tubules. George (1952) studied the processes of digestion and absorbtion of fat in various bivalves, producing evidence which conflicted somewhat with the earlier results of Yonge. He found a free extracellular lipase in the lumen of the stomach as mentioned above, and although he did observe ingestion and digestion of fat droplets by wandering phagocytes, he found little evidence of the transport of fat droplets through the gut wall by phagocytes, and he did not judge this to be a significant contribution to the processes of digestion.

It is generally understood that digestion in bivalves is confined to the stomach and digestive diverticula, and that digestion and absorbtion do not occur in the mid-gut or hind-gut, but Reid (1964) has found evidence of digestion in the mid-gut of both *Lima* and *Mya*. Phagocytes may often be found in the lumen of the mid-gut. Graham (1931) found evidence of lipolytic and amylolytic activity in the mid-gut of *Ensis*, and judged that this may be attributable to the presence of phagocytes there. Any experiments intended to determine the presence or absence of extracellular enzymes in any part of the lumen of the gut would have to be specially devised to exclude phagocytes from the enzyme/substrate mixtures, and to avoid bursting any phagocytes.

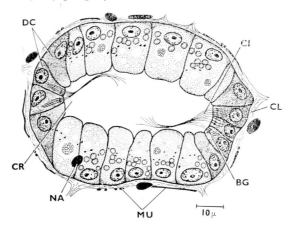

FIG. 87. Transverse section through a tubule of the digestive diverticula of *Venerupis pullastra*, showing the disposition of two tracts of darkly stained ciliated cells in the tubule. For interpretation of lettering, see pp. 263–4. (Originally published in Owen, 1955, *Quart. J. micr. Sci.* **96**, 522, fig. 5.)

We may now turn our attention to the functions served by the tubules of the digestive diverticula, having already explained the methods by which an appropriate selection of the stomach contents is conveyed thither. Yonge (1926) showed that these tubules consisted of two kinds of cells and judged that the smaller, dark-staining cells which lay in groups in the walls of the tubules were probably young cells, which would give rise to new digestive cells as and when required. Owen (1955, 1956), however, concluded that the smaller dark-staining cells were not young digestive cells, but that two contrasted types of cell were present (see Fig. 87). The small dark-staining cells lie in two or more strips which extend continuously through the length of the tubule and meet at its apex. As in the Nuculidae these cells bear long, languidly beating flagella, but they differ from those of the Nuculidae in bearing several such flagella per cell.

The other type of cell in the tubules may be termed the digestive cells; these are larger, vacuolated, and lack cilia or flagella. They are capable of four functions. First, they can ingest small particles. Particles of iron saccharate were ingested by the digestive cells of the tubules of *Mya* and of *Ostrea*, their presence within the cells being demonstrated by a modification of the prussian blue reaction (Yonge, 1923, 1926, 1926a). Owen (1955) disposed of a theory that the particles entered the cells by some purely physical means that had been mistaken for phagocytosis by demonstrating the ingestion of particles of titanium dioxide, no particles of which were less than $0.5\ \mu$ and most of which were $1.0\ \mu$ in diameter. Second, the digestive cells have the power of digesting small particles by means of intracellular enzymes. Third, they can absorb soluble matter such as is produced by amylolytic digestion in the lumen of the stomach. Indeed, these cells provide the only absorbtive area in the gut. Fourth, they can accumulate unwanted material at their distal ends and then nip off these ends as minute spherules which are then discharged via the intestinal groove to the mid-gut. Opinions differ as to whether this is solely an excretory function or whether this is also a source of extracellular enzymes in the lumen of the stomach. In all probability conditions differ in different genera, and typically these spherules may only make an insignificant contribution to extracellular digestion in the gut. *Lasaea*, however, appears to be exceptional. In this genus feeding and digestion occur in phases in relation to the state of the tide, and 2–3 hours after the cessation of feeding the contents of the digestive cells are nipped off and discharged into the stomach. This apparently ends the life of the cell, and Morton (1956) regards this as a form of holocrine secretion; the spherules presumably burst in the stomach and provide the relatively powerful extracellular enzymes which initiate the *next* cycle of digestion (see Fig. 88). The production of such spherules are the only grounds on which the digestive cells could be regarded as being secretory in bivalves. It is not yet quite clear how these spherules are discharged from the lumen of the digestive tubule. Owen (1955) showed that the tubules are surrounded by a sheath of collagen, outside which lies a delicate basketwork of circular and longitudinal muscle fibres. Owen suggests that the circular muscle fibres may be concerned with tonic balancing of internal fluid pressure, while the longitudinal muscle fibres may serve to empty the tubules. Morton observed such contractions of the tubules of *Lasaea* with the expulsion of the contents into the stomach.

The most important enzymes which have been demonstrated in extracts of the digestive diverticula are amylases. Thus Yonge (1923, 1926) found an amylase in *Ostrea* which had an optimum activity at pH 5·5 at 44·5°C. Starch, glycogen, sucrose, maltose, lactose, and two glucosides were reduced by extracts from the diverticula, and a very similar condition was found in *Mya*. Chromatography has revealed the presence of several distinct enzymes (Reid, 1964). Ganapati and Nagabhushanam (1956) found a strong amylase in the digestive diverticula of *Martesia*, which could reduce various carbohydrates including maltose, sucrose, lactose, and raffinose. It is not certain whether a distinct enzyme would be required for the reduction of each of these substances. George (1952) found a lipase in extracts of the digestive diverticula of *Venus mercenaria* and of *Ensis directus*, but this was weaker than a comparable enzyme he extracted from the crystalline style. He failed to find a lipase in the digestive diverticula of *Crassostrea virginica* or of *Atrina rigida*. A weak protease was found in the digestive diverticula of *Ostrea edulis*, and it exhibited two distinct optima, one at pH 3·7 and the other at pH 9·0 (Yonge, 1926a). Similarly, Ganapati and Nagabhushanam found a weak protease with optima at pH 4 and pH 8·5 in the digestive diverticula of *Martesia*. In contrast the digestive diverticula of *Mya* contain a strong protease (Yonge, 1923).

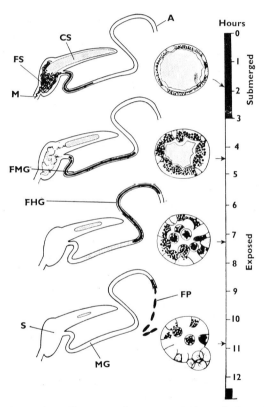

FIG. 88. *Lasaea rubra*, diagrammatic representation of the relationship between periodicity in the digestive system and the tidal regime. During the short period of submergence by the tide the crystalline style (CS) is well developed, the stomach (S) is full of food (FS), and waste material is beginning to be passed into the anterior end of the mid-gut; absorbtion is occurring in the cells lining the tubules of the digestive diverticula. Shortly after the animal is exposed by the falling tide the style begins to diminish in size; the cells of the digestive diverticula are fully loaded. Finally, the gut becomes empty and the digestive diverticula are discharging. For interpretation of other lettering, see pp. 263–4. (Originally published in Morton, 1956, *J. mar. biol. Ass. U.K.* **35**, 583, fig. 8.)

Reid (1964) made a comparison of the digestive systems of *Lima* and of *Mya* and found them to be significantly different. In *Lima*, which appears to fit the traditional theory, digestion of polysaccharides occurs largely in the stomach, and the digestion of fats and proteins occurs intracellularly in the digestive diverticula. In *Mya*, on the other hand, the emphasis seemed to be on extracellular digestion. This comparative study by Reid draws attention to the fact that digestive systems are evolving independently in the various phylogenies of bivalves, and that as yet we have reliable data on too few examples to permit generalisation.

The occurrence of a cellulase has been demonstrated in *Teredo* in a number of different ways, the earlier literature having been reviewed by Potts (1923), who found that one part of the digestive diverticula of *Teredo* is highly modified for the purpose of digesting small fragments of wood. The left posterior lobe of the digestive diverticula differed from the remainder in possessing larger lobules with wider lumina and with thinner walls composed of flattened cells. This part of the diverticula, which was white in appearance, was described by Potts as the digestive part, and its cavities were filled with fragments of wood. Phagocytes lie in the lumen of this digestive part of the diverticula and these phagocytes can put out pseudopodia and ingest wood fragments. Bits of wood are also ingested by the digestive cells forming the walls of the tubules. The other lobes of the digestive diverticula are brown in colour; they resemble the diverticula of other bivalves, but their digestive cells were not seen to contain wood fragments although these were present in the lumen. It seems clear that the specialised part of the diverticula contain an intracellular cellulase, by means of which the ship-worm is able to derive nourishment from the wood through which it drills. Lazier (1924) showed that whereas wood fragments in the stomach and in the stomach caecum would stain freely with Delafield's haematoxylin, those in the mid-gut did not stain so readily. Since this stain is a selective stain for cellulose, this suggested that much cellulose had been extracted from the wood before it was passed to the mid-gut. Dore and Miller (1923) analysed the wood in which *Teredo* was living, and also the wood which was passed through the gut of the ship-worm, and showed that in the digestive process some 80 per cent of the cellulose, and a considerable amount of the hemicellulose, had been extracted from it. Greenfield and Lane (1953) found that the cellulase extracted from the post-caecal part of the body of *Teredo* showed optimal activity at pH 5·6–6·7.

In general in the Bivalvia the products of digestion are stored in the form of glycogen and fat.

The function of the mid-gut and hind-gut is chiefly to consolidate the faeces. This is effected by secretion of mucus into the lumen and by a change in the pH which increases the viscosity of the mucus. Typically it is only necessary to make the faeces sufficiently firm to prevent them from fouling the mantle cavity. In the Tellinacea the mid-gut is notably longer in those species which take great quantities of mud into the mantle cavity. In such bivalves as *Scrobicularia*, *Macoma*, *Abra*, etc., the inhalant siphon collects food material from the surface of the substratum, which is usually muddy. Here it is important that the faeces should retain their form for a long period after they have been extruded to ensure that the same material is not collected again by the inhalant siphon in the near future. It is for this reason, doubtless, that the mid-gut is especially long in such forms.

The feeding mechanisms of bivalves do not include qualitative selection of the food material, and much may be swallowed at the mouth which is not digestible. For this reason one cannot judge the basic food requirements of a bivalve from an examination of gut contents. Naked flagellates will be very quickly digested and will leave no trace, whereas indigestible organisms may remain easily recognisable. Nelson (1933) found enormous numbers of a small nematode, *Chromadora*, in the stomach of *O. edulis*, and thought that these were being digested by some undetermined enzyme which could penetrate the cuticle of the nematode. It is far more probable that the *Chromadora* were collected and swallowed inadvertently, were dying quickly in the acid stomach contents, and were undergoing auto-digestion. Conversely, the sorting mechanisms of the stomach are not 100 per cent perfect, and food organisms may chance to pass quickly through the stomach and be discovered undigested, and perhaps unharmed, in the mid-gut or in the faeces. The occurrence of such an organism in the mid-gut or faeces does not prove that it is not acceptable food to the bivalve in question.

The Tridacnidae probably feed to a certain extent in the normal way upon material collected by the ctenidia, swallowed at the mouth and digested by the extracellular amylase of the crystalline style, followed by intra-cellular digestion in wandering phagocytes and in the epithelium of the digestive diverticula. The crystalline style of a 3-ft long *Tridacna derasa* was 34 cm long and 0·5 cm in maximum width, which suggests that this organ is not without importance. The digestive diverticula, however, are greatly reduced in numbers although they are normal in histological structure. The main source of food in the Tridacnidae is derived from the symbiotic association with zoo-xanthellae which has had so marked an effect upon the morphology of the family. The zoo-xanthellae only occur within phagocytes which lie in great numbers in blood spaces in the mantle. The zoo-xanthellae are "farmed" in the superficial tissues of the inner lobe of the mantle margin, being particularly abundant around the lens-like hyaline organs which may transmit sunlight deeper into the mantle tissues (but see p. 132). The zoo-xanthellae contain a small assi-milation product and a small pyrenoid, but have a relatively large accumu-lation of starch; the cell wall apparently lacks cellulose, so there is no hindrance to the ultimate digestion of the plant cell by the phagocyte. The zoo-xanthellae in the mantle tissues are always found to be undigested, whereas vast numbers of zoo-xanthellae in various stages of digestion lie in phagocytes around the gut and in the spaces among the digestive diverticula. At some stage in the history of these symbiotic plant cells they are evidently transported by the phagocytes to the vicinity of the gut, to be digested there by the phagocytes. Transport of symbionts from the mantle to the region of the gut must be rapid, for phagocytes containing plant cells were very rarely found in samples of blood taken from the heart. This additional source of food and method of digestion is doubtless responsible for the great size that can be reached in this family; *T. derasa* is the largest bivalve ever to have occurred in the history of the world, and this great size could not have been gained solely by the orthodox ctenidial feeding mechanisms of bivalves. Symbiosis also occurs in the Horse-shoe Clam, *Hippopus*, but the condition is less advanced here than in *Tridacna*. *Hippopus* possesses no hyaline organs, it has fewer zoo-xanthellae, fewer phagocytes in the viseral mass, and the digestive diverticula are correspondingly better developed than in *Tridacna*.

Gastropoda

The Gastropoda are remarkable for the diversity of their feeding mechanisms, and for the wide variety of foods on which they subsist. As is indicated in Chapter 2, the main classificatory units of the Gastropoda cannot be distinguished from one another on the basis of feeding habits and food preferences. The Prosobranchia in particular display almost every conceivable food preference, while the Opisthobranchia are nearly as diversified in their requirements. It seems clear that the Gastropoda underwent an explosive radiation at a very early stage in their evolutionary history, this explosion largely concerning food selection and methods of feeding. Both the Opisthobranchia and the Pulmonata must have originated in very ancient times from an early and unspecialised prosobranch ancestry, and all three divisions of the class have radiated on the basis of feeding habits, with some measure of parallel evolution.

The three great divisions of the Gastropoda are not as clearly distinguished from each other as one might expect on the concept of three monophyletic lineages of great antiquity. To mention only a few examples, Fretter and Graham (1949) have recommended transferring the Pyramidellidae from the Prosobranchia to the Opisthobranchia; Fretter (1943) considers that *Onchidella* is best regarded as an opisthobranch which has paralleled the Pulmonata in acquiring a lung, while Morton (1955) shows that consideration of the stomach of *Onchidella* suggests pulmonate affinities; again, the taxonomic position of *Siphonaria* is rather a problem, for this marine limpet does not conform entirely satisfactorily with either the

Opisthobranchia or the Pulmonata (Morton, 1955). It therefore seems possible that the Opisthobranchia and the Pulmonata are polyphyletic groups and that these might be linked with the Prosobranchia by early offshoots of problematical affinity.

On these grounds it seems simplest and most advisable to ignore here the main classificatory subdivisions of the Gastropoda, but to distinguish between herbivores and carnivores, and to draw attention to specialised features where these occur.

Herbivores

Undoubtedly the primitive condition, which is found in archaeogastropod prosobranchs, some primitive opisthobranchs, and some primitive pulmonates, is the digestion of minute particles of plant origin which have been collected in various ways by the radula. In the archaeogastropod *Patella* the food consists mainly of unicellular algae and diatoms, with only rare fragments of a larger seaweed. As this material is drawn into the buccal cavity by the radula this is lubricated by admixture of secretions from the four salivary glands, which do not contain any enzymes. The mixture passes backwards in a centrally sited food-groove which is flanked by a series of lateral pouches of the fore-gut. These oesophageal pouches pour on to the food a secretion containing an amylase which shows optimal activity at pH 6·2 at 30°C. With the exception of this amylase there is a general lack of extracellular enzymes in the lumen of the gut, where healthy diatoms and other algae may be found. There is a valve in the anterior part of the mid-gut which probably serves to close the mid-gut and so to guide food entering from the fore-gut into the single duct leading to the digestive diverticula. The digestive diverticula are the only site for the ingestion of such experimental food material as iron saccharate, dogfish blood corpuscles, and *Nitzschia*. Extracts of the digestive diverticula contain a weak protease which has two optima at 30°C, one at pH 5·6 and the other at pH 8·2; this protease is intracellular in action. Indigestible waste matter is excreted by the digestive diverticula in a mucoid string and this is compacted into a faecal rod, together with material passing directly backwards from the stomach, in the loops of the mid- and hind-gut. Posterior to the orifice of the duct from the digestive diverticula the gut does not display any digestive or absorbtive function, and no phagocytes are to be found there. The great length of this part of the gut is for the purpose of consolidation of the faeces, a matter of the greatest importance in view of the need to avoid fouling the mantle cavity (Graham, 1932).

Although intracellular digestion is generally regarded as a primitive feature, it is now known that various primitive members of the phylum Mollusca, e.g. the Polyplacophora, the Nuculidae, and the Fissurellidae, all exhibit extracellular digestion. The stomach of *Scutus breviculus* (Fissurellidae) appears to show secondary simplifications associated with the increased importance of extracellular digestion. Much of the interior of the stomach is not ciliated but is covered by a cuticular sheath. Ciliary sorting mechanisms are lacking and material is transported from the stomach to the digestive diverticula by muscular compressions of the stomach wall. The tubules of the digestive diverticula comprise three types of cell: absorptive, excretory, and flagellated. Digestion is apparently exclusively extracellular, there being no evidence of ingestion of particles by the cells of the digestive diverticula (Owen, 1958).

The archaeogastropod *Tegula funebralis* feeds by scraping up the film of minute organisms growing on the surfaces of rocks and by browsing on inter-tidal algae. A considerable proportion of the food material consists of structural carbohydrates, and little is yet known of the extent to which molluscs can hydrolyse such energy-rich polysaccharides. Galli and Giese (1959) investigated the digestive enzymes of the fore-gut and of the mid-gut of *T. funebralis* and, at the same time, they cultured many microorganisms found in the gut of *Tegula* and investigated their individual digestive capacities. They found the extracts of fore-gut, including the salivary glands, to contain highly active amylase and laminarase, whilst alpha- and beta-glucosidases were also present. Extracts of the mid-gut, including the digestive diverticula, contained highly active amylase and laminarase, and also a cellulase, an alginase, a fucoidase, and an iridophycinase. Clearly *Tegula* "has a remarkable arsenal of digestive enzymes with which to attack the structural carbohydrates which comprise the bulk of its food material". Of the eighteen strains of bacteria which were isolated from the gut of *Tegula*, only four proved active in utilising structural and other carbohydrates present in algae, including alginic acid, carrageenin, and agar. However, test counts of bacteria in the gut of *Tegula* indicated that such micro-organisms were present only in small numbers, and it is unlikely that they would contribute significantly to the digestion of food material in the gut of *Tegula*.

Some very primitive microphagous herbivores possess a style-sac which is conjoined with the anterior part of the mid-gut. This sac contains a rod of stiff, viscous mucus which is studded with faecal matter and which is continuous posteriorly with the faecal string. This rod projects into the stomach where it aids digestion mechanically by stirring the contents, since it is rotated by the ciliated epithelium of the style-sac. This "protostyle", as it is termed by Morton (1953), differs from a crystalline style in lacking any adsorbed digestive enzymes. Morton considers that from such a primitive condition a true crystalline style has arisen twice by parallel evolution, once in the Bivalvia and once in some of the Prosobranchia.

In such primitive prosobranchs as feed continuously on minute particles, the stomach possesses a ciliated sorting area of ridges and grooves by means of which coarse particles are rejected into the gut, and the anterior end of the mid-gut has become specialised to form a style-sac containing a crystalline style. The structural and functional similarity with the stomach of some filibranch bivalves is striking. The mesogastropod *Aporrhais* is such a case, the stomach having well-developed ciliary sorting mechanisms, a gastric shield, and a large crystalline style. *Aporrhais* is a specialised detritus feeder, selecting only material of plant origin, but it does not possess a cellulase (Yonge, 1937).

The closely related *Pterocera* feeds exclusively by nibbling the finest algal growths, the algal fragments being passed via the crop into the stomach together with secretions from the salivary glands. Extracellular digestion occurs in the crop and in the stomach, an amylase and a glycogenase being derived both from the salivary glands and also from the crystalline style. *Pterocera* is a highly specialised herbivore since it possesses a powerful cellulase, which operates extracellularly in the lumen of the stomach, having an optimum activity at pH 5·85 which approximates to the conditions of the stomach contents. The source of this enzyme was not determined. The cellulase reduces the cell walls of plant material to glucose, and at the same time liberates the cell contents for digestion by other enzymes. A weak extracellular lipase which operates in the stomach is probably derived from the digestive diverticula.

The ciliary sorting area passes coarse particles to the mid-gut for disposal, and passes the finest particles to the two ducts from the digestive diverticula. There must be some mechanism for conveying fluid from the stomach into the lumina of the digestive diverticula, and this may possibly be by a counter-current entering the ducts comparable to that found in bivalves by Owen (1955), or by changes in tonus of the muscle fibres enveloping the stomach and digestive diverticula. Absorbtion probably occurs exclusively within the digestive diverticula, and the amylase and glycogenase found in extracts of the diverticula are probably intracellular enzymes (Yonge, 1932).

A crystalline style occurs in four super-families in the Prosobranchia, namely the Strombacea, e.g. *Aporrhais* and *Pterocera* discussed above, the Calyptraeacea, the Rissoacea, and a few families in the Cerithiacea, e.g. the Turritellidae and the Melaniidae. In all these cases the presence of the crystalline style is associated with the absence of oesophageal glands (the proteolytic secretions of the oesophageal glands, where these are present, would prevent the development of a crystalline style). Fretter (1962) concludes that the four super-families mentioned above have a common ancestry.

The amphibious, freshwater mesogastropod, *Pomacea canaliculata*, feeds on the leaves of *Pistia* and *Vallisneria*, swallowing large pieces of leaf, together with sand and grit, and these are triturated in a part of the stomach which is specialised to form a gizzard. The salivary glands supply a strong amylase, the secretory cells of the digestive diverticula provide a strong cellulase, and digestion is exclusively extracellular, in the stomach (Andrews, 1965).

The marine slug *Onchidella*, which was generally regarded as a pulmonate, is now thought by Fretter (1943) to be an opisthobranch, as mentioned above. *Onchidella* rasps food off rock surfaces and consumes pieces of alga up to 1·0 cm long, diatoms, *Polystomella*, pieces of sponge, detritus, and sand. This material is passed along the gut mainly by muscular action, and the stomach contents are digested at leisure during high tide, the diatoms forming a large part of the diet. Apparently digestion is entirely extracellular, a highly advanced condition, and the soluble products of digestion are passed into the digestive diverticula. There is no ingestion of particles by the cells of the digestive diverticula. Absorbtion of soluble matter occurs only in the digestive diverticula, and after the absorbtive phase there is a phase of excretion by both the digestive and the excretory cells. Although *Onchidella* is mainly herbivorous, there is apparently no trace of a cellulase, and much of the plant material which is swallowed must pass through the gut undigested (Fretter, 1943).

The sea hare, *Aplysia*, feeds by cutting short lengths of fine algal fronds, which are passed back into the crop together with secretions of the salivary glands. These glands contain two types of gland cells, one of which probably secretes mucus. An extract of the salivary gland contains a strong amylase which displays optimal activity at pH 5·6, and also a weak protease. The outstanding feature of the gut of *Aplysia* is the great development of the anterior part which comprises an oesophagus, a crop, a gizzard, and a filter chamber, all of which have great freedom of movement in the body cavity. As a result of considerable muscular activity the contents of the gut are churned forwards and backwards and all the extracellular enzymes which operate in the lumen are mixed together.

There is only very slight trituration of the lengths of algal frond in the buccal cavity, but peristaltic contractions at the posterior end of the crop force the weed against the numerous teeth, which are polygonal with broad bases. As these teeth converge on the fragments of weed these are compressed and finally crushed, liberating the cell contents and exposing them to the action of enzymes. Within the crop the food receives digestive juices which are regurgitated from the stomach and which were derived from the digestive diverticula. These include a strong amylase, sucrase, lactase, maltase, and pectinase, and also a cellulase which is surprisingly weak. An extracellular lipase occurs in the lumen of the gut and is probably supplied by the digestive diverticula. The protease, which was presumably supplied by the salivary glands, had two optima, one at pH 2·8–3·4 and the other at pH 8·4–9·4.

The digestive glands were considered to include as many as four types of cell, two of which were secretory, a third being absorbtive and the last being solely excretory. An alternative interpretation is that fewer cell types are present, but that these individually pass through two or more contrasted phases of activity, each phase being characterised by certain cytological features. None of these cells was seen to ingest particles, and it is concluded that digestion is wholly extracellular. However, a glycogenase was found in extracts of the diverticula but not in the contents of the gut, and this may be a tissue enzyme. The ducts of the digestive diverticula open into the stomach adjacent to a ciliary sorting area; material excreted from the digestive diverticula is conveyed by this sorting area into a caecum of the stomach where it is consolidated into a faecal rod. This faecal rod is passed into the mid-gut, where it combines with undigested matter passed back from the stomach, and material from these two sources is combined in the mid-gut to form faeces (Howells, 1942).

Among the Opisthobranchia, various ascoglossan nudibranchs such as *Elysia, Limapontia, Actaeonia,* and *Hermaea* have adopted the habit of feeding on seaweeds by slitting open the plant cells and sucking out their contents. Owing to the fluid and highly digestible nature of the food, the alimentary canal is modified and greatly simplified. The cell contents are sucked up by alternate expansions and contractions of the buccal mass and, in order to force the secretions of the salivary glands through the muscular walls of the buccal mass, the salivary duct bears a contractile vesicle near . its distal end. The liquid food, mixed with salivary gland secretions, is carried backwards through the oesophagus by ciliary action; in *Elysia* and *Hermaea* a muscular pouch at the posterior end of the oesophagus serves to pump this mixture into the stomach and into the ducts of the digestive diverticula. The stomach is small and simple, and is deficient in ciliary sorting mechanisms, and the ducts to the digestive diverticula have wide orifices. Movement of food material is mainly by muscular action, contraction of the stomach walls will force food into the ducts of the digestive diverticula, and the latter may aid in the process by dilating and sucking in the contents of the stomach. Such ciliary currents as remain are chiefly concerned with passage of waste material to the mid-gut for elimination as faeces. Owing to the nature of the food there is very little faecal matter. Since there is no mantle cavity there is little need for consolidation of the faeces and, in consequence, the hind-gut is relatively short and is deficient in gland cells. In *Limapontia capitata* and in *L. cocksi* the rectum is very short and the anus lies high up on the dorsal surface of the body; the faeces are fluid and are ejected clear of the body. It is interesting to note, however, that in *L. depressa* and also in *Alderia modesta* the rectum is longer and the anus is almost terminal: both these species occur terrestrially, in salt marshes, and this elongation of the rectum is clearly a secondary adaptation to the habitat to ensure that the faeces do not soil the body (Fretter, 1941; Gascoigne, 1956). The closely related *Tridachia crispata* closely resembles *Elysia* in its processes of feeding and digestion. Here the tubules of the digestive diverticula are apparently differentiated into two parts, the proximal part of the tubules being absorbtive while the distal ends of the tubules are secretory. Contraction of muscle fibres which ramify through the body will force down the secretions of the digestive diverticula so that these will mix with the food in the stomach. At the same time waste material will be forced out of the tubules into the stomach. Although symbiotic zooxanthellae are present, lying free in the connective tissue, there is no reason to think that they are consumed as they are in the bivalve *Tridacna* (Yonge and Nicholas, 1940).

The Pulmonata comprise two sub-orders, the aquatic Basommatophora, which occur in fresh waters or inter-tidally on the sea shore, and the terrestrial Stylommatophora. The Pulmonata are generally regarded as highly specialised herbivores, and Karrer and Illing (1925) demonstrated a cellulase with optimal activity at pH 5·28 in the case of *Helix*. Myers and Northcote (1958) have listed all the digestive enzymes which have been detected in the digestive tract of *H. pomatia*, and emphasised the roles played by cellulase and lipase. Commenting that a carbohydrase is usually specific with reference both to the linkage and to the monosaccharide constituents of the polymer it splits, Myers and Northcote concluded that *H. pomatia* possesses no less than twenty distinct carbohydrases. An intercellular proteinase was also present, but was relatively weak in action which is not surprising since this snail is herbivorous. Nevertheless, many pulmonates are also carnivorous, even cannibalistic, on occasion.

One of the most primitive members of the Basommatophora is the tiny *Otina otis* which occurs inter-tidally on European shores in the vicinity of high-water mark of neap tides. *Otina* feeds like a limpet by rasping the rocks with its radula. The stomach of *Otina* is small and thin-walled, and it bears two digestive diverticula. A small muscular pouch on one side of the stomach serves as a simple gizzard. Muscular contractions force fluid out of the food mass in the lumen of the stomach and into the digestive diverticula. The tubules of the digestive diverticula comprise many digestive cells, which ingest small particles, and fewer excretory cells. The digestive diverticula exhibit distinct phases of ingestion and of excretion. "Fragmentation phagocytes" nipped off the tips of the digestive cells pass down into the lumen of the stomach and provide a supply of free extracellular enzymes in the stomach, but digestion is primarily intracellular within the digestive diverticula (see Fig. 89). Indigestible material in the stomach combines with material excreted by the digestive diverticula to form a protostyle which rotates, due to ciliary activity, and is gradually passed backwards into the mid-gut (Morton, 1955a).

Among the Basommatophora the aquatic *Lymnaea stagnalis* cuts suitable sized pieces of plant material and swallows these without much trituration. Sand in the gut is essential to digestion, serving for trituration of the gut contents; sand is most abundant in the gizzard, but also occurs in the crop and in the recurrent passage of the pylorus. If deprived of sand, *L. stagnalis* loses its appetite and eventually dies. Secretions from the salivary glands, containing an amylase and also a tryptic enzyme, are poured on to the food as it is swallowed, and extracellular digestion commences in the post-oesophagus. The food is thoroughly comminuted in

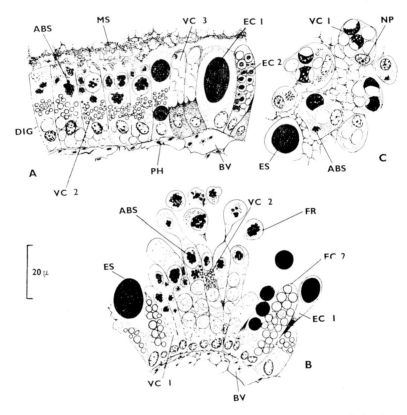

Fig. 89. *Otina otis*. A, epithelium of the digestive gland at the phase of absorption. B, the same at the stage of fragmentation and excretion. ABS, material absorbed from the stomach, in vacuoles; DIG, digestive cells. EC1, EC2, two types of excretory cell. Note the process of fragmentation of the tips of the digestive cells. C, a mass of fragmentation phagocytes in the lumen of the stomach. For interpretation of other lettering, see pp. 263–4. (Originally published in Morton, 1955a, *J. mar. biol. Ass. U.K.* **34**, 130, fig. 8.)

the crop, gizzard, and in the anterior part of the retrocurrent passage of the pylorus, with the aid of the sand grains normally found in these parts of the gut, and extracellular digestion is continued throughout this process. An extracellular cellulase is also operative during the process of trituration. Peristaltic contractions and also violent alternating pulsations force the fluid contents of the crop, pylorus, and "hepatic" ducts to and fro in

swirling currents, and strong ciliary currents in the vestibule to the digestive diverticula beat outwards, preventing the entry of coarse particles.
The pylorus acts as a filter chamber, allowing only finely divided particles
and the soluble products of extracellular digestion to enter the digestive
diverticula (see Fig. 90). Movement of the contents of the digestive diverti-

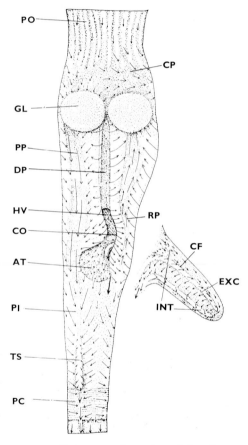

FIG. 90. *Lymnaea stagnalis appressa.* A portion of the alimentary canal has
been slit open along the mid-ventral line in order to display the ciliary sorting currents of the post-oesophagus (PO), the crop (CP), the gizzard, the pylorus, the hepatic vestibule (HV), the atrium (AT), and the anterior part of the
pro-intestine (PI). The ciliary currents on the inner surface of the caecum are
also shown. For interpretation of other lettering, see pp. 263–4. (Originally
published in Carriker, 1946, *Biol. Bull. Woods Hole* **91**, 101, figs. 3 and 9.)

cula is effected by the thin sheath of muscle fibres which invests the follicles.
There is some digestion by free amoebocytes in the lumen of the gut. The
digestive cells in the follicles of the digestive diverticula are capable of

ingesting small particles, as was demonstrated by feeding specimens with a mixture of egg white and lamp black (carbon), particles of the latter being found within the digestive cells. The final stages of digestion are therefore intracellular. Fats and carbohydrates can be absorbed by the walls of the pylorus.

The digestive diverticula contain lime cells, the function of which may perhaps be to buffer the contents of the stomach. The digestive cells of the diverticula form vacuoles, one per cell, in which indigestible material is accumulated. The digestive cells seem to conform to a common rhythm, first undergoing a digestive phase and then an excretory phase. These vacuoles are discharged into the lumen of the duct, where they combine to form a mucoid string which is passed outwards to the stomach. A similar mucoid string of indigestible matter is discharged at the same time from the stomach caecum, and at this time the pulsations of the stomach wall are arrested. These mucoid strings combine in the mid-gut together with the loose material passed out from the gizzard, to form the faeces. In the mid-gut water is absorbed and mucoid secretions are poured on to the faeces. Mucus being an amphoteric protein, the slightly alkaline conditions in the mid-gut will increase its viscosity and so aid in the consolidation of the faeces (Carriker, 1946).

Morton (1955) has studied various very primitive genera in the family Ellobiidae in the sub-order Basommatophora. These have a number of primitive structural features which indicate their origin from the Proso-branchia. *Leucophytia bidentata* occurs in crevices and under stones on the sea shore at and below high tide mark. The normal form of *Ovatella myosotis* is terrestrial on salt marches while a sub-species occurs intertidally. Finally, two species of *Carychium* occur terrestrially at inland sites. In all cases particles of plant origin are picked up from the substratum, admixed with mucoid secretions from the supra-pedal gland. Further supplies of mucus are poured on to the food from buccal mucous glands, and also from the salivary glands in the case of *Leucophytia*. A second type of gland cell in the salivary glands resembles glands found in the oesophagus, and these may supply extracellular enzymes. The stomach comprises three parts, a thin-walled anterior chamber, a posterior chamber with thick muscular walls, which serves as a gizzard, and a caecum which may contain a ridged and grooved ciliary sorting area. The gizzard is lined by a tough cuticle. The two masses of the digestive diverticula open into the stomach either by two separate ducts or by one common duct. Regular peristaltic waves in the wall of the anterior chamber convey food mixed with mucus into the gizzard, where intermittent contractions of the circular muscle compress the contents. There are no ciliary sorting mechanisms, the chief function of cilia being to hinder the entry of particles into the ducts of the digestive diverticula. However, in *Ovatella* a strong ciliary rejection tract runs from the mouth of each of the ducts from the digestive diverticula to the mid-gut, and these tracts remove waste matter discarded by the diverticula.

Muscular contractions of the gizzard squeeze fluid and finely divided particles out of the contents and, since the orifice of the mid-gut is closed at this time, this material is forced up the ducts to the digestive diverticula. Final contractions of the gizzard compact the remaining debris into a faecal rod which protrudes into the mid-gut. The sphincter at the mouth of the mid-gut nips off small portions of this faecal rod one at a time. The tubules of the digestive diverticula are composed of digestive and of excretory cells, and the tubules display distinct phases of activity. During the absorbtive phase numerous small vacuoles in the digestive cells become filled with material absorbed from the lumen of the tubule. After digestion is completed there is a phase of fragmentation when non-nucleate fragments are nipped off from the tips of the cells and are discarded into the stomach. These discarded fragments may provide enzymes for preliminary extracellular digestion in the lumen of the stomach, the final stages being absorbtion and intracellular digestion only effected by the digestive cells of the digestive diverticula.

Waste material from the diverticula is passed to the mid-gut where it joins debris from the gizzard to be formed into loosely bound faeces. Secretions from the abundant mucus glands aid in forming a continuous faecal rope which is pushed towards the anus partly by ciliary action and partly by peristalsis. Wandering amoebocytes are more common in the mid-gut than in the stomach, but are particularly abundant in the hind-gut, and are presumably scavenging the gut contents (Morton, 1955).

Among the Stylommatophora, *Helix* lacks the ciliary sorting mechanisms which are found in *Lymnaea*, and to a lesser extent in the Ellobiidae. *Helix* also lacks a muscular gizzard, this deficiency being made good by the presence of a more powerful cellulase than that possessed by *Lymnaea*.

In general in the Stylommatophora most of the extracellular enzymes are supplied by the salivary glands.

In the slugs *Arion* and *Agriolimax* and in the snail *Helix*, it was found that in the digestive diverticula the digestive cells and the lime cells absorbed phosphorus and iodine from the food; phosphorus was also absorbed by the epithelium of the mid-gut and, to a lesser extent, by the epithelium of the crop. These facts were ascertained by feeding these pulmonates with lettuce leaves on which measured doses of P^{32} and of I^{131} had been allowed to dry; the labelled phosphorus and iodine were respectively in the forms of sodium di-hydrogen phosphate, and sodium iodide (Fretter, 1952). The labelled iodine was absorbed by the digestive cells and by the lime cells of the diverticula (Fretter, 1952).

Carnivores

It is easy to envisage that among the ancestral gastropods, which fed by rasping small particles of food off rock surfaces, the habit of feeding chiefly or wholly on sponges would inevitably arise at an early stage in various independent lineages. The adoption of this simple carnivorous habit may have required only very slight changes in the processes of feeding and digestion. Among the Prosobranchia, the archaeogastropod *Diodora* is such an example. Here digestion is probably largely extracellular by enzymes derived from the oesophageal pouches and perhaps also from the digestive diverticula. Digestion of the cells of the sponge material must be a comparatively simple matter, but the elimination of masses of sharp sponge spicules without damage to the epithelium of the gut is a major consideration. These spicules are compounded into a faecal rod which superficially resembles a crystalline style, and this extends backwards through the mid-gut. Such a faecal rod may be distinguished from a crys· talline style by its mode of formation, the direction of its movement through the gut, and by the lack of absorbed extracellular enzymes (Graham, 1939).

In another lineage, the Opisthobranchia, various dorids feed on sponges, as, for example, *Jorunna tomentosa* and *Archidoris pseudoargus*, both of which feed on the crumb-of-bread sponge, *Halichondria panicea*. These two dorids differ to a surprising extent in the basic features of their digestive processes. *A. pseudoargus* swallows large lumps of sponge, up to 2 mm³, and has no means of breaking these down mechanically. The salivary secretions contain amylolytic and lipolytic enzymes, and extracellular digestion commences in the buccal cavity. The animal feeds voraciously and the stomach becomes distended with food which is mixed with enzymes, the latter being derived chiefly from the digestive diverticula. The digestive diverticula contain three types of gland cells and these undergo three distinct phases of secretion, absorbtion, and ingestion, and, finally, of extrusion. (The term extrusion is regarded as preferable to "excretion", the latter term being best applied to the elimination of nitrogenous wastes.) When digestion is well advanced, the sponge spicules tend to fall to the bottom of the stomach. The products of extracellular digestion are passed into the digestive gland, partly by muscular contractions of the stomach wall and partly by ciliary activity. The final stages of digestion are intracellular, in the cells of the digestive gland (Forest, 1953).

The digestive processes of *J. tomentosa* have been studied in detail, and are very different from those of *A. pseudoargus* (Millott, 1937). *Jorunna* has no salivary glands, and the food particles are mixed with mucus secreted from glands in the oesophageal walls. When the animal is feeding actively the food coalesces into a large mass around which a copious secretion of mucus protects the oesophagus from damage by the spicules of the sponge. The digestive diverticula open into the mid-gut by three wide-mouthed ducts into which the food material can enter with ease. Ciliary action directs food material into the first of these ducts, and many particles are ingested either by phagocytes or by the digestive cells in the tubules. Material which is not ingested is then returned to the mid-gut, and it may then enter another duct and be exposed once more to phagocytosis within this second lobe of the digestive diverticula. Waste material passing out of the ducts from the digestive diverticula is not isolated in any way from potential food material, which suggests that the mechanics of the mid-gut system are at a low level of organisation.

Material which is finally discarded from all the lobes of the digestive diverticula, in which sponge spicules will predominate, is passed into a caecum of the mid-gut. Ciliary currents direct this material to the head of the caecum where it is revolved and compacted with freshly secreted mucus into a bolus. Muscular contractions of the caecal wall knead this bolus and this is probably ejected into the intestine by peristaltic contractions.

Digestion in *Jorunna* is exclusively intracellular, there being no free enzymes in any part of the gut. In this respect the opisthobranch *Jorunna* is more primitive even than the archaeogastropod, *Diodora*. The phagocytes are confined to the lumen of the digestive diverticula and of the midgut, thereby differing from the phagocytes of the Bivalvia, which are essentially blood cells that can wander through all the tissues of the body. The phagocytes are an effective way of increasing the surface area for ingestion, and they are probably able to engulf many particles that are too large for ingestion by the cells of the digestive diverticula. The occurrence of ingestion by phagocytes and by digestive cells was experimentally demonstrated by means of iron saccharate and fish blood corpuscles. The intracellular enzymes of the digestive diverticula include two proteases, one with optimal activity in an acid medium, and the other with an optimum in an alkaline medium; a lipase was found, and possibly also an amylase. Since the powerful protease is exclusively intracellular, it may be presumed that the carnivorous habit had arisen from an earlier herbivorous mode of life, and this very primitive carnivorous condition may be compared with that occurring in the Septibranchia.

The mesogastropods *Cerithiopsis* and *Triphora* also feed on sponges, but show greater specialisation to this habit. *Cerithiopsis* prefers to feed on the sponge *Hymeniacidon sanguinea*, and since it sucks up the deeper tissues by means of its proboscis, it tends to avoid the dense layer of spicules at the surface of the sponge. The salivary glands are enlarged, a copious supply of their secretions lubricates the jaws and the radula, and may contain an extracellular enzyme. In *Cerithiopsis* the greatly enlarged left salivary gland contains a majority of a type of gland cell which does not secrete mucus, while in *Triphora* both glands are large, and both contain three kinds of gland cells. In both genera the stomach is small, unspecialised, receives two ducts from the digestive diverticula, and serves as a crop in which food lies while undergoing at least partial digestion by extracellular enzymes. The interior of the stomach does not bear any protective chitinous or cuticular lining, but is only protected from the sharply pointed sponge spicules by a thin sheet of mucus (Fretter, 1951).

Turning to the eolid gastropods we encounter still more specialised carnivores, for many of these are stenophagous and feed exclusively on particular species of hydroid or other coelenterates. In *Aeolidia*, *Cratena*, and *Facelina*, a copious supply of mucus from the pedal gland is poured over the food before it is eaten, and this provides adequate lubrication of the fore-gut. The salivary gland has two lobes, an anterior lobe which consists largely of mucus gland cells, and a posterior lobe in which a different type of gland cell predominates. This second type of gland cell may perhaps produce an extracellular enzyme. Movement of the food in most parts of the gut is effected by muscular contractions, but there are also ciliary tracts in the stomach, mid-gut, and ducts to the diverticula. Food is churned to and fro in the ducts of the digestive diverticula, and is passed into the cerata and out again. The digestive diverticula secrete extracellular enzymes which operate in the lumen of the stomach; these include an amylase with optimal activity at pH 6·4 at 30°C, and a protease with two optima at 30°C, one at pH 4·0, and the other at pH 6·2. The stomach contents were found to be at pH 6·4. In addition to these, the digestive diverticula possess an intracellular glycogenase which does not occur in the lumen of the gut. Lime cells in the digestive diverticula may serve to control the pH of the stomach contents.

The soluble products of extracellular digestion are absorbed partly by the cells of the stomach wall, by the cells of the digestive diverticula, and to a lesser extent by the cells lining the ducts of the latter. Small particles can be ingested by the cells of the diverticula. After a meal much food may pass into the intestine, where no digestion or absorbtion can occur, and this is a complete waste of the food. After the animal has had a meal and digestion is far advanced, for about a week excretory spherules continue to accumulate in the cells of the digestive diverticula, and in due course these spherules are nipped off and are discharged into the stomach as waste matter. The faeces are not consolidated, this being unnecessary since the animal does not possess a mantle cavity (Graham, 1938).

Aeolis pilata feeds upon the hydroid *Pennaria tiarella* which possesses no less than six different kinds of nematocysts. Of these all are digested except for the microbasic mastigophores which are passed undamaged through the stomach and digestive diverticula and are concentrated in the cnidosacs at the tips of the cerata of the sea slug. If the stomach contents of *A. pilata* are examined shortly after a meal it is found that all the nematoblasts have been eroded and most of the nematocysts lie naked and undischarged in the lumen of the stomach. The stomach contents are passed via gland ducts into the digestive diverticula which invade the cerata, and it was found that within 35 minutes after feeding all the nematocysts had left the stomach. Within the cerata each digestive tubule is subdivided into three sections—a basal digestive section which communicates via a slender ciliated canal with an apical cnidosac. Of all the food material which enters the basal digestive sections of the cerata, only the nematocysts are allowed to pass through the ciliated canals into the cnidosacs. Here all types of nematocyst are ingested and all except the microbasic mastigophores are digested. The latter accumulate, and are oriented with their opercular ends directed towards the lumen of the cnidosac (Kepner, 1943).

The nematocysts stored in the cerata can be ejected from the cnidosacs through pores at the tips of the cerata, and large numbers of these nematocysts are liable to explode at the time of ejection. Should a predacious fish take an eolid into its mouth it is therefore liable to be stung on the sensitive lips, or within the mouth, by the exploding nematocysts. This unpleasant experience would cause the fish to reject the eolid, unharmed, from its mouth.

Many eolids do not feed on coelenterates, and accordingly do not possess this defensive mechanism. In such cases protection from predators may be afforded alternatively by distasteful secretions from special glands which occur in the cerata. This subject has been extensively investigated and reviewed by Edmunds (1966).

Among the nudibranchs, *Calma glaucoides* is a most unusual and highly specialised carnivore which feeds on the eggs of various shore fishes. The contents of the eggs are highly digestible and in consequence the gut of *Calma* has become blind, there being no anus and no intestine (Evans, 1922). Suitable eggs are only available during the breeding seasons of the fish concerned (*Lepadogaster*, *Blennius*, etc.), and *Calma* gorges on eggs at this time and stores the yolk for long periods in a particularly large gastric sac which has been developed for this purpose (Rowett, 1946).

As a good example of a predacious carnivore, we may consider *Philine aperta*, which feeds on diatoms, foraminifera, small molluscs, and other organisms living in sub-littoral sands. The prey is swallowed whole, together with a certain amount of sand. The posterior part of the oesophagus is modified to form an anterior crop, a gizzard, and a posterior crop, the exit from the latter being guarded by a sphincter muscle. The walls of the gizzard are highly muscular and are lined internally by three calcareous plates, each of which has a boss on its inner surface. A sheet of muscle surrounding the gizzard operates these plates by means of which food is crushed. Small molluscs, such as the snail *Hydrobia*, may remain in the gizzard for many hours while the shell is being smashed. During all this time the food is subjected to the action of the enzymes contained in the secretion of the salivary glands, including an amylase and a glycogenase which show optimal activity between pH 6·3 and pH 7·0 at 30°C. From time to time small jets of fluid and finely divided particles are allowed to pass through the sphincter into the stomach, where the above-mentioned extracellular digestion continues. The stomach contents are at about pH 6·2–6·4. The stomach is set on one side, embedded in the digestive diverticula, so coarse and indigestible matter can be passed directly from the posterior crop to the mid-gut without entering the stomach and causing damage there. The stomach is not concerned with sorting and with disposal of coarse matter, and is considerably simplified in accordance with this, and with extracellular digestion. When coarse material is passed to the mid-gut by peristalsis, the orifice leading to the stomach is closed by circular muscles and by muscular lobes which guard the opening.

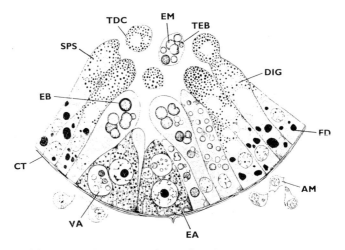

Fig. 91. *Philine aperta*. A representative portion of a transverse section through a tubule of the digestive gland. The digestive cells (DIG) contain droplets of fat (FD) near their bases; the tips of the digestive cells (TDC) may be cut off and liberated into the lumen of the tubule. There are excretory cells of two kinds (EA, EB), and those of type B also cut off their tips (TEB) and liberate these into the lumen of the tubule, so disposing of excretory masses. For interpretation of other lettering, see pp. 263–4. (Originally published in Fretter, 1938, *Trans. roy. Soc. Edinb.* **59**, 606, fig. 5.)

    The digestive diverticula contain three types of cells, of which the digestive cells are the most abundant. These can ingest particles into vacuoles and can also extrude spherules of secretory or excretory matter. The two remaining types of cell are both excretory and one of these, which accumulates lime, may be responsible for adjustments to the pH of the stomach contents (see Fig. 91). The digestive diverticula secrete a lipase and a protease into the stomach, the latter having two optima at pH 6·0 and at pH 8·5. Thus digestion is principally extracellular, though amoebocytes become active and phagocytosis is displayed by these and also by the digestive cells. The digestive cells of the digestive diverticula pass through distinct phases of ingestion and secretion, followed by an excretory phase of the other two types of cell. In the midgut the pH is raised to 8·4, which increases the viscosity of the mucus and so aids the consolidation of the faeces. Mucus from the hypobranchial gland aids in the passage of faeces through the mantle cavity (Brown, 1934; Fretter, 1938).

The Pyramidellidae are parasitic, feeding on the body fluids of a variety of invertebrate hosts by means of a long and delicate proboscis. The ducts of the two salivary glands unite and pass centrally down a stylet which opens above the mouth and in the centre of a sucker at the tip of the proboscis. The purpose served by the salivary secretion is not known. A muscular sac on the course of the gut, the buccal pump, serves to draw body fluids from the host into the stomach of the snail. The stomach is small and is extremely simple, being indistinguishable from the ends of the ducts of the digestive diverticula. This simplification is correlated with the simple nature of the food, which requires little manipulation or digestive action (Fretter, 1949). Many terrestrial slugs and snails are carnivorous and must be equipped with an efficient extracellular protease.

Conclusions

Correlated with the wide variety of feeding mechanisms, and of food preferences, the Gastropoda display considerable variety in their digestive processes, and it is not possible to offer an account of one single "typical" digestive process. The most primitive conditions seem to concern *intermittent* feeding on particles of rather a coarse nature, scraped off rock surfaces. Here the secretions of the salivary glands serve for lubrication, and they lack digestive enzymes. Oesophageal pouches are present and may supply an amylase, and there is no crystalline style. In more advanced prosobranchs, e.g. in many of the Taenioglossa, there is *continuous* feeding on minute particles and the stomach is adapted to sorting of particles and a crystalline style has developed. The presence of the style with its amylase is correlated with the loss of the oesophageal pouches. The appearance of the crystalline style in these comparatively advanced prosobranchs indicates that it has arisen independently from that of the Bivalvia. This is a remarkable case of parallel evolution.

The salivary glands may develop an amylase in more advanced herbivores, and a protease in carnivores. Similarly, in carnivores the oesophageal pouches may develop a free protease.

Primitively, digestion was wholly intracellular, in the digestive diverticula, the stomach achieving considerable complexity where it is concerned in sorting particulate material. With the advent of extracellular digestion in the more advanced gastropods, the digestive diverticula became an important source of extracellular enzymes, ciliary sorting mechanisms in the stomach became less important, and the movement of the contents of the gut was effected primarily by muscular activity.

Among herbivores the development of an extracellular cellulase was particularly important for browsers on sea weeds and on land vegetation, and in particular this must have provided a major impetus in the evolution of the Pulmonata.

Among some carnivores an interesting feature is the specialisation of the fore-gut into a number of distinct chambers including, for example, a crushing gizzard. The development of a powerful extracellular protease enabled gastropods to exploit the carnivorous mode of life very thoroughly.

Cephalopoda

Although all the Cephalopoda are carnivorous, the class exhibits considerable variety in mode of life and in feeding habits. The familiar *Octopus* is a benthic hunter associated with rocky shores, which lurks in some hollow or crevice in the rocks and emerges from this den to attack crabs which venture near. The octopus crawls over the rocks by means of the suckers on its arms, and attacks small moving objects such as crabs, throwing itself upon the crab and enveloping it with the membrane which joins the bases of the arms. The crab is bitten and killed, and is carried back to the den to be consumed, the discarded shells of the prey accumulating on a midden in the vicinity of the den.

The cuttle-fish, *Sepia*, is also associated with the sea bed, being found primarily over a sandy bottom, and often in association with a canopy of seaweeds such as the eel grass *Zostera*. *Sepia* feeds on crabs crustaceans such as prawns which occur in these plant associations, the prawn being recognised by its size and its movement. The cuttle-fish may cruise slowly in the water just above the bottom, disturbing this with jets of water from its funnel; any prawn which is buried in the sand and which is flushed out by these jets from the funnel, will endeavour to bury itself again, and will be recognised by these movements. Alternatively, a cuttle-fish may adopt a suitable camouflage and then rest upon the bottom, partly buried by sand thrown up by undulations of its lateral fins. In such a position the cuttle-fish may lie in wait for passing prawns, the two long tentacles being shot out suddenly to grasp the prey.

In contrast to these benthic forms, the squids *Loligo* and *Alloteuthis* are pelagic hunters which spend their entire lives swimming in midwater and they never approach the sea bed. Squid feed on fish, crustacea, and also on smaller squid; fish are seized behind the head, which is bitten off and then dropped. The fish is bitten into small pieces by the two horny jaws, the radula serving as a conveyor belt to pass the meat back into the oesophagus. The alimentary canal of the fish is not eaten, but hangs in the water until the anus is reached, when the rectum is bitten off and the alimentary canal is allowed to fall. Similarly the posterior vertebrae and the tail are rejected (Bidder, 1950).

These various examples differ from one another in occupying different habitats, in having different modes of life, and by capturing different kinds of prey by different techniques. They also evidently differ in their degrees of appetite, for *Octopus* will attack and kill another crab when it has only recently fed, or when it still has a crab held under its interbrachial mem.- brane, while *Sepia* may not feed for some considerable time after having had a meal. The duration of digestion is also very variable: digestion of a meal takes 18 hours in *Octopus*, 12 hours in *Sepia*, and only about 6 hours in *Loligo* (Bidder, 1950). It follows that there will be other differences, perhaps very important ones, in the digestive processes of the different subdivisions of the Cephalopoda, and it is urgent that generalisations should not be made at the present time on the digestive processes of the Cephalopoda as a whole from observations on individual genera.

In the squids, *Loligo* and *Alloteuthis*, digestion is entirely extracellular, and is well suited by its great rapidity to the active mode of life of these pelagic animals. Fish, crustacea, or smaller squid, are held by the suckers on the arms and are bitten, being poisoned by injection of secretions from the posterior salivary—or poison—glands. The poison is tyramine. The function of the anterior salivary glands is not known for certain, but it may be to lubricate the food before it is swallowed. The oesophagus is highly muscular and small pieces of meat are passed to the stomach by peristalsis, the interior of the oesophagus being protected from damage by fish bones, etc., by a continuous cuticular lining. The highly muscular stomach is lined by a soft and distensible cuticular sheath which protects it from damage by bones and scales when the contents are being churned about. Prior to the meal a supply of pancreatic secretion accumulates in the caecum, from which it is transferred to the stomach on the arrival of food masses. The food stays in the stomach for $1\frac{1}{2}$–2 hours, during which time it is subjected to gastric digestion under the influence of enzymes in the secretions of the "pancreas". The violent churning action of the stomach wall accelerates this process.

When this preliminary gastric phase of digestion is well advanced, some of the fluid contents of the stomach, together with small particles in suspension, is allowed to pass through a sphincter from the stomach to the caecum. No solid matter is allowed to enter the more delicate caecum, and the coarse, indigestible, skeletal residues are retained in the stomach for the time being. The secretions of the "liver" are allowed to accumulate in the tubules and ducts of that organ, there being a sphincter muscle at the head of each of the main "hepatic" ducts. On the admission of fluid food into the caecum, this sphincter is temporarily relaxed, and some of the "hepatic" secretion is passed down into the caecum, where extracellular digestion is completed in the following 3 or 4 hours. During this period the sphincter between the stomach and the caecum is closed, and the contents of these two organs are isolated from each other. The sphincter muscles at the head of the "hepatic" ducts are also closed, preventing the liberation of further "hepatic" secretion, while "pancreatic" secretions can still pass down the lower parts of the ducts and enter the stomach. As a result of this ingenious arrangement gastric digestion of more recently swallowed meat can proceed in the stomach while the digestion of an earlier meal is still being completed in the caecum.

The caecum may be considered to comprise two main parts. Firstly, there is an anterior spirally coiled part which receives the ducts from the two parts of the digestive gland ("liver" and "pancreas"), and which communicates both with the stomach and with the intestine. This part of the caecum bears on its walls a system of radiating pleats which centre upon the orifice of the intestine. The grooves between these pleats lead to the intestinal orifice, and ciliary currents on the pleats and in the grooves collect particulate material and bind it in mucus. This waste material is then passed in a series of mucoid strands into a deep ciliated groove which passes along the wall of the intestine towards the anus. All such particulate material is rejected. Secondly, there is a simple, highly extensible, caecal sac which has smooth walls bearing a ciliated epithelium. The walls of the caecal sac are muscular and are capable of great distension and contraction. Ciliary currents on its walls serve to circulate the fluid contents of the sac.

Prior to a meal the sphincter muscle at the head of each hepatic duct is closed, retaining "hepatic" secretion in the gland; "pancreatic" secretion accumulates in the sac of the caecum, which contracts to expel the secretion into the stomach when food masses are swallowed. A sphincter muscle between the stomach and the caecum can be closed in order to isolate the contents of the two chambers, or can be relaxed to allow the fluid products of gastric digestion to pass into the caecum. A valve at the orifice of the intestine into the caecum prevents the untimely passage of fluid from the caecum to the intestine, while permitting mucoid strands of waste particles to be passed out to the intestine in a deep ciliated groove. Absorbtion of the soluble products of digestion occurs in the walls of the caecal sac, and also in the intestinal walls, but nowhere else. Fluid can be passed out from the caecum and be held in the intestine while absorbtion is proceeding there. Finally, at the end of the digestive process, by adjustment of the opening of the stomach into the caecum, and the valve at the intestinal orifice, hard indigestible matter such as fish scales, crustacean limbs, etc., can be passed directly from the stomach into the intestine, thereby avoiding risk of damage to the soft and unprotected walls of the caecum. Manipulation of the various sphincters and valves which control the flow of enzymes, and the movement of the contents of the gut from one part of the gut to another, is presumably under the control of the sympathetic system, of which the splanchnic ganglion lies near at hand. The walls of the intestine and the rectum are protected by a liberal flow of mucus, which invests the mass of waste material.

Both parts of the digestive gland are concerned solely with secretion—and perhaps with excretion. The contents of the caecum are not passed up the hepatic ducts to the gland, and the epithelia of the gland do not absorb the soluble products of digestion nor do they ingest small particles.

The digestive gland consists of two distinct parts, the "liver" which forms a compact mass on each side of the oesophagus, and the "pancreas" which is a diffuse gland investing the walls of the ducts which drain the "liver", and discharging into these ducts. The "liver" is pervaded by a system of blood capillaries, and the organ serves partly as a source of digestive enzymes and partly as a storage organ for food reserves. The epithelium of the tubules contains only one type of cell with distinct immature and mature phases. There is histological evidence that these cells produce two different types of secretion, which mingle in the hepatic ducts. As has been reported for various other molluscan types, the swollen tips of these secreting cells are nipped off into the lumen of the tubule. The "pancreas" also contains only one kind of cell, which is secretory. Secretion by these cells seems to be rhythmic, with long resting periods, and the products of secretion accumulate in the caecal sac (Bidder, 1950).

It is possible that the epithelium of the pleated collecting organ in the caecum may be secretory, and the secretion may be an enzyme. In *Sepia* these ciliated leaflets were demonstrated to produce an activator (Romijn, 1935).

Key to the Lettering on the Figures

A	Anus.	CI	Cilia.
AA	Anterior adductor muscle.	CL	Collagenic layer.
AB	Aortic bulb.	CO	Caecal opening.
ABS	Material absorbed from the stomach, in vacuoles.	CP	Crop.
		CR	Crypt.
AM	Amoebocyte.	CS	Crystalline style.
AO	Aorta.	CT	Connective tissue.
AP	Appendage of palp.	DC	Darkly staining cells.
AT	Atrium.	DDD	Opening of duct to the digestive diverticula.
AVV	Auriculo-ventricular valve.		
BCI	Ingested blood corpuscle.	DH	Dorsal hood.
BCL	Blood corpuscle in lumen of digestive tubule.	DIG	Digestive cell.
		DP	Dorsal passage.
BCP	Blood corpuscle in process of being ingested.	EA	Excretory cell type A.
		EB	Excretory cell type B.
BG	Basal granules.	EC1	Excretory cell type 1.
BM	Basement membrane.	EC2	Excretory cell type 2.
BV	Blood vessel.	EM	Excretory masses from excretory cell type B.
C	Ctenidium.		
CF	Caecal fold.	EO	Excretory organ.
CG	Chitinous girdle.	ES	Excretory spheres.

ETY	Extension of major typhlosole into opening of duct to the digestive diverticula.
EX	Exhalant water current.
EXC	Excurrent tubule.
EXD	Exhalant portion of main duct to digestive diverticula.
F	Foot.
FD	Fat droplet.
FHG	Food in hind-gut.
FMG	Food in mid-gut.
FP	Faecal pellets.
FR	Fragmented tip of digestive cell, constricted off into the lumen.
FS	Food in stomach.
G	Gonad.
GL	Lobe of gizzard.
GS	Gastric shield.
HG	Hypobranchial gland.
HV	Hepatic vestibule.
IG	Intestinal groove.
IN	Inhalant water current.
IND	Inhalant portion of main duct to the digestive diverticula.
INT	Incurrent tubule.
L	Labial palps.
LC	Left caecum.
LP	Left pouch.
M	Mouth.
MD	Main duct of digestive diverticula.
MF	Mantle fusion.
MG	Mid-gut.
MS	Mucus from the stomach.
MT	Minor typhlosole.
MU	Muscle fibres.
NA	Nucleus of amoebocyte.
NBC	Nucleus of ingested blood corpuscle.
NP	Nucleus of "phagocyte".
NT	Nucleus of tubule cell.
O	Oesophagus.
OC	Older, vacuolated cells of tubules of the digestive diverticula.
OIG	Origin of intestinal groove.
OMG	Opening of mid-gut.
P	Small particles in suspension.
PA	Posterior adductor muscle.
PC	Pellet compressor.
PG	Pedal gape.
PH	Phagocytic blood cell.
PI	Pro-intestine.
PO	Post-oesophagus.
PP	Procurrent passage.
PRM	Posterior retractor muscle.
PSA	Posterior sorting area.
R	Rectum.
RC	Right caecum.
RM	Rim of orifice of oesophagus into stomach.
RP	Retrocurrent passage.
RT	Ciliated rejection tract.
S	Stomach.
SA	Ciliated sorting area
SD	Secondary ducts of digestive diverticula.
SPS	Spherules of secretory material.
SS	Style sac.
T	Tubules of digestive gland.
TDC	Tip of digestive cell, cut off from epithelium.
TEB	Tip of excretory cell type B, cut off from epithelium.
TO	Tongue of major typhlosole.
TS	Typhlosole.
TY	Major typhlosole.
V	Ventricle.
VA	Large vacuole in excretory cell of type A.
VC1 VC2	Vacuoles in digestive cells.
VC3	Vacuole in excretory cell at the absorbing phase.
W	Wall of stomach caecum.
YC	Crypts of young cells in tubules of digestive diverticula.

Reference List

ALLEN, J. A. (1958) On the basic form and adaptations to habitat in the Lucinacea (Eulamellibranchia), *Phil. Trans.* B, **241**, 421–84.

ALLEN, J. A. and TURNER, J. F. (1974) On the functional morphology of the family Verticordiidae (Bivalvia) with descriptions of new species from the abyssal Atlantic. *Phil. Trans. B*, **268**, 401–536.

ANDREWS, E. B. (1965) The functional anatomy of the gut of the prosobranch gastropod *Pomacea canaliculata* (D'Orb.) and of some other pilids, *Proc. zool. Soc. Lond.* **145**, 19–36.

BIDDER, A. M. (1950) The digestive mechanism of the European squids *Loligo vulgaris*, *L. forbesi*, *Alloteuthis media* and *A. subulata*, *Quart. J. micr. Sci.* **91**, 1–44.

BROWN, H. H. (1934) A study of a tectibranch gasteropod mollusc, *Philine aperta* (L.), *Trans. roy. Soc. Edinb.* **58**, 179–210.

CARRIKER, M. R. (1946) Observations on the functioning of the alimentary system of the snail *Lymnaea stagnalis appressa* Say., *Biol. Bull. Woods Hole* **91**, 88–111.

CHUANG, S. H. (1959) The structure and function of the alimentary canal in *Lingula unguis* (L.) (Brachiopoda), *Proc. zool. Soc. Lond.* **132**, 283–311.

COE, W. R. (1948) Nutrition, environmental conditions, and growth of marine bivalve molluscs, *Sears Foundation: J. Mar. Res.* **7**, 586–601.

DINAMANI, P. (1964) Feeding in *Dentalium conspicuum*, *Proc. malac. Soc. Lond.*, **36**, 1–5.

DORE, W. H. and MILLAR, R. C. (1923) The digestion of wood by *Teredo navalis*, *Univ. Calif. Publ. Zool.* **22**, 383–400.

EDMUNDS, M. (1966) Protective mechanisms in the Eolidacea (Mollusca, Nudibranchia), *J. Linn. Soc. (Zool.)* **46**, 27–71.

EVANS, T. J. (1922) *Calma glaucoides*, a study in adaptation, *Quart. J. micr. Sci.* **66**, 439–55.

FOREST, J. E. (1953) On the feeding habits and the morphology and mode of functioning of the alimentary canal in some littoral dorid nudibranchiate molluscs, *Proc. Linn. Soc. Lond.* **164**, 225–35.

FRETTER, V. (1937) The structure and function of the alimentary canal of some species of Polyplacophora (Mollusca), *Trans. roy. Soc. Edinb.* **59**, 119–64.

FRETTER, V. (1938) The structure and function of the alimentary canal of some tectibranch molluscs, with a note on excretion; *Trans. roy. Soc. Edinb.* **59**, 599–646.

FRETTER, V. (1941) On the structure of the gut of the ascoglossan nudibranchs, *Proc. zool. Soc. Lond.* **110**, 185–98.

FRETTER, V. (1943) Studies in the functional morphology and embryology of *Onchidella celtica* (Forbes and Hanley) and their bearing on its relationships, *J. mar. biol. Ass. U.K.* **25**, 685–720.

FRETTER, V. (1948) The structure and life history of some minute prosobranchs of rock-pools: *Skeneopsis planorbis* (Fabricius) *Omalogyra atomus* (Philippi), *Rissoella diaphana* (Alder), and *Rissoella opalina* (Jeffreys), *J. mar. biol. Ass. U.K.* **27**, 597–632.

FRETTER, V. (1949) The structure and mode of life of the Pyramidellidae, parasitic opisthobranchs, *J. mar. biol. Ass. U.K.* **28**, 493–532.

FRETTER, V. (1951) Observations on the life history and functional morphology of *Cerithiopsis tubercularis* (Montagu) and *Triphora perversa* (L.), *J. mar. biol. Ass. U.K.* **29**, 567–86.

FRETTER, V. (1952) Experiments with P^{32}, and I^{131} on species of *Helix*, *Arion* and *Agriolimax*, *Quart. J. micr. Sci.* **93**, 133–46.

FRETTER, V. (1962) Interrelations of monotocardian gastropods, *Proc. 1st European Malac. Congr.*, pp. 55–59.

FRETTER, V. and GRAHAM, A. (1949) The structure and mode of life of the Pyramidellidae, parasitic opisthobranchs, *J. mar. biol. Ass. U.K.* **28**, 493–532.

GALLI, D. R. and GIESE, A. C. (1959) Carbohydrate digestion in a herbivorous snail, *Tegula funebralis*, *J. exp. Zool.* **140**, 415–40.

GANAPATI, P. N. and NAGABHUSHANAM, R. (1956) Digestive enzymes of *Martesia striata* Linn., *Curr. Sci.* **25**, 156–7.

GASCOIGNE, T. (1956) Feeding and reproduction in the Limapontiidae, *Trans. roy. Soc. Edinb.* **63**, 129–151.

GEORGE, W. C. (1952) The digestion and absorbtion of fat in lamellibranchs, *Biol. Bull. Woods Hole* **102**, 118–27.

GRAHAM, A. (1931) On the morphology, feeding mechanisms and digestion of *Ensis siliqua* (Schumacher), *Trans. roy. Soc. Edinb.* **56**, 725–51.

GRAHAM, A. (1932) On the structure and function of the alimentary canal of the limpet, *Trans. roy. Soc. Edinb.* **57**, 287–308.

GRAHAM, A. (1938) The structure and function of the alimentary canal of aeolid molluscs, with a discussion on their nematocysts, *Trans. roy. Soc. Edinb.* **59**, 267–307.

GRAHAM, A. (1939) On the structure of the alimentary canal of style-bearing prosobranchs, *Proc. zool. Soc. Lond.* **109**, 75–112.

GRAHAM, A. (1949) The molluscan stomach, *Trans. roy. Soc. Edinb.* **61**, 737–78.

GREENFIELD, L. J. and LANE, C. E. (1953) Cellulose digestion in *Teredo*, *J. biol. Chem.* **204**, 669–72.

HOWELLS, H. H. (1942) The structure and function of the alimentary canal of *Aplysia punctata*, *Quart. J. micr. Sci.* **83**, 357–97.

JORGENSEN, C. B. (1955) Quantitative aspects of filter feeding in invertebrates, *Biol. Rev.* **30**, 391–454.

KARRER, P. and ILLING, H. (1925) In Yonge, C. M., 1932.

KEPNER, W. A. (1943) The manipulation of the nematocysts of *Pennaria tiarella* by *Aeolis pilata*, *J. Morph.* **73**, 297–312.

LAVINE, T. H. (1946) A study of the enzymatic and other properties of the crystalline style of clams: evidence for the presence of a cellulase, *J. cell. comp. Physiol.* **28**, 183–95.

LAZIER, E. L. (1924) Morphology of the digestive tract of *Teredo navalis*, *Univ. Calif. Publ. Zool.* **22**, 455–74.

MANSOUR, K. and ZAKI, F. G. (1947) The digestive diverticula of *Unio prasidens* as organs of secretion, *Proc. Egypt. Acad. Sci.* **2**, 38–44.

MATHERS, N. F. (1970) *Studies on the physiology of feeding and digestion in bivalve molluscs.* Ph.D. Thesis, University of London.

MEEUSE, B. J. D. and FLUEGEL, W. (1958) Carbohydrate-digesting enzymes in the sugar gland juice of *Cryptochiton stelleri* Middendorff (Polyplacophora, Mollusca), *Arch. néerl. Zool.* **13**, 301–13.

MILLAR, R. H. (1955) Notes on the mechanism of food movement in the gut of the larval oyster, *Ostrea edulis*, *Quart. J. micr. Sci.* **96**, 539–44.

MILLOTT, N. (1937) On the morphology of the alimentary canal, process of feeding, and physiology of digestion of the nudibranch mollusc *Jorunna tomentosa* (Cuvier), *Phil. Trans.* B, **228**, 173–217.

MORTON, J. E. (1953) The functions of the gastropod stomach, *Proc. Linn. Soc. Lond.* **164**, 240–6.

MORTON, J. E. (1955) The functional morphology of the British Ellobiidae (Gastropoda, Pulmonata) with special reference to the digestive and reproductive systems, *Phil. Trans.* B, **239**, 89–160.

MORTON, J. E. (1955a) The functional morphology of *Otina otis*, a primitive marine pulmonate, *J. mar. biol. Ass. U.K.* **34**, 113–50.

MORTON, J. E. (1956) The tidal rhythm and action of the digestive system of the lamellibranch *Lasaea rubra*, *J. mar. biol. Ass. U.K.* **35**, 563–86.

MORTON, J. E. (1959) The habits and feeding organs of *Dentalium entalis*, *J. mar. biol. Ass. U.K.* **38**, 225–38.

MYERS, F. L. and NORTHCOTE, D. H. (1958) A survey of the enzymes from the gastrointestinal tract of *Helix pomatia*, *J. exp. Biol.* **35**, 639–48.

NELSON, T. C. (1918) On the origin, nature and function of the crystalline style of lamellibranchs, *J. morph.*, **31**, 53–111.

NELSON, T. C. (1925) Recent contributions to the knowledge of the crystalline style of lamellibranchs, *Biol. Bull. Woods Hole* **49**, 86–99.

NELSON, T. C. (1933) On the digestion of animal forms by the oyster, *Proc. Soc. exp. Biol. N.Y.* **30**, 1287–90.

NEWELL, B. S. (1953) A cellulolytic enzyme in *Ostrea edulis*, *J. mar. biol. Ass. U.K.* **32**, 491–5.

OWEN, G. (1953) On the biology of *Glossus humanus* (L.) (*Isocardia cor* Lam.), *J. mar. biol. Ass. U.K.* **32**, 85–106.

OWEN, G. (1955) Observations on the stomach and digestive diverticula of the Lamellibranchia. I. The Anisomyaria and Eulamellibranchia, *Quart. J. micr. Sci.* **96**, 517–37.

OWEN, G. (1956) Observations on the stomach and digestive diverticula of the Lamellibranchia. II. The Nuculidae. *Quart. J. micr. Sci.* **97**, 541–67.

OWEN, G. (1958) Observations on the stomach and digestive gland of *Scutus breviculus* (Blainville), *Proc. malac. Soc. Lond.* **33**, 103–14.

OWEN, G. (1961) A note on the habits and nutrition of *Solemya parkinsoni* (Protobranchia, Bivalvia), *Quart. J. micr. Sci.* **102**, 15–21.

POTTS, F. A. (1923) The structure and function of the liver of *Teredo*, the shipworm, *Proc. Camb. phil. Soc.* **1**, 1–17.

PURCHON, R. D. (1955) The structure and function of the British Pholadidae (rock-boring Lamellibranchia), *Proc. zool. Soc. Lond.* **124**, 859–911.

PURCHON, R. D. (1956) The stomach in the Protobranchia and Septibranchia (Lamellibranchia), *Proc. zool. Soc. Lond.* **127**, 511–25.

PURCHON, R. D. (1957) The stomach in the Filibranchia and Pseudolamellibranchia, *Proc. zool. Soc. Lond.* **129**, 27–60.

PURCHON, R. D. (1958) The stomach in the Eulamellibranchia; stomach type IV, *Proc. zool. Soc. Lond.* **131**, 487–525.

PURCHON, R. D. (1960) The stomach in the Eulamellibranchia; stomach types IV and V, *Proc. zool. Soc. Lond.* **135**, 431–89.

PURCHON, R. D. (1960a) Phylogeny in the Lamellibranchia *Proc. Cent. and Bicent. Congr. Biol. Singapore 1958*, 69–82.

PURCHON, R. D. (1971) Digestion in filter feeding bivalves—a new concept. *Proc. malac. Soc. Lond.*, **39**, 253–62.

REID, R. G. B. (1964) The digestive system in the Bivalvia, Ph. D. thesis, University of Glasgow.

REID, R. G. B. (1965) The structure and function of the stomach in bivalve molluscs, *J. Zool.* **147**, 156–84.

ROMIJN, C. (1935) In Bidder, A. M., 1950.

ROSEN, B. (1949) Proteases in the digestive gland of lamellibranchs, *Ark. Kemi* **1**, 205–11.

ROWETT, H. G. C. (1946) A comparison of the feeding mechanisms of *Calma glaucoides* and *Nebaliopsis typica*, *J. mar. biol. Ass. U.K.* **26**, 352–7.

YONGE, C. M. (1923) Studies on the comparative physiology of digestion. I. The mechanism of feeding, digestion, and assimilation in the lamellibranch, *Mya. Brit. J. exp. Biol.* **1**, 15–63.

YONGE, C. M. (1926) The digestive diverticula in lamellibranchs, *Trans. roy. Soc. Edinb.* **54**, 703–18.

YONGE, C. M. (1926a) Structure and physiology of the organs of feeding and digestion in *Ostrea edulis*, *J. mar. biol. Ass. U.K.* **14**, 295–386.

YONGE, C. M. (1928) Structure and function of the organs of feeding and digestion in the septibranchs *Cuspidaria* and *Poromya*, *Phil. Trans.* B, **216**, 221–63.

YONGE, C. M. (1928a) Feeding mechanisms in the invertebrates, *Biol. Rev.* **3**, 21–76.

YONGE, C. M. (1930) The crystalline style of the mollusca and a carnivorous habit cannot normally coexist, *Nature, Lond.* **125**, 444–5.

YONGE, C. M. (1932) Notes on feeding, and digestion in *Pterocera* and *Vermetus*, with a discussion on the occurrence of the crystalline style in the Gastropoda, *Sci. Rep. Gr. Barrier Reef Exped.* **1**, 259–81.

YONGE, C. M. (1936) Mode of life, feeding, digestion and symbiosis with zooxanthellae in the Tridacnidae, *Sci. Rep. Gr. Barrier Reef Exped.* **1**, 283–321.

YONGE, C. M. (1937) The biology of *Aporrhais pes-pelecani* and *A. serresiana*, *J. mar. biol. Ass. U.K.* **21**, 687–704.

YONGE, C. M. (1939) The protobranchiate mollusca, a functional interpretation of their structure and evolution, *Phil. Trans.* B, **230**, 79–147.

YONGE, C. M. (1946) Digestion of animals by lamellibranchs, *Nature, Lond.* **157**, 729.

YONGE, C. M. (1949) On the structure and adaptations of the Tellinacea, deposit-feeding Eulamellibranchia, *Phil. Trans.* B, **234**, 29–76.

YONGE, C. M. and NICHOLAS, H. M. (1940) Structure and function of the gut and symbiosis with zooxanthellae in *Tridachia crispata* (Oerst.) Bgh., *Publ. Carneg. Inst.* **517**, 287–301.

CHAPTER 6

REPRODUCTION

Synopsis

Primitively molluscs were of opposite sexes, the gametes were discharged into the sea water via the nephridia, fertilisation was random, and the small eggs developed into trochophore larvae. The reproductive function of the adult was limited to the emission of large numbers of gametes, and no protection or special provision was made for the developing embryos. This condition is retained in various modern lineages. In other lineages there have been specialisations of one kind or another.

In the Monoplacophora two pairs of gonads discharge via the third and fourth pairs of nephridia, and fertilisation is presumably at random in the sea water. It is suggested that the Mollusca are metamerically segmented animals which, except for the Monoplacophora, have only one genital segment. In the Scaphopoda and in most of the Polyplacophora there is only one, unpaired gonad, the gametes are shed into the sea and development is external. Most of the Aplacophora are specialised as hermaphrodites which discharge gametes via the pericardium and the nephridia. In most of the Bivalvia the reproductive system is simple, the sexes are separate, gametes are discharged via gonad ducts into the mantle cavity and thence into the surrounding water, and fertilisation and development are external. Certain bivalves show specialised features such as hermaphroditism, sex reversal, incubation of developing young, etc. In the Cephalopoda sexes are separate, the single gonad only having one gonad duct; courtship is followed by copulation and spermatophores are transferred to the female by one arm of the male—the hectocotylus—which is modified to some extent. In the female nidamental and accessory nidamental glands provide secondary coverings to the eggs.

The Gastropoda show considerable diversification as regards their reproductive systems. In the Prosobranchia some members, in the Archaeogastropoda, retain the primitive ancestral condition; in more advanced types the right kidney has aborted save for the renal contribution to the genital duct and the genital duct is carried across the floor of the mantle cavity either as a ciliated and glandular groove or by the enclosure of this groove to form the pallial gonad duct.

In most of the higher prosobranchs the male has a penis and after copulation the fertilised eggs receive nourishment from the albumen gland at the inner end of the pallial oviduct and a protective coat from the capsule gland which comprises the distal part of the pallial oviduct. As regards the complexity of their reproductive systems, the Neritacea are more advanced than other members of the Archaeogastropoda, and this raises a difficult question in connection with their systematic position. In some mesogastropods the male has lost the penis and in the female the pallial oviduct has secondarily re-acquired the form of an open ciliated gutter into which water-borne spermatozoa or spermatozeugmata are drawn from the inhalant water stream. The spermatozeugmata are compound structures consisting of a large oligopyrene sperm bearing large numbers of normal, eupyrene sperms attached to its tail; these modifications are attributed to lateral compression of the mantle cavity which would make normal copulation a matter of great difficulty. Some prosobranchs exhibit protandric hermaphroditism, e.g. *Crepidula*, and functional hermaphroditism is found throughout the Opisthobranchia and Pulmonata, copulation usually effecting reciprocal fertilisation. In some forms a triaulic condition has evolved, there being a vagina as well as an oviduct and a vas deferens. This triaulic condition has been achieved independently in the Neritacea. Hypodermic impregnation occurs in some species, e.g. *Limapontia capitata*.

Although many molluscs retain the primitive unisexual state, hermaphroditism is typical of the Aplacophora, the Opisthobranchia, the Pulmonata, the bivalve order Anatinacea, and in many smaller categories scattered in otherwise unisexual groups. The mechanism governing the transition between unisexuality and hermaphroditism must therefore be very delicately poised. It is concluded that the young individual has the latent capacity to develop either or both sets of sexual organs, and suppression of one or other of these sets is determined by its genetical constitution under given environmental conditions. A genetical or an environmental change could therefore lead to a transition from the unisexual to the hermaphrodite condition, or vice versa. Thus hermaphroditism could emerge wherever it ultimately proves to have survival value. Functional hermaphroditism is the outcome when ovary and testis develop and mature synchronously. If the rate of development of the testis is accelerated as compared with that of the ovary, then the result is protandric hermaphroditism, e.g. *Crepidula fornicata*, *Teredo navalis*, and *Ostrea edulis*. Very favourable environmental conditions in the early stages of growth may cause the eclipse of the initial male phase in a protandric hermaphrodite, and the reproductive pattern of a given species may differ from one locality to another, e.g. *Tivela stultorum* and *O. virginica*. In some gastropods which are protandric hermaphrodites tactile association with other individuals of the same species may influence the duration of the male phase, e.g. *C. nivea*. In some species there is regular alternation of sex, e.g. in *O. edulis* there is believed to be one male and one female phase each year. The mechanism of sex change is not yet known. It is difficult to see in what way a species may benefit from protandry, or from simultaneous hermaphroditism. Among the Bivalvia, *Xylophaga dorsalis* is unique in possessing a receptaculum seminis by means of which eggs can be fertilised as they leave the oviduct. Parthenogenesis is known in few examples, e.g. *Paludestrina* (*Hydrobia*) *jenkinsi*, and as a result of this capacity one individual of this species could establish itself and found a new colony unaided.

Where eggs and sperm are broadcast and fertilisation is the result of random contact in the sea water there is nevertheless some measure of co-ordination, so that large numbers of individuals, of both sexes, spawn at the same time. In high latitudes this may be an environmental factor such as the achievement of a threshold temperature. Thus *O. edulis* spawns at 15–16°C, and *Hiatella* spp. breeds at temperatures below 12°C. The spawning of *O. edulis* is influenced by lunar periodicity, and occurs at the period of spring tides. In contrast, *Patella vulgata* spawns in periods of strong winds and protracted heavy surf. In *O. virginica* the spermatozoa carry a hormone, diantlin, which causes relaxation of the adductor muscle, enlargement of the ctenidial ostia, and increased ciliary activity on the ctenidia. The presence of spermatozoa in the inhalant water stream of a female therefore induced an increased rate of pumping of water by the ctenidia, after which ovulation commenced. Similarly, the presence of fresh oyster eggs in the mantle cavity of a male specimen of *O. virginica* caused it to emit a steady stream of spermatozoa within 30 seconds of the eggs touching the ctenidia. Thus extrinsic and/or intrinsic factors serve to concentrate the spawning of a majority of the animals in the locality into a short period of time. A high concentration of gametes in the water ensures the success of the random process of fertilisation and the production of large numbers of larvae, which suffer very heavy mortality in the period prior to the acquisition of sexual maturity. *Bankia gouldi* may be unique in the Bivalvia in practising a process of copulation by passing the exhalant siphon into the inhalant siphon of a neighbouring specimen perhaps for the purpose of transport of spermatozoa.

Where the eggs are retained in the parental mantle cavity, or pallial oviduct, until after fertilisation they are usually provided with nourishment and/or a protective coat. In the Gastropoda a penis has probably evolved independently in a number of lineages; it is naturally sited on the right side of the head aligned with the exhalant water stream. An open ciliated groove on the floor of the mantle cavity carried the spermatozoa up to the side of the penis and it is natural that in some lineages this groove should ultimately become enclosed to form the pallial vas deferens which penetrates the penis to its tip. Concurrent with these developments a ciliated gutter on the floor of the mantle cavity of the female would aid in transport of foreign spermatozoa to the site of the genital aperture, and with the development of a penis in the male, this ciliated gutter became enclosed to form a pallial oviduct. The pallial oviduct accommodates the penis during copulation, and provides nourishment and a protective coat to the fertilised eggs.

In the Prosobranchia there is much variation in the degree of provision of food and protection to the eggs. In the most primitive condition the unfertilised eggs and the sperm are broadcast, and the embryo develops to a free swimming trochophore, e.g. *Patella*; in a more advanced condition the embryo only hatches as a veliger larva, e.g. *Littorina littorea*, while in the most advanced condition the young emerge from the capsule as small crawling snails, the planktonic distributive larval stage having been eclipsed, e.g. *L. littoralis*. In some cases a number of eggs are enclosed together in a common capsule and the first embryos to develop eat the remainder and hatch at the veliger stage, e.g. *Theodoxus fluviatilis*, or as crawling snails, e.g. *Nucella lapillus*. In some sedentary forms the capsules of developing eggs are protected beneath the shell of the mother, e.g. *C. fornicata*, or are deposited in specially protected places, e.g. *Trivia monacha*. Such functional advances in the provision for and care of the developing embryos have been pursued independently in innumerable lineages, as seen in Table 5 (pp. 302–4). Where the eggs receive more nourishment and more protection, they are generally fewer in number and there is a lower mortality rate than elsewhere. An advantage is gained by those species which have been able to extend their range and colonise otherwise inaccessible habitats by virtue of special provision for the developing embryos; thus the Neritacea have been able to invade fresh waters, as also have several lineages of bivalves in which the eggs are incubated, e.g. *Sphaerium*, *Anodonta*, etc. The percentage of prosobranchs with pelagic larvae decreases with increasing latitude in the North Atlantic, and in the coastal waters of east Greenland none remain. These northern waters lack adequate food to support planktotrophic larvae, and suppression of the larval phase has permitted certain prosobranchs, e.g. *Acmaea rubella*, and certain bivalves, e.g. species of *Astarte* and *Macoma* to colonise the coasts of east Greenland. Some marine bivalves incubate eggs in their ctenidia, e.g. *O. edulis*, and *Teredo navalis*; in the case of the freshwater genus *Sphaerium* there is evidence that food is passed from the maternal tissues to the embryos, which hatch as young crawling adults.

Many gastropods protect the egg capsules by retaining these within the mantle cavity until the embryos hatch, and it is not a great step from this condition to that of ovo-viviparity where the developing embryos are retained in the pallial oviduct till they hatch as veligers, e.g. *Ianthina janthina* or as crawling snails, e.g. *Planaxis sulcatus*. Many freshwater prosobranchs are ovo-viviparous, e.g. *Viviparus viviparus*, and *Hydrobia ventrosa*. The Octopoda are noted for the protection of developing eggs; the female *Octopus* remains on guard over the eggs for 2 months or more, brushing away debris or potential predators with her arms, and ventilating the eggs with squirts of water from her funnel. The female *Argonauta* carries the developing eggs within her secondary shell until they hatch.

Self-fertilisation has been recorded for the pulmonate snail *Lymnaea columella*, and for *L. stagnalis appressa*, and is presumed to occur in those colonies of *Bulinus contortus* in which the genetical trait of aphallism is predominant. Among the Bivalvia self-fertilisation may be practised as a rare exception by *Xylophaga dorsalis*, which is a protandric hermaphrodite, and which accumulates a store of spermatozoa in a receptaculum seminis close to the opening of the oviduct. It is difficult to demonstrate beyond all doubt the occurrence of parthenogenesis, but this seems to occur in the fresh water snail *Paludestrina (Hydrobia) jenkinsi;* continental specimens of this species are brackish in habit, whilst those found in Britain are fresh water, and appear to be tetraploid. *P. jenkinsi* was little known in Britain prior to 1889, but since that date it has appeared at many freshwater localities, spread perhaps by industrial barges, and aided by its parthenogenetic mode of reproduction.

At the time of metamorphosis the behaviour of the larva in some cases is likely to increase the prospect of survival; thus the larva of *Adalaria proxima* will only settle and metamorphose in the presence of the polyzoan *Electra pilosa*, which is its chief source of food as an adult. Again, oyster larvae only swim directly upwards and tend to settle on the shady undersides of overhanging objects, where such are available; the larva explores an area before it finally metamorphoses and attaches itself, and this behaviour will ensure that there will be adequate space for initial growth by the spat. On the other hand, in some species the settlement of spat is indiscriminate, over a wide vertical range, and only those which settle in an appropriate zone are likely to grow to maturity. In fresh waters the larvae of the Unionidae, and of the Mutelidae, emerge from the parent, and attach themselves to the skin of appropriate species of fish, and develop there as ecto-parasites; the larvae probably exhibit chemo-taxis when approaching the fish, and the ecto-parasitic period supplies an important distributive phase in the life history.

In the Cephalopoda, the development of the gonad of *Octopus* is controlled by the central nervous system. In the Mediterranean the male octopus becomes sexually mature at about 300 g body weight, but in the female sexual maturity is delayed till a body weight of about 1000 g is reached; up to this weight light falling on the retinae causes stimulation of two sub-pedunculate lobes in the supra-oesophageal brain mass. The sub-pedunculate lobes in turn inhibit two optic glands sited on the dorsal surface of the optic tracts. When one of the optic glands is no longer inhibited thus it enlarges, and emits a larger quantity of a gonadotrophic hormone which stimulates the ovary to mature with rapidity. Control of maturation of the testis seems to be comparable, but to operate at a much lower threshold. The mechanism which permits maturation of the gonads under normal conditions is not yet known, but it probably concerns a change in photoperiod. The delay in onset of sexual maturity may be beneficial in favouring those octopuses which excel in discriminating between favourable and unfavourable situations, and comparisons are drawn with similar systems of control of maturation of the gonads in crustacea, in insects, and in vertebrates.

In advanced members of the Gastropoda, and in the Cephalopoda, copulation may be preceded by courtship, e.g. sexual excitation by use of the dart in the snails *Helix* and *Cepaea*. Courtship displays may also serve to provide sexual recognition signals. Thus the mature male *Octopus*, which is smaller than the female, signals its sex by displaying the disproportionately large suckers near the base of its second and third arms, and thereby avoids an attack by the female. Courtship display between two *O. horridus* involved the establishment of a strong pattern of vertical stripes on that side of the male which was visible to the female, and this display was followed by mating. In *Sepia* courtship display by a mature male establishes its sex and its mature condition and this constitutes a threat to other mature males. If the other *Sepia* is a female, or an immature male, it fails to display, and a temporary association is set up between the pair. The male swims above its partner, frequently stroking it on the back with its arms, and the male may attempt copulation with its partner at any time. The sexual display of a male *Sepia* is a marked black and white zebra pattern on the upper surface of the visceral mass; colour changes in *Sepia* are effected by three types of chromatophore which are controlled by nerve fibres arising in special chromatophore lobes in the brain.

The reproductive process concerns all aspects of the replacement of the present generation by its successor. Accordingly we have to consider under this heading not only the various expressions of sexuality of the adult individual, but also courtship, copulation, egg laying, parental protection of eggs and young, developmental stage at emergence of the young, aspects of larval behaviour, and any other matters which may have some bearing on the proportions of offspring which survive to reach maturity. This can best be done separately under these various headings. First, however, let us designate what is likely to be the aboriginal reproductive pattern in the Mollusca, from which other patterns have arisen as deviations. In some cases these deviations supply a major feature in the adaptive radiation of the phylum, e.g. hermaphroditism in the opisthobranch and the pulmonate Gastropoda; in other cases deviations from the norm may be assembled to indicate probable progressive steps in an evolutionary process, e.g. the protection of eggs and early developmental stages in prosobranch Gastropoda; yet other deviations may be minor end points in adaptive radiation which possibly solve a functional problem for one particular species occupying one particular habitat, such as larviparity in some species of oyster e.g. *Ostrea edulis*.

The Primitive Molluscan Reproductive System

It is suggested as a basis for discussion that the sexes were separate, that there were no accessory sexual organs and no special genital ducts, and that the gametes were discharged via the nephridia into the sea water. The eggs were numerous, small, and lacking in yolk. Cleavage and gastrulation occurred rapidly while drifting in the sea, and the organism hatched as a planktonic trochophore larva. The parental generation was solely concerned with the emission of large numbers of gametes for which no protection was provided.

The Reproductive System

Neopilina galatheae (Monoplacophora) exhibits metamerism in many of the organ systems including the muscles, the nerve connectives, the nephridia, the ctenidia, the auricles of the heart, and the gonads. In the male there are clearly two pairs of testes, which pass their products by two gonad ducts into the third and fourth pairs of nephridia and thence to the exterior. The female system is probably identical, but the division between the first and the second pair of ovaries has not been established with certainty. The problem is to decide whether this metamerism has been derived from a segmental plan of construction possessed by the remote, common pre-molluscan ancestor, or whether it is a secondary specialisation confined to the Monoplacophora. This remains unresolved, but I myself think the former theory to be simpler, and therefore to be more readily acceptable. On this basis I judge that the gonad of the remote ancestral mollusc was probably a paired, segmental structure, confined to an exceedingly small number of reproductive segments. Possibly all early molluscan lineages other than that leading to the Monoplacophora had only one single reproductive segment, and therefore only one pair of gonads. In those molluscs in which there is only a single unpaired gonad, this may have been produced either by fusion of the originally paired gonads or by the loss of one of these gonads.

In the Monoplacophora the only feature of special interest here is the presence of two genital segments. Eggs found in the nephridia were without shells or secondary membranes, and fertilisation is presumably external (Lemche and Wingstrand, 1959).

The Scaphopoda retain a primitive reproductive system, there being a single unpaired gonad in the roof of the mantle cavity. A single gonad duct leads from the anterior end of the gonad into the right excretory organ—a condition which also occurs in the more primitive prosobranch gastropods. The animals have separate sexes, the gametes are shed into the water, and development is external.

In the Polyplacophora the sexes are separate. The gonad is paired in one genus, *Nuttallochiton*, but in the remainder there is a single median gonad which is thought to have been produced by fusion of the left and right gonads. The gonad lies anterior to the pericardium and communicates with the lateral pallial groove by means of paired gonad ducts which do not join the nephridia. In some genera the eggs are fertilised in the lateral pallial groove—presumably by spermatozoa which have been drawn in with the inhalant water current—and are retained in the pallial groove during the early stages of development, but in other cases the fertilised eggs are laid singly or in strings, and are abandoned.

The majority of the Aplacophora are hermaphrodite, the ovotestis discharging gametes into the pericardium—which is, of course, a remnant of the coelome. The gametes are voided from the pericardium by the nephridia, which are somewhat specialised as adjuncts of the reproductive system. A caecum on the proximal wall of each nephridium may serve as a sperm reservoir; the distal ends of the two nephridia fuse to form a median pouch with glandular walls where the fertilised eggs are enclosed in a capsule; copulation is known to occur in some species. *Chaetoderma* is unusual in being unisexual, with a single median gonad, and with simple, unmodified nephridia.

The reproductive system remains simple and unspecialised in most of the Bivalvia. Typically the sexes are separate and the gametes are discharged into the mantle cavity by gonad ducts which open very close to the excretory pores. In the majority of cases the eggs are fertilised by random contact with spermatozoa in the water, and develop into free-swimming bi-valved veliger larvae. Certain bivalves show specialised features such as functional hermaphroditism, sex reversal, incubation of developing embryos and young larvae within the ctenidium, and even development of a part of the ctenidium into a special brood pouch or marsupium. These specialisations will be discussed individually under appropriate headings later in this chapter.

The sexes are separate in the Cephalopoda, there being a single median gonad at the posterior end of the visceral mass. Ova, or sperm, are passed into the perivisceral coelome, and are discharged into the mantle cavity by a single gonad duct which lies on the left side of the visceral mass. In the male the gonad duct is extremely complicated and is the site of manufacture of spermatophores. In the female the ventral surface of the visceral mass bears paired nidamental glands, and paired accessory nidamental glands, secretions from which provide secondary coverings to the fertilised eggs. The male has no penis, and yet mating occurs, the spermatophores being transferred by one of the arms of the male.

Finally, we have to consider the Gastropoda. Here, in the Archaeogastropoda we find examples which exhibit extremely simple conditions approximating closely to those postulated for the primitive ancestral mollusc.

In addition, however, we find more advanced conditions in the Mesogastropoda and Neogastropoda, and functional hermaphroditism throughout the Opisthobranchia and Pulmonata. It is therefore necessary to pay considerably closer attention to the Gastropoda than to any of the other molluscan classes (see Fig. 92).

In the Archaeogastropoda there is a single gonad which is either an

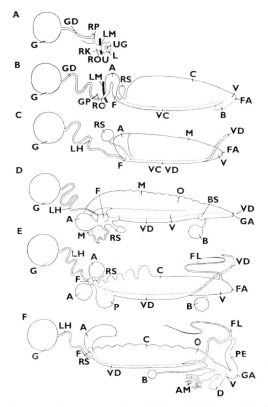

FIG. 92. Diagrammatic representations of the genital ducts of various gastropods. Broken lines represent incomplete separation of channels. A, female specimen in the Trochidae. B, female specimen of *Nucella*. The remainder are hermaphrodite specimens of C, *Actaeon*; D, *Aplysia*; E, *Onchidella*; F, *Helix*. The gonad (G) discharges via a gonad duct (GD), or little hermaphrodite duct (LH) which passes into a large hermaphrodite duct. In the latter the vas deferens (VD) is an open groove in the floor of the larger female duct. The female part of the large hermaphrodite duct is specialised into a series of sections including an albumen gland (A), a mucus gland (M), a capsule gland (C), a receptaculum seminis (RS), and a bursa copulatrix (B). The male and female systems may open together or separately. In the pulmonates there may be a flagellum (FL) a dart sac (D), and a special mucus gland (AM). For other lettering, see p. 327. (Originally published in Fretter, 1946, *J. mar. biol. Ass. U.K.* **26**, 345, fig. 7.)

ovary or a testis and which passes its products via a gonad duct into the right nephridium, and thence to the mantle cavity. In *Diodora* and in the Trochidae the gonad duct opens into the reno-pericardial duct, in *Patella* and in *Haliotis* the gonad duct opens directly into the kidney, while in other Docoglossa the gonad duct opens into the ureter. The gametes therefore pass through a duct which is of dual origin, the proximal part being derived from the gonad and the distal part from the nephridium (Fretter, 1946). Retention of this condition would seem to hinder the inception of internal fertilisation by copulation; development of the accessory structures necessary for internal fertilisation can only occur after the genital pathway has been isolated from the products of excretion.

In the Mesogastropoda and in higher gastropods the right excretory organ has disappeared except for that part of it which contributes to the genital duct—the renal oviduct of the female—the renal vas deferens of the male. In the Mesogastropoda a further relic of the right excretory organ may be found in the gono-pericardial duct, which occurs in the female only, and which links the renal oviduct with the pericardium. Its purpose is not known, but it may serve some essential function since it appears at sex reversal from the male to the female phase in *Calyptraea* and in *Crepidula* (Fretter, 1946). After dissociation of the excretory and reproductive systems in this way there is increased prospect of the emergence of internal fertilisation by copulation.

In the Mesogastropoda the gonad duct of the female has been extended across the floor of the mantle cavity alongside the similarly extended rectum by the development of a third component to the genital duct. This distal extension of the female duct was probably effected, firstly, by the development of a ciliated and glandular groove crossing the floor of the mantle cavity and, secondly, by the fusion of the ridges bounding this groove, so converting it into a closed ciliated and glandular duct. Since the distal component of the female duct has presumably been derived in this way from the floor of the mantle cavity, it is generally known as the pallial oviduct. Thus the female duct comprises firstly the oviduct *sensu stricto*, then the renal oviduct which is now the sole remnant of the right nephridium, and, finally, the pallial oviduct.

In the Mesogastropoda the ova are fertilised at the inner end of the pallial oviduct by spermatozoa which have been conveyed to this point in a variety of ways, and the main functions of the pallial oviduct are to surround the fertilised eggs with a supply of nourishing material and then with a protective coat or capsule. The inner end of the pallial oviduct is specialised to form an albumen gland for the former purpose, while the distal three quarters is developed into a capsule gland. In some, e.g. *Littorina*, and the Neogastropoda, there is a bursa copulatrix adjacent to the orifice of the pallial oviduct, for the reception of spermatozoa from the penis of the male. The spermatozoa are then transported inwards along a ciliated groove in the floor of the pallial oviduct to a receptaculum seminis sited at the inner end of the pallial oviduct. Spermatozoa are liberated from the receptaculum as and when required for the purpose of fertilisation of eggs arriving from the renal oviduct.

Pedal glands may participate in the reproductive process; in the neogastropod *Nucella lapillus* the egg capsule is passed to the ventral pedal gland, by which it is moulded into its final shape and firmly fixed to the substratum; in the opisthobranch *Onchidella celtica* the egg masses are similarly shaped in the pedal mucus gland (Fretter, 1941, 1943).

In some mesogastropods the male has a long penis which penetrates the pallial oviduct to deposit spermatozoa either near or in the receptaculum seminis. In such cases, e.g. *Crepidula, Calyptraea, Capulus,* and *Trivia,* there is no bursa copulatrix and no ciliated sperm-conducting groove in the pallial oviduct, neither of these structures being required (Fretter, 1953). Alternatively, in some examples the ventral sperm-conducting groove has been cut off from the pallial oviduct to form a separate duct which accommodates the penis during copulation, this new duct being known as the vagina, e.g. in *Paludestrina jenkinsi* and *Theodoxus fluviatilis.* The inner end of the vagina passes to the receptaculum seminis, which communicates with the inner end of the pallial oviduct.

In some mesogastropods the growth form of the shell results in lateral compression of the mantle cavity to such an extent that the introduction of a penis during copulation would occlude the mantle cavity and interfere with the processes of respiration and feeding. In such examples the male lacks a penis, the pallial oviduct of the female is secondarily simplified into an open ciliated and glandular groove and the spermatozoa are drawn into this groove via the inhalant water stream. Prosobranchs may have two kinds of spermatozoa. Firstly, there are the normal gametes, or eupyrene spermatozoa. Secondly, there are oligopyrene spermatozoa of various forms, which may reach enormous size. An oligopyrene sperm transports thousands of eupyrene sperms which are oriented on its tail with their heads embedded in the cytoplasm. Such compound structures are called spermatozeugmata. These can traverse considerable distances and they obviate the need for copulation as they can swim into the open pallial oviduct of the female, e.g. *Cerithiopsis* and *Clathrus* (Fretter, 1946, 1951, 1953). *Ianthina janthina* has no locomotory organs and no copulatory organs, and Wilson and Wilson (1956) judge that fertilisation is effected by such spermatozeugmata which swim across the relatively large spaces between individual snails in the swarm. Spermatozeugmata have been found in all parts of the closed oviduct of *Janthina britannica.* There is no copulation and the relatively huge spermatozeugmata swim to the female and are transported up the glandular pallial groove to the site of fertilisation, e.g. in *Turritella communis* (Fretter, 1946, 1951, 1953). Other examples in which the shell is coiled in a tight spiral, the mantle cavity is constricted, and the pallial oviduct is secondarily open throughout its length, include *Cerithiopsis, Triphora, Clathrus, Bittium,* and *Balcis* (Fretter, 1953).

The genital duct of the male similarly consists of two sections in the Archaeogastropoda, but of three sections in the Mesogastropoda and the Neogastropoda. The proximal section, derived from the gonad, serves as a vesicula seminalis and the spermatozoa are stored here. The second section is the renal vas deferens, and this commences with a sphincter muscle which only permits spermatozoa to leave the vesicula seminalis at the appropriate time. In the Archaeogastropoda the spermatozoa are released from the renal vas deferens into the mantle cavity and are emitted in the exhalant water stream. In the Mesogastropoda and Neogastropoda, due to the development of a closed pallial oviduct in which the eggs are encapsulated, internal fertilisation is necessary and in the male a penis has been developed on the right side of the head. In the most primitive condition the seminal fluid is conveyed from the mouth of the renal vas deferens to the penis by means of a ciliated groove on the floor of the mantle cavity and up the side of the penis. During copulation the ridges bordering this ciliated groove will approximate above the groove, so converting it temporarily into a closed tube, e.g. *Trichotropis, Littorina, Crepidula, Capulus,* and even the cypraeid *Simnia* (Fretter, 1951; Graham, 1954).

In *Littorina*, and in the secondarily open groove of *Turritella,* glandular strips on either side of the groove provide prostatic secretions. In the more advanced Mesogastropoda, and in the Neogastropoda, the ridges bordering this groove in the floor of the mantle cavity have fused above the groove, with the formation of a permanent tubular pallial vas deferens which penetrates the penis and opens at its tip. The pallial vas deferens has glandular walls and forms a prostate. In the higher Mesogastropoda and in the Neogastropoda the proximal end of the prostate opens into the mantle cavity either by a dorsal slit or by a slender ciliated duct. The purpose of this is to provide a safety valve through which spermatozoa and prostatic secretions can easily be ejected, permitting the penis to be suddenly withdrawn if the animal is disturbed during copulation, so reducing the risk of rupturing the internal organs by backflow of excessive quantities of fluid. Where the shell is a helicoid spiral with a small apical angle, the mantle cavity is deep and is laterally constricted, as mentioned earlier for the female system. In males of such species, e.g. *Cerithiopsis* and *Triphora,* the pallial vas deferens is an open groove and lacks great prostatic development, and there is no penis; both apyrene and eupyrene spermatozoa are produced; sperm and prostatic secretion is emitted in a cloud which does not disperse much, and this is sucked in through the inhalant siphons of females in close proximity to the spawning males. Only the eupyrene spermatozoa have been found in the female reproductive system (Fretter, 1951).

Some members of the Archaeogastropoda resemble the Mesogastropoda in possessing a closed prostatic tube crossing the mantle cavity, and a penis on the right side of the head, e.g. the Neritacea,* including *Nerita* and *Theodoxus*. A similar parallelism is found in the female system of *Calliostoma zizyphinum* which possesses a closed pallial oviduct.

Some prosobranchs exhibit protandric hermaphroditism and the reproductive system is naturally complicated by the development of both male and female components. In *Crepidula* and *Calyptraea*, for instance, the gono-pericardial duct begins to develop during the preparations for sex reversal from the male to the female phase (Fretter, 1946). *Omalogyra atomus* is a minute protandric hermaphrodite with quite remarkable modifications to the reproductive system. In the spring young specimens pass through a male phase when the gonad duct serves as a vesicula seminalis. There is no penis. A bursa copulatrix opens into the prostatic pallial vas deferens, and the opening of the bursa is developed into a long, slender, muscular tube which lies within the pallial vas deferens, pointing inwardly. It is thought that this provides a device whereby the bursa can become filled with native spermatozoa during the male phase. The method of copulation is not known, but it seems possible that the long slender tubular orifice of the bursa may be reversed in position so that it can be protruded through the genital aperture and used as a penis. It is not at all clear how this reversal of the tube could be effected. The organs of the female system are present at the same time as the male organs, but do not reach maturity until later. Distally the pallial oviduct and vas deferens join to form a hermaphrodite duct. In the summer the ovary is greatly developed and the animal now enters the female phase. Individuals which develop from eggs laid by summer spawners fail to develop the male system fully, although a testis with ripe spermatozoa is produced, but pass quickly into a functional female phase. There is a succession of generations in the summer months and self-fertilisation may possibly occur (Fretter, 1948, 1953).

In some Mesogastropoda and Neogastropoda special areas of the reproductive system have acquired the capacity to destroy superfluous or senile spermatozoa; these are ingested by the epithelial cells which proceed to digest the spermatozoa intracellularly. This is known in the male system in *Omalogyra atomus*, *Ocenebra erinacea*, and *Buccinum undatum*. It occurs in the female system in *Cerithiopsis tubercularis*, *Trivia monacha*, *Nucella lapillus*, *Ocenebra erinacea*, and *Buccinum undatum*, and in the last-named species the ingesting area can also ingest superfluous yolk. The engulfing of spermatozoa may also occur in the Opisthobranchia, e.g. *Onchidella celtica*. It is curious that species within the same genus may differ in possessing or in lacking this capacity, thus it is present in *T. monacha*, but absent in *T. arctica* (Fretter, 1941, 1943, 1946, 1948). Senile spermatozoa might be

* For such reasons Wilbur and Yonge (1964) have withdrawn the Neritacea from the Archaeogastropoda and raised this group to ordinal status while giving recognition to relationship between the Neritacea and the Archaeogastropoda as amended. For brief discussion see Chapter 10, pp. 515–16.

able to penetrate the ovum, thereby blocking entry by any other sperma-
tozoon, and yet be unable to initiate normal development. If this were true,
then the capacity to destroy superfluous spermatozoa would clearly be
advantageous. However, Coe (1942) has shown that in *Crepidula onyx* the
spermatozoa can remain functional for at least a year after the last copulation.

In all the Opisthobranchia, and all the Pulmonata, there is a condition of
functional hermaphroditism, and normally there is reciprocal fertilisation
between pairs of copulating individuals. *Aplysia* is unusual in developing
the habit of copulating in chains, in which the first individual acts only as
female to the second, but the second individual acts as male to the first,
and as female to the third, etc. This apparently sensual pattern of behaviour is less efficient than the usual opisthobranch and pulmonate habit of
reciprocal fertilisation of pairs of individuals, for in *Aplysia* the last individual in the chain does not become fertilised. The achievement of functional hermaphroditism in the Opisthobranchia and Pulmonata provided the
climax to the development of the reproductive system. Much diversity in
form has been revealed in published descriptions of the reproductive systems of many different genera, e.g. the parasitic Pyramidellidae (Fretter and
Graham, 1949), the tectibranch *Philine* (Brown, 1934), *Actaeonia* (Gascoigne, 1956), *Onchidella* (Fretter, 1943), the Ellobiidae (Morton, 1955),
Oxychilus, (Rigby, 1963), etc. It does not seem necessary to investigate such
diversity of form here, and I would simply draw attention to the ultimate
development in some examples of three distinct genital ducts to the exterior. These are the vas deferens which passes to the penis, the oviduct by
means of which fertilised eggs are laid, and the vagina which has become
separated off from the oviduct for reception of the penis and passage of
foreign spermatozoa to the head of the pallial oviduct. This "triaulic"
condition has already been encountered in the Neritacea, e.g. *Nerita* and
Theodoxus, which as members of the archaeogastropod Prosobranchia are
unusually advanced in this respect*.

Nevertheless, it is interesting to note the occurrence of hypodermic impregnation in *Limapontia capitata*, in which the bursa copulatrix lies immediately below the body wall but does not open to the exterior. During
copulation hypodermic impregnation occurs at this point, the slender,
gently curved penial style being stabbed through the body wall and into
the bursa of the partner, and spermatozoa with prostatic secretion are deposited in the bursa. It is striking that *L. depressa* differs in possessing an
open bursa copulatrix, and here the penial style is adapted for retaining its
position in the bursa during copulation, having a distinctly angular contour, and bearing four short lateral spines (see Fig. 93). The occurrence of
notable differences in the genital armature in two closely related species is

* See the footnote on p. 281.

doubtless of importance in establishing and maintaining the two separate species (Gascoigne, 1956). In the sacoglossan opisthobranch *Elysia maoria* there is no vagina and fertilisation is by hypodermic impregnation in spite of the fact that the penis is not armed with a style. The turgor pressure of the penis is evidently sufficient to ensure penetration of the body wall of the partner. Prostatic secretion has been found in the haemocoele subsequent to copulation, but no spermatozoa were found there and it is not clear how the spermatozoa injected into the haemocoele pass thence into the bursa copulatrix (Reid, 1964). Hypodermic impregnation is also known in *Alderia modesta*.

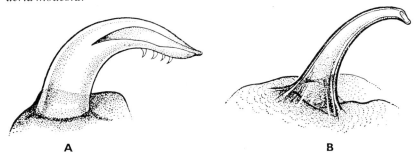

A   **B**

Fig. 93. The penial styles of A, *Limapontia depressa*, and B, *L. capitata*. In *L. depressa* the bursa opens to the exterior by a vaginal duct into which the penial style is inserted, the curvature of the style and the array of small hooks may serve to hold the penis in position during copulation. In *L. capitata* the bursa is closed to the exterior, copulation is by hypodermic impregnation, and the long slender penial style is suitably shaped for this purpose. (Originally published in Gascoigne, 1956, *Trans. roy. Soc. Edinb.* **63**, 132, fig. 7.)

The two British cowries *Trivia monacha* and *T. arctica* are very closely related, and were originally grouped together as one species, *T. europaea*. One of the few differences between the two species is the size and shape of the penis, that of *T. arctica* being larger and having a broader, leaf-like shape. This difference is clearly of great importance in preventing cross-breeding, and so maintaining these two species (Fretter, 1946). A somewhat similar condition is found in the Pulmonata in the genus *Cepaea*. The geographical ranges of *C. hortensis* and *C. nemoralis* overlap considerably, and populations of the two species are sometimes found occupying the same habitat. Under natural conditions the two species are prevented from cross-breeding by small differences in their courtship behaviour, and also by structural differences. These pulmonate snails possess a dart which is used for reciprocal stimulation during courtship. *C. hortensis* is rather smaller, and has a slightly curved dart, and is likely to be put off, or even damaged, by the more vigorous movements and straight dart of *C. nemoralis*. By this means the two species are evidently maintained in spite of the fact that there is some overlap in their ecological preferences, and that mixed populations of the two species frequently occur (Diver, 1940).

Expressions of Sexuality

Many molluscs retain the primitive unisexual state. This is true for the classes Scaphopoda, Polyplacophora, Monoplacophora, and Cephalopoda, and it is also true for the majority of Bivalvia. Yet there are major and minor exceptions, for hermaphroditism is found in the majority of the class Aplacophora, in all the Opisthobranchia and Pulmonata, in the bivalve order Anatinacea and families Tridacnidae (Yonge, 1936), Erycinidae and Montacutidae (Popham, 1940), Sphaeriidae (Monk, 1928), and also in many smaller categories scattered in predominantly unisexual groups, e.g. *Pecten, Modiolaria, Entodesma,* and *Poromya* among the Bivalvia. While most species of *Pecten* are functional hermaphrodites, some are unisexual, e.g. *P. magellanicus*; conversely, among the predominantly unisexual genus *Cardium,* a few species are hermaphrodite, e.g. *C. crassum* (Lebour, 1938) and *C. ciliatum* (Ockelmann, 1958). There is a strong suggestion of protandry with sex reversal in the pholad *Martesia striata* (Ganapati and Nagabhushanam, 1953). Again, some hermaphrodite specimens have been found in various species in the Unionidae, e.g. *Anodonta grandis* and *A. cygnea* (Schalie and Locke, 1941). Examples of hermaphroditism occur in the prosobranch Gastropoda, e.g. *Valvata* and *Velutina. Patella coerulea* and *P. vulgata* are probably protandric hermaphrodites, whereas *P. aspera* and *P. depressa* are apparently unisexual (Dodd, 1956).

It necessarily follows that the mechanism for the determination of sex must be very delicately poised, so that there is no great difficulty in effecting a transition from unisexuality to hermaphroditism and vice versa. We have to conclude that the young developing individual must have the latent capability of developing either or both sets of sexual organs, and that unisexuality in molluscs is achieved by suppression of one or other of the two possible sets of sexual organs. We may surmise that the suppression of either the male or of the female complement of organs is effected through the influence of the chromosomes possessed by an individual, under particular environmental conditions. We now have to qualify our definition of the primitive molluscan reproductive pattern by stating that the primitive unisexual condition is superimposed on an underlying totipotency which remains latent and undisturbed. This provides a form of insurance for the future for, by adjustments in the developmental machinery, hermaphroditism may emerge and may become consolidated where it is found to have survival value (Purchon, 1951).

Let us suppose that functional hermaphroditism is achieved when there is a complete lack of interference with the rates of development of the male and the female complements of the reproductive system; these then lie side by side and may reach maturity synchronously. We have good examples of this condition in the Opisthobranchia, the Pulmonata, the Anatinacea, and in *Pecten maximus* among the pseudolamellibranch bivalves.

Let us suppose that, as a result of changes in the gene complex, the rates of development of the ovary and the testis are varied, that of the testis being accelerated and that of the ovary being retarded. This would produce a protandric hermaphrodite which is a functional male in its youth and a functional female when it reaches final maturity. Well-known examples of this condition are the Teredinidae, or ship-worms, and the Calyptraeidae, e.g. *Crepidula fornicata*. Under this condition all individuals that survive to maturity should first pass through a male phase and should then enter a female phase. We must expect some variation in the genotype and therefore some variation in the potency of the internal forces influencing the rates of development of the ovary and the testis. These internal forces may, moreover, be dependent on environmental variables for the degree of their expression. In other words we must expect a population of protandric hermaphrodites to exhibit some degree of phenotypic variation in expression of sexuality. This may explain why Coe speaks of "true males" in *Teredo navalis*, and in *Bankia setacea*, which retain the male phase long after other individuals have undergone sex reversal to the female phase, and of "protogynous individuals" of *T. navalis* which appear to become functional females without ever having undergone an initial male phase. The rate of development of ship-worms in warm seas is as remarkable as their productivity. *T. navalis* may become a fully functional male within a month from settlement, in the warm season, and may then proceed to spawn as a female at an age of only 8–10 weeks (Coe, 1941, 1945).

In the protandric oyster *Ostrea virginica* rapid growth at a very early stage may eclipse the initial male phase in some specimens. Thus favourable environmental conditions influence the expression of sexuality in this species, increasing the number of specimens which first spawn as females. For every 100 which first spawned as males, the following numbers first spawned as females in the localities cited: Newhaven Harbor, 3–12; Milford (Bed), 7–25; West Sayville, 31; Delaware Bay, 41; Beaufort, N. Carolina, 37–48 (Coe, 1938). The effect of the environment upon the expression of sexuality in protandric hermaphrodites is also well illustrated in the Calyptraeidae by studies on the effects of isolation on the development of very young specimens of certain species of *Crepidula* (see Fig. 94). When very young specimens of *C. plana* were isolated in circulating sea-water aquaria in the laboratory about 30 per cent developed an apparently functional male reproductive system, many others developed some rudiments of the male system, while others showed no signs whatsoever of the male system. When similar young specimens were isolated in the harbour with free access of normal sea water, as many as 68 per cent developed an apparently normal and functional male system while 32 per cent showed no signs of the male system. When comparable young specimens were raised in company with females of the same species, as many as 90 per cent of the individuals developed a functional male phase. In comparable ex-

periments with *C. nivea* (= *C. nummaria*) only 9 per cent developed into
the male phase in isolation in aquaria, some 42 per cent developed into
the male phase when isolated in the sea, and 90 per cent developed as
males when raised in company with females of the same species. In these
two examples it seems clear that the higher percentage of individuals

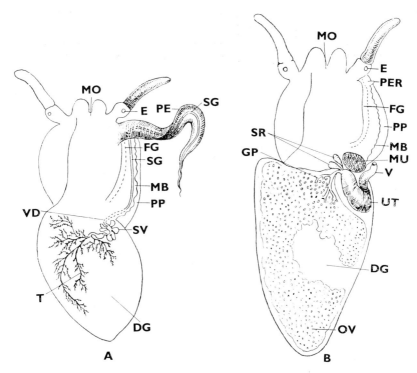

Fig. 94. Protandric hermaphroditism in the prosobranch *Crepidula onyx*. A,
the reproductive system in the male phase. B, the reproductive system in the
subsequent female phase. In the male phase the testis (T) passes spermatozoa
to the seminal vesicles (SV) which are drained by the vas deferens (VD).
Spermatozoa are passed up the penis (PE) in a superficial sperm groove (SG)
during copulation. In the female phase the ovary (OV) passes ova to the
"uterus" (UT). Spermatozoa are stored in the seminal receptacles (SR), only
two out of the twelve of these being shown in the figure. The rudiment of the
penis (PER) is still visible on the right side. For interpretation of other lettering, see p. 327. (Originally published in Coe, 1953, *J. exp. Zool.* **122**, 7, fig. 2.)

which attained the male phase in isolation in the sea, as compared with
those isolated in aquaria, is attributable to the richer food supply in natural
sea water. The increase in the percentage of individuals which achieved
the male phase when reared in association with females of the same species
is clearly due to that association; the underlying cause may be tactile

stimuli received first by the tentacles and later on by the penis of the young male. Not only does a larger proportion of individuals achieve the male phase when reared in association with females, but also the male phase is maintained for a far longer period if this association continues. In *C. onyx* a male may grow to almost the same length as the underlying female before it begins sex reversal to the female phase (Coe, 1938a, 1953).

Comparison of different species of *Crepidula* shows that environmental influence—especially association with a female—is of greater importance in the establishment of the male phase in some species than in others. Thus in *C. nivea* only 42 per cent of young individuals attained the male phase in the absence of female specimens; in *C. plana* a larger proportion—68 per cent achieved the male phase in isolation; in *C. onyx* as many as 90 per cent achieved the male phase in spite of being isolated; in *C. williamsi* and *C. norrisiarum* "nearly all" young specimens developed as males when isolated from females of the same species. On the basis of this evidence it seems reasonable to conclude that there may be considerable variation in the potency of the intrinsic factors which underline the phenotypic expression of sexuality in these protandric hermaphrodites; within a population this may lead to the production of apparently anomalous "true males" and "protogynous individuals"; between species or between genera, this variability could initiate a return to unisexuality wherever the circumstances favoured this condition.

The return to unisexuality from a previous hermaphrodite state would require the development of two alternative conditions in the gene complex. One condition would cause diminution of the early, potentially male phase, as a result of which the individual would pass quickly to the later female phase and would retain this for the remainder of its life. Some evidence for this hypothesis may be found in the "Quahog", *Venus mercenaria*, in which 98 per cent of the individuals from Long Island Sound, Connecticut, were found to pass through a juvenile male phase during the first year of life, after which the population comprised definitive male and female individuals in about equal proportions (Loosanoff, 1937). The other genetical condition would augment the early male phase until this filled the whole life of the individual, and in consequence the individual would never reach the theoretically possible subsequent female phase. Evidence for this might be sought in the development of ova in the gonads of senile males, but so far as I am aware, no such evidence is yet available.

It is suggested that the primitive unisexual condition postulated for the Mollusca was attained in this way, by the suppression of the organs of one sex or the other by genetical means. If this hypothesis approximates to the truth, then it is evident that a condition of functional hermaphroditism or protandric hermaphroditism could be attained in any unisexual molluscan stock as a result of changes either in the gene complex or in the environment. It has seemed necessary to go to these lengths to seek to

find a possible basis on which hermaphroditism could have arisen sporadically on numerous occasions in the phylum.

In view of the preponderance of unisexual types in the Prosobranchia
and in the Bivalvia, it seems natural to think of hermaphroditism as a
specialised feature meeting some special functional requirement. *Tivela
stultorum* is almost invariably unisexual in San Diego, but is hermaphrodite on the Pacific coast (Coe, 1945). Again, *O. virginica* is stated to be
unisexual along the Atlantic coast of North America, but to be a protandric hermaphrodite in the region of the Gulf of Mexico (Burkenroad, 1931).
Coe (1941) states that in this species the juvenile male phase is often reduced or omitted under conditions favourable to rapid growth; it may be
presumed that under such favourable conditions rapid growth would
carry the genetical females so quickly through the potentially male phase
that the testis would not have time to differentiate. Although *O. virginica*
is potentially a protandric hermaphrodite, a population might therefore
prove to be uni-sexual in an environment favourable to rapid early
growth. It is unprofitable to consider whether this is an example of transition from uni-sexuality to hermaphroditism or vice versa. All that is
important is that transition from one condition to the other is possible,
and depends on environmental conditions.

Sex reversal in some of the higher Mesogastropoda not only concerns
change in the nature of the gonad, and modifications to the genital duct,
but also involves reduction and loss of the penis, e.g. *Crepidula*, *Crucibulum*,
and *Trichotropis* (Graham, 1954).

Some protandric hermaphrodites are not confined to a single act of sex
reversal from male to female, but may follow this by further changes of
sex. In some cases this may involve a regular alternation of sex, but this
is not always the case. In *O. edulis* the change of sex from female to male
is rapid, and automatic. Ninety-seven per cent of specimens begin to
differentiate into the male phase very soon after having spawned as a
female. After the individual has spawned as a male, however, it does not
necessarily undergo sex reversal to enter a female phase. The course of
events is probably dependent on both intrinsic and extrinsic factors, e.g.
whether the act of spawning completely cleared the follicles of the gonad
or not, whether the animal carried an abundant store of food reserves,
and whether the spawning occurred early or late in the breeding season
(Orton, 1927, 1933). Cole (1942) concluded that in *O. edulis* change of
sex proceeds with regularity, there generally being one male and one female
phase in each year.

It remains for us to consider how the condition of hermaphroditism
with rhythmic changes of sexual phase can be equated with the more
straightforward condition of protandric hermaphroditism which has
already been discussed. It is easy to envisage protandry as an example of
heterochrony, the male phase being confined to the early part of the life

history of the individual. If this were the whole story, then there could be no question of putting the clock back, and a protandric hermaphrodite which survived for an unusually long time would be obliged to continue as a female in successive breeding seasons. On the other hand, sex reversal may not be determined solely by the rates of developmental processes, but may depend partly or wholly on metabolic processes. The production of spermatozoa may exhaust the resources of the animal for the production of nucleo-proteins and thereby supply a trigger mechanism for initiating sex reversal to the female phase. The act of spawning as a female might put the clock back from the biochemical point of view, under certain circumstances, if the primary oocytes had been effectively discharged, if the animal carried adequate food reserves, etc. On such a basis it seems possible to regard rhythmic sex reversal as a natural development of protandric hermaphroditism, and one that might be expected to occur in a long-lived animal which exhibits sex reversal.

One further manifestation of sexuality remains to be considered, namely parthenogenesis. This has been demonstrated for *Paludestrina* (*Hydrobia*) *jenkinsi*, a small taenioglossan prosobranch snail which occurs in brackish and tidal waters and also in fresh water. In this species there are no traces of the occurrence of dwarf males, nor are there any signs of protandry. Ova are present at most seasons, and these are passed into the pallial oviduct, which serves as a brood pouch. The embryo is attached to a large mass of yolky macromeres, and is contained in a hyaline capsule, and a fully grown snail may carry from 35 to 45 developing embryos in its brood pouch. There is no free-swimming larval stage, and the young emerge as miniature adults with strong powers of adhesion. This species is able to withstand considerable changes in salinity, e.g. from 17 parts per 1000 to fresh water (Robson, 1923).

It is generally understood that hermaphroditism does not remove the necessity for cross-fertilisation, and it is difficult to see in what way a bivalve species might benefit by the possession of simultaneous hermaphroditism, or protandric hermaphroditism, or rhythmic sex reversal. In some cases the gametes of one sexual phase have been completely discharged before those of the succeeding phase begin to mature, and self-fertilisation is impossible. In some simultaneous hermaphrodites, e.g. *Pecten maximus*, the two types of gametes are spawned in alternate clouds apparently without self-fertilisation. In *Rissoella diaphana*, a hermaphrodite prosobranch, the ova are obliged to pass down the hermaphrodite duct through the mass of spermatozoa which are stored in that part of the duct which serves as a vesicula seminalis (Fretter, 1948). The eggs are not fertilised as the spermatozoa are presumably in an inactive state due to the lack of prostatic secretions. *Xylophaga dorsalis* is unique among the Bivalvia in possessing a receptaculum seminis which lies in the suspensory membrane of the ctenidium, very close to the orifice of the genital duct.

The receptaculum seminis becomes packed with spermatozoa which are oriented with their heads against the surface of the epithelium. These are presumably native spermatozoa which were passed directly from the genital aperture into the receptaculum. It is not known whether these spermatozoa are used to fertilise eggs liberated at the next spawning, but if so, this would be a unique occurrence among bivalves as far as is known at the present time (Purchon, 1941).

Among opisthobranch and pulmonate Gastropoda, all of which are simultaneous hermaphrodites, and in any other hermaphrodite molluscs which experience copulation, a small benefit can be seen from possessing the hermaphrodite state. Firstly, *any* chance meeting of two mature individuals can lead to a successful mating, whereas only 50 per cent of chance meetings of adult unisexual individuals would permit of copulation; this would be beneficial in any species in which the populations are of low density. Secondly, where hermaphrodite animals indulge in reciprocal copulation, *both* individuals become fertilised. This is clearly beneficial to any species which is lacking in defences and subject to predation, for both individuals are fertilised and the maximum advantage is taken of the time spent in a defenceless condition while copulating.

Paludestrina jenkinsi possibly benefits in two ways from parthenogenetic reproduction. Firstly, it may be able to withstand the greatest rigours .of estuarine regions by withdrawing into its shell at a period when the rate of change of salinity is unduly high; there is no danger of being caught at such a time and exposed to osmotic hazards during a protracted period of copulation. Secondly, a *single individual* has the capacity to establish the species in a new habitat. There are very few references to the occurrence of this species in fresh waters in England before 1889, and the species has apparently established itself in considerable numbers at a wide variety of localities comparatively suddenly towards the end of the nineteenth and fairly early in the twentieth century. The first areas in which *P. jenkinsi* became established seem to have been industrial districts and canalised waterways. Since the species has no free-swimming larva, it seems probable that it has been transported to new localities while attached to the bottoms of boats. Under these conditions the capacity of a single individual to reproduce by parthenogenesis has obvious survival value (Robson, 1923).

Spawning

In the primitive condition the ova and the spermatozoa were emitted in the exhalant water current and fertilisation occurred at random in the sea water. This is still true for the Monoplacophora, the Scaphopoda, the majority of the Polyplacophora and the Bivalvia, and some of the Aplacophora. It is also true for some of the most primitive prosobranch gastropods. It need not be supposed, however, that such processes of spawning and fertilisation are entirely fortuitous, for the condition of the individual and the maturation of the gonads are dependent on the richness of the food supplies which in turn are dependent on the climate. In temperate, boreal, and arctic seas there are marked seasonal changes in temperature and illumination, and the increase in food supplies and in general metabolism in spring and early summer will inevitably impose a limited period of peak breeding activity. In this process there may be two contrasted features: firstly, a period of increasing metabolic activity with growth and maturation of the gonads; secondly, there may be some trigger mechanism, which is sometimes a threshold temperature, which initiates the act of spawning. Nelson (1928) noted the following spawning temperatures for various species of bivalves:

10–12°C   *Mytilus edulis, Mya arenaria.*
15–16°C   *Ostrea edulis, O. lurida, Pecten irradians, Teredo navalis.*
20°C   *Gryphaea (O.) virginica.*
24–25°C   *Venus mercenaria, Mytilus recurvus.*

When the ambient temperature rises to the stated level, spawning is initiated and with the high concentration of gametes in the sea water a maximal level of fertilisation is presumably achieved. Hunter (1949), showed that *Hiatella* spp. breed at temperatures *below* 12°C (see Fig. 95).

In tropical seas there may be very little annual variation in temperature, and no marked seasonal changes in abundance of food so far as is known at present. Even under these nearly uniform conditions spawning may not be spread over the whole year; thus Yonge (1936) reported that *Hippopus hippopus* bred in mid-summer at Low Islands on the Great Barrier Reef of Australia; the majority of specimens which were studied spawned rather suddenly in January, though some breeding occurred from December to March. The stimulus which initiates spawning in such a case is not known. Lunar periodicity is considered to supply the stimulus to commence spawning in *Pecten maximus* (Mason, 1958). In *O. edulis* Orton (1926) found that in July, August, and September there were frequency peaks in the numbers of *young* larvae in the gills immediately after the full moon, while *mature* larvae showed peaks in abundance a week later. Knight-Jones (1952) found that data on the breeding of *O. edulis* at Burnham-on-Crouch in 1948 showed fortnightly peaks in the abundance of planktonic larvae, and these peaks—with rare exceptions—occurred about 8 days after spring tides (see Fig. 96). *O. edulis* is larviparous, and the eggs and young developing larvae are retained in the spaces of the ctenidium for about 8 days, so the original act of spawning coincided approximately with the spring tides. Korringa (1947) recorded large numbers of larvae to be liberated about 10 days after full and new moon, which indicated that spawning occurred 1 or 2 days after the spring tides.

Spawning in *Patella vulgata* has been observed to coincide with the occurrence of winds of over 20 knots, so the stimulus to spawn in this case may be provided by continued rough seas and protracted heavy surf (Orton, Southward, and Dodd, 1956).

In the oviparous oyster *Gryphaea (Ostrea) virginica* spawning may be induced by the presence of eggs or of sperm in the inhalant stream,

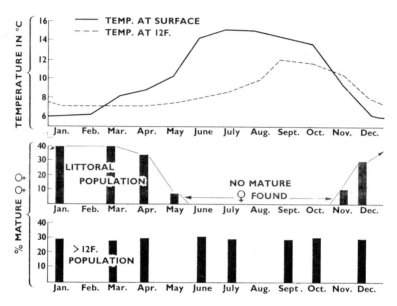

FIG. 95. Reproduction of *Hiatella* spp. in the Clyde Sea Area. In the lower part of the figure histograms show the percentage of individuals with mature ova in the population at different times of the year. Temperatures at the surface of the sea ——————, and at 12 fathoms, — — — —, are plotted graphically above. It is found that in the population of *Hiatella* living at 12 fathoms 30 per cent of the females are mature at all seasons, and that the temperature does not exceed 12°C at any time. In contrast, few or no mature females are found in the littoral population during the summer months when the surface waters exceed 12°C. (Originally published in Hunter, 1949, *Proc. roy. Soc. Edinb.* **63**, 273, fig. 1.)

so that once spawning has been initiated in a few individuals, the spawning of the whole population tends to be synchronised. Nelson and Allison (1940) found that the sperm carry a hormone, which they named "diantlin", which causes relaxation of the adductor muscle, enlargement of the ostia of the ctenidia, increased ciliary activity on the ctenidia, and, as a consequence of these changes, increased rate of pumping of water by the ctenidia. Nelson and Allison extracted from the spermatozoa an acid-soluble protein and also an alkali-soluble nucleo-protein. By titrating solutions of these they obtained a conjugated protein which was as effective as fresh spermatozoa in causing the various effects mentioned above. When a ripe female *G. virginica* is stimulated by the addition of spermatozoa to the surface of the ctenidium, it commences rhythmic and deep contractions and relaxations of the adductor muscle at intervals of approximately 30 seconds, and after about 9 or more

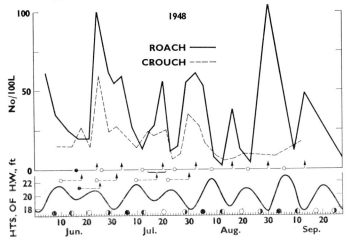

FIG. 96. Lunar periodicity in the abundance of recently liberated oyster larvae in the rivers Roach —————— and Crouch — — — — in 1948. Phases of the moon and heights of high tide are shown at the foot of the figure. The developing embryos are incubated in the ctenidia for about 8 days after spawning before they are liberated into the tideway. Many of the peaks of abundance of recently liberated larvae can be linked with spring tides about 8 days earlier, as is indicated by arrows, and by broken arrows. However, two of the peaks seem to be associated with neap tides. (Reproduced by permission of the Controller of *H.M.* Stationery Office, from Knight-Jones, 1952, Min. of Agr. and Fish., *Fish Invest.*, Ser. II, vol. 18, no. 2, page 12, fig. 4.)

minutes it commences to spray out fan-shaped clouds of eggs into the water. If a ripe male *G. virginica* is stimulated by the addition of fresh oyster eggs on the surface of the ctenidium, it remains quiescent with the shell valves gaping widely, and spermatozoa are emitted in a steady stream within 30 seconds. When the eggs have been removed from the surface of the ctenidium by ciliary activity the stimulus to spawn is apparently removed, and accordingly the emission of spermatozoa ceases.

Where the population of a species is very dense the concentration of gametes in the supernatant water may be very high, and a very high percentage of ova may be fertilised. The native oyster, *O. edulis*, may spawn about 1 million eggs per season by the time it has reached maturity at 4 years of age, while *O. virginica* has been calculated to produce up to

100 million eggs per spawning period (Cole, 1941). Kristensen (1957) studied a population of *Cardium edule* on tidal flats in the Dutch Wadden Sea, and estimated the productivity of ova per square metre. The intertidal population had been completely destroyed by the very severe winter of 1946–7, and the area had been heavily re-colonised the following year by larvae derived from adults living in deeper water nearby. In subsequent years the settlement of cockle larvae was very poor, and Kristensen was able to follow the progress of the population derived from the 1947 settlement for a number of years. The average density of this population per square metre was recorded as follows:

1948	150	1951	20
1949	60	1952	15
1950	35	1953	6

The average size was determined for each of these year classes, and the average number of eggs produced by cockles at each of these sizes was also determined. In this way Kristensen was able to calculate the total number of ova produced by this population of cockles over its whole life span of 6 years as approximately 5 million per square metre. It must be presumed that from this output of 5 million ova (and an infinitely larger number of spermatozoa) only two or three hundred larvae might survive to metamorphose and settle, and only about a half of these would survive to the following season. The mortality rate is exceedingly high, but the future of the species is assured by the very large numbers of ova produced.

Even when the gametes are all distributed at random in the supernatant water the reproductive process is not entirely fortuitous, for extrinsic and/or intrinsic factors may serve to concentrate the spawning of a majority of individuals into a narrow period of time. Where the population is very extensive, or is very densely crowded, the output of gametes is astronomically high. The annual cycle of seasons may necessarily determine a limited period of peak breeding; extrinsic factors such as the lunar or tidal cycle or the temperature of the water, or intrinsic factors such as sensitivity to the presence of sperm or ova in the water, may provide a trigger mechanism which initiates maximal spawning action.

Reproduction by the random union of gametes in the supernatant water can only work effectively if the provision of these gametes is both copious, and highly co-ordinated. Here nature is so bounteous that a 999.9 per cent wastage is of no consequence. Not only is it of no consequence, but it is essential that this wastage *should* occur; this particular form of reproduction necessitates the excess production of gametes, and the excess *must* be eliminated if the population is to remain stable. There is, of course, a sufficiency of pelagic and benthic carnivores to ensure this drastic reduction in numbers.

Specimens of *Bankia gouldi* living in test panels at the Harbor Island Test Station, Wilmington, North Carolina, have been observed to exhibit a remarkable form of siphonal activity at irregular intervals. On such occasions the siphons of large numbers of these ship-worms writhed continuously until neighbouring shipworms united in pairs, the exhalant siphon of one individual being thrust into the inhalant siphon of another specimen. In a test panel showing about 100 pairs of siphons as many as 75 exhalant siphons were seen entering the inhalant siphon of a different specimen, but on no occasion did an exhalant siphon enter the inhalant siphon of the same specimen. While the transparent siphons of neighbouring individuals were thus associated, a minute amount of a "somewhat transparent fluid" was seen to be ejected spasmodically from the exhalant siphon of the one specimen into the inhalant siphon of the other. It seems possible that what was observed was an occasion of co-ordinated sexual activity in which the exhalant siphon was being used as an intromittent organ—which, if proven, would be a unique event in the class Bivalvia. Use of the exhalant siphon as an intromittent organ could only be established if the transference of spermatozoa from one individual to another by this means were demonstrated beyond all doubt. The development of an intromittent organ would obviously be of great survival value to species of shipworm living in isolated pieces of drift wood (Turner, 1966).

Special Provision for the Fertilised Ova

The previous section on "spawning" was confined to consideration of those molluscs in which the ova and the spermatozoa were discharged in clouds into the sea water. In a more advanced condition the spermatozoa are received in the mantle cavity of the female, where the eggs are fertilised before being liberated into the sea, or being laid on the surface of the substratum. Characteristically such eggs receive some special provision from the mother in the form of additional food material, or some form of protective covering. Fertilisation of the ova within the mantle cavity is known to occur in some of the Polyplacophora and in a variety of Bivalvia, the sperm being drawn into the mantle cavity in the inhalant water stream. A few specialised bivalves deposit eggs on the substratum instead of emitting them into the plankton. In *Turtonia minuta*, one of the smallest British bivalves, a few large yolky eggs are produced. These, having been fertilised, are laid in a gelatinous capsule with a tough outer skin which is attached to the byssus and thereby to the substratum. The capsule is probably secreted by the glandular margin of the mantle lobes (Oldfield, 1955) (see Fig. 97). Again, Ockelmann reports that in coastal waters of East Greenland several species of *Astarte* and *Macoma* lay eggs with a mucoid and adhesive surface membrane, and these are probably attached to the substratum. Further, *Modiolaria discors* and *M. nigra* spawn egg strings and attach these to the substratum (Ockelmann, 1958). These specialisations may have enabled *Astarte*, *Macoma*, and *Modiolaria* to become established in these northern coastal waters which are possibly deficient in suitable food for planktotrophic larvae.

Internal fertilisation occurs in the great majority of the Gastropoda, but most of these are relatively mobile, and there is generally some measure of association—usually copulation—for breeding purposes. Let

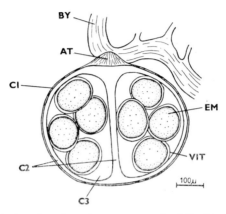

Fig. 97. Vertical section through an egg capsule of the bivalve *Turtonia mi-nuta*. The capsule possesses outer, middle, and inner layers (C1, C2, C3 respectively) and is attached (AT) to the byssus (BY). The capsule contains a number of embryos (EM), each of which is enclosed in its own vitelline membrane (VIT). (Originally published in Oldfield, 1964, *Proc. malac. Soc. Lond.* **36,** 84, fig. 2.)

us digress for a moment to consider how the Gastropoda came to acquire a penis situated on the right side of the body at a considerable distance from the original genital aperture. In forms with only one ctenidium the respiratory current passes through the mantle cavity from left to right. The spermatozoa are therefore passed out on the right-hand side. It seems possible that at an early stage in the evolution of prosobranchs individuals may have congregated in small clusters prior to spawning, which would increase the density of spermatozoa received in the inhalant current. If individuals next acquired the capacity to recognise a member of the opposite sex they would be able to associate in pairs and the male could station itself so that the exhalant sperm-laden water current was directed at the mouth of the mantle cavity of the female. The development of a ciliated pathway from the opening of the vas deferens, crossing the floor of the mantle cavity, would serve to concentrate the outgoing stream of spermatozoa. Delivery of spermatozoa into the mantle cavity of the female would be rendered still more efficient if the ciliated pathway were extended on to a projection from the body of the male. It is natural, therefore, that a penis bearing a lateral ciliated seminal groove should develop on the right side of the male. Such a condition occurs in *Littorina*, *Calyptraea chinensis*, *Crepidula unguiformis*, *Capulus ungaricus*, and *Cypraea* spp., in which the vas deferens still opens at the posterior end of the mantle cavity (Fretter, 1946).

At the same time, the development of a ciliated pathway on the floor of the mantle cavity of the female, on which the cilia beat inwards towards the genital aperture, would aid in carrying the spermatozoa to the point where the ova are to be fertilised. The embedding of this ciliated pathway in a deep groove would isolate the incoming spermatozoa from the exhalant current of the female, and the fusion of the epithelia superficial to this groove would complete the process of formation of the pallial oviduct. The pallial oviduct serves not only to guard the inward journey of the spermatozoa but also to nourish and to encapsulate the freshly fertilised ova. With the development of a closed pallial oviduct in the female, it is natural that the evolving penis of the male should become more highly specialised as an intromittent organ, with a sub-central closed vas deferens instead of a superficial ciliated seminal groove. This penis would deliver spermatozoa either into a bursa copulatrix, or directly into a receptaculum seminis sited close to the point at which the ova are to be fertilised.

In the majority of gastropods the penis is pedal in origin but in the Ampullariidae, e.g. *Pila*, the penis and penis sheath are pallial structures (Jutting, 1956) while in the Neritidae the penis is described as cephalic (Fretter, 1965). It therefore seems that a penis has evolved independently in more than one lineage in the Gastropoda, pedal, pallial, and cephalic penes being analogous, not homologous, organs.

This line of evolution with the production of a penis has been put into reverse in some prosobranchs in which the shells have a narrow apical angle and the mantle cavity is deep and laterally constricted. In such animals the intrusion of a penis during copulation would seriously interfere with the processes of feeding and respiration, and accordingly there has been a secondary simplification with the loss of the penis. The spermatozoa are emitted in a cloud into the sea water, and the suspension of spermatozoa is drawn into the mantle cavity of the female in the inhalant water current, e.g. *Turritella*, *Cerithiopsis*, *Triphora*, etc. (Fretter, 1951).

From the point of view of special provisions which are made for the fertilised ova, the prosobranch Gastropoda offer a variety of conditions worthy of special attention. The Prosobranchia are divided systematically into three orders, of which the Archaeogastropoda is the most primitive. In some of these both the ova and the spermatozoa are discharged into the sea water, where fertilisation occurs, and the embryo soon hatches as a free-swimming trochophore larva. The trochophore is the most primitive larval form in the Mollusca, and this condition is very close to that postulated for the remote ancestral mollusc; e.g. *Patella vulgata*; certain of the Trochidae, such as *Trochus niloticus* and *Gibbula* spp.(Lebour, 1937); and *Haliotis tuberculata* (Crofts, 1937). A slightly more advanced condition is found in the limpet *Patelloida tessulata*, in which the eggs are laid in gelatinous layers which are attached to the substratum. Here again the embryo hatches as a free-swimming trochophore larva. In the Neritidae there is an unusual "crystal sac" which opens into the dorsal side of the rectum, from which it receives sand grains, diatom cases, etc. from the faeces. These materials are passed from the crystal sac into the capsule gland, where they are incorporated as strengthening materials into the walls of the egg capsules (Fretter, 1946).

Turning to the Mesogastropoda, a second order of prosobranch gastropods, we find that two winkles, *Littorina neritoides* and *L. littorea*, both send their eggs out singly to float among the plankton, but this is not to be confused with the ancestral condition for these eggs have already been fertilised and they are enclosed in a transparent egg shell, from which the embryo hatches as a free-swimming veliger larva. Some other members of the Mesogastropoda lay eggs in gelatinous layers attached to the substratum, from which the embryo hatches as a veliger larva, e.g. *Lacuna vincta* and *Bittium reticulatum*.

The occurrence of a free-swimming veliger larva in a few marine pulmonates, e.g. *Melampus* in the Ellobiidae, the Siphonariidae, and the Gadinidae, indicates that these are very primitive forms which lie close to the prosobranch ancestry of the Pulmonata (Morton, 1955a).

The provision of a little more food material to eggs which are laid in a gelatinous sheet will allow the embryo to develop to a more advanced stage before hatching, and thereby to delete the planktonic larval stage from the life history. The animal emerges from the egg as a minute crawling snail. This very advanced condition is found even in some of the Archaeogastropoda, e.g. *Diodora apertura* and *Calliostoma zizyphinum*, and also in some of the Mesogastropoda, e.g. *Littorina littoralis*, *Calyptraea*, and *Trichotropis*.

Alternatively, a number of eggs, each enclosed in its own egg membrane, may be grouped together and enclosed in a tough capsule which is secured to the substratum. Normally all of the eggs hatch, and emerge as veliger larvae through a sort of plug or lid near the apex of the capsule (see Fig. 98). In some the number of eggs per capsule is unusually large, e.g. *Lamellaria perspicua*, in which there may be from 1000 to 3000 eggs per capsule. In *Balcis alba* the relatively huge egg capsules contain some hundreds of eggs, which hatch as veligers (Lebour, 1935). In *Crepidula fornicata* there are about 250 eggs per capsule, which is a more usual figure, but in many cases there are very few indeed, e.g. *Aporrhais pes-*

pelecani, in which there are from one to three eggs per capsule. In many species the number of eggs per capsule is partly dependent on the size of the parent individual, for older and larger specimens may produce measurably larger capsules. In *Urosalpinx cinerea* a specimen of 16·5 mm shell length produces an average of 4·7 eggs per capsule, while a specimen of 29·1 mm shell length produces an average of 11·5 eggs per capsule (Carriker, 1955).

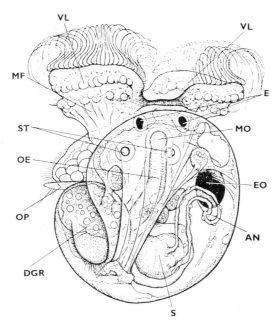

FIG. 98. A veliger larva of the nudibranch *Stiliger niger.* The transparent shell is sinistrally coiled; there is a system of fine retractor muscle fibres (MF) in the basal parts of the velar lobes (VL); the two eyes (E) are well defined. The median, ciliated oesophagus (OE) passes back to the stomach (S) and the twisted rectum opens at the anus (AN) on the right side of the body. The excretory organ (oval, black, EO) lies close to the anus. The paired statocysts (ST), each with a single statolith, can be seen in the base of the foot. The right lobe of digestive diverticula (DGR) occupies the apex of the shell. For interpretation of other lettering, see p. 327. (Originally published in Rasmussen, 1951, *Vidensk. Medd. dansk naturh. Foren. Kbh.* **113**, 231, fig. 17.)

In some species the egg capsules are attached to the substratum or to the shells of other specimens of the same species, e.g. *Hydrobia ulvae.* In some species a number of capsules are deposited close together, and the group of capsules is protected from predators by being covered by the shell of the mother at all times until the eggs hatch, e.g. *C. fornicata* and *C. plana.* This habit could only be expected in a sedentary gastropod which feeds passively by water filtration. Many mesogastropods feed upon a substratum of sponge or compound ascidian, and some of these cut a hole in the food substrate and deposit each capsule singly in such a hole. The top of the capsule projects slightly above the surface of the substratum, and one may see the developing larvae within swimming in a clear fluid as though in an aquarium. Such examples include *L. perspicua* and the cowrie *Trivia monacha*. The vase-shaped egg capsule of *Trivia* is buried in the tissues of a compound ascidian such as *Polyclinum* or *Botryllus*, with its slender neck protruding slightly above the surface, the capsule being moulded into place in the hole by the ventral pedal gland of the cowrie (Fretter, 1946).

The predacious carnivore *Natica* hunts for its prey in the surface layers of sandy beaches and therefore has no firm substratum on which to attach its egg masses; instead, these form nearly flat semicircular, circular, or spiral "collars" which lie on the surface of the sand. Each collar contains many hundreds of separate egg spaces in which the eggs develop, and from which they escape as veliger larvae in the case of *N. poliana*. Most species of the genus *Conus* deposit very large numbers of small eggs in a mass of egg capsules which are attached in a cluster on the underside of a boulder. There may be from 30,000 to $1\frac{1}{2}$ million eggs in such a cluster of egg capsules, and the young emerge as planktonic veligers. In very few species of *Conus* there are fewer, larger eggs, and the planktonic phase is either very brief or is non-existent. Thus in *C. pennaceus* there are about eighty eggs per capsule and a total of about 3500 eggs per egg mass, and when the young emerge they almost immediately adopt the benthic mode of life (Kohn, 1961, 1961a).

Some members of the order Neogastropoda produce capsules from which all of the ova normally hatch as veliger larvae, e.g. *Nassarius reticulatus*, in which each capsule may contain from 50 to 100 eggs.

In a still further advanced condition many eggs are laid in each capsule, but only a few eggs hatch; the eggs do not possess individual protective membranes and those embryos which develop most rapidly are able to consume the remainder, which therefore serve as a food store. The embryos which survive may hatch as veliger larvae which spend a period swimming in the plankton, as in the archaeogastropod *Theodoxus fluviatilis* and in the mesogastropod *Natica catena*. In many of the Neogastropoda there is a far larger provision of "nurse eggs", and the greater food supply permits a longer period of development within the capsule. Accordingly, the survivors emerge as young snails which crawl on the substratum (see Fig. 99), e.g. *Nucella lapillus* and *Buccinum undatum*.

These and some other examples culled from the literature have been set out in tabular form (Table 5). It is obvious from a glance at this table that there is very great variation in detail in the form of protection and in the provision of nourishment for the developing embryos in the Prosobranchia,

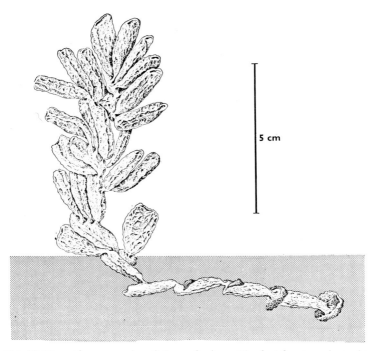

5 cm

Fig. 99. Mass of egg capsules of *Conus figulinus*. The first five capsules to be formed contain no eggs, and are modified to provide anchorage in the sand. The remaining capsules contain eggs and project above the surface of the sand. In other species of *Conus* the egg capsules are attached to a firm substratum by an adhesive basal disc. (Originally published in Kohn, 1959, *Ann. Mag. nat. Hist.*, ser. 13, **11**, 318, fig. 3.)

and that the tabulation adopted is highly artificial. Some species could possibly have been placed differently in the present table, and the table itself could have been drawn up differently, with the production of somewhat different groupings of species. Nevertheless, certain general principles can be drawn from this compilation of data, as follows:

1. The most primitive conditions are to be found among members of the Archaeogastropoda.

2. The suppression of the free-swimming larval phase can be effected comparatively easily by the provision of more food for the developing embryo. The young emerge at the crawling stage in *Gibbula tumida*, *Littorina littoralis*, *Lacuna pallidula*, and *Crepidula williamsi*, whilst other species in these genera produce free-swimming trochophore or veliger larvae. The five species of the genus *Littorina* show an extreme range from emission of planktonic eggs which liberate veliger larvae, to viviparity with the emergence of crawling young. The stage at

TABLE 5. *Reproductive processes in the Prosobranchia*

	Archaeogastropoda	Mesogastropoda	Neogastropoda
Eggs set free singly into the plankton.			
(a) Hatching as trochophores	*Haliotis tuberculata* *Patella* spp. *Patina pellucida* *Trochus niloticus* *Gibbula cinerarea* *G. umbilicalis* *Monodonta lineata*		
(b) Hatching as veligers		*Littorina neritoides* *L. littorea*	
Eggs laid in gelatinous layers attached to the substratum.			
(a) Hatching as trochophores	*Patelloida tessulata*		
(b) Hatching as veligers		*Lacuna vincta* *Bittium reticulatum* *Lacuna pallidula* *Littorina littoralis*	
(c) Hatching at crawling stage	*Diodora apertura* *Calliostoma zizyphinum* *Gibbula tumida* *Cantharidus* spp.		
Eggs laid in capsules, all eggs hatch as veligers.			
(a) Capsules attached to substratum or shells of other specimens		*Turritella communis* *Aporrhais pes-pelecani* *Hydrobia ulvae* *Cingula semistriata* *Rissoa* spp.	*Nassarius reticulatus* *Mangelia nebula* *Philbertia gracilis*

(b) Egg "collars" lie free upon the surface of the sand	*Natica poliana*	
(c) Egg capsules placed in a hole cut in the substratum of food material	*Cerithiopsis tubercularis* *Trivia monacha* *Lamellaria perspicua* *Capulus ungaricus* *Crepidula fornicata* *C. nivea* (= *nummaria*)	
(d) Capsules are guarded by being covered by the shell of the parent		*Nucella lapillus* *Buccinum undatum* *Colus islandicus* *Neptunea despecta*
Many eggs laid in capsules; only few hatch, remainder being eaten.		
(a) Capsules laid on surface of substratum; hatch as veligers	*Natica catena*	
(b) Capsules laid on surface of substratum; hatch at crawling stage	*Theodoxus fluviatilis*	
Few eggs laid in capsules, all hatch at crawling stage.		
(a) Capsules are abandoned	*Natica clausa* *Onoba striata* *Cingula fulgida* *C. semicostata* *C. cingillus* *Rissoella diaphana* *Calyptraea* spp. *Crepidula williamsi*	
(b) Capsules are guarded by being covered by the shell of the parent		*Urosalpinx cinerea* *Ocenebra erinacea* *Conus pennaceus*

	Archaeogastropoda	Mesogastropoda	Neogastropoda
Viviparous forms.			
(a) Young emerge as veligers		*Littorina angulifera*	
(b) Young emerge at crawling stage			
(1) Marine	*Acmaea rubella*	*L. saxatilis* *Planaxis sulcatus*	
(2) Fresh water		*Hydrobia ventrosa* *Viviparus viviparus* *Paludestrina jenkinsi*	

which the young emerge from the egg therefore has no systematic value. This is confirmed for the Opisthobranchia by Rasmussen (1951) who found that in *Brachystomia rissoides* there may be numerous small eggs which hatch as swimming veligers, or fewer large eggs which hatch at the crawling stage.

3. Generally speaking, where the eggs receive a greater amount of food material, and a greater measure of protection, the eggs are fewer in number, and there is a much lower mortality rate.

4. It seems natural to suggest that the examples quoted can be roughly graded as being "primitive", "advanced", and "highly advanced" according to the degree of provision afforded to the developing embryos and according to the stage of development at hatching. This is only meaningful if by these special means some species have enjoyed and are enjoying "greater success" than others. How can this "success" be measured? Presumably not actuarially by compilation of prospects of survival, for in general each species manages to maintain its numbers regardless of the details of its reproductive processes. Presumably not on metabolic grounds, for the production of few and large eggs, well protected by a stout capsule, may demand as many calories as the production of innumerable, minute, and unprotected eggs. But there may be justification in regarding some forms as "highly advanced", and the whole sequence as exhibiting some degree of evolutionary progress, if some of the more advanced forms have been able to enter habitats which would otherwise have been denied to them. This certainly seems to be true for the Neritacea,* e.g. *Theodoxus fluviatilis*, which as members of the Archaeogastropoda are precocious in their reproductive processes, and which have managed to invade fresh waters. It is also true of *Acmaea rubella*, another archaeogastropod, for Thorson (1944) has shown that no prosobranchs with planktonic larvae are to be found on the coasts of East Greenland. *A. rubella* has doubtless only managed to become established in East Greenland by virtue of the reproductive speciality of ovoviviparity.

Studies on the prosobranch faunas of various localities in the North Atlantic have shown that the percentage of prosobranchs which have pelagic larvae decreases with increasing latitude until in East Greenland exhaustive studies of the plankton in all months of the year have failed to detect any pelagic prosobranch larvae. Success in penetration of prosobranchs into the colder regions of the North Atlantic is therefore correlated with the capacity to suppress the planktonic larval phase of the life history. This is doubtless concerned with a progressive deficiency in boreal and sub-arctic seas of the minute phytoplankton organisms on which plankto-

* See the footnote on p. 281.

trophic larvae would depend for their growth and development (Thorson, 1940, 1941, 1944). Suppression of the planktonic larval phase can be accomplished by the provision of larger quantities of food material within the egg capsule, for the developing embryos do not emerge from the capsule until they have consumed all the food material available within the capsule. Examination of the capsules and developing embryos of one particular species of opisthobranch, *Brachystomia rissoides* (Pyramidellidae) shows that the young individual may emerge from the capsule at different stages of development at different localities. It is not known whether this is an early stage in speciation or is a reversible response to the environment.

Mating also occurs in the Cephalopoda in spite of the absence of a penis. The spermatozoa are compacted into spermatophores of complex structure, and these are passed from the male to the female during the mating process by one particular arm of the male. This arm may be slightly, or considerably, modified for this purpose, and is known as the "hectocotylus". In *Sepia* the spermatophores are placed in the genital bursa of the female, a shallow mid-ventral pocket in the interbrachial web, and here the ova are fertilised. In *Octopus* the hectocotylus of the male is inserted into the mantle cavity of the female. In extreme forms, e.g. the Paper Nautilus, *Argonauta argo*, the hectocotylus is broken off and left hanging from the mantle cavity of the female. After autotomy of the hectocotylus, the male can generate a replacement (Pelseneer, 1906).

With the further exception of the presence of a penis in the male, or male phase, in the prosobranch Gastropoda, sexual dimorphism is rare in the Mollusca. *Xylophaga* is unique in the Bivalvia in possessing a large foliaceous glandular organ which hangs in the suprabranchial chamber, suspended from the posterior adductor muscle in the male phase. It is of unknown function, but may be concerned with the passage of native spermatozoa into the receptaculum seminis. It is well developed in the male phase, but diminishes in size during sex reversal and is absent from the female phase (Purchon, 1941). The development of a part of the ctenidium into a marsupium in which the developing young are incubated is correlated with a modification to the shape of the shell valves to accommodate this marsupium in the ancient freshwater Lampsilinae. This modification in the shape of the shell is found only in the female, and in the genus *Truncilla* the difference is so considerable that males and females were originally accorded different specific names (Walker, 1917). In the Cephalopoda the male of some genera can be easily distinguished by the highly modified hectocotylus. The octopod *A. argo* is peculiar in that there is a dwarf male, while the female possesses a thin, papery, secondary shell which encloses her body and within which the eggs are incubated. The dorsal arms of the female are expanded into broad flat webs which secrete and also hold the shell (Pelseneer, 1906).

Parental Care of the Eggs and Developing Young

Not only can the embryos be aided by the parent by special provision of sustenance, but they can be guarded, or incubated, by the parent until the young individuals hatch. The Whelk *Neptunea despecta* often attaches clusters of egg capsules to its own shell, and carries them about, though it may alternatively deposit them on the substratum (Thorson, 1955) (see Fig. 100). The prosobranch *Clanculus bertheloti* has a shell which is orna-

FIG. 100. The whelk *Neptunea despecta* may deposit its egg capsules on the surface of a rock, but the capsules are often attached to the shell of an adult whelk, and are carried about, thereby possibly gaining some protection from predators. Each capsule originally contains up to 5000 eggs, but most of these are consumed either as eggs or as embryos and only three or four young snails survive to emerge at the crawling stage. (Originally published as Dr. Gunnar Thorson's Christmas card for 1955.)

mented with spiral grooves within which the eggs are deposited and covered with a thin layer of mucus. There are no copulatory organs, but individuals of opposite sexes must associate closely for breeding purposes, for eggs are deposited on the shells both of males and of females. The young snails develop in the grooves of the parental shell and with the protective veil of mucus which covers most of the shell (Thorson, 1950) (see Fig. 101).

Various species of *Crepidula* lay their egg capsules in clusters and then protect these clusters by covering them with the shell, e.g. *C. fornicata*, and *C. walshi*. This is also true of *Calyptraea pellucida*, *Crucibulum scutellatum*, and *Amalthea lissa* in the Iranian Gulf (Thorson, 1940). Some sedentary Mesogastropoda attach their egg capsules to the inner side of the shell, as in the Vermetidae, and the young emerge at the crawling stage. Thus in

Serpulorbis and in *Novastoa* there is a deep mid-dorsal slit in the mantle wall in the female only, permitting attachment of the egg capsules to the inner surface of the shell. In *Pyxipoma* there is a special brood chamber which opens to the exterior just below the right margin of the foot and this may contain from 100 to 150 small embryos (Morton, 1951). Some genera of Gastropoda attach the capsules to their body, as in the protandric hermaphrodite *Capulus*, and also in *Hipponyx*. Thus *C. ungaricus* carries a single egg capsule, the base of which is attached to the propodium. *H. antiquatus* carries a bunch of six to eight capsules which more or less

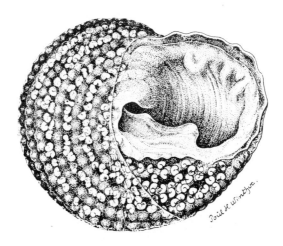

FIG. 101. Protection of the young by the prosobranch *Clanculus bertheloti*. The snail is withdrawn into its shell, which is here seen from below. Spiral grooves in the surface of the shell contain large numbers of developing embryos which are held in position by a thin veil of mucus covering the greater part of the shell. Eggs and developing young occur on the shells of both male and female individuals. (Originally published as Dr. Gunnar Thorson's Christmas card for 1950.)

fill the mantle cavity, and must obstruct the respiratory current somewhat. In *H. antiquatus* the slender bases of the egg capsules are attached to a perforated calcareous plate which is sited in a depression in the surface of the propodium (Yonge, 1953) (see Fig. 102). In these various ways the eggs are guarded by the shell of the maternal parent at all times until they hatch.

From this it is not a far step to ovoviviparity, in which the eggs are retained in the pallial oviduct during their development, and the young individuals emerge either as veligers or as young crawling snails according to the amount of food material available to them within the egg capsule. Wilson and Wilson (1956) observed a specimen of *Ianthina janthina* to liberate a number of veliger larvae in captivity, but it is not known whether they would have been retained to a later stage of development under natural conditions. Lebour (1945) observed veliger larvae being liberated from the Bermudan mangrove winkle, *Littorina angulifera*, whereas in the European *L. saxatilis* the young remain protected within the maternal body until they reach the crawling stage. In the Iranian Gulf, Thorson (1940) observed that in *Planaxis sulcatus* the embryos break out of the egg membrane, but remain in the "uterus" (= pallial oviduct) where they

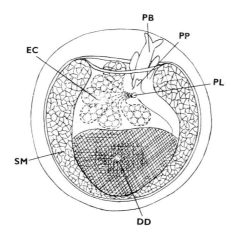

FIG. 102. *Hipponyx antiquatus*. Ventral view of the animal, showing a group of egg capsules (EC) lying within the mantle cavity. There are six to eight egg capsules each of which contains about fifty eggs. The capsules are attached by slender stalks to a perforated calcareous plate (PL) which is itself attached to the body of the animal in a depression below the propodium (PR). For interpretation of other lettering, see p. 327. (Originally published in Yonge, 1953, *Proc. Calif. Acad. Sci.* **28**, 17, fig. 8.)

feed on other developing eggs and finally emerge as young crawling snails at a much greater size (see Fig. 103). This is a most interesting observation, for in New Caledonia this same species was observed to produce free swimming veliger larvae, which is probably the characteristic condition. Ovoviviparity occurs rather surprisingly in the archaeogastropod *Acmaea rubella*, and it is probably as a result of this capability that *A. rubella* has been able to colonise the shores of East Greenland (Thorson, 1944).

Many freshwater prosobranchs are ovoviviparous, including the family Viviparidae, e.g. *Bellamya javanica*, in which the pallial oviduct may contain from 15 to 20 young. In the family Thiaridae also, many species are ovoviviparous, e.g. *Brotia* spp.. *Melanoides torulosus* broods eggs in the pallial oviduct, from which the young emerge as veligers (Jutting, 1956). *Viviparus viviparus* and *Hydrobia ventrosa* are two well-known freshwater gastropods in British waters which exhibit ovoviviparity. There is little doubt that the possession of ovoviviparity must have been a major factor in the invasion of fresh waters by many of these examples. Some terrestrial pulmonates are viviparous, e.g. *Partula*, which may on an average contain about two embryos (Crampton, 1916).

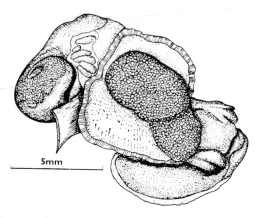

5mm

FIG. 103. *Planaxis sulcatus*. A mature female specimen which has been removed from its shell. The margin of the mantle has been reflected to expose the terminal part of the pallial oviduct, which is packed with about 300 embryos. (Originally published in Thorson, 1940, *Dan. Sci. Invest. Iran*, Pt. II, p. 163, fig. 1A.)

Various bivalves are known to brood their eggs and developing young in spaces in the ctenidium. These include the oysters *O. edulis* and *O. lurida*. In the former species the larvae are retained only for about 8 days before being liberated as planktonic veligers. Among the ship-worms, *Teredo navalis* incubates its larvae in the suprabranchial chamber, but *T. norvegica* and *T. megotara*, two other British species, do not (Lebour, 1946). Other marine bivalves which are known to incubate their young in the suprabranchial chamber include *Kellia suborbicularis, Lasaea rubra, Galeomma turtoni, Mysella bidentata* (Lebour, 1938), and *Transenella tantilla* (Hansen, 1953). In the case of *Cardium elegantulum* the eggs are retained in two brood pouches in the *infra-branchial* chamber, these brood pouches being formed from the margins of the mantle lobes (see Figs. 104 and 105). Another bivalve which broods eggs in the infra-branchial chamber is *Thracia myopsis* (Ockelmann, 1964). In *Galeomma turtoni* the young are liberated as free-swimming larvae, as also in *K. suborbicularis* which has unusually large veligers. In the other examples quoted above, the young hatch as miniature adults. *Nucula delphinodonta* is unique in developing a brood chamber which is plastered on to the posterior end of the shell, externally. The walls of this chamber appear to be formed from secretions of the hypobranchial gland, strengthened by adherence of small particles

from the substratum. From 20 to 70 eggs are brooded in this external chamber. Apparently all other species of *Nucula* liberate their eggs into the sea
water (Drew, 1901). Species of *Cardita* in South California brood their

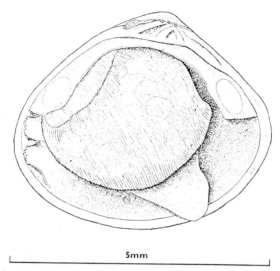

5mm

FIG. 104. *Transenella tantilla*, seen from the right side after removal of the
right shell valve and right mantle lobe. In this venerid the brood is protected
in the space between the inner demibranch and the visceral mass, and the
brood can be seen by transparency through the demibranch. All stages of development from newly laid eggs to young ready to leave the mother may occur in the same brood. (Originally published in Hansen, 1953, *Vidensk. Medd.
dansk naturh. Foren. Kbh.* **115**, 314, fig. 1.)

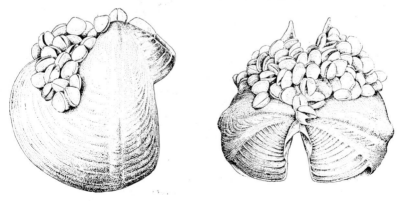

FIG. 105. Brood protection in the deep water wood-boring bivalve *Xylophaga
panamensis*. About fifty young specimens are attached to the postero-dorsal
surface of the shell, and it is concluded that the eggs are fertilised and the
young are brooded within the burrow occupied by the mother. (Originally
published in Knudsen, 1961, *Galathea Rept.* 5, 172, fig. 10.)

young in the spaces in the inner demibranch as a rule, though the embryos may sometimes also extend into the outer demibranch. Ten specimens of *C. ventricosa* showed a range of 14–93 young, with a mean of about 37 em-

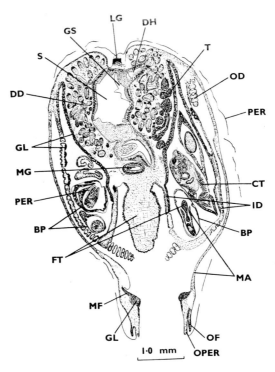

FIG. 106. Transverse section through the freshwater bivalve *Sphaerium* sp. showing well-developed young individuals lying within the inner demibranch (ID). The fertilised eggs develop to an advanced stage within a brood pouch (BP) the outer wall of which is attached by a thick stalk to the epithelium of the ctenidium (this stalk is not shown here). Sections through the young individuals show characteristic adult features such as the foot (FT), the mantle lobes (MA), the periostracum (P), etc. In the parent the testis (T) is only developed on one side. For interpretation of other lettering, see p. 327. (Freehand drawing magn. ×15, of a section which was cut and stained by Miss. E. Turner.)

bryos per adult specimen. When the young are liberated they may remain attached to the parent by byssus threads (Jones, 1963).

The incubation of eggs and young developmental stages is characteristic in the bivalves of fresh waters, but *Dreissena polymorpha* still produces a free-swimming veliger larva. In *Sphaerium notatum* fertilisation probably occurs in the autumn and the young are retained in the brood pouch throughout the winter, and are liberated as miniature adults (Monk,

1928) (see Fig. 106). In the unionid *Anodonta cygnea* an enormous number of glochidia larvae are incubated in the spaces of the outer demibranch.

The Octopoda are noted for the care that is given to protecting the developing eggs. The female of *Argonauta argo* carries the egg mass within the secondary shell until they hatch. Species of the genus *Octopus* lay large numbers of eggs in sheltered places, attached to the substratum, and the mother octopus takes up her station nearby and remains on guard until the eggs hatch. Batham (1957) observed a mass of about 7000 eggs which were laid in captivity by a specimen of *Octopus maorum*. The female remained close to the egg sheet, and ventilated them by slow radial sweeps of two or three of the tentacles. This would prevent settlement of debris on the eggs, and would ensure an adequate oxygen supply for the eggs. At the same time these arms serve for tactile observation of the entry of any potential predators into the area. A specimen of the snail *Scutus breviculus*, which approached and ate some of the eggs, was soon detected and was then gently pushed away from the area by the arms of the octopus. Shortly before the eggs began to hatch, the mother octopus changed her behaviour and began to squirt water over the eggs with her siphon. The eggs hatched in about 80 days. Throughout this long brooding period the mother octopus failed to respond in the normal way to the presentation of crabs, and although these were eventually eaten—probably at night—she fed much less than usual. It was possibly for this reason that she died shortly after the eggs hatched, but it is not clear whether this is the usual procedure.

Self-fertilisation and Parthenogenesis

In the vast majority of molluscs reproduction necessarily involves cross-fertilisation between two individuals; this is true even for hermaphrodite species. In only a very few examples, all of which are gastropods, have single individuals been known to reproduce either by self-fertilisation or by parthenogenesis. The occurrence of self-fertilisation has been established under rigorous experimental conditions by rearing individuals from isolated eggs, and by obtaining viable eggs from these isolated individuals. This has been done for the pulmonate snail *Lymnaea columella*, a self-fertilising line of this species having been reared through ninety-three generations extending over 20 years (Coulton and Pennypacker, 1938). Self-fertilisation has also been demonstrated in *L. stagnalis appressa* by Crabb (1927) who considers that here self-fertilisation is a normal process, and that cross-fertilisation seldom occurs in this species. Basch (1959) isolated freshly laid eggs of the freshwater limpet *Ferrissia shimekii* and reared the young individuals which hatched from these eggs. Three such individuals, reared from the egg in isolation, themselves produced viable eggs when they were only 4 or 5 weeks old. Basch was not able to obtain a further generation of snails from these individuals which had been produced by self-fertilisation. It is not known whether these species would necessarily resort to self-fertilisation under natural conditions. If an extremely young and virgin snail happened to be completely isolated in nature by some strange circumstance, it might be able to establish the species in the new locality by self-fertilisation, but this would obviously be a very unusual occurrence. If the individual was only a little older it would probably already have mated and any young produced in isolation would have been derived from normal cross-fertilisation.

The possibility of the occurrence of self-fertilisation has been raised for the summer generations of *Omalogyra atomus*; here normal spermatozoa are to be found in the testis and in the seminal vesicles, but the full complement of male organs is not developed and copulation cannot occur. Self-fertilisation certainly seems to be possible (Fretter, 1948).

The terrestrial pulmonate *Bulinus contortus* is remarkable in that some specimens may lack a penis. It has been shown that in some localities aphallic specimens of this species are rare, whilst in other localities aphallic specimens are abundant, or even predominant. The aphallic condition is apparently the outcome of the genetical constitution of the individual. In colonies of *B. contortus* in which aphallism is predominant it may be presumed that self-fertilisation is the principal mode of reproduction (de Larambergue, 1939, 1941).

The bivalve *Xylophaga dorsalis* is a protandric hermaphrodite, and is unique among bivalves in possessing a receptaculum seminis which lies in the suspensory membrane of the ctenidium, very close to the genital aperture. This becomes filled with oriented spermatozoa while the animal is still in the male phase. It is not known for certain whether the receptaculum is filled with native spermatozoa or whether foreign spermatozoa are collected from the inhalant water current by the ctenidia and passed in some way to the receptaculum seminis. On the whole it seems far more probable that the receptaculum contains native spermatozoa which are passed straight into it from the genital aperture. If this supposition is correct, then the purpose served by the receptaculum is to enable the animal to reproduce by self-fertilisation (Purchon, 1941). Perhaps *Xylophaga* is normally induced to spawn by the presence of foreign spermatozoa in the inhalant water current, but in the absence of such stimulation it can use the stored native spermatozoa as a last resort to ensure fertilisation.

It is far more difficult to demonstrate the occurrence of parthenogenesis. One has not only to show that individuals are able to breed in isolation, but also to exclude the possibility that the animal is a protandric hermaphrodite and has utilised spermatozoa formed and stored in a very early male phase. Robson (1923) collected specimens of *Paludestrina* (*Hydrobia*) *jenkinsi* from twenty-three localities in the British Isles and in Denmark, some of these being brackish and others being tidal waters. Specimens were collected over a period of 4 years, during which time Robson found no dwarf male specimens and no trace of an early male phase. An unmodified spermatheca is present, which suggests that parthenogenesis has only been adopted comparatively recently by this species. *P. jenkinsi* was apparently little known in Britain prior to 1889, but subsequent to that date it appeared in appreciable numbers comparatively suddenly at a wide variety of freshwater localities. Since the first areas to be invaded by this species were either industrial districts, or areas served by canals, it is probable that specimens of *P. jenkinsi* may have been transported to new localities on the bottoms of barges, etc. The capacity of a single individual to establish itself in a new locality by parthenogenesis may account for the sudden spread of this species at about the end of the nineteenth century. Continental specimens of *P. jenkinsi* seem to be restricted to brackish waters and to be diploid, whereas British representatives occur in fresh waters and appear to be tetraploid. The tolerance of British populations of *P. jenkinsi* to fresh water may possibly stem from the tetraploid condition (Sanderson, 1940). A second example of normal parthenogenetic reproduction is found in another prosobranch snail, *Campeloma rufum*, in which male specimens are unknown (Mattox, 1938).

Larval Behaviour and Settlement

The prospect of survival of the individual to maturity is aided by the adoption of appropriate behaviour by the larva when seeking out a suitable habitat and selecting a good point at which to undergo metamorphosis. A very few examples will suffice to illustrate the point. Wilson (1937) has shown that in certain polychaet worms the post-larva is able to delay metamorphosis if at first it does not encounter a suitable substratum for metamorphosis and settlement. It is probable that many other larval forms are similarly able to delay metamorphosis when necessary and thereby increase their chances of finding a suitable substratum after a further period of planktonic life. Thus Thompson (1958) has shown that the larva of the dorid *Adalaria proxima* will only settle and metamorphose in the presence of the polyzoan *Electra pilosa*, which is the source of its food in adult life. The veliger larva has a short obligatory planktonic phase of only 1 or 2 days when it tends to swim upwards, and then it passes through a phase of searching which may last from a week to a fortnight, during which time the larva is seeking a substratum of *E. pilosa*. The larva will metamorphose in the vicinity of *E. pilosa* even when separated from the polyzoan by a screen of fine bolting silk, which suggests that the larva detects the scent of the polyzoan and that its final approach is by chemo-taxis.

Oyster larvae only swim vertically upwards, and therefore naturally tend to encounter the under surfaces of overhanging objects. Cole and Knight-Jones (1939) found that the larvae of *Ostrea edulis* preferred to settle on the undersides of sheets of ground-glass which had been painted black on the upper surface, as compared with similar sheets of unpainted glass, i.e. the larva preferred to settle at shady sites. Moreover, larger numbers of larvae settled during day-time, when shady spots could be selected, than at night when this was not possible. Having encountered a substratum, the larva may crawl over it for several minutes, presumably testing the surface for the slight roughness and the degree of shade that are required. If the larva has difficulty in finding a suitable substratum, swimming alternating with crawling may continue for several days; the capacity to delay metamorphosis has definite survival value. However, this capacity to select a suitable spot for metamorphosis is not necessarily a universal one. Quantitative studies in the intertidal zone have frequently suggested that immense numbers of larvae of certain species (e.g. *Tellina fabula*) may settle indiscriminately over a wide vertical range, and only those which accidentally fall within the optimum horizontal zone will survive to maturity.

When an oyster larva is crawling on a suitable substratum the distances crawled in various directions gradually become less and finally the larva ceases to crawl and the shell is rocked backwards and forwards while a drop of byssal cement is exuded from the foot on to the substratum. The larva then settles on one side in this drop of cement and in a few minutes the cement has hardened and the oyster is fixed once and for all to the substratum. The exploratory crawling phase is necessary to ensure that there is a sufficient area of surface to meet the immediate needs of the larva for early growth. Under natural conditions the larvae of *O. edulis* are gregarious and show a preference for settlement on fresh oyster "clocks"—i.e. the attached lowermost valves of recently deceased oysters, which they may recognise by a chemo-tactile sense. These habits will tend to maintain the high density of the oyster colony that is necessary for survival. The larvae of *O. lurida* have similarly been shown to prefer downward facing horizontal surfaces for settlement.

The glochidia larvae of the Unionidae, when liberated from the suprabranchial chamber of the mother, are attracted to certain species of freshwater fish, and attach themselves to the skin of the fish as ecto-parasites. Similarly, the haustorius larvae of the African mutelid *Mutela bourguignati* are attracted to the cyprinid fish *Barbus altianalis radcliffi* and generally become attached to the fins of the fish (see Figs. 107 and 108). In such examples the larvae doubtless exhibit chemo-taxism in approaching the host fish to which they become attached (Fryer, 1959, 1961). In the Unionacea generally there is this important ecto-parasitic distributive phase in the life history.

Maturation of the Gonads in the Octopoda

Studies on the life of various gastropods and bivalves have indicated that reproduction is a seasonal event, dependent on external factors such as temperature and availability of food. Hypertrophy of the gonads could be

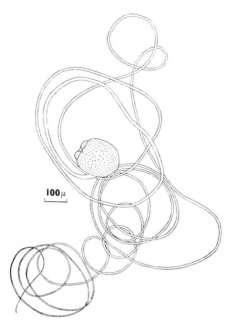

FIG. 107. Mature, free-living larva of *Mutela bourguignati*, showing the full length of the uncoiled tentacle. (Originally published in Fryer, 1961, *Phil. Trans.* B, **244**, 264, fig. 10.)

taken simply to represent a seasonally accumulating balance in the metabolic profit and loss account, there being little evidence to indicate that development of the gonad may be under either neural or hormonal influence. Various gastropods and bivalves exhibit sex reversal and it seems probable that this must be effected by some internal machinery, but any such mechanism remains undiscovered. The bivalve *Cuspidaria* exhibits slight sexual dimorphism as regards glands associated with the genital apertures of male specimens (Grobben, 1892), and the bivalve *Xylophaga* possesses a peculiar glandular organ in the supra-branchial cavity, which develops in association with the male phase (Purchon, 1941). Lemche (1955) has found evidence of neurosecretion in the tectibranch gastropod *Cyclichna cylindracea*. The functions of these various glands remain unknown.

In the Cephalopoda, on the other hand, it has been clearly established that the gonads of *Octopus* are under the influence of the central nervous system (Wells and Wells, 1959; Wells, 1964). In the Mediterranean the male *Octopus* appears to be sexually mature at a body weight of only 300 g, when it already has a hectocotylised third right arm and spermatophores in its genital duct, and it will attempt to mate even at this small size. In contrast, the female *Octopus* remains sexually immature until it has exceeded a body weight of some 1000 g. Up to about this body size the

FIG. 108. Bud and base of the stalk of a well-developed but still totally enclosed haustorial larva of *Mutela bourguignati*, in ventro-lateral view. The rudiments of the ctenidia (CR) can be seen anteriorly on either side of the foot
(FT). The mantle (MA) sends a prolongation (MP) down the stalk to the
haustorial tube. HLC, haustorial larval cuticule. (Originally published in
Fryer, 1961, *Phil. Trans.* B, **244**, 277, fig. 40.)

ovary is small and inactive and remains constant at about 1/500th the
body weight, and in all probability female *Octopus* of less than 1000 g
body weight do not breed.

Wells and Wells (1959) have shown that up to a body weight of about
1000 g maturation of the gonad of the female *Octopus* is inhibited by a pair
of sub-pedunculate lobes situated postero-dorsally in the supra-oesophageal part of the brain (see Fig. 109 and also Fig. 142, p. 431). Brain lesions
involving the loss of even only one of the sub-pedunculate lobes resulted
in rapid growth and maturation of the ovary, this change being recognisable within 4 or 5 days. During the first 3 days after operation there
was no significant increase in the weight of the ovary, but 15 days after

suitable brain lesions, and about twice as long after section of all the
optic nerves, the ovaries were found on post-mortem dissection to be
significantly larger than in equivalent but unoperated animals. In addition
to the ovary, the oviduct and oviducal gland were also enlarged. It was
further noted that there was conspicuous enlargement of the optic gland
on the operated side of the animal. The optic glands are sited distally on
the optic stalks and are small and pale yellow in unoperated females of
less than 1000 g body weight. After suitable brain lesions, or after section of the optic nerves, the optic gland on the operated side swells to
ten times its original size and assumes a bright orange colour.

Operations which result in enlargement of the gonad invariably also
cause hypertrophy of one or both optic glands; conversely, any operation
which leads to enlargement of one of the optic glands also causes enlargement of the gonad. If in addition to such an operation the optic

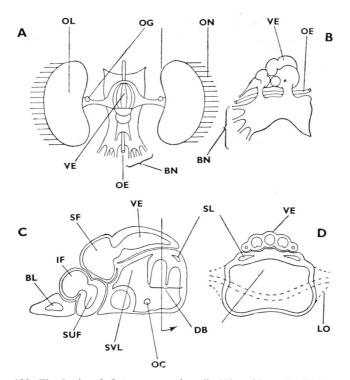

FIG. 109. The brain of *Octopus* seen dorsally (A) and laterally (B). An enlarged sagittal section of the supra-oesophageal part of the brain (C), and a
transverse section through the posterior part of the supra-oesophageal mass
(D) are also given. Note the optic glands (OG) and the sub-pedunculate lobes
(SL) which together control the onset of maturation of the gonad. For interpretation of other lettering, see p. 327. (Originally published in Wells,
1964, *Bull. nat. Inst. Sci. India* **27**, 63, fig. 1.)

glands are also removed, then there is no consequential enlargement of the gonad. This indicates that the development of the gonad is influenced both by the subpedunculate lobes of the brain and also by the optic glands. In females of under 1000 g body weight the sub-pedunculate lobes of the brain inhibit development of the optic glands, the secretions of which are necessary for initiating the development of the ovary. The decontrolled optic gland(s) supplies a gonadotrophic hormone which may be the same in the male as in the female, for no histological differences have been noted between the optic glands of male and of female *Octopus*. Data on the male *Octopus* are less conclusive; evidently the very low level of secretion by the small, pale optic glands is sufficient to reach the low threshold for response by the testis. Within the size range of specimens available it was not possible to produce such dramatic differences between operated and unoperated male animals as it was for the females described above. Nevertheless suitable brain lesions in male specimens resulted in enlargement of the optic gland(s) and in increase in the size of the testis, while extirpation of the optic glands caused diminution of the bulk of the testis. It seems reasonable to conclude that development and maturation of the gonads is mediated in the same way in both sexes, the only important difference being that the threshold of response of the testis is much lower than that of the ovary.

The nerve supply to the optic glands arises in the supra-oesophageal brain mass in the general region occupied by the sub-pedunculate lobes and the dorsal basal lobes, as has been shown by following the courses of nerve fibres which have been caused to degenerate. Sectioning the optic tract *proximal* to the site of the optic gland therefore interrupts the innervation of the gland by the sub-pedunculate lobe and consequently terminates the inhibiting effect of the latter (see Fig. 110). Section of the optic tract proximal to the site of the optic gland accordingly has precisely the same effect on the gland and on the gonad as does extirpation of the sub-pedunculate lobe. Complete unilateral section of all the optic nerves, or alternatively section of the optic tract *distal* to the site of the optic gland, causes the optic gland to develop, but in both cases the development is more leisurely than in those operations which interrupt the nerve supply from sub-pedunculate lobe to optic gland. It seems probable, therefore, that the optic gland is doubly innervated, receiving excitatory impulses from the retina as well as overriding inhibitory commands from the sub-pedunculate lobe. The inhibitory centre in the sub-pedunculate lobe is itself dependent on visual stimulation, for as noted above, complete section of the optic nerves, or of the optic tract distal to the optic gland destroys its inhibitory action.

There is as yet no clear indication as to how the female *Octopus* is ultimately released from the inhibitory action of the sub-pedunculate lobes, but since this inhibition is itself directly dependent on impulses received from the retina and optic lobes, it seems probable that the inhibition may be overridden under natural conditions by external circumstances, such as a change in photoperiod.

The optic gland is universal in dibranchiate cephalopods, but is absent from *Nautilus*, and it is therefore possible that this system of regulation of the reproductive cycle may be broadly applicable to all the Dibranchiata.

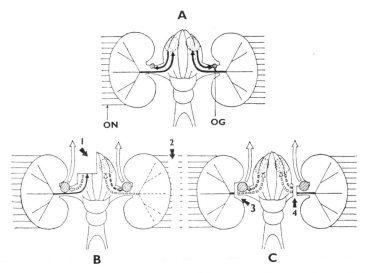

FIG. 110. Diagrammatic representation of the system of hormonal control of the onset of sexual maturity in *Octopus*. Dorsal view of the brain. A, in a normal, immature animal. Arrows indicate nerve pathways from the retina to the sub-pedunculate lobe, and from the latter to the optic gland which remains small. B, the consequences of two operations are indicated, either of which leads to enlargement of the optic gland, liberation of a gonadotrophic hormone, and development of the ovary. On the left, operation 1 is removal of one sub-pedunculate lobe; on the right, operation 2 is section of all the optic nerves on one side of the body. C, two further operations which similarly lead to enlargement of the ovary are indicated; on the left, operation 3 is section of the optic tract distal to the optic gland; on the right, operation 4 is section of the optic tract proximal to the optic gland. (Originally published in Wells, 1964, *Bull. nat. Inst. Sci. India*, **27**, 74, fig. 9.)

Wells and Wells (1959) draw attention to the interesting parallelism between the optic gland system of the Cephalopoda, the X organ and sinus gland complex in the eyestalks in Crustacea, the corpus allatum in the Insecta, and the pituitary system in the Vertebrata. There are close functional similarities in these four major animal groups in the mechanisms adopted for regulating the onset of sexual maturity, involving the perception of light and an endocrine organ closely associated with the central nervous system. There is a particularly close analogy between the condition in the Cephalopoda and that in the higher Vertebrata, both of which are able to modify their behaviour on the basis of previous experience. Wells and Wells (1959) suggest that in such animals a long post-embryonic period of development is desirable to enable the individual to gain the full benefit of its ability to discriminate between favourable and unfavourable situations. Delay in the onset of sexual maturity, as effected in the female *Octopus* by the optic gland/sub-pedunculate lobe system, is therefore advantageous. Many experiments have shown that there is considerable individual variation in *Octopus* in the ability to discriminate between favourable and unfavourable situations. A protracted adolescent period subjects *Octopus* to natural selection on the basis of "intellectual" ability. Those individuals which are slow to learn important discriminations, or which fail to establish long-lasting memories of these discriminations, are more liable to be eliminated before they have had a chance to reproduce; those individuals which are quick to learn and which establish lasting memories of unfavourable situations are more likely to survive and to breed. The establishment in the Dibranchiata of a mechanism which delays the onset of sexual maturity is therefore highly relevant to the development in these animals of the "higher" brain centres responsible for learning and memory. Wells (1964) suggests that from a functional standpoint the Cephalopoda and the Vertebrata have undergone convergent evolution, the optic gland of the former and the pituitary of the latter supplying different but closely comparable solutions to the same functional problem.

Courtship Behaviour and Mating

In the most primitive forms the ova and spermatozoa are independently dispersed and fertilisation is effected at random in the super-natant water. A more advanced condition is the association of adult individuals usually in pairs, though sometimes in chains, for the purpose of copulation and internal fertilisation. In the most advanced conditions species which are characterised by copulation may exhibit highly specific forms of courtship behaviour which are relevant in some way to the reproductive processes. In its simplest form courtship may involve sexual excitation leading directly to copulation, e.g. the reciprocal use of the dart by many pulmonate snails such as *Helix* and *Cepaea*.

Alternatively, courtship display may serve for the recognition of the sexes of the two individuals when they meet (see Fig. 111). The male *Octopus vulgaris* is usually appreciably smaller than the female with which it will mate, and when male and female octopuses meet it is necessary for the former to establish its male sex and so avoid being attacked by the female. The suckers on the arms of octopuses increase in diameter from the base of the arm outwards as far as the seventh or eighth sucker and they then diminish regularly in diameter towards the tip of the arm. On the second and third arms of the male the largest sucker in the series is disproportionately large and this is a reliable means of recognising individuals of the male sex after sexual maturity has been reached. When male and female octopuses meet the former identifies itself as being of the male sex by displaying these large suckers. The second and third arms on the side nearest the female octopus are turned

FIG. 111. Courtship display by a male octopus; on the second and third arms of the male the largest of the first seven or eight suckers is disproportionately large. (Originally published in Packard, 1961, *Nature, Lond.* **190**, 736–7, fig. 1.)

upwards over the head, thus exposing the suckers, and at the same time the smaller suckers beyond the eighth in the series are contracted, so emphasising the size of the seventh and eighth suckers (Packard, 1961). J. Z. Young observed courtship display and mating by *O. horridus* in shallow water at Raffles Light, Singapore, and reported that the male kept the female in his right visual field and displayed a strong pattern of vertical stripes on the left side of the body and on the left arms which were visible to the female. There was a strong band of colour along the posterior border of each arm and this extended up on to the head; the skin on the top of the head was raised into two pronounced horns above the eyes. This display was maintained while the male stretched out his third right arm and inserted its tip into the mantle cavity of the female, which was about 18–24 in. distant. This position was held for about 10 minutes and, after a short interval, copulation was repeated for a further 10 minutes (Young, 1962). During copulation spermatophores are understood to be conveyed along the groove between the suckers on the hectocotylised arm and are deposited in the mantle cavity of the female. The pair of octopuses observed by Young were previously lying under the same large intertidal coral boulder, and it is probable that they were already established as a mated pair. Both the sucker display, which advertises the male sex of the octopus at first encounter with a female, and the striped pattern exhibited by the male immediately before and during copulation, may be regarded as components of the courtship display.

Drew (1911) observed the sexual activities of the squid *Loligo pealii* in aquaria at Woods Hole. One form of copulation was preceded by a period of unusual, excited swimming movements which may constitute a form of courtship display. The male then swam below the female and turned its arms upwards to grasp the visceral mass of the female; the left ventral arm of the male was swept quickly past the funnel in order to collect spermatophores, and it was then inserted for only about 10 seconds into the mantle cavity of the female. At the ejaculatory end of the spermatophore there is a filament which is evidently jerked during the transfer of spermatophores to the female, so initiating ejaculation of the sperm reservoir within. The sperm reservoir is surrounded with a sticky cement by means of which it becomes attached to the body of the female. Sperm reservoirs deposited within the mantle cavity near the oviducal opening continue to liberate active spermatozoa into the mantle cavity for up to 2 days, and eggs are fertilised within the mantle cavity. Mantle cavity copulation only occurred with females which had swollen nidamental glands and bright red accessory nidamental glands, and which were nearly ready to deposit their eggs. In an alternative form of copulation the two squid face each other at a distance of about 30 cm, and then with no apparent preliminaries the male darts at the female and the two become attached head to head with their arms interlocked. The left ventral arm of the male is swept past the funnel to collect spermatophores and these are then deposited between the two ventral arms of the female. The sperm reservoirs become attached on the inner side of the outer buccal membrane of the female, and as the spermatozoa are emitted from the sperm reservoirs they accumulate in a deep pit, the receptaculum seminis, which lies nearby in the buccal membrane of the female. Within the receptaculum seminis the spermatozoa are stored in an inactive state. In *L. pealii* the modification to the hectocotylised left ventral arm of the male consists of a glandular pleated ridge between the suckers near the tip of the arm.

In *Sepia* courtship display seems to serve two purposes; firstly, it serves to establish the sex of the individual, and, secondly, it serves to set up a non-permanent association between two individuals, usually one male and one female. The display does not seem to be excitatory since it does not necessarily lead on to copulation, and its main function seems to be to establish a bond between two individuals and thereby to enable copulation at some later time. In this respect there is some similarity with the courtship displays of birds, except that in the latter the bond is a lasting one, perhaps for one breeding season only, but sometimes for life. Due to the ephemeral nature of the association between two sepias there can be no mutual recognition of the individual identity of the partners.

Mature male specimens of *Sepia* display when they encounter another *Sepia*, and thereby announce their own sex and sexually mature condition. If the newcomer is a male it displays too, and the less vigorous of the two will swim away. If the newcomer is a female, or an immature male, it will fail to display and a temporary bond may be set up between the two, the non-displaying member being treated as a female. The colour scheme of a *Sepia* is determined by interaction of a deep-seated layer of iridocytes which reflect light and a superficial layer of chromatophores of three main colour types. The distribution of the iridocytes on the back of the animal is in alternate thick and thin bands; during courtship display by the male *Sepia* the chromatophores overlying the thick bands of iridocytes are contracted and so these areas appear pale, while the intervening areas are dark due to the shallowness of the deep-seated layer of iridocytes and to the expansion of the superficial chromatophores. The courtship display of the male is therefore in the form of a marked black and white zebra pattern on the upper surface of the visceral mass. In addition the upper surface of the head of the male assumes a deep red coloration. The brilliant system of stripes is also shown on the dorsal side of the fourth ventral arm of the male. In addition the arms are extended sideways somewhat, one of the fourth ventral arms in particular being extended far out to the side, and the eye on that side of the body is opened very widely. Details of the chromatophores and of the various colour changes and colour patterns exhibited by *Sepia* are given by Holmes (1940).

The resting coloration of *Sepia* of both sexes is a grey-brown, due to partial expansion of all the chromatophores. The coloration of the female during courtship is similar to that of the courting male except that the pattern is far less pronounced, the stripes being alternately dark brown and a grey-white, and the head is not a deep red colour.

The mature male displays as described above towards any intruding *Sepia*, and if the latter is a mature male it will display in return; the display constitutes a threat towards other males, and one of the males will be driven away. Females and immature males do not respond by giving the typical male display and are therefore not driven away but may be accepted as a partner. Having established such a partnership the male swims constantly above its partner and often strokes its back with its arms. If a third cuttle-fish now appears the male will display vigorously and will chase away any intruding mature male *Sepia*. If the intruder is a female there may be a change in partnership by the male, and all such partnerships seem to be short-lived.

The males are indifferent to motionless sepias, and will desert their partners if these cease to move for a time; the female does not make any positive response to display by a male, and is identified as a female by virtue of its failure to display in reply. For this reason the male *Sepia* seems unable to discriminate between a mature female and an immature male. Having formed a pair with another specimen, the male may at any time attempt to copulate with it, seizing it by the head, and changing the position of its arms until the axis of its body is in line with that of its partner. Any attempts to copulate with an immature male must fail since the latter fails to make the correct responses. When copulating with a female the latter throws back her arms as soon as her head has been encircled by the arms of the male. The male holds the head of the female with the first three pairs of arms, it approximates its funnel to the base of the fourth ventral, hectocotylised arm, and passes spermatophores into the longitudinal groove in the base of this arm. The funnel is then withdrawn, and the base of the hectocotylised arm is stretched forwards between the ventral arms of the female and the spermatophores are passed to the bursa copulatrix in the oral membrane of the female. This pose is held for about 2 minutes, the distal part of the hectocotylised arm hanging down freely in the water. The spermathecae discharge their burden of spermatozoa, and these presumably swim towards and into the seminal receptacle of the female, where they are stored. The seminal receptacles comprise a number of chambers which are surrounded with muscle fibres the contractions of which will in due course presumably discharge the spermatozoa for the fertilisation of eggs (Tinbergen, 1939).

Binocular vision is only possible for a small part of the whole field of vision of cuttle-fish, and is used by *Sepia* when stalking and attacking its prey. During courtship, on the other hand, a male *Sepia* keeps its partner in the centre of the field of vision of one eye.

The elliptical eggs are fully formed in the mantle cavity, they are passed between the arms for fertilisation, and they are then deposited one at a time at selected laying sites. The eggs of *Sepia* are not brooded (Tinbergen, 1939).

Perhaps the most spectacular of all forms of courtship is that displayed by certain arboreal slugs. These hermaphrodite animals launch themselves into space in pairs, suspended from a branch by a thin but viscous strand of mucus secreted by their pedal glands, and mating is accomplished while suspended in mid-air in this manner.

Key to the Lettering on the Figures

A	Albumen gland.	MB	Border of mantle.
AM	Additional mucous gland.	MF	Muscle fibres.
AN	Anus.	MG	Mid gut.
AT	Attachment of egg capsule to byssus.	MO	Mouth.
		MP	Prolongation of the mantle which runs down the stalk to the haustorial tube.
B	Bursa copulatrix.		
BL	Buccal lobe.		
BN	Brachial nerves.	MU	Right shell muscle.
BP	Brood pouch.	O	Oviduct.
BS	Bursa seminalis.	OC	Optic commissure.
BY	Byssus.	OD	Outer demibranch.
C	Capsule gland.	OE	Oesophagus.
C1		OF	Outer fold of mantle margin.
C2	Outer, middle, and inner layers of the egg capsule.	.OG	Optic gland.
C3		OL	Optic lobes.
CR	Rudiment of the ctenidium.	ON	Optic nerves.
CT	Ctenidium.	OP	Operculum.
D	Dart sac.	OPER	Origin of periostracum.
DB	Dorsal basal lobe of brain.	OV	Ovary.
DD	Digestive diverticula.	P	Prostate.
DG	Digestive gland.	PB	Proboscis.
DGR	Right lobe of digestive gland.	PE	Penis.
DH	Dorsal hood.	PER	Periostracum.
E	Eye.	PL	Plate to which stalks from egg capsules are attached.
EC	Egg capsules.		
EM	Embryo.	PP	Propodium.
EO	Excretory organ.	RK	Right kidney.
F	Site of fertilisation.	RO	Renal oviduct.
FA	Female genital aperture.	RP	Reno-pericardial duct.
FG	Food groove.	RS	Receptaculum seminis.
FL	Flagellum.	S	Stomach.
FT	Foot.	SF	Superior frontal lobe of brain.
G	Gonad.	SG	Sperm groove.
GA	Common genital aperture.	SL	Sub-pedunculate lobe of brain.
GD	Gonadial duct.	SM	Shell muscle.
GL	Gland cells.	SR	Seminal receptacles.
GP	Gono-pericardial duct.	ST	Statocysts.
GS	Gastric shield.	SUF	Subi-frontal lobe of brain.
HLC	Haustorial larval cuticle.	SV	Seminal vesicles.
ID	Inner demibranch.	SVL	Sub-vertical lobe of brain.
IF	Inferior frontal lobe of brain.	T	Tests.
L	Glandular lips.	U	Ureter.
LG	Ligament.	UT	Uterus.
LH	Little hermaphrodite duct, of gonadial and renal origin.	V	Vagina.
		VC	Ventral sperm channel.
LM	Inner limit of mantle cavity.	VD	Vas deferens.
LO	Level of optic stalks and commissure.	VE	Vertical lobe of brain.
		VIT	Vitelline membrane.
M	Mucous glands.	VL	Two lobes of velum.
MA	Mantle lobe.		

Reference List

BASCH, P. F. (1959) Studies on the development and reproduction of the fresh water limpet, *Ferrissia shemeckii* (Pilsbry), *Trans. Amer. Micr. Soc.* **78**, 269–76.

BATHAM, E. J. (1957) Care of eggs by *Octopus maorum*, *Trans. roy. Soc. N.Z.* **84**, 629–38.

BROWN, H. H. (1934) A study of a tectibranch gasteropod mollusc, *Philine aperta* (L.), *Trans. roy. Soc. Edinb.* **58**, 179–210.

BURKENROAD, M. D. (1931) Sex in the Louisiana oyster, *Ostrea virginica, Science* (2), **74**, 71–72.

CARRIKER, M. R. (1955) Critical review of biology and control of oyster drills *Urosalpinx* and *Eupleura*, *Fishery Prod. Rep. Fish. Wildl. Serv. U. S.* **148**, 1–150.

COE, W. R. (1938) Primary sexual phases in the oviparous oyster *(Ostrea virginica)*. *Biol. Bull. Woods Hole* **74**, 64–75.

COE, W. R. (1938a) Influence of association on the sexual phases of gastropods having protandric consecutive sexuality, *Biol. Bull. Woods Hole* **75**, 274–85.

COE, W. R. (1941) Sexual phases in wood-boring molluscs, *Biol. Bull. Woods Hole* **81**, 168–76.

COE, W. R. (1942) The reproductive organs of the prosobranch mollusc *Crepidula onyx* and their transformation during the change from male to female, *J. Morph.* **70**, 501–12.

COE, W. R. (1945) Development of the reproductive system and variations in sexuality in *Pecten* and other pelecypod molluscs, *Trans. Conn. Acad. Arts. Sci.* **36**, 673–700.

COE, W. R. (1948) Variations in the expression of sexuality in the normally protandric gastropod *Crepidula plana* Say., *J. exp. Zool.* **108**, 155–69.

COE, W. R. (1953) Influence of association, isolation and nutrition on the sexuality of snails of the genus *Crepidula, J. exp. Zool.* **122**, 5–20.

COLE, H. A. (1941) The fecundity of *Ostrea edulis, J. mar. Biol. Ass. U.K.* **25**, 243–60.

COLE, H. A. (1942) Primary sex phases in *Ostrea edulis, Quart. J. micr. Sci.* **83**, 317–56.

COLE, H. A. and KNIGHT-JONES, E. W. (1939) Some observations and experiments on the setting behaviour of larvae of *Ostrea edulis, J. Cons. int. Explor. Mer* **14**, 86–105.

COULTON, H. D. and PENNYPACKER, M. (1938) The results of twenty years of self-fertilisation in the snail *Lymnaea columella* Say, *Amer. Nat.* **68**, 129–36.

CRABB, E. D. (1927) The fertilisation process in the snail *Lymnaea stagnalis appressa*, *Biol. Bull. Woods Hole* **53**, 67–98.

CRAMPTON, H. E. (1916) Studies on the variation, distribution and evolution of the genus *Partula*; the species inhabiting Tahiti, *Publ. Carneg. Instn.* **228**, 3–313.

CROFTS, D. R. (1937) The development of *Haliotis tuberculata*, with special reference to organogenesis during torsion, *Phil. Trans.* B, **228**, 219–68.

DIVER, C. (1940) The problem of closely related species living in the same area; in *The New Systematics*, ed. J. Huxley, pp. 303–28, Oxford University Press.

DODD, J. M. (1956) Studies on the biology of limpets. III. Hermaphroditism in the three British species of *Patella, J. mar. biol. Ass. U.K.* **35**, 327–40.

DREW, G. A. (1901) The life history of *Nucula delphinodonta* (Mighels), *Quart. J. micr. Sci.* **44**, 313–91.

DREW, G. A. (1911) Sexual activities of the squid *Loligo pealii* (Les.) I. Copulation, egg-laying and fertilisation, *J. Morph.* **22**, 327–59.

FRETTER, V. (1941) The genital ducts of some British stenoglossan prosobranchs, *J. mar. biol. Ass. U.K.* **25**, 173–211.

FRETTER, V. (1943) Studies in the functional morphology and embryology of *Onchidella celtica* (Forbes and Hanley) and their bearing on its relationships, *J. mar. biol. Ass. U.K.* **25**, 685–720.

FRETTER, V. (1946) The genital ducts of *Theodoxus, Lamellaria* and *Trivia* and a discussion on their evolution in the prosobranchs, *J. mar. biol. Ass. U.K.* **26**, 312–51

FRETTER, V. (1948) The structure and life history of some minute prosobranchs of rock pools: *Skeneopsis planorbis* (Fabricius), *Omalogyra atomus* (Philippi), *Rissoella diaphana* (Alder) and *Rissoella opalina* (Jeffreys), *J. mar. biol. Ass. U.K.* **27**, 597–632.

FRETTER, V. (1951) Observations on the life history and functional morphology of *Cerithiopsis tubercularis* (Montagu) and *Triphora perversa* (L.), *J. mar. biol. Ass. U.K.* **29**, 567–86.

FRETTER, V. (1951a) Some observations on the British Cypraeids, *Proc. malac. Soc. Lond.* **29**, 14–20.

FRETTER, V. (1953) The transference of sperm from male to female prosobranch, with reference also to the Pyramidellidae, *Proc. Linn. Soc. Lond.* **164**, 217–24.

FRETTER, V. (1955) Observations on *Balcis devians* (Monterosato) and *Balcis alba* (Da Costa), *Proc. malac. Soc. Lond.* **31**, 137–44.

FRETTER, V. (1965) Functional studies of the anatomy of some neritid prosobranchs, *J. Zool.* **147**, 46–74.

FRETTER, V. and GRAHAM, A. (1949) The structure and mode of life of the Pyramidellidae, parasitic opisthobranchs, *J. mar. biol. Ass. U.K.* **28**, 493–532.

FRYER, G. (1959) Development in a mutelid lamellibranch, *Nature, Lond.* **183**, 1342–3.

FRYER, G. (1961) The developmental history of *Mutela bourguignati* (Ancey) Bourguignat (Mollusca: Bivalvia), *Phil. Trans.* B, **244**, 259–98.

GANAPATI, P. N. and NAGABHUSHANAM, R. (1953) Sex changes in a wood-boring mollusc, *Martesia striata* Linn., *Curr. Sci.* **22**, 345–6.

GASCOIGNE, T. (1956) Feeding and reproduction in the Limapontiidae, *Trans. roy. Soc. Edinb.* **63**, 129–151.

GRAHAM, A. (1954) The anatomy of the prosobranch *Trichotropis borealis* Broderip and Sowerby, and the systematic position of the Capulidae, *J. mar. biol. Ass. U.K.* **33**, 129–44.

GROBBEN, C. (1892) Beiträge zur Kenntnis des Baues von *Cuspidaria* (*Naera*) *cuspidata* Olivi nebst Betrachtungen über das System der Lamellibranchiaten, *Arb. zool. Inst. Univ. Wien* **10**, 101–46.

HANSEN, B. (1953) Brood protection and sex ratio of *Transennella tantilla* (Gould), a Pacific bivalve, *Vidensk. Medd. dansk naturh. Foren. Kbh.* **115**, 313–24.

HOLMES, W. (1940) The colour changes and colour patterns of *Sepia officinalis* L., *Proc. zool. Soc. Lond.* A, **110**, 17–35.

HUNTER, W. R. (1949) The structure and behaviour of *Hiatella gallicana* (Lamarck) and *H. arctica* (L.) with special reference to the boring habit, *Proc. roy. Soc. Edinb.* **63**, 271–89.

HUXLEY, J. (1942) *Evolution, the Modern Synthesis*, Allen and Unwin, London.

JONES, G. F. (1963) Brood protection in three Southern Californian species of the pelecypod *Cardita*, *Wasmann J. Biol.* **21**, 141–8.

JUTTING, W. S. S. VAN BENTHEM (1956) Systematic studies on the non-marine mollusca of the Indo-Australian archipelago. V. Critical revision of the Javanese freshwater gastropods, *Treubia* **23**, 259–477.

KNIGHT-JONES, E. W. (1952) Reproduction of oysters in the rivers Crouch and Roach, Essex, during 1947, 1948 and 1949, *Fishery Invest. Lond.* **18**, 1–48.

KNUDSEN, J. (1961) The bathyal and abyssal *Xylophaga* (Pholadidae, Bivalvia), *Galathea Rept.*, **5**, 163–209.

KOHN, A. J. (1959) Ecological notes on *Conus* (Mollusca: Gastropoda) in the Trincomalee region of Ceylon. *Ann. and Mag. nat. Hist.* **13**, 309–20.

KOHN, A. J. (1961) Spawning behaviour, egg masses and larval development in *Conus* from the Indian ocean, *Bull. Bingham Oceanogr. Coll.* **17**, 3–51.

KOHN, A. J. (1961a) Studies on spawning behaviour, egg masses, and larval development in the gastropod genus *Conus*. Part I. Observations on nine species in Hawaii, *Pacif. Sci.* **15**, 163–79.

KORRINGA, P. (1947) Relations between the moon and periodicity in the breeding of marine animals, *Ecol. Monogr.* **17**, 347–81.

KRISTENSEN, I. (1957) Differences in density and growth in a cockle population in the Dutch Wadden Sea, *Arch. néerl. Zool.* **12**, 351–453.

LARAMBERGUE, M. DE (1939) Étude de l'autofécondation chez les gastéropodes pulmonés: récherches sur l'aphallie et la fécondation chez *Bulinus (Isadora) contortus* Michaud, *Bull. biol. Fr. Belg.* **123**, 19–231.

LARAMBERGUE, M. DE (1941) Races aphalliques et euphalliques de *Bulinus contortus*, *Int. Conf. Genet. Edinb. 1939*, pp. 185–6.

LEBOUR, M. V. (1935) The larval stages of *Balcis alba* and *B. devians*, *J. mar. biol. Ass. U.K.* **20**, 65–70.

LEBOUR, M. V. (1937) The eggs and larvae of the British prosobranchs, with special reference to those living in the plankton, *J. mar. biol. Ass. U.K.* **22**, 105–66.

LEBOUR, M. V. (1938) Notes on the breeding of some lamellibranchs from Plymouth and their larvae, *J. mar. biol. Ass. U.K.* **23**, 119–44.

LEBOUR, M. V. (1945) The eggs and larvae of some prosobranchs from Bermuda, *Proc. zool. Soc. Lond.* **114**, 462–89.

LEBOUR, M. V. (1946) The species of *Teredo* from Plymouth waters, *J. mar. biol. Ass. U.K.* **26**, 381–9.

LEMCHE, H. (1955) Neurosecretion and incretory glands in a tectibranch mollusc, *Experientia* **11**, 320–2.

LEMCHE, H. and WINGSTRAND, K. G. (1959) The anatomy of *Neopilina galatheae* Lemche 1957, *Galathea Rep.* **3**, 9–71.

LOOSANOFF, V. L. (1937) Development of the primary gonad and sexual phases in *Venus mercenaria* Linnaeus, *Biol. Bull. Woods Hole* **72**, 389–405.

MASON, J. (1958) A possible lunar periodicity in the breeding of the scallop, *Pecten maximus* (L.), *Ann. Mag. nat. Hist.* ser. 13, **1**, 601–2.

MATTOX, N. T. (1938) Morphology of *Campeloma rufum*, a parthenogenetic snail, *J. Morph.* **62**, 243–61.

MONK, C. R. (1928) The anatomy and life history of a freshwater mollusc of the genus *Sphaerium*, *J. Morph.* **45**, 473–503.

MORTON, J. E. (1951) The structure and adaptations of the New Zealand Vermetidae, Parts 1, 2, and 3, *Trans. roy. Soc. N.Z.* **79**, 1–51.

MORTON, J. E. (1955) The functional morphology of the British Ellobiidae (Gastropoda, Pulmonata) with special reference to the digestive and reproductive systems. *Phil. Trans.* B, **239**, 89–160.

MORTON, J. E. (1955a) The evolution of the Ellobiidae with a discussion on the origin of pulmonates, *Proc. zool. Soc. Lond.* **125**, 127–68.

NELSON, T. C. (1928) On the distribution of critical temperatures for spawning and for ciliary activity in bivalve molluscs, *Science* **67**, 220–1.

NELSON, T. C. and ALLISON, J. B. (1940) On the nature and action of diantlin, a new hormone-like substance carried by the spermatozoa of the oyster, *J. exp. Zool.* **85**, 299–338.

OCKELMANN, W. K. (1958) The zoology of East Greenland: marine Lamellibranchiata, *Medd. Grønland*, **122**, 5–256.

OCKELMANN, K. W. (1964) Spreading phases in marine bivalves, Symposium on the Mollusca, 1964, Zoological Society of London (unpublished).

OLDFIELD, E. (1955) Observations on the anatomy and mode of life of *Lasaea rubra* (Montagu) and *Turtonia minuta* (Fabricius), *Proc. malac. Soc. Lond.* **31**, 226–49.

OLDFIELD, E. (1964) The reproduction and development of some members of the Erycinidae and Montacutidae (Mollusca, Eulamellibranchiata). *Proc. malac. Soc. Lond.* **36**, 79–120.

ORTON, J. H. (1926) On lunar periodicity in spawning of normally grown Falmouth oysters (*O. edulis*) in 1925, with a comparison of the spawning capacity of normally grown and dumpy oysters, *J. mar. biol. Ass. U.K.* **14**, 199–225.

ORTON, J. H. (1927) Observations and experiments on sex-change in the European oyster (*O. edulis*). Part I. The change from female to male, *J. mar. biol. Ass. U.K.* **14**, 967–1045.

ORTON, J. H. (1933) Observations and experiments on sex-change in the European oyster (*O. edulis*). Part IV. On the change from male to female, *J. mar. biol. Ass. U.K.* **19**, 5–53.

ORTON, J. H., SOUTHWARD, A. J. and DODD, J. M. (1956) Studies on the biology of limpets. II. The breeding of *Patella vulgata* L. in Britain, *J. mar. biol. Ass. U.K.* **35**, 149–76.

PACKARD, A. (1961) Sucker display of *Octopus*, *Nature, Lond.* **190**, 736–7.

PELSENEER, P. (1906) *Mollusca*: *A Treatise on Zoology*, vol. V., ed. E. Ray Lankester, A. and C. Black, London.

POPHAM, M. L. (1940) The mantle cavity of some of the Erycinidae, Montacutidae and Galeommatidae, with special reference to the ciliary mechanisms, *J. mar. biol. Ass. U.K.* **2 4**, 549–87.

PURCHON, R. D. (1941) On the biology and relationships of the lamellibranch *Xylophaga dorsalis* (Turton), *J. mar. biol. Ass. U.K.* **25**, 1–39.

PURCHON, R. D. (1951) Hermaphroditism, *Gazette of King Edward VII Medical Soc. University of Malaya* **2**, 1–7.

RASMUSSEN, E. (1951) Faunistic and biological notes on marine invertebrates. II. The eggs and larvae of some Danish marine gastropods, *Vidensk. Medd. dansk naturh. Foren. Kbh.*, **113**, 201–49.

REID, J. D. (1964) The reproduction of the sacoglossan opisthobranch *Elysia maoria*, *Proc. zool. Soc. Lond.* **143**, 365–93.

RIGBY, J. E. (1963) Alimentary and reproductive systems of *Oxychilus cellarius* (Müller) (Stylommatophora), *Proc. zool. Soc. Lond.* **141**, 311–59.

ROBSON, G. C. (1923) Parthenogenesis in the mollusc *Paludestrina jenkinsi*, *Brit. J. exp. Biol.* **1**, 65–78.

SANDERSON, A. R. (1940) Maturation in the parthenogenetic snail *Potamopyrgus jenkinsi* Smith, *Proc. zool. Soc. Lond.* A, **110**, 11–15.

SCHALIE, H. VAN DER, and LOCKE, E. (1941) Hermaphroditism in *Anodonta grandis*, a freshwater mussel, *Occ. Pap. Mus. Zool. Univ. Mich.* **432**, 1–7.

THOMPSON, T. E. (1958) The natural history, embryology, larval biology and post-larval development of *Adalaria proxima* (Alder and Hancock) (Gastropoda, Opisthobranchia), *Phil. Trans.* B, **242**, 1–58.

THORSON, G. (1940) Studies on the egg masses and larval development of Gastropoda from the Iranian Gulf, *Dan. Sci. Invest. Iran*, Pt. II, 159–238.

THORSON, G. (1941) Marine Gastropoda Prosobranchiata, *The Zoology of Iceland* **4**, 1–150.

THORSON, G. (1944) The Zoology of East Greenland: Marine Gastropoda Prosobranchiata, *Medd. Grønland* **121**, 3–181.

THORSON, G. (1950) Christmas card.

THORSON, G. (1955) Christmas card.

TINBERGEN, L. (1939) Zur Fortpflanzungsethologie von *Sepia officinalis* L., *Arch. néerl. Zool.* **3**, 323–64.

TURNER, R. D. (1966) *A Survey and Illustrated Catalogue of the Teredinidae*, Museum of Comparative Zoology, Harvard University, Cambridge, Mass.

WALKER, B. (1917) The method of evolution in the Unionidae, *Occ. Pap. Mus. Zool. Univ. Mich.*, No. 45.

WELLS, M. J. (1964) Hormonal control of sexual maturity in cephalopods, *Bull. nat. Inst. Sci. India* **27**, 61–77.

WELLS, M. J. and WELLS, J. (1959) Hormonal control of sexual maturity in *Octopus*, *J. exp. Biol.* **36**, 1–32.

WILBUR, K. M. and YONGE, C. M. (1964) *Physiology of Mollusca*, voc. **1**, Academic Press. New York and London.

WILSON, D. P. (1937) The influence of the substratum on the metamorphosis of *Notomastus* larvae, *J. mar. biol. Ass. U.K.* **22**, 227–43.

WILSON, D. P. and WILSON, M. A. (1956) A contribution to the biology of *Ianthina janthina*, *J. mar. biol. Ass. U.K.* **35**, 291–305.

YONGE, C. M. (1936) Mode of life, feeding, digestion and symbiosis with zooxanthellae in the Tridacnidae, *Scient. Rep. Gt. Barrier Reef. Exped.* **1**, 283–321.

YONGE, C. M. (1952) Studies on Pacific Coast molluscs. V. Structure and adaptation in *Entovalva saxicola* (Baird) and *Mytilimeria nuttallii* Conrad, with a discussion on evolution within the family Lyonsiidae (Eulamellibranchia), *Univ. Calif. Publ. Zool.* **55**, 439–50.

YONGE, C. M. (1953) Observations on *Hipponyx antiquatus* (Linnaeus), *Proc. Calif. Acad. Sci.* **28**, 1–24.

YOUNG, J. Z. (1962) Courtship and mating by a coral reef octopus (*O. horridus*), *Proc. zool. Soc. Lond.* **138**, 157–62.

CHAPTER 7

DISTRIBUTION OF MOLLUSCS

Synopsis

The distribution of species and of higher categories depends on the antiquity of the group, its powers of dispersal, its adaptability to environmental variables and the effects, if any, of isolation. Some stenothermal taxonomic groups are restricted to warm seas, e.g. the Tridacnidae and the Conidae. In contrast some eurythermal groups are very widely distributed, e.g. the Limidae and the Naticidae. The only molluscan classes to have invaded fresh waters are the Gastropoda and the Bivalvia; some of these freshwater molluscs are of great antiquity and are extensively adapted to the rigours of freshwater life, especially as regards the reproductive system. These ancient freshwater forms commonly comprise a whole family or order, e.g. the Unionacea and the Sphaeriacea, which are cosmopolitan and show no signs of relationship with any marine orders. The very large number of genera in the Unionacea indicates the great success of this initial colonisation of fresh waters by bivalves. In contrast, freshwater elements of more recent origin are more restricted in distribution, are less obviously adapted to freshwater life, and may be closely related to marine forms, e.g. *Egeria* which occurs only in West African rivers, and belongs to the marine family Donacidae.

The terrestrial pulmonate genus *Partula* comprises 112 species which inhabit Pacific islands extending over a huge area, in general each species being found only on one island. There is no hint as to the origin of the genus on any adjacent land mass. Some islands support many species of *Partula*, these often occurring in the same valley and even under the same leaf. It is postulated that speciation was partly due to geographic isolation on islands after subsidence of a larger land mass, and partly to ecological isolation, and that the final stages of subsidence caused good species to be driven together centripetally into the same valley systems. Speciation in such instances will have been aided by the general lack of predators on the islands. The widespread occurrence of *Partula* on Pacific islands supports the theory of the origin of coral islands by subsidence and not by submarine elevation.

The limestone hills of Malaya are especially interesting on account of their gastropod faunas, which total over a hundred different species of terrestrial prosobranchs and pulmonates; some of these are endemic to a single hill while some others occur on two, three, or more hills. The sides of these hills are generally precipitous and the hills are separated from each other by soils of a different constitution which are not likely to favour migration of snails from one hill to another. Some hills support more than one species in the same genus and clearly these sympatric species must have evolved elsewhere, perhaps in ecological isolation over a larger area centring on the hill, and were finally driven to coexistence on the hill as the result of submergence. Statistical analyses show that the distribution of species of snail on these hills is random, and closely fits expectation according to the Poisson distribution. However, few of these snails have been studied alive and perhaps not all are good species.

Studies on the non-marine molluscan faunas of the East Indies reinforces the view that there is here no single zone of discontinuity as suggested by "Wallace's line", but that the whole archipelago is a region of transition from one zoo-geographical area to another. Three tabulations of the distributions of selected terrestrial and freshwater prosobranchs and pulmonates clearly demonstrate faunistic affinities running through the archipelago. The islands of this archipelago must at one time have been interconnected by land bridges which have long since disappeared due to general subsidence of the land.

Molluscs of the rocky shore demonstrate ecological succession from high- to low-water mark, each species being most abundant in a certain horizontal stratum, some of these strata being narrow and clearly defined and others being broader and less striking. The upper and lower limits of distribution of each species arise from interplay of a number of environmental factors, including degree of exposure to wave impact, duration of exposure, and of immersion by the sea, maximum and minimum air and sea temperatures, the presence or absence of competitors, predators, and food sources. The criteria differ from one species to another. The movements of motile species may contribute to the final pattern of distribution of the adults. Thus post-larvae of *Littorina littorea* settle sub-littorally and young individuals gradually move inshore. *Hydrobia ulvae* survives but will not float at the low salinity of 3·5 parts per 1000 and therefore does not permit itself to be carried further up an estuary on the tidal stream than the 3·5 per 1000 isohaline. Where the tidal range is high and there is some protection from surf the scheme of zonation is primarily dependent on the vertical range of the tide, but on open oceanic beaches the tidal range is small and is far outweighed by the splash zone and the latter predominates in determining the pattern of inter-tidal zonation. The potential vertical range of a species can only be judged from study of many contrasted localities, due to the possibility of the existence of local limiting factors. The zonation of molluscs on various rocky shores are described in detail. Tests of the food preferences of *L. obtusata* (= *L. littoralis*) showed that the inter-tidal zonation of this species was in accordance with its preferences for various species of sea weed. Difficulties in the identification of species of *Patella* are discussed, and the distributions of four European species of *Patella* are reviewed and illustrated.

The molluscan elements of the infauna of shores of sand and of sandy mud are similarly subject to horizontal zonation. The zonation of each species may be influenced by the degree of exposure to surf, the extent of the tidal range, the composition of the soil and rapidity of drainage, the presence of organic matter and micro-organisms, the presence of predators, etc. The influence of these various factors can only be evaluated if all are studied concurrently, and comparisons are made of zonation schemes at various localities. Deposit-feeding species, e.g. *Tellina* spp. and *Scrobicularia*, tend to favour sheltered beaches where the shore profile is gentle, where there is an abundance of silt, organic matter, and micro-organisms. Filter-feeding bivalves are dependent on other properties of the soil and predominate at favourable levels where there is shelter from exposure by wave action, etc. The nature of the soil may be more important than the tidal level, and where the sand is approximately constant in composition at all tidal levels *Tellina tenuis* was found from high water to low water and beyond; *T. tenuis* was most abundant at the lowest levels where spat can settle in the largest numbers, but the largest specimens were found on the upper half of the beach. The distribution of molluscs on various shores of sand, and of sandy mud are described in detail.

The abundance of the blue rayed limpet, *Patina pellucida*, per unit area of sub-littoral weed has been shown to depend on the strength of the water currents, this limpet being most abundant in areas of moderate current strength; it is shown that the strength of the current may influence the limpet in a number of different ways and this is by no means a simple environmental factor.

Small molluscs may be accidentally transported by various agencies; *Anodonta* was once reported to have fallen in numbers during torrential rain; *Patella* spp. have been transported by tidal currents, buoyed up by attached algae. The terrestrial prosobranch *Pomatias* has been seen attached to the leg of a flying bumble-bee, and freshwater bivalves have been seen similarly attached to the beetle *Dytiscus*. The Upland Plover apparently places many specimens of *Physa* in its plumage prior to migration, for use as rations on the journey, the snail adhering to the plumage of the bird by the sticky slime secreted by the snail. The distributions of many molluscs have been greatly extended by man; *Crepidula fornicata* and *Urosalpinx cinerea* were introduced into British coastal waters probably with consignments of *Ostrea virginica*, and are now serious pests of oyster beds. *Helix aspersa* was deliberately introduced into the U.S.A. and elsewhere as a source of food, and the dispersal of *Achatina fulica* (illustrated in Fig. 129) was partly a matter of supposed scientific interest, and partly a matter of provision of additional food. In contrast, *Gonaxis* was used very cautiously to control *Achatina* on small islands in the Pacific. The cutting of the Suez Canal permitted interchange between Mediterranean and Red Sea faunas and *Pinctada vulgaris* became established in the Mediterranean. The dispersal of the freshwater *Dreissena polymorpha* through the river systems of western Europe as a result of commercial activity in rivers and canals is especially interesting, for prior to the nineteenth century *Dreissena* was confined to an area near the Black, Caspian, and Aral seas. What may be the long-term consequences of the erection of dams at Aswan, and on the Volta in Ghana, on the snail vectors of schistosomiasis? Is it possible to control these snail vectors by biological methods, e.g. by introduction of suitable sciomyzid flies, the larvae of which feed on snails?

A brief account is given of studies on the snail *Cepaea nemoralis*, which exhibits balanced polymorphism as regards colour and banding of the shell. In some areas there is a correlation between the nature of the vegetation and the various colour forms of the snail, the most abundant types being cryptically coloured. This is attributed to predation by thrushes, rabbits, and rodents, on the more conspicuous shells. Yellow and banded shells tend to predominate in short turf, while unbanded, brown or pink shells predominate in dense deciduous woodlands where undergrowth is sparse. On some chalk downs certain phenotypes predominate over large areas regardless of the vegetation, and the reasons for these "area effects" are not yet known, but the low level of visual predation by thrushes on open downland may be relevant.

The present-day geographic distribution of species and of higher taxonomic groupings of Molluscs is dependent on a number of factors including primarily the antiquity and region of origin of the group, the efficiency of the type as a biological machine, the adaptability of the type to environmental variables, and the effects of geographic isolation. We also have to consider the mobility of the group—and in particular the degree of dispersal effected by the larval phase—major climatic changes during the geological history of the earth, changes in the alignments of the oceans and land masses, and passive dispersal resulting partly from the activities of man.

Within the geographic range of the group there may be discontinuities which are imposed by various ecological factors including the nature of the substratum, excessive exposure, e.g. to wave action, to desiccation, to extremes of temperature, etc., and restrictions in distribution of the basic sources of food. The geographic range of the group may in some cases be subdivided into a number of mutually exclusive areas each of which is populated by a different representative of the taxonomic group. Alternatively, the area may be populated by a series of representatives, the geographic ranges of which overlap to greater or lesser extent.

The geographic distribution of molluscs is interesting in itself as affording one aspect of the viability of the various molluscan phylogenies, it is interesting in its application to the broad problems of zoo-geography, and it provides useful material for contemplation of the principles of organic evolution.

The Mollusca originated in the seas, in the shallow littoral zone, where their ranges were not generally restricted by major physiographical barriers. Some taxonomic groups are markedly stenothermal and are restricted either to warm waters, or to cold waters, according to the temperature of the seas in which they originated, e.g. the Tridacnidae and the Conidae, both of which are confined to warm seas. Other taxonomic groups of molluscs are eurythermal and suffer no such restriction, e.g. the Limidae and Veneridae among the Bivalvia, the Trochidae and Naticidae in the Gastropoda. Such eurythermal types are cosmopolitan and are represented in tropical and in temperate waters. If a marine family or order is eurythermal it should experience negligible geographical restriction, and the distribution of the group through the seas and oceans, and the number of constituent genera, provide meaningful indices of the general efficiency and vigour of the phylogeny. On this basis the family Veneridae, with twenty-nine genera, is very successful. Examination of the littoral zone in different latitudes shows that in general the same niches are present and that they are often exploited by representatives of the same genus. It seems to the writer that the main faunistic differences between temperate and tropical shores are: (a) the tropical shores are enriched by a diversity of stenothermal warmth-loving forms, e.g. the Tridacnidae, and the Conidae which are never found in temperate waters (see Figs. 112 and 113); (b) there is notably greater degree of speciation on tropical shores.

Molluscs in Fresh Waters

The only molluscan classes to have invaded freshwater systems are the Gastropoda and the Bivalvia. These freshwater faunas are rigorously restricted to their respective freshwater systems by geographical and physiological barriers, for they cannot migrate over land, nor can they return to the sea and become dispersed along the coast. The distributions of these freshwater forms were originally determined by the disposition of the land mass and the configuration of the river systems in which they evolved; their geographic distributions could only be moderated thereafter by changes in the outline of the land mass—by elevation, subsidence, or perhaps by continental drift, and by changes in alignment of watersheds with consequential changes in composition of river systems. Passive distribution, e.g. by attachment to the legs of migratory water birds, may have occurred to a limited extent, but it seems unlikely that this has been responsible for any major change in the geographic distribution of the higher taxonomic categories. The molluscs of fresh waters can be roughly divided into two groups, namely those of great antiquity and those which are more recent additions to

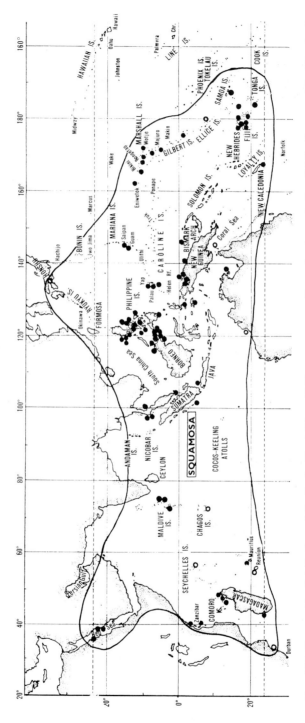

FIG. 112. Geographic distribution of the Scaly Giant Clam, *Tridacna squamosa*. *T. maxima* has a similar distribution except that it extends considerably further eastwards across the Pacific. The six recent species in the family Tridacnidae are confined in their distribution to the tropical Indian, and western Pacific Oceans. (Originally published in Rosewater, 1965, *Indo-Pacific Mollusca*, **1**, 383, plate 286.)

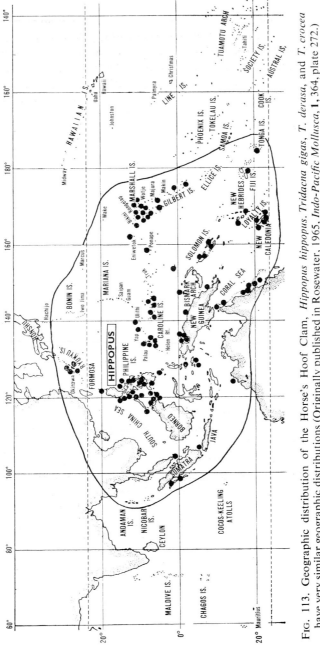

Fig. 113. Geographic distribution of the Horse's Hoof Clam, *Hippopus hippopus*. *Tridacna gigas*, *T. derasa*, and *T. crocea* have very similar geographic distributions (Originally published in Rosewater, 1965, *Indo-Pacific Mollusca*, **1**, 364, plate 272.)

the faunas of fresh waters. The former can generally be recognised by three criteria: they are often most highly modified, both structurally and functionally, to meet the rigours of the freshwater environment, this concerning in particular the reproductive system. They commonly comprise a whole family or order, their relationship with marine families and orders being obscure. Their distribution may be cosmopolitan. In contrast the more recent increments to the faunas of fresh waters are less highly modified in form and function; they may be individual freshwater genera belonging to families the other members of which are marine or estuarine; they are of more limited distribution (see Fig. 76, p. 202).

Among the Bivalvia the Unionacea and the Sphaeriacea are outstanding examples of freshwater categories of the greatest antiquity. In the Unionacea the ctenidium is modified to carry the developing embryos, a special organ, the marsupium, being developed in some cases. The life history is much modified, there being no free swimming veliger stage, and the glochidium larva (Unionidae) or haustorius larva (Mutelidae) is ecto-parasitic on the skin of fish, thereby providing an alternative distributive phase. The Unionacea are cosmopolitan, occurring in all continents, but they comprise four constituent families of which the Mutelidae and the Etheridae are confined to the southern hemisphere. In the Mesozoic a warm water Tethys ocean separated a northern land mass from a southern land mass. Subsequent land movements and marine transgressions caused rearrangements of the land masses and their constituent faunas, thus complicating the picture considerably. Thus North America joined with South America, enabling Nearctic elements to migrate southwards into South America, and Neotropical elements to migrate northwards into North America. McMichael & Hiscock (1958) have postulated the invasion of Australia ,and New Zealand by mutelids derived from South-east Asia.

The constituents of the order Unionacea must have become fully established in the freshwater systems of the world even before the Mesozoic since they occur in all continents. The Unionacea differ from all other orders of bivalves in having generated a relatively enormous number of genera (see Fig. 76, p. 202). We may presume that this was because they were probably the first bivalves to enter and colonise fresh waters, and they encountered no competition and probably negligible predation. They therefore enjoyed an unique evolutionary opportunity, they exploited it to the full, and they underwent a major phase of adaptive radiation in the freshwater systems of the world.

If we conceive the Unionacea as a monophyletic freshwater stock with representatives in the freshwater systems of all continents, what are we to think of the mode of origin of this phylogeny from a marine ancestry? To be a strictly monophyletic freshwater order the Unionacea should have originated *from a single interbreeding population* of marine bivalves which became adapted first to estuarine conditions and then to freshwater conditions. This single stock would then have been dispersed over the entire system of freshwaters by changes in configuration of the land, changes in watersheds, etc., which enabled these bivalves to spread into other river systems. This seems to be extremely improbable. The alternative possibility is that an extremely widely dispersed marine ancestral stock gradually invaded estuarine waters over a wide geographical range. These innumerable isolated estuarine stocks, having comparable genetical constitutions, continued the process in parallel and eventually became established in fresh waters at innumerable localities, and became adapted to life in freshwaters in much the same way, both anatomically and physiologically. They would have formed a number of distinct, but probably closely related species and genera, each with its own restricted geographic distribution. This seems less improbable. The stock is still monophyletic but the common ancestor was marine and rather more remote, there was much parallel evolution and the actual entry into and acclimatisation to fresh waters was polyphyletic. This is a question on which there is still room for speculation.

The Sphaeriacea are also an extremely ancient freshwater order of bivalves with a cosmopolitan distribution. The genera *Sphaerium* and *Pisidium* are themselves cosmopolitan, occurring even in South America, Australia, and New Zealand. The reproductive process is highly specialised, a small number of larvae being developed in the spaces of the ctenidium until they are eventually released as miniature adult individuals. The larvae are apparently nourished in some way from the maternal tissues while they are retained in the ctenidium. The Unionacea and the Sphaeriacea are the only bivalves to be found in high mountain lakes; the genus *Pisidium* in particular has a remarkable distribution, including lakes in the arctic circle and relict lakes at high altitudes, e.g. Lake Titicaca at nearly 4000 m above sea level in South America.

It is curious, to say the least, that *Pisidium* should have undergone world wide distribution over a period of hundreds of millions of years apparently without generating any taxonomic units above the level of species—and many so-called species of this genus are of doubtful status. Is this a bradytelic phylogeny, i.e. one which is evolving extremely slowly, like the brachiopod *Lingula* (Simpson, 1953)? Perhaps *Pisidium* is remarkably well adapted to its environment and little improvement in efficiency is possible. The environment may be generally rather uniform, with a lack of selective predation. In other words, for *Pisidium* the organism–environment relationship may be very stable, in which case the situation will lack evolutionary impetus. Alternatively, the minute *Pisidium* is probably a paedomorphic form, like the tiny *Lasaea* and *Turtonia* of the rocky shore. Such minute forms have to be exceedingly economical with the small space available within the protection of the shell valves. There is just enough room for the organs of feeding, digestion, excretion, and reproduction. There is little room to spare even for connective tissues, and as a paedomorphic form the shell may be committed to the relatively simple form and contour of the dissoconch.

In contrast with the cosmopolitan distribution of the Sphaeriacea and the Unionacea, which indicates their extreme antiquity, each of these orders has a component family which is confined to the southern regions, viz. the Cyrenoidea and the Etheridae respectively. The more restricted distribution of these two families suggests that they are of more recent origin than the orders to which they are assigned; their distribution was restricted by geographical isolation. The Cyrenoidea occur in Australia, the Philippines, West Africa, and Central America. Similarly, the genera of the Etheridae occur in South America, tropical Africa, Madagascar, and India. These distributions, like that of the Dipnoi, indicate that these southern lands were once joined together to form a huge southern land mass, "Notogaea". The typically marine family Donacidae has given rise to two brackish and freshwater genera, *Iphigenia* and *Egeria*, of which the former occurs in rivers of West Africa and also in rivers on both coasts of South and Central America. From this and much other accumulated evidence there can be no doubt that the southern continents must once have been joined by land, perhaps during a warmer climate via Antarctica. Perhaps they were once connected directly, but subsequently separated by the process of continental drift postulated by Wegener (1924). After 40 years of controversy informed opinion now mainly supports the theory of continental drift. Measurements of rock magnetism indicate the probable latitude and climate in which rocks were formed, subject to stability of the earth's magnetic field, thereby adding support to the theory. Postulated convection currents within the earth's mantle may cause upward flow beneath the mid-oceanic ridges and an outward drag beneath the continents (as yet unproven). The subject was exhaustively reviewed in a recent symposium (Royal Society, 1965).

Molluscs of Islands and Archipelagos

The distribution of the pulmonate genus *Partula* in the Pacific is of especial interest from various points of view. Due partly to the effects of geographic isolation on oceanic islands, the genus contains as many as 112 species, many of which have numerous well-defined varieties. *Partula* occurs over an enormous area of Oceania stretching from the Austral, Society and Marquesas Islands in the east, through Polynesia, Melanesia, and Micronesia, while a single species reaches as far west as Talauer in the East Indies. Mead (1961) has shown that in about a quarter of a century of isolation on various islands in the Pacific the snail *Achatina fulica* has generated a number of distinct forms which are restricted to certain islands. Had we been ignorant of the recent colonisation of these islands by *A. fulica* it is conceivable that the various island forms of this species might

TABLE 6. *Distribution of 112 species of the pulmonate genus*
Partula *on the islands of Oceania (Crampton, 1916, 1925, 1932)*

Region	Island group	Numbers of species	
Polynesia	Society	44	
	Marquesas	6	
	Austral and Cook	2	
	Samoan	8	
	Fiji, Rotuma, and Tonga	3	63
Melanesia	New Hebrides and Santa Cruz	17	
	Solomon	11	
	New Ireland, New Britain, and Admiralty	5	
	New Guinea and Louisiade	4	37
Micronesia	Pelew	3	
	Caroline	4	
	Marianas	4	11
East Indies	Talauer	1	

have been accorded sub-specific, or even specific status. We cannot exclude the possibility that *Partula* was similarly distributed from island to island in prehistoric times, perhaps as much as 1000 years ago. We do not know the origins and movements of the long extinct people responsible for the archaeological remains on Easter Island; they may have come from the western Pacific where other unexplained archaeological remains have been found. We can assume these people to have been engaged in shipping, and they might have been responsible for the distribution of one or a few species of *Partula*. Populations of *Partula* could have undergone much diversification in the ensuing period of isolation. We cannot, therefore, be certain that all these are good species. In the lack of relevant evidence, however, I shall assume that they are good species.

Judging from the present day distribution of species of *Partula* (see Table 6) it seems that the headquarters of the genus lies in the Society Islands. Since no species of *Partula* occurs on any major land mass, there is no indication of the source from which the islands of Oceania were originally colonised by these snails.

Partula cannot survive more than a very brief contact with sea water, and the idea that it achieved its present wide distribution by transportation on drifting vegetation can be safely dismissed. The distances are far too great, and no terrestrial pulmonate could pass through the surf when making landfall, or traverse the arid and saline coastal strip to reach the humid valleys of the interior where alone these snails can thrive. Similarly, we may perhaps dismiss the possibility that *Partula* has mainly been inadvertently dispersed by man, e.g. among loads of fruit, vegetables, and other merchandise; had this occurred at all recently, neighbouring islands would generally possess one or more species of *Partula* in common, which is not the case. It is true that *P. gibba* occurs on Guam, Tinian, and Saipan in the Marianas. Although extensive plateaux of coral limestone on these islands indicate that elevation has also occurred, it is possible that these three islands were formed by subsidence of an extensive land mass, from which each island inherited its stock of *P. gibba*. Alternatively, in this particular case *P. gibba* may have been accidentally transported from Guam to Tinian and to Saipan by human agency. Another exceptional case is *P. hyalina* which occurs on Tahiti in the Society Group and also in the Cook and the Austral groups. This is only explicable by parallel evolution, by long continued subsidence of a very extensive parental land mass, or, more improbably, by passive transport by human agency. With these two exceptions, it seems that all other species of *Partula* are individually confined in their distribution to a single island. Even where islands are closely adjoining, and one might expect snails to be inadvertently moved from one island to another in canoe loads of fruit, vegetables, timber, etc., there is no sign of this ever having occurred.

Peake (1968) records 31 families of terrestrial molluscs on the Pacific islands and assigns these to four groups according to geographic distribution and supposed centres of dispersal. The Pacific group comprises four families, including the Partulidae, which are endemic to or have their greatest diversity on the high islands of the central Pacific, and a fifth family on the Hawaiian islands. This group has no affinity with any existing major land mass; it is an ancient assemblage which presumably evolved in the Pacific on some large land mass which has subsequently disappeared—due to systematic formation and destruction of the ocean floor? From such an origin *Partula* must have spread by "island-hopping". The greater part of the dispersal of the molluscan fauna of the Pacific islands is not attributable to land bridges since any such bridges could only have lain along island arcs, at right angles to the main direction of dispersal.

Speciation in *Partula* cannot have been solely determined by geographic isolation, for a number of islands possess several species of *Partula*, and some of these species are subdivided into distinct varieties. *Partula* requires the humid conditions found in luxuriant vegetation in sheltered valleys with high and regular rainfall, such as occur on the "high" islands of volcanic origin. With very rare exceptions found on "low" islands of coral origin in the Cook and Austral groups, *Partula* is generally confined to "high" islands. On these "high" islands the ancient and long extinct

TABLE 7. *Distribution of 10 species of* Partula *on the island of Moorea (Society Group) and of 3 species of* Partula *on Guam (Mariana Group) (Crampton, 1925, 1932)*

		Number of species per valley					
		0	1	2	3	4	Total
Number of valleys studied	Moorea	—	12	22	15	7	56
	Guam	—	10	21	8	—	39

volcanic cones have been deeply eroded by rainfall and typically each island is deeply dissected into a series of valleys radiating from the highest inland crags. *Partula* cannot survive in the more arid coastal strip, nor can it survive on the exposed ridges which separate the valleys. Particularly in the interior, where they are often delimited by precipitous cliffs, valleys with their constituent snail populations tend to be more or less isolated from their neighbours. Consequently these snail populations seem to be breeding, and evolving, in isolation in these valleys. Some species are restricted in their distribution to a single valley or to a small group of adjacent valleys. Other species occupy many, or even all of the valleys of an island, and may be divided into as many as six or more sub-species each of which has its own distinct geographical range in a group of valleys which are usually contiguous. Conversely, individual valleys may be occupied by a single species of *Partula*, or by 2, 3, or even 4 species of *Partula* (see Table 7).

Not only may three or four species of *Partula* inhabit the same valley, but representatives of all these species may be found attached to the undersurface of the same banana, ginger, or caladium leaf.

Such groups of sympatric species were compared by Crampton, who made careful statistical analyses of individual populations, taking into account various dimensions and proportions of the shell, and the incidence of sinistrality and dextrality, etc., and the investigations showed that in no case was there any suggestion of close relationship between these sympatric species. In other words, a species which was restricted in its distribution to one or two valleys only had not evolved from any of the species with which it now coexists. These species must have evolved elsewhere and must have acquired adequate barriers to interbreeding before they came to coexist in the present valley systems. At first sight speciation in *Partula* was initiated by geographic isolation on oceanic islands and was continued by geographic isolation within individual valleys or groups of valleys, an ubiquitous species breaking up into subspecies and then into distinct species of more restricted range. This may well be true for some species, but statistical analysis of populations of sympatric species of *Partula* indicates clearly that for these groups of sympatric species different origins must be sought. Perhaps these sympatric species first arose as the result of ecological isolation on a larger land mass which carried a greater variety of terrestrial habitats. Widespread subsidence would reduce the area of the island, destroying some of these habitats and gradually driving the various populations of different species of *Partula* centripetally into the heads of the valley systems where they are found at the present day. There is now no interbreeding between these sympatric species, for adequate barriers would have been established during this earlier epoch of ecological isolation.

The species of *Partula* in the Society Group of islands can be divided into three main groups according to their habits. One of these groups is generally found only on the ground, though the snails may sometimes climb a few feet up the trunks of trees. A second group is arboreal during the daytime, when the shells are sealed and at rest on the underside of a leaf, but snails of these species descend to the ground at night in order to feed. The third group is exclusively arboreal, and these species remain high up in the trees. In all cases these snails feed on decaying plant material and on fungal mycelia and never on the living tissues of angiosperms. It seems clear that speciation in the genus *Partula* has been effected by a combination of factors including geographic isolation on oceanic islands, geographic isolation in restricted valley systems, and ecological isolation in selected habitats. Speciation has been greatly aided by the general lack of predation, the only birds on the Society islands, for example, being either frugivorous or insectivorous.

The number of species of *Partula* per island in the Society Group bears

TABLE 8. *Relation between the area of an island and the number of resident species of* Partula *in the Society Group* (*Crampton,* 1916)

Island	Area in square miles	No. of species of *Partula*
Boraboa	8	1
Huahine	19	5
Tahaa	32	5
Moorea	40	10
Raiatea	60	21
Tahiti	350	8

an interesting relationship with the area of the island, which may serve as an indicator of the diversity of habitats once offered on that island, and the longevity of the island (Table 8). The only exception is the island with the largest area and this may perhaps be explained in terms of the "Sewall Wright effect", the larger island having larger populations of snails which would tend to evolve less rapidly.

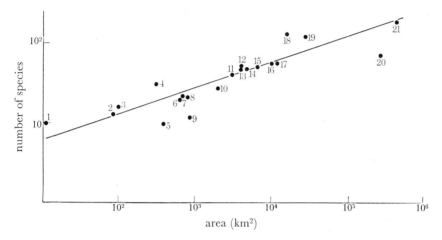

FIG. 114. The relationship between the area of an island and the number of molluscan species occurring on it. The islands selected represent a large area of the western Pacific from Sumatra to Samoa. 1, Krakatau; 2, Nissan; 3, Treasury; 4, Florida; 5, Tanna; 6, Rennell; 7, Tonga; 8, Vate; 9, Erromanga; 10, Manus; 11, Samoa; 12, Malaita; 13, Esperito Santo; 14, San Cristobal; 15, Guadalcanal; 16, Viti Levu; 17, New Ireland; 18, New Caledonia; 19, New Britain; 20, Java; 21, Sumatra. The data from Krakatau were not used in calculating the regression line. (Originally published in Peake, 1969, *Phil. Trans. B.*, 255, p. 287, figure 19.)

Another example of great interest in this connection is provided by the snails of the limestone hills of Malaya. These hills arise abruptly—usually precipitously—from surrounding country often with a soil of an entirely different constitution. It is thought that the snails of these limestone hills are generally as effectively isolated by the surrounding alluvial soils as are the faunas and floras of oceanic islands by the surrounding sea. Each closely associated group of limestone hills possesses its own endemic species of snails, some of which are shared by a number of hills in the group, while others are found only on one single hill.

Recent geological investigations have shown that the limestone hills are outcrops belonging to at least four separate groups, of which the Permian exposures are the most widespread. In some areas, e.g. in Perlis, there are parallel ranges of limestone hills which are only a few miles apart, yet these ranges can never have formed parts of a continuous limestone exposure since they have different geological origins and have always been separated by Carboniferous shales and quartzites estimated to be about 5000 ft in thickness. In some regions the limestone hills arise vertically from an extensive and relatively flat bench of more ancient limestone which bears a comparatively thin layer of residual soil and river gravels, which *may* provide little hindrance to migration of calcicolous snails along the principle, north–south axis of the underlying bench. In other areas this underlying limestone bench is covered to various depths by an alluvium deposited during an extensive marine incursion. Due to the very low calcium content this alluvium would seriously restrict, but not necessarily prevent, migration of calcicolous snails between neighbouring limestone hills. In some areas the underlying limestone deposits are lenticular in shape, and are isolated from one another. Thus for various good reasons, the limestone hills of Malaya are now recognised as comprising ten major divisions, no pair of which can ever have been joined by a continuous limestone surface (Paton, 1961).

The snails of these limestone hills include both prosobranch and pulmonate examples. Very few species have ever been seen alive, which is not surprising under the circumstances; many hills are in remote situations and can only be approached by long walks, leaving only short periods for collecting. Moreover the sides of the hills are generally precipitous, and often virtually unscalable. The collection of leaf litter at the foot of these cliffs, under boulders, etc., by a method described by Tweedie (1961), is an effective way of obtaining large numbers of empty shells of the species which occur on the hill. Repeated visits to the same hill have yielded comparable results, which suggests that the sampling method is not unreliable. Due to the high solubility of calcium salts, the minute size of most of these snails, and the generally humid conditions, it seems probable that these empty snail shells are the remains of specimens which only died recently and are not the accumulation of centuries. If this is true, it follows that the limestone hills are the sites of large and flourishing populations of these snails. Some species, e.g. of *Cyclophorus* and of *Hemiplecta*, also occur sparingly in the surrounding country, while in Sumatra species of *Discartemon*, *Sinoennea*, and *Opisthostoma* have been found in the soil in the jungle. It is possible, therefore, that some species of these snails are not confined to the limestone hills, but are specially conspicuous at these sites through occurring there in considerable abundance. Some hills are only a few hundred yards, or perhaps a mile apart in the same valley system and these might exchange species occasionally when living specimens happen to be carried off in flood waters.

Tweedie (1961) has tabulated the distribution of over 100 species of snails on 28 Malayan limestone hills (see Fig. 115). As many as 70 species are known each from one hill only, while the remaining species occur on

2, 3, 4, or more hills. Three species of *Diplommatina* occur on 8, 12, and 13 hills respectively, one species of *Paraboysidia* occurs on 10 hills, while *Gyliotrachela hungerfordiana* is found on 19 out of the 28 hills investigated. The study includes approximately equal numbers of species of Proso-branchia (*Diplommatina* and *Opisthostoma*) and of Pulmonata (*Boysidia, Discartemon, Gyliotrachela, Hypselostoma, Oophana, Paraboysidia,* and *Sinoennea*).

The study poses many difficult questions, one of the most important of these being the way—or ways—in which the limestone hills may have originally acquired their snail faunas, since these hills are now known never to have have been joined together in a single widespread limestone plateau. Perhaps the nine genera in question were originally widely distri-buted and were not dependent on the presence of lime, and the present presumed calcicolous habit may have developed on innumerable occasions, subsequent to the establishment of the individual species on the various hills. Secondly, we should inquire where speciation occurred. At first sight the faunas of these limestone hills seem to have behaved like the

FIG. 115. The distribution of the limestone hills of Malaya, which are popu-lated by species of prosobranch and of pulmonate snails. Each hill is indicated by an arabic numeral.(Redrawn from the original published in Tweedie, 1961, *Bull. Raffles Mus.* **26,** 62, fig. 1.)

faunas of oceanic islands, new species having emerged on each isolated hill by genetical "drift", as the result of geographic isolation. This is not necessarily always true. On two separate occasions the genus *Sinoennea* has produced a pair of species which occur only on that single hill. *Diplommatina* has produced a pair of species on one hill and three species on another, and the same is true for *Opisthostoma*. There are no special grounds for supposing that these pairs, and triplets of species arose in each case by divergence from a common ancestor on that hill. We do not know for any of these hills whether the snails of any one genus occur within the same ecological habitat and whether their ecological niches are quite distinct. Indeed, this would be one of the most important questions to settle in any future field investigation of this problem. It seems possible that the various species evolved in an earlier period of ecological isolation over a larger area, and became concentrated on what are now the limestone hills during a period of subsidence and marine incursion. Having evolved into good species while ecologically isolated the groups of species could then live side by side in harmony when imprisoned on one of our present day limestone hills.

We also have to ask whether we are dealing in all cases with "good" species? As none of these species has yet been exhaustively studied alive under controlled conditions in the laboratory, we cannot yet completely exclude the possibility that some of the observed variation may be only phenotypic. Berry (1962, 1963) has studied living specimens of four Malayan species of Prosobranchia (Cyclophoridae) in the genera *Diplommatina* and *Opisthostoma*, and of two Malayan species of Pulmonata (Streptaxidae) in the genera *Oophana* and *Sinoennea*. In all six cases Berry found that a new rib was added at the mouth of the shell once very 24 hours, usually at night, under normal living conditions with sufficient moisture. If the snails were allowed to become dry the new ribs were of reduced size and were crowded together. If the specimens remained dry, growth soon ceased. In the case of *Opisthostoma retrovertens*, assuming that the habitat remains sufficiently moist during the growing season and that one new rib is added to the shell every day, Berry concluded that the animal would become fully grown in about 110–124 days. There are probably some differences in the micro-climates endured by the various species which are endemic to particular hills, or groups of hills, and some of these "species" may ultimately prove to be no more than phenotypic variants. This may be revealed by rearing stocks of many different "species" under uniform conditions in the laboratory. In some cases it may be found that differences between "species" arise from an underlying ecocline. The shells of many of these land snails are ornamented with ribs which are laid down as flanges at the mouth of the shell at regular intervals as the shell grows, as described above. The appearance of these shells is therefore influenced by the degree of development of these ribs, and by the size of the spaces between them, both of these criteria probably being subject to phenotypic variation.

Analysis of the data supplied in tabular form by Tweedie (1961) shows that the numbers of species of snails found on the limestone hills of Malaya are distributed at random, and conform to the Poisson distribution. This is an important form of distribution which is characteristic of events which may occur at random either 0, 1, 2, 3, 4,..., n times on any given occasion. The frequency of occurrence for each successive point in the above progression is determined by the formula

$$e^m = 1 + m + \frac{m^2}{2!} + \frac{m^3}{3!} + \frac{m^4}{4!} + \ldots + \frac{m^n}{n!},$$

where m is the arithmetic mean, and $4! = 4 \times 3 \times 2 \times 1$. Whereas the normal distribution of a variable has two parameters, the arithmetic mean and the standard deviation, the Poisson distribution has only one parameter,

TABLE 9. *Distribution of 70 species of Malayan limestone hill snails, each of which occurs on only one limestone hill, on a total of 28 such hills: comparison of field data with results expected from application of the Poisson theorem. $m = 2 \cdot 5$; $6! = 6 \times 5 \times 4 \times 3 \times 2 \times 1$*

Number of species per hill	Poisson distribution	Frequency	
		Observed	Expected
0	1	2	2·30
1	m	5	5·75
2	$\dfrac{m^2}{2!}$	9	7·18
3	$\dfrac{m^3}{3!}$	6	5·98
4	$\dfrac{m^4}{4!}$	3	
5	$\dfrac{m^5}{5!}$	2 ⎫	6·79
6	$\dfrac{m^6}{6!}$	0 ⎬ = 6	
7	$\dfrac{m^7}{7!}$	1 ⎭	
Totals	e^m	.28	28·0

the arithmetic mean, which also equals the variance (Fisher, 1941). Tweedie recorded seventy species of snail, some prosobranch and some pulmonate, each of which was known to occur only on one single limestone hill. Of the 28 limestone hills studied, some possessed no such endemic species, while others possessed 1, 2, or more of these species of snails. In Table 9 the observed frequency distribution is compared with that expected according to the Poisson distribution (see Fig. 115). There is close agreement between the observed and the expected results $x^2 = 0.6902$; $P = >0.5$; and the hypothesis is highly significant. This clearly indicates that the occurrence of any one species on a particular limestone hill is a random event, completely uninfluenced by the presence or absence of any other species. We may further conclude that interspecific competition is negligible or non-existent even on those hills where a number of species of the same genus occur side by side.

In addition to the 70 species of snail each of which occurs only on one single hill, there are also 33 further species each of which occurs on two or more hills. With regard to these 33 species it is possible to compare each limestone hill with each other hill in turn, recording the number of species of snail which are common to each pair of hills. Since there are 28 limestone hills each hill can be compared with 27 other hills, and there will be a total of $27 \times 14 = 378$ such comparisons. In many cases two selected hills possess no species of snail in common; in other cases a pair of hills may have one, two, or more, species of snail in common. These

TABLE 10. *Distribution of 33 species of Malayan limestone hill snails each of which occurs on two or more such hills.*

Number of species common to two selected hills	Frequency	
	Observed	Expected
0	111	107·11
1	112	118·89
2	70	65·98
3	24	24·41
4	4 ⎫	
5	3 ⎬ = 8	8·61
6	1 ⎪	
7	0 ⎭	
Totals	325	325·00

comparisons provide us with a numerical estimate of the "degree of faunistic relationship" exhibited by any pair of limestone hills, with a range from 0 to 7. Statistical analysis shows that there is an excess of occasions where two hills share as many as 5, 6, or 7 species of snail. Some of these are probably due to the subdivision of a larger limestone hill by erosion into two daughter hills, and the investigation suggests that hills 8, 9, and 10 in Perak should be regarded in this light. After grouping these three hills as a single locality the data conform to a Poisson distribution (Table 10), which indicates that there is no significant zoo-geographical discontinuity in the area under consideration. The distributions of the individual species vary at random, depending on such factors as their antiquity, their region of origin, their reproductive capacity, micro-habitat, and susceptibility to passive distribution.

In a long series of studies on the non-marine mollusca of Malaysia and the East Indies it has been shown that both terrestrial and aquatic proso-branchs and pulmonates have undergone extensive speciation in this vast archipelago, and many of the islands bear endemic species (Jutting, 1941–63a; Laidlaw and Solem, 1961). Here, however, the pattern of distri-bution differs strikingly from that of *Partula* described previously, for very many species are common to a pair of adjacent islands such as Su-matra and Java, or Java and Bali, etc., while many species range widely through the whole area as shown in Tables 11, 12, 13. Thus the freshwater prosobranch *Thiara scabra* ranges from Mauritius through the Indo-Malayan region and even reaches some of the islands in the Pacific Ocean. Again, the terrestrial pulmonates *Amphidromus perversus* and *A. inversus* have very extensive distributions through the Indo-Malayan region. The species in question are now effectively isolated by sea water, but their present distributions demonstrate beyond all doubt that the various islands of the East Indies have at some earlier date been interconnected by land bridges. This area is of great zoo-geographical interest; Alfred Russel Wallace drew a sharp distinction between the faunas of the Indo-Malayan and the Australian Regions, the well known "Wallace's line" passing through the narrow strait between the islands of Bali and Lombok, and thence northwards between Borneo and Celebes. Variations on this proposal were advanced by Weber and by T. H. Huxley in attempts to correct the short-comings of the original boundary line, but these modi-fications were no more satisfactory. Wegener (1924) postulated that Austra-lia had drifted away from Antarctica to the north-east, carrying with it its indigenous fauna, and striking the Indo-Malayan island chain in the vicin-ity of New Guinea. This theory supports the view that these two zoo-geographical regions had acquired their faunas from different sources and that the closely spaced chain of Indo-Malayan islands had thereafter permitted a certain degree of mingling of the two faunas according to the

TABLE 11. *Distribution of selected species of terrestrial prosobranch gastropods in the Far East.*
(Data from W. S. S. van Benthem Jutting, 1940–63a)
The table does not purport to be exhaustive, and the absence of a record for a particular locality does not necessarily mean that the species does not occur there

	Malaya	Sumatra	Bangka	Biliton	Borneo	Philippine Isles	Celebes	Moluccan Isles	Java	Bali	Sumbawa	Sumba	Flores	Timor	Amboina	Aru Isles	New Guinea	Kei Isles
Geophorus oxytropis						x	x	x					x	x	x	x	x	x
Cyclophorus perdix including sub-species	x	x	x		x				x	x								
Leptopoma perlucidum	x	x	x	x		x	x	x	x	x	x	x	x	x	x	x	x	x
Diplommatina javana									x	x			x					
Sulfurina parva		?				x		x							x	x		

mobility of their various constituents. Be that as it may, we now recognise that there must have once been land bridges between the islands of the East Indies, these land bridges having been lost as the result of a rise in the level of the sea—or a subsidence of the land mass. No special significance can be attached to the strait between Bali and Lombok, or to any other strait in this area, as a barrier between two terrestrial and fresh water faunas of contrasted origin. On the contrary, we now regard the Indo-Malayan archipelago as a region of transition, and this view is amply supported by the known distribution of the terrestrial and freshwater Gastropoda.

In the case of *Partula* isolation on oceanic islands led to speciation, and each island colonised by *Partula* carries its own endemic species of this snail. Why did not isolation lead inexorably to a similar degree of speciation in the terrestrial and freshwater snails of the Indo-Malayan archipelago? Firstly, the relative change in water level probably only occurred comparatively recently, and the various island populations have not been

TABLE 12. *Distribution of selected species of freshwater prosobranch gastropods in the Far East.* (Data from W. S. S. van Benthem Jutting, 1940–63a)

The table does not purport to be exhaustive, and the absence of a record for a particular locality does not necessarily mean that the species does not occur there

	Mauritius	Seychelles	India	Ceylon	Burma	Cambodia	Malaya	Sumatra	Bangka	Borneo	Philippine Isles	Celebes	Moluccan Isles	Java	Bali	Sumbawa	Sumba	Flores	Amboina	New Guinea	North Australia	Solomon Isles	Fiji	New Caledonia
Brotia costula			×		×		×	×	×					×										
Thiara scabra	×	×	×		×		×	×		×	×	×	×	×	×	×	×	×	×	×			×	×
T. rudis		×	×		×			×			×		×	×	×		×		×					
T. amarula	×			×		×		×			×		×	×					×	×				
T. cancellata			×					×			×	×	×	×	×		×	×	×	×				
Melanoides rusticus								×					×	×	×	×	×	×		×				
M. tuberculatus			×		×		×	×	×	×	×	×	×	×	×	×	×	×	×	×	×			
M. torulosus			×		×			×					×	×			×		×	×		×		
Pila ampullacea							×	×	×	×		×		×										

TABLE 13. *Distribution of selected species of terrestrial and of freshwater Pulmonata in the Far East.* (Data extracted from van Benthem Jutting, 1940–63a, and from Laidlaw and Solem, 1962) The table does not purport to be exhaustive, and the absence of a record for a particular locality does not necessarily mean that the species does not occur there.

	Annam	Thailand	Malaya	Sumatra	Bangka	Borneo	Celebes	Philippine Isles	Java	Bali	Lombok	Sumba	Flores	Timor	Moluccan Isles	Aru Isles	New Guinea	Caroline Isles	S. E. Australia	Some Polyne- sian Isles
Gastrocopta pediculus ovatula				×				×	×			×		×			×	×	×	×
Amphidromus inversus	×	×	×	×	×	×	×													
Amphidromus perversus				×		×	×		×	×										
Microcystina nana				×					×	×		×	×							
Trochomorpha froggatti				×		×	×	×	×			×			×	×	×			

	Macrochlamys amboinensis	Prosopeas achatinaceum	Helicarion albellus	Lymnaea rubiginosa	Gyraulus convexiusculus
		×			
	×				×
	×				
	×			×	×
		×		×	
				×	
				×	×
		×	×	×	×
		×	×	×	×
	×	×	×	×	×
				×	×
			×	×	×
		×		×	
		×		×	
	×	×	×	×	×
				×	×
				×	×

isolated for a sufficiently long period of time. This is, of course, self-evident, but many endemic species *have* developed on individual islands. There must be some special reason why certain species, such as those listed in Tables 11, 12, and 13, did not similarly subdivide and produce numerous daughter species, one endemic to each island in the area. The reason probably is that many of the islands are comparatively large, and in certain species of snail the interbreeding populations are presumably also very large. Under such conditions there will be greater stability, and little likelihood of evolution by genetical "drift". Again, there may perhaps be very little variation in the micro-climates of these freshwater and terrestrial snails, and consequently there may be little or no measurable adaptive evolution, or phenotypic diversification. Under such circumstances a species may remain unchanged over a very wide geographical range in spite of isolation on numerous islands, while speciation may continue in other closely related lineages if there is adaptation to different ecological niches, or if the size of the interbreeding units are kept small by some special circumstance.

Molluscs of the Littoral Zone

We can distinguish between the geographic range and the geographic distribution of any animal, for no animal is uniformly distributed through-out its geographical range; there are gaps in the distribution of a species where the edaphic, biotic, or climatic conditions are entirely unsuitable, while there are concentrations of the population wherever the habitat is particularly congenial. Whereas in freshwater and in terrestrial molluscs such discontinuities generally subdivide the species into innumerable isolated inbreeding populations, this is not necessarily true for the great majority of marine molluscs. In many marine molluscs there may be free emission of water-borne gametes and/or a planktonic distributive larval phase which may override any restrictions apparently imposed by sudden changes in the substratum. In spite of the contrast between marine and other molluscs in the extent to which species are dissected into isolated interbreeding populations, there seems to be no fundamental difference in the degree of generation of species or in the possible sizes of the geographic range of species. In marine, as in freshwater and in terrestrial molluscs, there are examples with exceptionally wide geographic ranges, and other examples where speciation has been rife and the geographic distributions are correspondingly restricted.

The littoral zone is particularly suitable for demonstrating the limitations imposed upon the distributions of various species of molluscs by environmental variables. Rocky, sandy, and muddy shores have their own characteristic molluscan faunae, and discontinuities in the extent of these faunae are immediately obvious at the most superficial inspection. These reflect the very broadest structural and functional adaptations to the habitat and the mode of life. Thus *Patella* and *Pholas* are characteristic of the rocky shore, while *Natica* and *Donax* are colonists of the sandy shore. Closer inspection of the rocky shore shows that the distribution of the various characteristic molluscan species is not dependent solely on the presence of a rocky substratum for attachment. Many gastropods and bivalves are restricted to a particular horizontal belt on the rocky shore, this belt being broad in some species and narrow in others. Whilst there is considerable overlapping of these distribution bands, a general scheme of horizontal zonation prevails and there is a distinct ecological succession as one passes from above high-water mark down towards low-water mark.

Morton (1954) has reported on an ecological succession as regards the penetration of rock crevices by a diversity of small organisms including several minute molluscs. This was demonstrated most effectively in an inter-tidal reef of slate with clefts which were penetrated most deeply by the prosobranch *Cingula cingillus*, the pulmonate *Leucophytia bidentata*, and by the bivalve *Lasaea rubra*. The pulmonate *Otina otis* occurred nearer to the mouths of the crevices where small specimens of various common rocky shore molluscs were also encountered.

The causes underlying the restriction of the occurrence of a species of gastropod or bivalve to a particular horizontal band on the shore are not fully understood. These restrictions arise from the interplay of a number of environmental factors including degree of exposure to wave impact, duration of wetting by immersion and by splashes from waves, the presence or absence of competitors, of predators or of food sources, maximum and minimum air and sea temperatures, and the duration of exposure to air. The last-mentioned factor may operate in different ways on different types of mollusc; thus in many filter-feeding bivalves and in browsing herbivorous gastropods the duration of exposure to air determines the remaining amount of time available for feeding, for feeding and digestion may here be almost continuous processes. In predacious carnivores, on the other hand, restriction of the hours available for feeding may be of little consequence for they may feed quickly and to repletion. Again, exposure to air brings the hazards of desiccation by sun and wind, of overheating, or of freezing; the distribution of some species may be limited by intolerance of a high summer temperature or of a low winter temperature, while other species may be limited by failure to reach a threshold temperature for breeding.

The distributions of animals may also be influenced by their own patterns of movement. Thus Smith and Newell (1955) have shown that the post-larvae of *Littorina littorea* settle sub-littorally, principally in the early summer, and young individuals gradually move inshore partly by wave action over the lower flatter part of the shore and partly by directed locomotion; the scheme of inter-tidal zonation that is characteristic of the adult winkles is achieved by the end of the first year of life. Thereafter individual periwinkles tend to remain at about the same level on the shore; when exposed by the ebbing tide those lying on flat wet sand exhibit a "light compass reaction" and crawl for a period towards the sun and then for a period in the opposite direction. Winkles exposed on vertical or nearly vertical surfaces, e.g. rocks, groynes, and piles, on the other hand, respond to gravitational influence; after a period of exposure and desiccation the winkles respond to wetting by crawling downwards for a time and after they have been immersed for some time they turn, crawl horizontally for a period, and then crawl upwards to the water line. Whether they are living on horizontal or on vertical surfaces, the winkles behave in such a way that they maintain approximately the same station on the sea shore (Newell, 1958). The snail *Hydrobia ulvae* spends a part of its time on the surface of the substratum and a part of its time floating suspended from the surface of the water. Under estuarine conditions the snail is lifted from the substratum by the rising tide and it is transported up the estuary passively by the tidal stream, and is then brought downstream again on the ebbing tide and is re-deposited on the substratum. *H. ulvae* does not pass further up an estuary than does water of 3·5 per 1000 salinity; with decreasing salinity fewer snails become active, and therefore fewer snails are floated and carried further upstream on a rising tide. The floating snail is sensitive to a decrease in salinity and if the salinity falls by a certain amount the operculum is closed and the snail drops to the bottom; the lower the salinity of the water by which the snail is floated, the more sensitive is the snail to a fall in salinity of the water. A limit is reached at about 3·5 per 1000 salinity, at which level *H. ulvae* is able to survive but will not float and therefore does not permit itself to be carried further upstream by tidal currents (Newell, 1964).

Where the tidal range is relatively high and there is considerable protection from surf, the scheme of horizontal zonation of intertidal molluscs is dependent primarily upon the vertical range of the tides. Where there is an increase in exposure to surf, as will be experienced when working round from a sheltered bay to an exposed headland, the increasing width of the splash zone becomes proportionately greater in importance and the whole scheme of horizontal zonation is shifted and extended upwards. Since both the tidal range and the degree of exposure to surf are subject to variation, it is not possible to reduce inter-tidal horizontal zonation at different localities to a common vertical scale of measurement. Moyse and Nelson-Smith (1963) have studied the vertical ranges of *L. neritoides* at fifteen selected localities in the vicinity of St. Anne's Head, Pembrokeshire, and have demonstrated the influence of wave action in extending the vertical distribution of this species upwards above high-tide mark, with outstanding clarity (see Figs. 116 and 117). In great shelter at the Slip Pier beach at Dale, *L. neritoides* does not reach any higher than mean high water of spring tides, whilst at St. Anne's Head this species reaches to 36 ft above

MHWS. We may conclude that from the biological point of view, as indicated by the presence of *L. neritoides*, 23 ft above chart datum at the sheltered Slip Pier beach at Dale is more or less equivalent to 59 ft above chart datum at the highly exposed St. Anne's Head. Height above chart datum has, in itself, no biological significance. Moreover, chart datum is not a constant for all localities, but is depressed in regions where the tidal range is greater than usual.

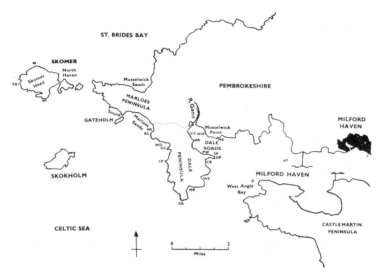

FIG. 116. Map of the coastline of west Wales showing the stations at which zonation studies were made on various littoral molluscs. Working around the coast from west to east the stations visited are: SK, Skomer Head. MS, Marloes Stack. RC, Red Cliff. WD, West Dale North. GC, Great Castle Head. LP, Long Point Stack. SA, St. Anne's Head. MB, Mill Bay North. WS, Watwick Bay South. CB, Castle Beach South. DB, Dale Point. SP, Slip Pier Beach. PW, Point Wood Beach. BR, Black Rock. CC, Cliff Cottage Beach. MW, Musselwick West. ME, Musselwick East. (Originally published in Moyse and Nelson-Smith, 1963, *Field Studies* **1** (5), 2, fig. 1.)

The vertical range of a species may be restricted locally, and even temporarily, by the presence or absence of another species which is competing for a foothold for settlement or for the same source of food. For this reason it is not possible to make a reliable judgement of the ultimate upper and lower limits of range of a littoral mollusc by inspection of only one or only a few localities. Moyse and Nelson-Smith (1963) have determined a mean distribution for *L. neritoides* with reference to the beaches studied by them. If the number and diversity of rocky beaches studied by these authors constitutes an adequate sample then their estimated mean distribution should serve as a reliable guide to the distribution of this species on other rocky shores in approximately the same latitude. Having determined in the same way the mean distributions of eleven other littoral molluscs on the same group of beaches in Pembrokeshire, Moyse and Nelson-Smith have provided a general scheme of inter-tidal zonation for these species (see Fig. 118). It must be appreciated that the vertical distributions of these various species are dependent on the interplay of a number of environmental variables, and it is unlikely that all species will respond in the same way to the same

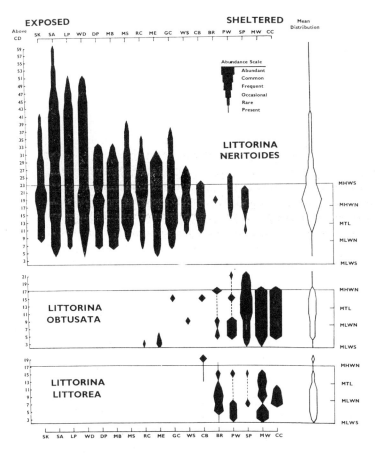

FIG. 117. Vertical distribution of three species of *Littorina* on rocky shores in west Wales. The initials at the head and at the foot of the graph indicate the names of the various stations and the locations of these stations are shown in Fig. 116. *L. neritoides* favours exposed situations, and extends higher above high water mark with increasing exposure, due to the proportionate increases in the extent of the splash zone. *L. obtusata* and *L. littorea* are unable to maintain their grip on exposed shores, and these species favour more sheltered sites. For lettering indicating tidal levels, see p. 394. (Originally published in Moyse and Nelson-Smith, 1963, *Field Studies* **1** (5), 18, fig. 8.)

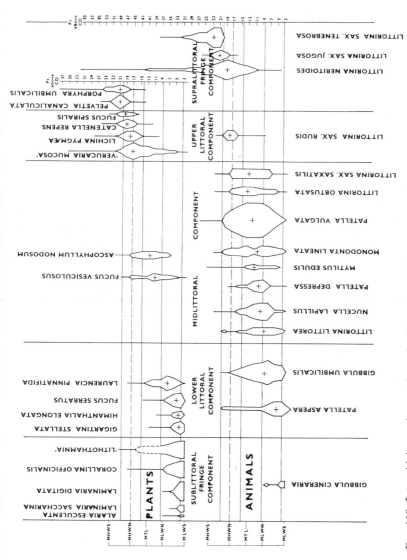

Fig. 118. Inter-tidal zonation on the rocky shore near Dale, Pembrokeshire. The vertical distribution of the common algae and lichens is given above to provide some indication of the habitats. The vertical distributions of twelve common molluscs are given below. For each species the "kite" diagram indicates the "mean distribution" which summates the variations in vertical distribution at all the different localities studied. The centre of abundance of each species is represented by a cross. For interpretation of lettering indicating tidal levels, see p. 394. (Modified from the original in Moyse and Nelson-Smith, 1963, *Field Studies* 1 (5), 27, fig. 14.)

variation in one particular variable. Although Skomer Head is at least as greatly exposed as is St. Anne's Head (Fig. 116), *L. neritoides* does not extend so far upwards above high-water mark at the former as it does at the latter station; this is because of the greater smoothness of the igneous rocks at Skomer Head and the lack of crevices providing protection against dislocation by wave impact. It follows that this scheme is a general statement

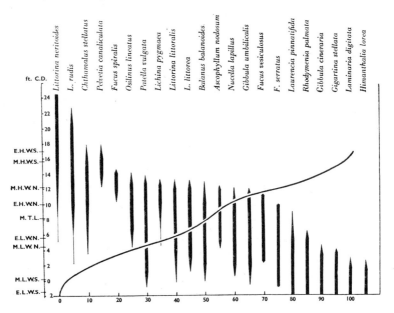

FIG. 119. Zonation of inter-tidal species on a rocky shore near Plymouth, south-west England. In addition to dealing with nine common molluscs, the vertical distributions of two barnacles, *Chthamalus stellatus* and *Balanus balanoides*, one lichen, *Lichina pygmaea*, and ten sea-weeds are also represented. The sinuous line indicates graphically the percentage exposure to air at all levels on the beach. It will be noted that only *Littorina neritoides* and *L. rudis* (= *L. saxatilis*) extend appreciably above high tide level into the splash zone above. For interpretation of lettering indicating tidal levels, see p. 394. (Originally published in Evans, 1947a, *J. mar. biol. Ass. U.K.* **27**, 209, fig. 13.)

of principle, and that minor variations from this scheme at other localities should occasion no surprise. Evans (1947) has provided a comparable representation of zonation on the coast near Plymouth (see Fig. 119).

Purchon and Enoch (1954) have studied the horizontal zonation of the littoral fauna and flora on rocks at Raffles Light in the vicinity of Singapore, in extremely sheltered waters (see Fig. 120). There, in calm weather the water level could be determined with an accuracy of ±2 in., while in "rough" weather the amplitude of the waves probably did not normally exceed 3 ft. Zonation was therefore determined almost exclusively by the range of the tide. Twentyone species of molluscs were studied and a clear pattern of horizontal zonation was established. Some species of gastropod showed great mobility, especially *L. undulata*, and *Tectarius malaccensis*. The taxisms exhibited by these snails were of assistance in maintaining their system of zonation. The tidal regime at Raffles Light was such that adjacent high tides differed noticeably in height, but this did not incommode

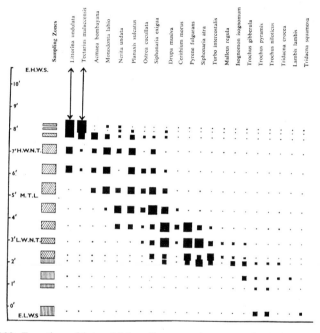

FIG. 120. Zonation of inter-tidal molluscs on the extremely sheltered rocky shore of Raffles Lighthouse, near Singapore. A dot indicates absence while black squares of four increasing areas represent rare, occasional, common, and abundant, respectively. For interpretation of lettering indicating tidal levels, see p. 394. (Modified from the original in Purchon and Enoch, 1954, *Bull. Raffles Mus.* **25**, 60, 61, fig. 4.)

these snails of the supra-littoral fringe, for they migrated downwards with the ebbing tide as long as they were able to remain on wet rocks. As the tide ebbed it gradually accelerated in its progress, and gradually the *L. undulata* and *T. malaccensis* were left behind at some rock crevice or other obstacle which could not be quickly passed. On the rising tide these snails ascended the rocks until at high tide they reached their zenith. The difference in height of two adjacent high tides, and the difference in height of high tide of spring and of neap tides, were of no significance to the two snails mentioned, for their disposition on the rocks was adjusted twice each day in accordance with the tidal levels attained. Doubtless other mobile littoral gastropods similarly maintained approximately the same level on the shore by rhythmic alternation of the appropriate taxisms. Certainly the two pulmonate limpets *Siphonaria atra* and *S. exigua* made foraging expeditions and each specimen returned to exactly the same spot, and established well-defined scars on the rocks at the points to which they "homed".

Batham (1956) gives a comprehensive review of inter-tidal zonation on a moderately sheltered rocky shore in New Zealand.

On the west coast of Africa, where the shores are pounded by the heavy Atlantic swell, the tidal range observed by Bassindale (1961) is only about 1·5 m, while the splash zone extends some 2·85 m above "high-water mark". On these shores horizontal inter-tidal zonation of the fauna and flora is dependent primarily upon the heavy surf, and the diurnal changes in tidal level are comparatively unimportant. This is emphasised by the contrasted conditions at those few localities which are well protected from wave action, and where the inter-tidal zone has accordingly shrunk to only 1·5 m. Of ten molluscan species figured by Bassindale for the rocky shore at Winneba, Ghana, the bivalves *Ostrea tulipa* (?) and *Brachyodontes puniceus* and the gastropods *T. granosus* and *Haminea orbignyana* were all restricted to zones appreciably above "high-water mark", while another gastropod, *L. punctata* extended over the whole splash zone of about 2·8 m and reached down as low as mean high water of neap tides. Bassindale described this as a "moderately exposed" rocky shore. Buchanan (1954) has reviewed the distribution of molluscs on the coast of Ghana, and has portrayed diagrammatically a scheme of horizontal zonation for ten species of gastropod and one species of polyplacophoran (see Fig. 121). The importance of the splash zone on West African beaches is emphasised by the high level achieved by the "keyhole limpet" *Fissurella nubecula*. This species may commonly be seen on bare, upwardly facing, but sea-wetted rocks high above the general level of the sea. In contrast the equivalent "keyhole limpet" *Diodora apertura* of British shores only occurs in deep shade on the under sides of boulders below low water mark of neap tides, and the same can be said of *D. singaporensis* in the Far East.

A few attempts have been made to correlate the ecological preferences of certain molluscan species in the general scheme of inter-tidal zonation, with the physiological attributes, and responses of those species. Thus Brown (1960) studied six species of littoral gastropods, in each case taking specimens from the top of their inter-tidal range, and determined their respective survival times at various relative humidities. For five species the survival time was correlated with the level occupied on the beach, the species living highest on the beach exhibiting the greatest endurance. The remaining species had a greater resistance to desiccation than would have been expected from its relative position on the shore, but this species had a lower resistance to high temperature than did the other five species. It seems, therefore, that the scheme of zonation of these six species corresponds with what was learned of their physiological attributes. Arnold (1957) compared the responses of specimens of the limpet *Patella vulgata* collected from different levels on the rocky shore. He found that specimens collected at high water of ordinary neap tides, i.e. at the upper limit of their

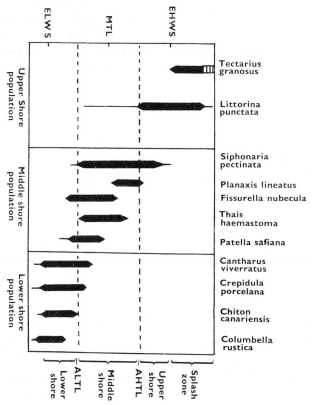

FIG. 121. Scheme of zonation of common inter-tidal molluscs on rocky shores in Ghana, West Africa. The broken lines represent the average levels of high and low water marks. For interpretation of lettering indicating tidal levels, see p. 394. (Originally published in Buchanan, 1954, *J. W. Afr. Sci. Ass.*, **1**, 33, fig. 1.)

range, showed greater tolerance to water of low salinity than did specimens collected from lower down the beach. Under laboratory conditions specimens from HWNT would give a positive response to wetting by 50 per cent sea water (i.e. of about 17 per 1000 salinity) and would begin to crawl about. These "high-level limpets" were deemed to be well adapted to life on the upper part of the beach, where they might often need to forage when the rocks are covered with water of low salinity during periods of rain.

The inter-tidal distribution of various herbivorous snails is probably influenced by their food preferences, the snails tending to accumulate in the zones occupied by the preferred seaweeds. Bakker (1959) subjected *L. obtusata* (= *L. littoralis*) to food-preference tests and found that when the snail had no alternative it was capable of consuming *Pelvetia canaliculata* although it does not colonise the *Pelvetia* zone on the sea shore. Large quantities of faeces were produced and it does not necessarily follow that *L. obtusata* was able to digest this weed. As the result of various large scale food-preference tests, Bakker concluded that for *L. obtusata* the order of preference for seaweeds was: (1) *Fucus vesiculosus*, (2) *F. spiralis*, (3) *Ascophyllum nodosum*, (4) *F. serratus*. It seems probable that the inter-tidal zonation of *L. obtusata* is at least partly determined by these food preferences.

Many genera of marine molluscs occur on both sides of the Atlantic and in some cases the same species occurs on European and on American shores. Colman (1932) examined populations of *L. obtusata* (= *L. littoralis*) from eleven localities on both sides of the Atlantic and was unable to detect any significant differences between the American and the European samples. In contrast the limpet genus *Patella* does not occur anywhere on the American side of the Atlantic Ocean, although it possesses a brief planktonic distributive phase with trochophore and veliger larvae, and one would expect its powers of dispersal to be somewhat comparable with those of *L. littorea* and to be decidedly superior to those of *L. saxatilis*. Species of *Patella* colonise the Atlantic coasts of Europe and Africa from the Lofoten Isles on the Norwegian coast to the Cape of Good Hope, more than one species commonly occurring at any one locality and within the same general habitat. As many as five species of *Patella* occur on European Atlantic shores and we shall now consider the distribution of these as a special exercise.

The identification of European species of *Patella* is no easy matter, due partly to confusion in nomenclature and partly to difficulty in finding entirely reliable criteria for the various species. Fischer-Piette (1935) was able to identify the various species of *Patella* with certainty for some stretches of coastline, but found difficulty elsewhere, where there seemed to be intergradation between the species. Regions of apparent intergradation were along the Basque coast of France and along the south coast of England in the vicinity of the Isle of Wight. Accordingly, in an address to the Linnean Society of London he presented the case for regarding the European species of *Patella* as a "syngameon" which in some areas had the form of a single highly variable species, while in other areas it comprised a number of distinct species (Fischer-Piette, 1938). Thus it was suggested that the high degree of intergradation between "species" at the Isle of Wight might be taken to signify that in this area little advance had yet been made locally in the subdivision of the basic stock into species, i.e. that in this area we are witnessing an early stage in speciation. Passing further west along the south coast of England there is a steady diminution in the degree of intergradation between "species" and this could be interpreted as meaning that as one passes further to the west the degree of speciation becomes more advanced.

Evans (1953) took large samples of limpets along the south coast of England from Torquay to Newhaven, a sample from Aberystwyth for comparison, and also a sample of *P. coerulea* from Marseilles. He made a preliminary identification of each specimen on the basis of external features of the living animal, and he then proceeded to re-identify the specimens on the basis of single characters, e.g. first on the external form of the shell, then on the internal appearance of the shell, then on the form of the pluricuspid radular teeth, etc. When studying a single character at a time in this way, he became convinced of the occurrence of transitional individuals which exhibited intergradation between one species and another. The evidence seemed at that time to exclude the possibility of attributing this intergradation to hybridisation between species. Evans therefore came to the tentative conclusion that in the region of the Isle of Wight there was an early stage in the separation of a basic stock into three types, with the final emergence of *P. aspera* and *P. depressa* as good species further to the west. This tentative conclusion seems to be at variance with an earlier statement in the same paper that out of a total of nearly 2000 limpets only three were wrongly identified during the preliminary identification on external appearance, and that all the specimens were definitely identified to one or another species when a number of characters were taken into account simultaneously. If these statements are true then there appears to be no basis for the argument that there is an initial phase in the emergence of species in the Isle of Wight area.

It seems to be simpler to suggest that we are concerned with three good species which exhibit a certain amount of phenotypic variation, this variation being greater in some parts than in other parts of their geographic ranges. The three species must have much in common genetically, and it is hardly surprising if phenotypic variation should lead to some overlapping in form. It is certainly curious that there should be a greater degree of variation in particular areas, such as in the neighbourhood of the Isle of Wight, than elsewhere. The fact that two of the species are close to the limits of their geographical range at this point may possibly be important in this respect. There is no fundamental reason why three good species must be easily separable by simple character differences; as in the present case it should suffice if specimens can be accurately determined to the species on consideration of a group of characters. Nevertheless, the chief difficulty here seems to be the failure to find a simple and reliable character difference by means of which the various European species of *Patella* can be distinguished.

More recently Fischer-Piette and Gaillard (1959) have claimed that the

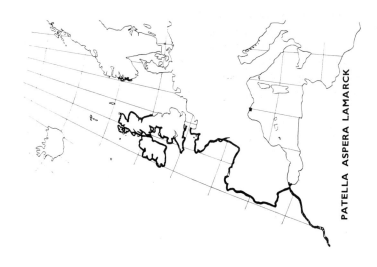

PATELLA ASPERA LAMARCK

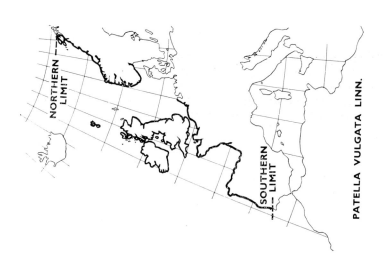

PATELLA VULGATA LINN.

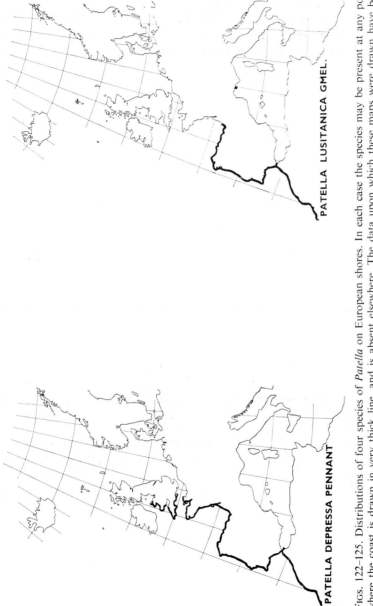

FIGS. 122–125. Distributions of four species of *Patella* on European shores. In each case the species may be present at any point where the coast is drawn in very thick line, and is absent elsewhere. The data upon which these maps were drawn have been obtained from many published sources discussed in the text and recorded in the reference list, and from various personal communications.

form of the first lateral tooth of the radula provides a sure means of identification of the five European species of *Patella*. It will be interesting to learn whether the application of this new criterion removes all difficulty in identification of limpets on the south coast of England. We will here work on the assumption that there are five distinct European Atlantic species of *Patella*. The geographic ranges, and habitats, of these species overlap to a considerable extent, and it is necessary to conclude that the various species originally evolved in geographic isolation from one another and that they have subsequently extended their ranges (see Figs. 122–125).

The ranges of the various species may be described in terms of several distinct "dimensions". Thus as regards tolerance of low temperatures *P. vulgata* outstrips all others, except perhaps *P. aspera*, and it extends as far north as Lofoten on the coast of Norway (Thorson, 1941). Kolstad (1959) found *P. aspera* on the west coast of Norway from Stavanger in the south to the islands west of Bergen in the north; in this area *P. aspera* is common only on wave-beaten sites on the outer islands and skerries. *P. depressa* Pennant only reaches northwards as far as Anglesey (Crisp and Knight-Jones, 1955), while *P. lusitanica* Gmel. barely passes the Spanish frontier in the Bay of Biscay (Fischer-Piette and Gaillard, 1959).

As regard tolerance of warmth, on the other hand, *P. vulgata* is least successful, occurring in abundance only as far south as Finisterre, and finally disappearing a little to the north of Cape St. Vincent, while the other species mentioned extend further south on to the coast of Africa.

Another relevant "dimension" concerns the degree of dilution of sea water that can be endured by limpets, e.g. in estuaries. Here again *P. vulgata* is more tolerant than any other European species of limpet. In the Bristol Channel *P. vulgata* extends up channel as far as Portskewett on the Welsh coast, and as far as Portishead on the English side (Bassindale, 1940; Purchon, 1957). In France, *P. vulgata* extends up the Rance estuary beyond Port Saint Jean as far as Plouer and Rochefort (Fischer-Piette, 1931). The other species of *Patella* do not occur in estuarine conditions.

As regards exposure to or shelter from surf, again *P. vulgata* is very tolerant although it is less abundant in very sheltered sites than it is at positions of considerable exposure (Moyse and Nelson-Smith, 1963) (see Fig. 126). Near the southern end of its geographical range *P. vulgata* may occur in the shelter of estuaries where it is lacking on the open coast nearby (Fischer-Piette and Gaillard, 1959). *P. aspera* is intolerant of very great shelter, but may occur over a range of conditions from extreme exposure to moderate shelter. While *P. aspera* thrives on extremely exposed sites, these are not necessary to it. *P. depressa*, on the other hand, is restricted to areas of moderately great exposure to surf (Evans, 1947, 1947a; Crisp and Southward, 1958).

As regards exposure to air, with the concomitant risks of desiccation by sun and wind, and exposure to extremes of temperature, once again

P. vulgata is more tolerant than the other species under consideration. *P. vulgata* may occur on dry sites up to about mean high water of neap tides (MHWNT), or even higher in situations which receive much wave splash. *P. depressa* is more limited in this respect, being most numerous up to mid tide level (MTL), but in pools it may reach to MHWNT or higher where there is much wave splash. *P. aspera* is characteristic of low-water mark, occurring in maximum numbers below mean low water of neap tides (MLWNT), from which level it apparently extends downwards into the sublittoral zone. *P. vulgata*, *P. depressa*, and *P. aspera* may

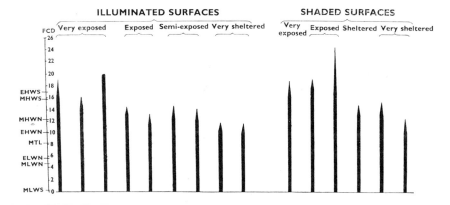

FIG. 126. The upper limit of the vertical range of the limpet *Patella vulgata* on rocky shores near Plymouth, south-west England. Comparison of different shores indicates that the upper level is depressed by exposure to sunshine and is raised in shade; the upper level is depressed in shelter from surf and is raised at sites exposed to surf. The scale on the left of the figure is in feet above chart datum. For interpretation of lettering indicating tidal levels, see p. 394. (Originally published in Evans, 1947a, *J. mar. biol. Ass. U.K.* **27**, 205, fig. 11.)

occur together in the same rock pool, and under such conditions *P. aspera* tends to be the dominant species.

Regarding immersion by the sea, *P. aspera* is evidently the most tolerant species, as was indicated in the previous paragraph. *P. vulgata* is slightly less tolerant, occurring only in reduced numbers below mean low-water mark of spring tides (MLWST), while *P. depressa* is the least tolerant of immersion since it does not occur below MLWST. *P. aspera* is the dominant limpet at the lowest levels on exposed rocky shores, and is also the dominant species in permanent immersion in rock pools (Evans, 1947, 1947a; Moyse and Nelson-Smith, 1963).

Passing eastwards along the shores of the English Channel, both *P. depressa* and *P. aspera* are moderately abundant as far as Ventnor and Culver Cliff on the Isle of Wight, and as far as Barfleur and the Isles of St. Marcouf on the coast of France. The ranges of both these species

terminate abruptly at these points, and the reasons for this are a matter
for speculation. The eastern half of the English Channel is less warm than
the western half, but there is no clear evidence that low winter temperatures east of Ventnor and Barfleur would be lethal to the adults, or that
the summer temperatures would restrict the fecundity of the limpets. The
shores of the eastern basin possess less frequent outcrops of the harder
rocks, and beaches of scouring materials are prevalent; however, there
are some suitable reefs to the east of Ventnor and Barfleur and yet these
lack *P. depressa* and *P. aspera*. At St. Catherine's Point, Isle of Wight,
the coast turns from a south westerly to a south easterly aspect, and at
Barfleur there is a change in aspect from north to east; these changes in
aspect are likely to be correlated with a reduction in shelter from the
weather, and a generally less equable climate may prevail. Finally, it
must be mentioned that at rocky headlands the tidal stream is likely
to be deflected from the adjacent coast, so that planktonic larvae are
carried out to sea. Where the planktonic phase is relatively brief, as it is
in *Patella*, such offshore currents caused by the configuration of the coast
line might well impose a limit to the geographical distribution of the species. Crisp and Knight-Jones (1955) showed that *P. depressa* extended as
far as Carmel Head at the north west tip of Anglesey, being common in
the relatively warm waters which bathe the west coast of Anglesey. The
species is rare at stations east of Carmel Head where the steep, northerly
facing shores are only slightly warmed by the sun. This change of aspect
is associated with a marked faunal discontinuity at Carmel Head.

Consideration of the distribution of the various Atlantic species of
Patella in relation to the several environmental "dimensions" outlined
above may perhaps offer some preliminary indications as to the isolating
mechanisms which permitted speciation. Thus *P. vulgata* may have evolved
in isolation inter-tidally on cold northern shores where inshore waters
may have been diluted by melting snow and ice. In contrast *P. lusitanica*,
for example, may have emerged as a species requiring more warmth than
could be tolerated by *P. vulgata*. Specific status having been achieved
in isolation, the two species could extend their geographical ranges and
come to overlap on the coasts of Portugal and Spain without interbreeding.
On the other hand, *P. aspera*, which is dominant on the lower part of the
shore, and which extends downwards into the sub-littoral, may have
originally emerged as a species which was adapted to life at greater depths
and may later have extended its range upwards on to the shore. This
could only have been achieved by virtue of effective genetical isolation,
e.g. by the adoption of a different breeding season from that of any other
sympatric species of *Patella*. Kolstad (1959) found that on the west coast
of Norway the breeding season of *P. aspera* (July–September) was slightly
earlier than that of *P. vulgata* (August–October). *P. depressa* may have
been a southern species which migrated northwards in post-glacial times,

reaching the coast of Britain while this was still continuous with that of the continent, but after Ireland had become separated from Britain. Orton and Southward found that *P. depressa* breeds some 4 months earlier than does *P. vulgata*, and Lewis (1964) records that in *P. depressa* the gametes are released from April to July, whereas in *P. vulgata* they are released from October till March, in English waters.

The failure of *P. depressa* to reach the coast of Ireland cannot be attributed to climatic or hydrographic rigours, for the species endures colder winter sea and air temperatures near the northern limit of its geographic range at Anglesey than obtain on the south-west coast of Ireland. The only likely explanation of the failure of *P. depressa* to reach Ireland is that the comparatively narrow St. George's Channel constitutes an effective barrier to dispersal. Tidal currents tend to run parallel to the coast line rather than to run directly out to sea, and it is unlikely that the comparatively brief planktonic distributive phase of *P. depressa* could cross the channel and establish a bridgehead for the species on the coast of Ireland. If this were true the colonisation of Irish shores by *P. vulgata* and *P. aspera* must have occurred at an earlier time or by a different route (Crisp and Southward, 1953).

If the narrow St. George's Channel is an effective barrier to the colonisation of Irish shores by *P. depressa*, how are we to account for the occurrence of species of *Patella* on the Azores which lie hundreds of miles away from the nearest European shores? Mr. Brian Morton, a member of the Chelsea College Expedition to the Azores, made extensive collections of limpets on São Jorge and selected specimens from these collections have been identified by Dr. Ronald G. Evans (personal communication) as *P. aspera* and *P. coerulea*. The presence of *P. coerulea* in the Azores is particularly interesting since this species has been regarded up to date as being confined to the Mediterranean. The Azores are volcanic; they stand on the mid-Atlantic ridge and have never been connected to Europe. These species of *Patella* must therefore have been transported to the Azores by shipping, presumably from a Mediterranean port.

From a consideration of characteristics of the various species of *Patella*, Dr. Evans (1958) considers that *P. vulgata* and *P. depressa* are very closely related and may have diverged from a common ancestor; he suggests that *P. coerulea* and *P. aspera* may be similarly related to each other.

The genus *Conus* is remarkable for the high degree of speciation which has occurred, and also for the extensive geographical ranges of many of the constituent species. Among these *C. chaldaeus* and *C. ebraeus* are very closely related and have comparable geographic ranges. They may coexist on the same reef, *C. chaldaeus* tending to predominate on the seaward side of the reef and *C. ebraeus* on the landward side of the reef. Both feed on polychaet worms, but they differ markedly in their food preferences, *C. chaldaeus* feeding exclusively on *Platynereis dumerilii*, and *C. ebraeus* feeding chiefly on *Perinereis helleri* in Hawaiian waters. Thus these two species of *Conus* occupy different micro-habitats and do not compete with each other for food (Fig. 26, p. 72). Kohn and Orians (1962) judge that speciation in animals is normally dependent on an initial period of geographic isolation, subsequent to which the two newly created species may extend their geographic ranges and come into contact. Kohn and Orians extend the term "sympatric" to include species whose ranges either overlap or coincide, and they suggest that sympatry is only possible if the environment is sufficiently diverse, in relation to the natural history of the species concerned, to permit more than one efficient means of exploitation of the environment by those species. While this view fits the example quoted above, it seems unnecessarily restrictive; there seems no reason to insist that sympatric species of *Patella*, previously discussed, must necessarily be exploiting the environment in different ways.

The infauna of sandy shores also exhibits horizontal zonation which is dependent on both the tidal range and the degree of exposure to surf. Those beaches which are exposed to the heaviest surf have a steep profile, the substratum tends to comprise chiefly coarse sand, there is a deficiency of silt and organic matter, and the infauna is impoverished. In contrast, more sheltered beaches have a more gentle profile and extend farther to seaward, the substratum is richer in silt and organic matter, and soil drainage is proportionately less efficient; the infauna of such sheltered beaches of sand, or muddy sand, tends to be rich both in numbers of species and in numbers of individuals. The molluscan element of the infauna of deposit beaches includes filter-feeding and deposit-feeding bivalves and gastropods, and also carnivorous and scavenging gastropods. It is immediately obvious that the distribution of deposit-feeding forms will be dependent primarily on the occurrence of sufficient silt and organic matter upon which to feed, and secondarily on the sheltered conditions which allow this material to accumulate locally. It is possible that such forms are also dependent on the relative durations of the periods of exposure to air and immersion in water. Due to the set of a local tidal current, one level on a beach may be more efficiently scoured than another; the local configuration of rocks may cause eddies which permit regular deposition of organic debris at certain levels on the beach and not on others. Accordingly it may be necessary to study the distribution of a species at several different localities before it is possible to discover the extent to which its inter-tidal distribution is influenced by the nature of the substratum and/or by the degree of exposure to air by tidal movements.

Filter-feeding forms are independent of the substratum for their food supply, but may be dependent on other properties of the soil. Thus the coarse sand of a steep, exposed beach is relatively easy to penetrate, but is much disturbed by surf; species lacking a well-developed foot for digging would be unable to re-bury themselves when washed out of the sand by surf, and these beaches are only suitable for species with powerful digging capabilities. Such a steep beach of coarse sand drains well, and while it may be necessary for the infauna to burrow deeply to avoid heating and desiccation during ebb tide, the interstitial water will be well oxygenated and respiration should present little difficulty. In contrast a soil composed of a mixture of coarse and fine particles will pack much more tightly, and if there is a high proportion of silt and organic matter, the soil will drain less readily and the interstitial water may become deficient in oxygen. Such substrates could not be tolerated by species which need to move with ease through the soil, and which are ill-adapted for respiring under such trying conditions.

Carnivorous and scavenging forms are dependent on an adequate supply of their prey and indirectly, therefore, on the soil conditions required by these prey organisms.

It follows, therefore, that in studying the inter-tidal distribution of the infauna it is necessary to take into account not only the position of the sample in relation to tidal level, but also the chemical and physical constitution of the soil from which the sample was taken. Any significant change in the abundance of one particular molluscan species can only be adequately evaluated if there is a significant change in only one of the three following: (1) position in relation to tidal levels; (2) physical or chemical constitution of the substratum; (3) composition of the remainder of the infauna. In many examples in the literature change in two or more environmental variables makes it impossible to isolate the principal cause of a noted limit in the distribution of a particular species. Thus at Loch Gilp, an arm of Loch Fyne on the Scottish coast, Stephen (1930) worked a series of stations from station 1 at high-water mark to station 8 at low-water mark and found that *Macoma balthica* and *Cardium edule* extended over the upper half of the beach from stations 1 to 5 and 1 to 6 respectively, while *Tellina tenuis* occupied the lower half of the beach from station 4 to station 8. *T. tenuis* overlapped with the other two species at stations 4, 5, and 6, but was only abundant at the bottom of the beach where *Macoma* and *Cardium* were excluded. Stephen showed by soil analysis and by consideration of conditions on other beaches that it was not tidal level, but the nature of the soil which led to the occurrence of two contrasted (but intergrading) bivalve associations at Loch Gilp. The *T. tenuis* association on the lower half of the beach was characterised by clean sand; the *Macoma/Cardium* association on the top half of the beach was characterised by the occurrence of black and dirty sand with poor aeration, beneath a thin layer of clean sand.

Newell (1965) has studied the distribution of the bivalve *Macoma balthica* and the prosobranch *Hydrobia ulvae*, which feed upon the micro-organisms which thrive in inter-tidal sedimentary deposits containing much silt, and has demonstrated that the abundance of both species is correlated with the abundance of fine particles in the subtratum (see Fig. 127).

Intensive studies of the infauna of the littoral and sub-littoral sand at Kames Bay, Isle of Cumbrae, in the Firth of Clyde, showed that the sand was of approximately the same composition at all levels on the beach and that *T. tenuis* extended uninterruptedly from high-water mark down to low-water mark and below this to a depth of about 3 fathoms. *T. tenuis* was most abundant near low-water mark, becoming progressively less abundant towards high-water mark. However, the size-frequency was in

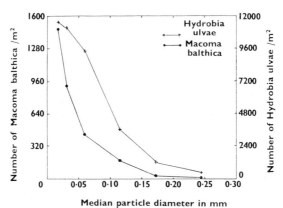

Median particle diameter in mm

FIG. 127. Distribution of the bivalve *Macoma balthica* and the prosobranch gastropod *Hydrobia ulvae*. For both species the numbers of individuals per square metre are plotted against median particle size, for a number of shore stations on the north coast of Kent. For *H. ulvae* each point is based on five samples, and for *M. balthica* on ten samples. It is seen that both of these deposit-feeding molluscs are more abundant at stations where the substratum is more muddy, and that abundance diminishes with increasing coarseness of grade of the substratum. (Originally published in Newell, 1965, *Proc. zool. Soc. Lond.* **144**, 30, fig. 2.)

inverse relation to abundance, the largest specimens being found at the higher levels. It seems that growth rates are higher on the higher levels of the beach, and that *T. tenuis* is well adapted as an occupant of the upper half of the littoral zone. It seems possible that the progressive increase in abundance of this species towards low-water mark could be accounted for by the fact that as one approaches low-water mark the surface of the sand is available for a progressively longer period on each tide for the settlement and metamorphosis of larvae. Whereas *T. tenuis* is adapted for life in the inter-tidal zone, *T. fabula*, on the other hand, is essentially an occupant of the sub-littoral zone, as is amply attested in Table 14.

The figures for the inter-tidal zone are derived from $\frac{1}{4}$ square metre samples; the data for the sub-littoral zone are derived from dredge hauls, and are not directly comparable (Stephen, 1929).

TABLE 14. *The abundance of 2 species of* Tellina
*at different levels inter-tidally and sub-littorally
at Kames Bay, Cumbrae*

Station	Inter-tidal (specimens per square metre)						Sub-littoral (specimens per 20 cm cube)				
	1 (HW NT)	1a	2	3	4	5 (LW ST)	2f	3 f	5 f	7 f	10f
Tellina tenuis	14	241	622	1180	2552	3388	327	29	—	—	—
T. fabula	—	—	—	1	7	25	26	101	144	134	10

Watkin (1942) made a detailed study of the macrofauna of inter-tidal sands of Kames Bay, taking 42 samples at 5-yard intervals from high-water mark to low-water mark, each sample consisting of 1 square foot dug to a depth of 6 in. The only bivalve he studied in detail was *T. tenuis*, and although he found markedly higher densities per square metre than did Stephen (1929) his figures show a continuous distribution from high-water mark of neap tides to below low-water mark, the density increasing progressively towards LWM. Brady (1943) worked a series of stations from low-water mark to high-water mark at Black Middens, Tynemouth, where the constitution of the soil was remarkably constant over the whole tidal range (fine sand 98·0 per cent or more, silt 0·7 per cent or less). The shore was extensively colonised by *T. tenuis* and by *M. balthica*, both species being most abundant in the middle of the beach and diminishing in numbers towards high-water mark and towards low-water mark. Brady reported an inverse correlation between the density of *M. balthica* and that of the polychaet *Scoloplos*. Clark (1955) studied the sub-littoral fauna of Kames Bay and confirmed that *T. tenuis* only extends down to about $2\frac{1}{2}$ fathoms at most, whilst *T. fabula* favours deeper water and is most abundant from $1\frac{1}{2}$ fathoms to $10\frac{1}{2}$ fathoms. In these deeper waters, however, the sandy substratum grades into fine mud and this change in the grade of the soil is doubtless an important factor in the ecological replacement of *T. tenuis* by *T. fabula*. Many other species of bivalves were encountered at these sub-littoral stations, and also the predacious carnivore *Philine aperta* which was common at all stations from $2\frac{1}{2}$ to 13 fathoms. In White Bay, which differed in being rather more exposed than Kames Bay, there was a comparable though not identical variety of bivalves, but *Philine* was replaced by *Natica alderi*, another predacious carnivore.

Ebling *et al.* (1948) studied the influence of the strength of the water currents on the differential distribution of the sub-littoral limpet *Patina pellucida* at Lough Ine, on the south coast of Eire. Lough Ine is connected with an arm of the sea by a stretch of rapids about 150 yards long, and diurnal tides enter and leave the lough via these rapids. Except for a few minutes at "high slack water" and again at "low slack water", a continual tidal stream passes through the rapids either inwards or outwards at all times. A considerable part of one side of the rapids is in the form of a vertical dry-stone wall and, since the channel gradually narrows, the water current steadily increases in strength towards the narrowest point in the rapids. *P. pellucida* occurs on the holdfasts, stipes, and fronds of the large weed *Saccorhiza bulbosa*. At the end of each period of slack water in the rapids the direction of the tidal stream reverses, the stipes and fronds of the seaweed rise and turn over and are then drawn out in the opposite direction by the newly established tidal stream. Thus the specimens of *Patina* are always oriented in the same way in relation to the water current except that for one part of the tidal regime they are upside down on the underside of the weed.

A series of stations was established along one side of the rapids and the rate of flow of water was determined at representative stages of the tidal cycle at these stations and also in mid-stream opposite each shore station. Water currents were measured subsurface, just above the canopy of seaweeds, and also at an intermediate depth, using a Watts meter. From these data it was possible to determine the current speed characterising each station in the canopy from which specimens of *Patina* were collected. Specimens of the weed *S. bulbosa* were then collected at ten stations in the rapids, and their populations of *Patina* were removed and counted. By arranging the stations in order of increasing current strength, the abun-

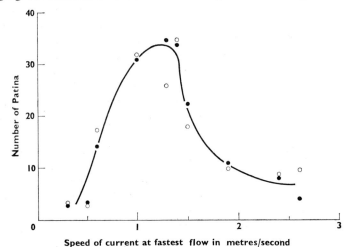

Fig. 128. Distribution of the limpet *Patina pellucida* on *Saccorhiza bulbosa* at various stations arranged in order of increasing speed of current. ● mean number of *Patina* per *Saccorhiza* plant; ○ number of *Patina* per 3 lb of *Saccorhiza*. (Originally published in Ebling *et al.*, 1948, *J. Anim. Ecol.* **17**, 240, fig. 14.)

dance of *Patina* was plotted against current strength, as in Fig. 128. It is seen that up to a speed of about 1·0 m/sec increased rate of flow of the tidal stream is matched by increased abundance of the limpet. Further increases in the rate of flow of the tidal stream are associated with progressive decrease in the abundance of *Patina*.

At first sight this may appear to be an uncomplicated relationship with a simple environmental variable; up to a speed of 1·0 m/sec, increased rate of flow is advantageous, while above this speed *Patina* may be dislodged from the surface of the weed by the strength of the current. It could alternatively be argued that there are here two interacting environmental effects: (1) Increased rate of flow means that a larger volume of water passes over each unit area of algal frond, bringing an increased number of planktonic larvae which can settle and metamorphose. If this alone is true, increased rate of flow up to 1·0 m/sec does not benefit the individual *Patina* in any way, and at low current speeds the abundance of *Patina* is no more than a measure of the volume of water passing the seaweed. (2) Above about 1·0 m/sec increased rate of flow is disadvantageous in that it dislodges either the newly settled larva, or the adult individual from its place on the surface of the weed. We have to concede, however, that the rate of flow of the tidal stream may exert an influence on the organism in many different ways, and that the graph is the summation of these various different effects. Thus at stations with very low current speeds the surface of the weed may become covered with sediment or with a growth of filamentous algae and thereby become unsuitable for settlement and metamorphosis of the larval *Patina*. At such stations the temperature may be higher, and there may be relevant deviations from "normal" in oxygen tension, carbon dioxide tension, and pH. Any one, or more of these variables may be significant to the young limpet. In contrast, where the current speed is higher the surface of the fronds of *S. bulbosa* remain free from growths of filamentous algae, and the physico-chemical qualities of the water will tend to "normality". At very high current speeds, especially where there is turbulence, the oxygen tension will be unusually high, which may be advantageous, but there may be mechanical difficulties with reference to feeding, and to movement on the surface of the weed. Even this analysis of the possible ways in which current speed may influence the abundance of *Patina* may well be far from complete.

The investigations at Lough Ine also showed that in contrast to *P. pellucida*, actively crawling snails such as *Bittium reticulatum*, *Gibbula cineraria*, and *Rissoa parva* were largely confined to situations where the current was very weak. This is probably due in part to the limited ability of these forms to resist dislodgement by moderate water currents. It is probable that the preference of these species for areas of calm water is partly due to the occurrence of sediment at such sites and to a growth of diatoms and filamentous algae on which the snails can browse. Other environmental factors dependent on low water current cannot be excluded from consideration.

Passive Distribution of Molluscs

Molluscs may be transported haphazardly from place to place in various ways, e.g. by birds, by insects, by wind, and by water currents. Evidence on this subject has been collated and reviewed by Rees (1965). There are many well authenticated records of the deposition on dry land of small fish which had previously been lifted high into the air in a waterspout, and it is a reasonable supposition that small molluscs may similarly be moved by tornadoes and high winds, though in most cases the event would pass unnoticed. Rees refers to the deposition of hundreds of *Anodonta anatina* in torrential rain at Paderborn in Germany in 1892. Clench (1954) studied the distribution of various insular species of the terrestrial pulmonate genus *Cerion* in Cuba, the Bahamas, and the Cayman Isles. He noted that after a recent hurricane a locality may be found to possess living specimens of a species not previously recorded there, and concluded that hurricanes may play a significant part in the random re-distribution of species.

Lucas (1954) reported that living specimens of three species of the prosobranch limpet *Patella* were transported to the Dutch coast by flotation in tidal currents. These specimens had developed on other shores and the upper surfaces of their shells bore substantial growths of sea weeds. The limpets had presumably been dislodged, perhaps by wave action during a storm, and had failed to re-attach themselves to the rock surfaces and they had then been carried off in the tidal current due to the buoyancy of the attached seaweed. Specimens of *P. aspera* (= *P. athletica* Bean) and *P. depressa* were transported to the coast of Holland through buoyancy provided by the attached fronds of *Himanthalea elongata*.

A living specimen of the marine snail *Littorina littoralis* was found in the bed of the river Endrick, which flows into Loch Lomond in Scotland, some 13 miles from the nearest colony of this species on the shore of the Clyde estuary. This can only be explained as the result of very recent passive transportation thither, probably by a sea gull (Hunter, Maitland and Yeoh, 1964).

Rees (1965) lists examples of bumble bees in flight with a specimen of the terrestrial prosobranch *Pomatias elegans* attached to one of the legs, and of the water beetle *Dytiscus marginalis* with the fresh water limpet *Ancylus fluviatilis* attached to the elytra or with the fresh water bivalves *Sphaerium* and *Pisidium* attached to a leg. In another paper Rees (1952) discusses the part played by Amphibia in the dispersal of fresh water bivalves.

Records of the transportation of small gastropods and bivalves attached to the feet and legs of aquatic birds are more abundant, and several examples are cited by Rees (1965). The snail *Vitrina pellucida* secretes rather sticky slime and has been found adhering to the plumage of various species of migrating birds. The most remarkable example is that of the Upland Plover, *Bartramia longicauda* which regularly carry from ten to thirty snails of the genus *Physa* under their wings on their arrival in Louisiana. The snails are only found on Upland Plovers which have recently arrived from the south. Crushed remains of the same snail may be found in the gizzard of the bird, and it seems probable that the bird deliberately places the snails under its contour feathers prior to migration, as rations for the journey. In such ways as these various molluscs may be carried to localities outside their normal geographic range and, if the new environment is suitable such passive transportation may sometimes result in the establishment of the species and extension of its range. More important than any of these natural processes of passive dispersal is dispersal due to the activities of man, to which a separate section of this chapter is devoted.

Distribution of Molluscs by Man

While man has been responsible for the extinction of a variety of vertebrates, his influence on the Mollusca has been otherwise. So far as I am aware no mollusc has suffered extinction at the hands of man, but many have had their geographic ranges greatly extended through his activities. In some cases there has been a deliberate movement of a mollusc and its liberation at a new site as a potential source of food, in a display of ill-guided experimentation or, more rarely, in an attempt at biological control of some pest. In other cases a mollusc has been moved inadvertently in merchandise, or perhaps attached to the sides of vehicles and ships. It is very difficult to envisage the ultimate consequences of the establishment of a mollusc at a new site far outside its original geographic range. However, there is one point on which we may be sure, namely that we shall probably have great difficulty in eradicating it or preventing it from spreading from the new site if its presence should ultimately prove to be undesirable.

One of the most interesting, and best annotated examples of extension of the geographic range of a mollusc by man's activities is that of "the Giant African Snail", *Achatina fulica* (see Fig. 129). In point of fact, this is not the only giant African snail. The family Achatinidae is endemic throughout Africa south of the Sahara, and the genus *Achatina* comprises perhaps as many as eighty valid species. The genus *Achatina* has been authoritatively reviewed by Bequaert (1950) who concluded that the ancestral home of *A. fulica* was an area on the East Coast of Africa, opposite Zanzibar, which is now occupied by the sub-species *A. fulica hamillei*. Bequaert considers this sub-species to represent the ancestral stock from which the insular sub-species *A. f. fulica* was derived. The ancestral stock was probably also endemic to coastal islands such as Zanzibar and Pemba. Bequaert could not finally decide as to whether *A. fulica* was truly endemic

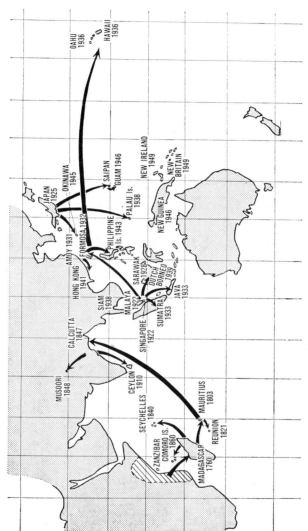

FIG. 129. Dispersal of the Giant African Snail, *Achatina fulica fulica*. The dates indicate the earliest published records of the occurrence of the snail at each locality. The hatched area on the coast of East Africa is considered to be the ancestral home of the snail. Heavy black arrows indicate the source of the introduced snail wherever this is known with some confidence. The data have been extracted from the review by Bequaert, 1950, *Bull. Mus. comp. Zool. Harv.* **105**, 3–216.)

to Madagascar or whether it had been introduced to this island by man. He is inclined to the view that the species was introduced to Madagascar by man some time prior to 1760 when it was first recorded as being common there. *Achatina* was introduced into Mauritius not long before 1800, and it soon became a serious pest. *A. fulica* reached Réunion a few years later, being reported there in 1821; in 1840 the snail was reported from the Seychelles, and it had reached the Comoro Islands by 1860. Specimens of *A. fulica* were deliberately taken from Mauritius to India and liberated in Calcutta in 1847, while further specimens were taken from Calcutta to Musoori in the foothills of the Himalayas in 1848. Specimens were liberated in Ceylon in about 1900, and it was reported to be abundant there by 1910.

It is not clear how the snail reached Malaysia, but it was probably imported by sea, being reported in Singapore, Johore, Perak, and Kelantan by 1922. The snail had arrived in Amoy in southern China by 1931, it was present in Siam by 1938, it had reached Hong Kong by 1941, and there can be little doubt that this rapid progress was due to deliberate importation by man, probably from Singapore. *Achatina* was introduced into Sarawak in about 1928, into Sumatra some time before 1933, Java and Dutch Borneo in about 1939, probably coming from Malaysia in all these cases. *Achatina* was imported from Singapore into Formosa in 1932, and thence to the Philippines during the Japanese occupation. Specimens of *Achatina* were also taken from Formosa to Hawaii in 1936 and thence to Oahu.

An unsuccessful attempt was made in 1925 to establish *Achatina* in Japan, and snails from Japan were transported to Guam, Saipan, and the Palau Islands in Micronesia, and also to Okinawa. *Achatina* reached New Guinea by 1946, and New Ireland and New Britain by 1949.

A few specimens were accidentally introduced into California from the Pacific theatre of war, and these were destroyed as soon as they were discovered. Due to the ensuing vigilance, the snail was intercepted on several other shipments from the Pacific area. Bequaert considers, however, that it is unlikely that *Achatina* could become established in California due to the climatic conditions. *Achatina* requires the ecological conditions of the humid tropics, and cannot endure frosts, or long continued dry weather. With the exception of Japan and California, *Achatina* has apparently become well established and has reached the status of a serious pest at all the sites into which it has been deliberately introduced, or into which it has naturally spread since it first "escaped" from Madagascar early in the nineteenth century. The chief reason why man has deliberately introduced *Achatina* into new localities was doubtless in order to make use of it as a source of food. It is hard to envisage *Achatina* as a delicacy, and equally hard to justify its distribution of the sake of adding a delicacy to the menu; however, one may well feel more sympathetic to those who saw in *Achatina* an opportunity to augment the impoverished diet of hungry peoples in the Far East. Be that as it may, the disadvantages of *Achatina* as a pest on agricultural land doubtless far outweigh its anvantages as an item of food wherever it is now well established. In some cases the snail was probably liberated in new localities as a matter of supposedly "scientific interest". In other cases, movement of the snail by land or sea may have been entirely fortuitous. While it is scarcely possible to make complete provision against human folly and simple mischance, it is now obvious that due precautions have to be taken to prevent another such disastrous outbreak by a potential molluscan pest. The reader should consult Mead (1961) for a very thorough study of the biology of *Achatina*, which includes detailed information on the attempted control of this pest by chemical and by many different biological agents.

Another pulmonate snail the distribution of which has been greatly extended at the hands of man is *Helix aspersa*. This european species has been successfully introduced into the United States, into South America, into South Africa, into Australia, and into New Zealand (Pilsbry, 1939).

It is pertinent to mention here the cautious use of the carnivorous streptaxid snail *Gonaxis kibweziensis* which was introduced into the uninhabited island of Agiguan, off Tinian in Micronesia, in an attempt to destroy, or at least to decimate the population of *Achatina* that had become established on that island. *Gonaxis* became well established on Agiguan, extended its range considerably beyond the original points of release, and contributed greatly to the observed reduction in abundance of *Achatina* in the area. As a result of this success, large numbers of *G. kibweziensis* were collected from Agiguan and were released on neighbouring islands including Tinian, Saipan, and Guam, and releases were also made on the islands of Rota, Ponape, and Truk. Specimens of *G. kibweziensis* were also shipped to Hawaii, to combat *Achatina* on the Hawaiian islands of Oahu and Maui, while some specimens of *Gonaxis* were shipped to California for experimental use there on a European snail pest (Anon.).

Various other snails owe their present wide distribution to man's activities, being transported with merchandise, with plants, etc. This is true of the snails *Lamellaxis clavulinus* and *L. gracilis* which have reached South-east Asia from East Africa and/or Madagascar. *Bradybaena similaris* and *Huttonella bicolor* are further tropical cosmopolites which seem to follow in the track of human settlements, occurring in coffee plantations, and other agricultural areas (van Benthem Jutting, personal communication).

Turning now to marine gastropods, there have been two most unfortunate, accidental additions to the British marine fauna in *Crepidula fornicata* and *Urosalpinx cinerea*. *C. fornicata* was apparently introduced accidentally from North America shortly before the end of the nineteenth century, together with oysters which were re-laid on British oyster beds. It was probably hoped that the introduced oysters would become established in British waters, but it is now known that *Ostrea virginica* requires a higher threshold temperature (20°C) for spawning than does the native *O. edulis* (15–16°C), and it was unlikely that the introduced oysters would be able to breed successfully. It is not known whether any attempt was made to cleanse the introduced oysters and to prevent the introduction of other marine species of unknown potentiality—probably not—but this would be an almost impossible task. Some specimens of *C. fornicata* evidently survived and became established. It is now known that *C. fornicata* may commence breeding at a threshold temperature of 10°C (= 50°F), and can breed for about $5\frac{1}{2}$ months in British waters, from April till September (Chipperfield, 1951). The eggs are incubated for 21–28 days, and the larvae spend perhaps a further month in the plankton, which permits wide dispersal. It is not surprising, therefore, that once introduced it has proved impossible to eradicate the pest. It is here to stay. *Crepidula* commences breeding earlier in the year than the oyster, its spat may take up spaces for settlement which would otherwise be available to the oyster, the chains of *Crepidula* shells are liable to smother the oysters, and *Crepidula* competes very successfully with the oyster for planktonic food.

The tingle, *U. cinerea* was presumably introduced from North America in the same way and at about the same time, but it was not discovered in British waters until 1928 (Orton and Winckworth, 1928). *Urosalpinx* prefers to feed on young oysters, the shells of which can be perforated quickly, and under experimental conditions the snail was found to consume 0·9 oyster spat per tingle per day. Breeding commences at a threshold temperature of 12–13°C, and each female may produce about 25 egg capsules, and a total of 300 crawling young in each season. A female may commence breeding in her second year, and may live and continue breeding for about seven more years. Fortunately the young leave the egg capsule at the crawling stage, and there is no planktonic distributive phase, which limits the rate at which new areas can be colonised by the snail. *U. cinerea* now occurs from the Humber to Dorset and the Isle of Wight, and also in the River Yealm in Devon. It is abundant in Essex in the rivers Blackwater, Colne, Crouch, and Roach. It is present on the oyster beds at Whitstable, but is not yet abundant there.

If *Urosalpinx* fed exclusively on oyster spat it would destroy about 59 spat per tingle per season. *Urosalpinx* may perhaps be controlled with great difficulty, but it certainly cannot be eradicated. The seriousness of the infection of oyster beds with this voracious pest cannot be overestimated. It is particularly frustrating to reflect that no useful purpose can have been served by the activities which led to the introduction of both *C. fornicata* and *U. cinerea* into British marine waters. It is important that we should learn from these lessons and ensure that such accidents do not occur again.

It is equally unfortunate that the drill *Ocinebra japonica* should have been introduced to the Pacific coast of America with the importation of Japanese seed oysters. *O. japonica* is now well established at various points along the Canadian and American Pacific coasts, and it feeds on various bivalves of commercial importance (Chew, 1960).

The bivalve *Petricola pholadiformis* was first recorded on the east coast of Britain in 1890, and it may have been accidentally introduced into this area together with *C. fornicata* and *U. cinerea*. From this initial bridgehead in European waters *P. pholadiformis* has gradually spread at a rate of about 10 km per annum as a result of its planktonic larval phase, aided by coastal currents; it reached the coast of Belgium in 1899, it extended to Dunkirk by 1906 and to the northern tip of Denmark by 1908. *P. pholadiformis* reached the Swedish coast by 1932 (Haas, vol. 2, p. 441). In British waters *Petricola* has spread northwards as far as Mablethorpe in Lincolnshire, and westwards as far as Poole Harbour (Duval, 1963). It is encouraging to report that at least one of the molluscs which has been introduced into British waters from North America is of potential commercial importance, namely the "Quahog", *Venus mercenaria*. The occurrence of this species in Southampton Water was only reported in 1961, but judging by size of shell and growth rings, it must have become established in Southampton Water before 1936. It was probably introduced accidentally, from North America, perhaps having been jettisoned from the kitchen of a transatlantic liner. The species has subsequently been introduced from America to Conway, North Wales, and to Burnham-on-Crouch, Essex, while samples of *V. mercenaria* from Southampton Water have been successfully re-laid at Poole Harbour, Dorset, and in the Newtown Estuary, Isle of Wight. It seems possible that this species may become capable of bearing planned commercial exploitation in British waters (Ansell *et al.*, 1964).

The distributions of various molluscs have been significantly altered by the industrial use of rivers and canal systems. Until about 1800 the bivalve *Dreissena polymorpha* was confined to an area surrounding the Black Sea, the Caspian Sea, and the Aral Sea. In about 1820, *D. polymorpha*, which possesses a well-developed byssus apparatus, "escaped" from its original geographical range by being carried about attached to the bottoms of barges and other vessels. *D. polymorpha* appeared in London by 1820, in Rotterdam by 1826, in Hamburg by 1830, and Copenhagen by 1840. It spread steadily up the Rhine, reaching the Danube via the Ludwig Canal; from Rotterdam *Dreissena* passed southwards into France and it had reached the south of France by 1865. Similarly, *Dreissena* spread steadily up the Elbe from Hamburg, and it also colonised the Weser and the Oder. In this way, though industrial influence, *D. polymorpha* colonised much of the fresh water systems of Western Europe (Haas, vol. 2, figs. 164 and 165).

The parthenogenetic snail *Paludestrina jenkinsi* occurs in both brackish and fresh waters, in Britain and on the Continent. Prior to 1889 there were very few records of this snail in Britain, but at about the turn of the century it appeared suddenly at a large number of localities. It first appeared in industrial areas and in canals, which strongly suggests that the dispersal of this snail was being effected by passive transport on the bottoms of barges (Robson, 1923).

Prior to the construction of the Suez Canal the molluscan faunas of the Red Sea and of the Mediterranean were well known, and had few species in common. Subsequent to the opening of the canal various molluscan species extended their range through the canal, from the Mediterranean into the Red Sea, and also vice versa. In some cases these species may have been aided by passive transport through attachment to the hulls of ships. *Pinctada vulgaris* entered and became firmly established in the Mediterranean, being abundant on the coast of Palestine, and it has extended westwards along the coast of Africa as far as Tunis and Algiers (Haas, vol. 2, p. 446).

Turning to the future, we should earnestly consider the possible consequences of the contruction of the high dams at Aswan, and on the River Volta in Ghana, with regard to the spread of disease. *Bulinus truncatus* is the snail vector of *Schistosoma haemotobium* in Egypt, and the subspecies *B. truncatus rohlfsi* and *B. globosus* are vectors of two strains of *S. haematobium* in Ghana. In Egypt there is in addition *S. mansoni*, which is transmitted by the snail *Biomphalaria boissyi*. These pernicious blood-flukes are already very extensively distributed and do untold damage through reduction of human vitality and consequent loss of commercial productivity. In Egypt in areas of only intermittent summer irrigation a 1 per cent infection of *B. truncatus* suffices to effect at least 50 per cent infection of the human population by *S. haematobium*. On completion of the Aswan high dam it is expected that there will be permanent irrigation over a far wider area, which will presumably result in a great increase in the range and abundance of the snail vectors of *Schistosoma*, with an inevitable rise in the incidence of this disease among humans. The potential advantages of the new dam may be heavily outweighed by the consequential spread of schistosomiasis unless there is a surprising advance in the application of medical malacology (Schalie, 1958, 1960). The same problem arises with respect to the Volta dam in West Africa. Water in abundance for industry, for inland fish culture, and for irrigation, will be a dubious blessing to Ghana if it also means a tidal wave of schistosomiasis. The Volta dam is expected to impound a lake with a surface area of 3275 square miles. Once freshwater pulmonates are established round the shores of such a vast lake, the prospect of eradication would be nil. The most important vector of schistosomiasis in Ghana is at present *B. globosus*. It might still be possible—though exceptionally difficult—to destroy this

snail in Ghana by a systematic treatment of the catchment area, commencing from the tops of watersheds and working slowly down all tributaries,
and using molluscicides such as copper sulphate, sodium pentachlorophen-
ate, or dinitrocyclohexylphenol (DCHP) (McCullough, 1959; McMullen
and Harry, 1958). The cost would be fantastic, but it would be more
worth while than some of the expenditure presently undertaken in some
"newly emergent" countries.

Alternatively, the snail vectors of schistosomiasis might be subjected
to biological control if a suitable predator could be produced. Consideration could be given in this respect to the dipteran family Sciomyzidae,
whose larvae attack and kill terrestrial or aquatic snails. The Sciomyzidae
occur mainly in the northern hemisphere, but some extend into moderately
warm latitudes. In all species for which the larvae have been reared these
larvae have been found to feed exclusively on terrestrial or on aquatic
snails (Knutson, Stephenson, and Berg, 1965). It seems probable that a
thorough review of the family might indicate a number of species which
are likely to feed on the snail vectors of *Schistosoma*, and which are likely
also to tolerate the prevailing ecological conditions. It might be necessary
to conduct large-scale breeding experiments in order to isolate particularly
suitable mutant stocks of flies which could then be released in areas troubl-
ed with schistosomiasis. The production of a "custom-built" new species
of sciomyzid for the biological control of schistosomiasis would be a major
genetical triumph.

Mead (1961) cites many examples of biological control of pulmonate
pests by many different kinds of insects, e.g. the "India Glowworm"
Lamprophorus tenebrosus, by crabs, by amphibians, e.g. the Central American toad *Bufo marinus*, by rats, and by birds. Mead draws attention to
the complex interrelations that can develop, sometimes with undesirable
consequences on the abundance of other serious pests. The introduction
of a new predator in an attempt at biological control of a pest should only
be attempted after much thought and careful preparation, and ideally
only on a completely isolated site such as a small uninhabited island from
which the introduced species cannot escape. If biological control is successful under such isolated conditions, only then should consideration be
given to releasing the introduced species on a more extensive land mass.

Studies on Populations of *Cepaea nemoralis*

The banded snail, *Cepaea nemoralis*, commonly exhibits a balanced
polymorphism with regard to the ground colour of the shell and to the
spiral bands of deeper pigmentation with which the shell is frequently
ornamented, and it seems probable that the heterozygous condition in
these respects is generally advantageous. In some areas which have been
studied closely there is a well defined correlation between the nature of

the vegetation and the relative abundances of different colour morphs in the various colonies of the snail, the most abundant morphs typically being cryptically coloured. Where there is a short turf there tends to be a predominance of yellow and heavily banded shells which will appear cryptically coloured against the background of blades of grass, shadows, and shafts of sunlight. Longer grass provides more effective cover and shade, and here one may find a higher proportion of effectively unbanded* shells. Whereas yellow shells may be common in grasslands, pink shells and brown shells tend to predominate in dense woodlands where the background is darker, and there is a general deficiency of low vegetation, and under such circumstances a high proportion of unbanded shells is to be expected. It seems probable that the relative abundance of cryptically coloured specimens may be attributable at least partly to the effects of visual selection by predators such as thrushes, rabbits, or rodents. It is often possible to test this proposition by making parallel collections of living specimens and also of empty, predated shells from the same habitat, and by making statistical comparisons of the relative abundances of cryptic and of conspicuous shells in the two samples. A higher proportion of conspicuous specimens in the sample of predated shells will suggest that the predator in question is exercising visual selection against those morphs. The effects of visual predation will depend upon the predator; thrushes can discriminate between yellow and other colours, whilst small mammals such as rabbits and rats, being colour-blind, are incapable of this discrimination but can exercise visual selection based on the tone of the shell. The effects of visual selection may change from season to season; in dense woodland, for example, yellow shells are conspicuous early in the year against a dark background of dead leaves, twigs, and earth, and there may be strong selection against yellow shells at this time of year. Later, as a carpet of woodland herbs develops, this selective disadvantage disappears, and predator pressure may operate against pink shells (Sheppard, 1952).

Studies on colonies of *C. nemoralis* in the vicinity of Oxford have successfully demonstrated that variation between colonies is here correlated with differences in habitat. Colonies from oak woods and from mixed deciduous woods were found to be characterised by a low percentage of yellow, and of banded shells; in contrast, colonies sampled from rough herbage and from hedgerows typically possessed a higher percentage of yellow and of banded shells (see Fig. 130) (Cain and Sheppard, 1954; Cain and Currey, 1963). Currey, Arnold, and Carter (1964) studied colonies of *C. nemoralis* in woodlands (11 samples) and in various open habitats (25 samples) and recorded the percentage of yellow shells and

---

* "Effectively unbanded" includes all shells which are likely to appear unbanded to a predator; traces of bands may be discerned on some of these shells when scrutinised carefully.

the percentage of effectively unbanded shells for each colony investigated. Their results, expressed in Table 15, demonstrate conclusively that in the area of country under investigation populations of *C. nemoralis* from woodlands tended to be less yellow, and more effectively unbanded, than populations from more open habitats.

Recent studies of populations of *C. nemoralis* on the Marlborough and Lambourn Downs (Cain and Currey, 1963) have shown that on these chalklands certain phenotypes predominate over large areas far exceeding

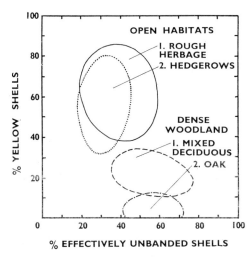

FIG. 130. Scatter diagram portraying the distribution of different colour forms of *Cepaea nemoralis* according to habitat, at various localities off the chalk and within 10 miles from Oxford. The ellipses represent the distribution of *the majority* of points plotted on an original, more detailed scatter diagram. It is seen that samples of shells taken from two types of "open" habitat tend to have a high proportion of yellow shells and rather fewer effectively unbanded shells. In contrast, samples of shells from two types of dense woodland tend to have fewer yellow shells and a higher proportion of banded shells. These differences are ascribed in this case to the effects of differential selection by predators. (Modified from fig. 7, p. 18, in Cain and Currey, 1963, *Phil. Trans.* B, **246.**)

the extent of a panmictic population of this snail, and quite regardless of the habitat. The transition from one such area to another may be marked by a remarkably abrupt change in the proportions of the phenotypes concerned, and it seems clear that the differential distribution of the various phenotypes cannot here be attributed to visual selection by known predators in relation to the background of vegetation. The more or less uniform distribution of shells of a particular colour or type of banding regardless of the nature of the vegetation has been defined non-committally as an "area effect" the causes of which remain as yet unknown. These

TABLE 15. *The relationship between colour and banding of shells of* Cepaea nemoralis *and habitat, in south Warwickshire. Each colony is recorded once for percentage of yellow shells and once for percentage of unbanded shells; for each of these variables the colonies are divided into two groups, those above and those below the median value* (Currey, Arnold, and Carter 1964)

	Number of colonies in	
	(a) Woods	(b) Open habitats
Colour of shell		
More yellow	1	17
Less yellow	10	8
Banding of shell		
More effectively unbanded	9	8
Less effectively unbanded	2	15

area effects cannot be attributed to any obvious environmental feature such as altitude, aspect, or geological formation.

For some unknown reason, on these downlands the heterozygous condition in respect of certain genes is evidently not favoured, and a balanced polymorphism is not maintained for these genes. The low level of visual predation by thrushes over extensive areas of downland may be an important contributory factor. In each area a particular genotype has emerged as the dominant form and the abrupt transition at the boundaries between adjacent areas suggests that interbreeding between contrasted harmonious gene complexes is relatively unsuccessful; the hybrid stocks are unable to establish themselves and to extend in competition against the more purely bred parent stocks on either side. If this were true competition would occur between a limited number of highly successful genotypes, and the most vigorous types would slowly gain ground. This could be the reason for the great size of the areas dominated by one particular morph of *C. nemoralis* (Goodhart, 1963). The establishment of a limited polymorphic system over a very large area of downland, and the lack of effective hybridisation with other forms on adjacent areas of downland, could be seen as a first step towards sub-speciation in *C. nemoralis*; if reinforced by genetical changes, isolation could become absolute and speciation would be effected.

It would be interesting to test the relative viabilities of contrasted morphs of *C. nemoralis* over a period of years by massive reciprocal transplantations into areas stripped as throughly as possible of their original snail populations. Would the transplanted populations hold their own, would they increase the area of land occupied, or would they be overwhelmed by the morph which characterises the surrounding country? The Marlborough and the Lambourn Downs would seem to be ideal for such an experiment.

Key to the Lettering on the Figures

AHTL	Average high tide level.	HWNT	High water of neap tides.
ALTL	Average low tide level.	LWNT	Low water of neap tides.
EHWN	Lowest level of high water of neap tides.	MHWN	Mean high water of neap tides.
EHWS	Extreme high water of spring tides.	MHWS	Mean high water of spring tides.
ELWN	Highest level of low water of neap tides.	MLWN	Mean low water of neap tides.
ELWS	Extreme low water of spring tides.	MLWS	Mean low water of spring tides.
FDC	Feet above chart datum.	MTL	Mid tide level.

Reference List

ANON. Biological control of *Achatina* by *Gonaxis*, *Pacif. Sci. Inf. Bull.* **7** (6), 11–12.

ANSELL, A. D., LANDER, K. F., COUGHLAN, J., and LOOSMORE, F. A. (1964) Studies on the hard-shell clam, *Venus mercenaria*, in British waters. I. Growth and reproduction in natural and experimental colonies, *J. appl. Ecol.* **1**, 63–82.

ARNOLD, D. C. (1957) The response of the limpet *Patella vulgata* L. to waters of different salinities, *J. mar. biol. Ass. U.K.* **36**, 121–8.

BAKKER, K. (1959) Feeding habits and zonation in some intertidal snails, *Arch. néerl. Zool.* **13**, 230–57.

BASSINDALE, R. (1940) Studies on the biology of the Bristol Channel, IV. The invertebrate fauna of the southern shores of the Bristol Channel. *Proc. Bristol Nat. Soc.*, 4 ser., **9**, 143–201.

BASSINDALE, R. (1961) On the marine fauna of Ghana, *Proc. zool. Soc. Lond.* **137**, 481–510.

BATHAM, E. J. (1956) Ecology of a southern New Zealand sheltered rocky shore, *Trans. roy. Soc. N.Z.* **84**, 447–65.

BEQUAERT, J. C. (1950) Studies in the Achatininae, a group of African land snails, *Bull. Mus. comp. Zool. Harv.* **105**, 3–216.

BERRY, A. J. (1962) The growth of *Opisthostoma* (*Plectostoma*) *retrovertens* Tomlin, a minute cyclophorid from a Malayan limestone hill, *Proc. malac. Soc. Lond.* **35**, 46–49.

BERRY, A. J. (1963) Growth and variation of the shell in certain Malayan limestone hill snails, *Proc. malac. Soc. Lond.* **35**, 203–6.

BRADY, F. (1943) The distribution of the fauna of some intertidal sands and muds on the Northumberland coast, *J. Anim. Ecol.* **12**, 27–41.

BROWN, A. C. (1960) Desiccation as a factor influencing the vertical distribution of some South African gastropods from intertidal rocky shores, *Portug. acta biologica*, B, **7**, 11–23.

BUCHANAN, J. B. (1954) Marine molluscs of the Gold Coast, West Africa, *J. W. Afr. Sci. Ass.* **1**, 30–45.

CAIN, A. J. (1953) Visual selection by tone of *Cepaea nemoralis* (L.), *J. Conch.* **23**, 333–5.

CAIN, A. J. and CURREY, J. D. (1963) Area effects in *Cepaea*, *Phil. Trans.* B, **246**, 1–81.

CAIN, A. J. and SHEPPARD, P. M. (1954) Natural selection in *Cepaea*, *Genetics*, **39**, 89–116.

CHEW, K. K. (1960) Study of food preferences and rate of feeding of Japanese oyster drill *Ocinebra japonica* (Duncker), *Fishery Prod. Rep. Fish. Wildl. Serv. U.S.* **365**, 1–27.

CHIPPERFIELD, P. N. J. (1951) The breeding of *Crepidula fornicata* (L.) in the river Blackwater, Essex, *J. mar. biol. Ass. U.K.* **30**, 49–71.

CLARKE, R. B. (1955) The sublittoral fauna of two sandy bays on the isle of Cumbrae, Firth of Clyde, *J. mar. biol. Ass. U.K.* **34**, 161–80.

CLENCH, W. J. (1954) The occurrence of clines in molluscan populations, *Syst. Zool.* **3**, 122–5.

COLE, H. A. (1942) The American whelk tingle, *Urosalpinx cinerea* (Say), on British oyster beds. *J. mar. biol. Ass. U.K.*, **25**, 477–508.

COLMAN, J. S. (1932) A statistical test of the species concept in *Littorina*, *Biol. Bull. Woods Hole* **62**, 223–43.

CRAMPTON, H. E. (1916) Studies on the variation, distribution, and evolution of the genus *Partula*: the species inhabiting Tahiti, *Publ. Carneg. Instn.* **228**, 3–313.

CRAMPTON, H. E. (1925) Studies on the variation, distribution, and evolution of the genus *Partula*: the species of the Mariana Islands Guam and Saipan, *Publ. Carneg. Instn.* **228 A**, 1–116.

CRAMPTON, H. E. (1932) Studies on the variation, distribution, and evolution of the genus *Partula*: the species inhabiting Moorea, *Publ. Carneg. Instn.* **410**, 3–335.

CRISP, D. J. and FISCHER-PIETTE, E. (1959) Répartition des principales espèces inter-cotidales de la côte atlantiques française en 1954–5, *Ann. Inst. oceanogr. Monaco*, **36**, 275–388.

CRISP, D. J. and KNIGHT-JONES, E. W. (1955) Discontinuities in the distribution of shore animals in North Wales, *Rep. Bardsey Bird Fld. Obs.*, 1954, pp. 29–34.

CRISP, D. J. and SOUTHWARD, A. J. (1953) Isolation of intertidal animals by sea barriers, *Nature, Lond.* **172**, 208.

CRISP, D. J. and SOUTHWARD, A. J. (1958) The distribution of intertidal organisms along the coasts of the English Channel, *J. mar. biol. Ass. U.K.* **37**, 157–208.

CURREY, D. J., ARNOLD, R. W. and CARTER, M. A. (1964) Further examples of variation of populations of *Cepaea nemoralis* with habitat, *Evolution* **18**, 111–17.

DAS, S. M. and SESHAPPA, G. (1948) A contribution to the biology of *Patella*: on popula-tion distribution and sex proportions in *Patella vulgata* Linnaeus at Cullercoats, England, *Proc. zool. Soc. Lond.* **117**, 653–62.

DUVAL, D. M. (1963) The biology of *Petricola pholadiformis* Lamarck (Lamellibranchi-ata, Petricolidae), *Proc. malac. Soc. Lond.* **35**, 89–100.

EBLING, F. J., KITCHING, J. A., PURCHON, R. D., and BASSINDALE, R. (1948) The ecol-ogy of the Lough Ine rapids with special reference to water currents. 2. The fauna of the *Saccorhiza* canopy, *J. Anim. Ecol.* **17**, 223–44.

ESLICK, A. (1940) An ecological study of *Patella* at Port St. Mary, Isle of Man, *Proc. Linn. Soc. Lond.* **152**, 45–58.

EVANS, R. G. (1947) Studies on the biology of British limpets, *Proc. zool. Soc. Lond.* **117**, 411–23.

EVANS, R. G. (1947a) The intertidal ecology of selected localities in the Plymouth neighbourhood, *J. mar. biol. Ass. U.K.* **27**, 173–218.

EVANS, R. G. (1953) Studies on the biology of British limpets, *Proc. zool. Soc. Lond.* **123**, 357–76.

EVANS, R. G. (1958) The genus *Patella* on the west coast of France, *J. Conchyliol.* **98**, 126–51.

FISCHER, R. A. (1941) *Statistical Methods for Research Workers*, 8th edn., Oliver and Boyd, London.

FISCHER-PIETTE, E. (1931) Sur la pénétration des diverses espèces marines sessiles dans les estuaires et sa limitation par l'eau douce, *Ann. Inst. Océanogr. Monaco* **10**, 217–41.

FISCHER-PIETTE, E. (1935) Systématique et biogéographie — les Patelles d'Europe et d'Afrique du Nord, *J. Conchyliol.* **79**, 5–66.

FISCHER-PIETTE, E. (1936) Sur la distribution de *Patella intermedia* et sur les répartitions dites capricieuses, *Livre Jubilaire de M. E. L. Bouvier*, Paris, pp. 27–32.

FISCHER-PIETTE, E. (1938) The concept of species and geographical isolation in the case of North Atlantic Patellas, *Proc. Linn. Soc. Lond.* **150**, 268–75.

FISCHER-PIETTE, E. (1941) Observations biométriques sur les Patelles de la Manche, *J. Conchyliol.* **84**, 300–6.

FISCHER-PIETTE, E. (1959) Contribution à l'écologie intercotidale du détroit de Gibraltar, *Bull. Inst. océanogr. Monaco* **1145**, 1–32.

FISCHER-PIETTE, E. and GAILLARD, J. M. (1959) Les Patelles au long des Côtes Atlantiques Ibériques et Nord Marocaines, *J. Conchyliol.* **99**, 135–200.

GOODHART, C. B. (1963) "Area effects" and non-adaptive variation between populations of *Cepaea* (Mollusca), *Heredity* **18**, 459–65.

HAAS, F. (1929–56) *Bronns Klassen und Ordnungen des Tier-reichs; Bivalvia*, Akademische Verlag, Leipzig.

HUNTER, W. R., MAITLAND, P. S. and YEOH, P. K. H. (1964) *Potamopyrgus jenkinsi* in the Loch Lomond area, and an authentic case of passive dispersal, *Proc. malac. Soc. Lond.* **36**, 27–32.

JUTTING, W. S. S. VAN BENTHEM (1941) Non-marine mollusca from the satellite islands surrounding Java, *Arch. néerl. Zool.* **5**, 251–348.

JUTTING, W. S. S. VAN BENTHEM (1948) Systematic studies on the non-marine mollusca of the Indo-Australian archipelago. I. Critical revision of the Javanese operculate land shells of the families Hydrocenidae, Helicinidae, Cyclophoridae, Pupinidae and Cochlostomatidae, *Treubia* **19**, 539–604.

JUTTING, W. S. S. VAN BENTHEM (1952) The Malayan species of *Opisthostoma* (Gastropoda, Prosobranchia, Cyclophoridae) with a catalogue of the species hitherto described, *Bull. Raffles Mus.* **24**, 5–62.

JUTTING, W. S. S. VAN BENTHEM (1952a) Systematic studies on the non-marine mollusca of the Indo-Australian archipelago. III. Critical revision of the Javanese pulmonate land snails of the families Ellobiidae to Limacidae, with an appendix on Helicarionidae, *Treubia* **21**, 291–435.

JUTTING, W. S. S. VAN BENTHEM (1953) Annotated list of the non-marine mollusca of the Moluccan islands Ambon, Haruku, Saparna and Nusa Laut. *Treubia* **22**, 275–318.

JUTTING, W. S. S. VAN BENTHEM (1956) Systematic studies on the non-marine mollusca of the Indo-Australian archipelago. V. Critical revision of the Javanese freshwater gastropods, *Treubia* **23**, 259–477.

JUTTING, W. S. S. VAN BENTHEM (1958) Non-marine mollusca of the island of Misool, *Nova Guinea* **9**, 293–338.

JUTTING, W. S. S. VAN BENTHEM (1958a) Landmollusken von Sumba, *Verh. naturf. Ges. Basel* **69**, 90–117.

JUTTING, W. S. S. VAN BENTHEM (1959) Non-marine mollusca of the North Moluccan islands Halmahera, Ternate, Batjan and Obi, *Treubia* **25**, 25–87.

Jutting, W. S. S. van Benthem (1959a) Catalogue of the non-marine mollusca of Sumatra and of its satellite islands, *Beaufortia* **7**, 41–191.

Jutting, W. S. S. van Benthem (1960) Some notes on land and freshwater molluscs of Malaya, *Basteria* **24**, 10–20.

Jutting, W. S. S. van Benthem (1962) Selection of lectotypes of non-marine Mollusca of New Guinea, described by Tapparone Canefri, and now preserved in the Museo Civico di Storia Naturale "Giagomo Doria" in Genoa, *Ann. Mus. Stor. nat. Genova* **73**, 1–16.

Jutting, W. S. S. van Benthem (1963) Non-marine mollusca of West New Guinea. Part I. Mollusca from fresh and brackish waters, *Nova Guinea, Zoology* **20**, 409–521.

Jutting, W. S. S. van Benthem (1963a) Non-marine mollusca of West New Guinea Part 2. Operculate land shells, *Nova Guinea, Zoology* **23**, 653–726.

Knutsen, L. V., Stephenson, J. W. and Berg, C. O. (1965) Biology of a slug-killing fly, *Tetanocera elata* (Diptera: Sciomyzidae), *Proc. malac. Soc. Lond.* **36**, 213–20.

Kohn, A. J. (1959) The ecology of *Conus* in Hawaii, *Ecol. Monogr.* **29**, 47–90.

Kohn, A. J. and Orians, G. H. (1962) Ecological data in the classification of closely related species, *Syst. Zool.* **11**, 119–27.

Kolstad, K. (1959) *Patella aspera* Lamarck new to Norway, *Nature, Lond.* **184**, 1886–7.

Laidlaw, F. F. and Solem, A. (1961) The land snail genus *Amphidromus*, a systematic catalogue, *Fieldiana, Zool. Mem.* **41**, 509–675.

Lewis, J. R. (1964) *The Ecology of Rocky Shores*, English Universities Press, London.

Lucas, J. A. W. (1954) Het genus *Patella* in Nederland (with an English summary), *Basteria* **18**, 36–40.

McCollough, F. S. (1959) The susceptibility and resistance of *Bulinus* (*Physopsis*) *globosus* and *Bulinus* (*Bulinus*) *truncatus rohlfsi* to two strains of *Schistosoma haematobium* in Ghana, *Bull. World Hlth. Org.* **20**, 75–85.

McMichael, D. F. and Hiscock, I. D. (1958) A monograph of the freshwater mussels (Mollusca: Pelecypoda) of the Australian Region, *Austr. J. mar. & Fish. Res.* **9**, 372–508.

McMullen, D. B. and Harry, H. W. (1958) Comments on the epidemiology and control of bilharziasis, *Bull. World Hlth. Org.* **18**, 1037–47.

Mead, A. R. (1961) *The Giant African Snail: A Problem in Economic Malacology*, University of Chicago Press, Illinois.

Morton, J. E. (1954) The crevice faunas of the upper intertidal zone at Wembury, *J. mar. biol. Ass. U.K.* **33**, 187–224.

Moyse, J. and Nelson-Smith, A. (1963) Zonation of animals and plants on rocky shores around Dale, Pembrokeshire, *Fld. Stud.* **1**, 1–31.

Nelson-Smith, A. (1964) Ph. D. thesis, University College of Swansea.

Newell, G. E. (1958) An experimental analysis of the behaviour of *Littorina littorea* (L.) under natural conditions and in the laboratory, *J. mar. biol. Ass. U.K.* **37**, 241–66.

Newell, R. (1964) Some factors controlling the upstream distribution of *Hydrobia ulvae* (Pennant) (Gastropoda, Prosobranchia), *Proc. zool. Soc. Lond.* **142**, 85–106.

Newell, R. (1965) The role of detritus in the nutrition of two marine deposit feeders, the prosobranch *Hydrobia ulvae* and the bivalve *Macoma balthica*, *Proc. zool. Soc. Lond.* **144**, 25–45.

Orton, J. H. and Southward, A. J. (1961) Studies on the biology of limpets. IV. The breeding of *Patella depressa* Pennant, *J. mar. biol. Ass. U.K.* **41**, 653–62.

Orton, J. H. and Winckworth, R. (1928) The occurrence of the American oyster pest *Urosalpinx cinerea* (Say) on English oyster beds, *Nature, Lond.* **122**, 241.

Paton, J. R. (1961) A brief account of the geology of the limestone hills of Malaya, *Bull. Raffles Mus.* **26**, 66–75.

PEAKE, J. F. (1969) Patterns in the distribution of Melanesian land Mollusca. *Phil. Trans. B*, **255**, 285–306.

PILSBRY, H. A. (1939) Land molluscs of North America (North of Mexico), *Acad. Nat. Sci. Philad.*, monogr. 3, 1–573.

PURCHON, R. D. (1957) Studies on the biology of the Bristol Channel. XVIII. The marine fauna of five stations on the northern shores of the Bristol Channel and Severn Estuary, *Proc. Bristol Nat. Soc.* **29**, 213–26.

PURCHON, R. D. and ENOCH, I. (1954) Zonation of the marine fauna and flora on a rocky shore near Singapore, *Bull. Raffles Mus.* **25**, 47–65.

PURCHON, R. D. and SOLARI, M. E. (1968) Studies on the distribution of species of prosobranch and pulmonate snails on the limestone hills of Malaya. *Proceedings of the Symposium on Mollusca, Ernakulam, India*, pp. 223–30.

REES, W. J. (1952) The role of amphibia in the dispersal of bivalve molluscs, *Brit. J. Herpet.* **1**, 125–9.

REES, W. J. (1965) The aerial dispersal of mollusca, *Proc. malac. Soc. Lond.* **36**, 269–82.

ROBSON, G. C. (1923) Parthenogenesis in the mollusc *Paludestrina jenkinsi*, *Brit. J. exp. Biol.* **1**, 65–78.

ROSEWATER, J. (1965) The family Tridacnidae in the Indo-Pacific, *Indo-Pacif. Moll.* **1**, 347–94.

ROYAL SOCIETY (1965) *A symposium on continental drift*, *Phil. Trans.* No. 1088, 1–323.

SCHALIE, H. VAN DER (1958) Vector snail control in Qualyub, Egypt, *Bull. World Hlth. Org.* **19**, 263–83.

SCHALIE, H. VAN DER (1960) Egypt's new high dam—asset or liability, *Biologist* **42**, 63–70.

SHEPPARD, P. M. (1952) Natural selection in two colonies of the polymorphic land snail *Cepaea nemoralis*, *Heredity* **6**, 233–38.

SIMPSON, G. G. (1953) *The Meaning of Evolution*, Oxford University Press.

SMITH, J. E. and NEWELL, G. E. (1955) The dynamics of the zonation of the common periwinkle (*Littorina littorea* (L.)) on a stony beach, *J. Anim. Ecol.* **24**, 35–56.

SOLEM, A. (1961) New Caledonian land and fresh water snails: annotated check list, *Fieldiana, Zool.* **41**, 415–501.

SOUTHWARD, A. J. and CRISP, D. J. (1954) The distribution of certain inter-tidal animals around the Irish coast, *Proc. R. Irish Acad.* **57**, 1–29.

SOUTHWARD, A. J. and DODD, J. M. (1956) Studies on the biology of limpets. I. The late J. H. Orton's work on *Patella*, *J. mar. biol. Ass. U.K.* **35**, 145–7.

STEPHEN, A. C. (1929) Studies in the Scottish marine fauna; the fauna of the sandy and muddy areas of the tidal zone, *Trans. roy. Soc. Edinb.* **56**, 291–306.

STEPHEN, A. C. (1930) Studies in the Scottish marine fauna; additional observations on the fauna of the sandy and muddy areas of the tidal zone, *Trans. roy. Soc. Edinb.* **56**, 521–35.

THORSON, G. (1941) *The Zoology of Iceland*; 4. *Marine Gastropoda, Prosobranchia*, 1–150.

THORSON, G. (1944) The zoology of East Greenland; Marine Gastropoda Prosobranchia, *Medd. Grønland* **121**, 3–181.

TURNER, R. D. (1966) *A Survey and Illustrated Catalogue of the Teredinidae*, Museum of Comparative Zoology, Harvard University, Cambridge, Mass.

TWEEDIE, M. W. F. (1961) On certain mollusca of the Malayan limestone hills, *Bull. Raffles Mus.* **26**, 49–65.

WATKIN, E. E. (1942) The macrofauna of the intertidal sand of Kames Bay, Millport, Buteshire, *Trans. roy. Soc. Edinb.* **60**, 543–61.

WEGENER, A. (1924) *The Origin of Oceans and Continents* (transl. by J. G. A. Skerl), Methuen, London.

CHAPTER 8

THE FUNCTIONS OF THE NERVOUS SYSTEM IN THE DIBRANCHIATE CEPHALOPODA

Synopsis

The basic adaptations of the Dibranchiata were to a free-swimming predacious mode of life, and this resulted in a high degree of cephalisation and highly efficient means of co-ordination of bodily activities. The respiratory demands of such large and active predators as *Sepia* and *Loligo* could not be met by the conventional, molluscan ciliary mechanisms, in place of which ventilation of the mantle cavity is by muscular contractions of the mantle wall. The mantle wall comprises an unique arrangement of intermingled circular and radial muscle fibres, and the efficiency of the system is such that it also subserves locomotion, the propulsive force being supplied by jets of water from the funnel. Gentle respiratory movements aid in slow cruising motion and these movements are maintained by nerve impulses travelling in the ordinary neurones in the stellar nerves. In the Decapoda, but not in the Octopoda, the stellar nerves also contain giant nerve fibres which are only activated when the animal is alarmed, and which initiate maximal contractions of the pallial muscle fibres. The giant fibre system comprises three series of neurones; there are two giant fibres of the first order, one on each side, lying in the central nervous system, and connected to each other by an interaxonic bridge which ensures bilateral response even to unilateral stimulation. These two giant fibres can be fired by visual, tactile, or gravitational disturbances; their axons synapse with a number of giant fibres of the second order which innervate the retractor muscles of the head and of the funnel as well as with a second-order fibre which passes to the stellate ganglion there to synapse with many giant fibres of the third order. The last-named giant fibres travel in the stellar nerves and innervate the pallial muscle fibres. Excitation of the giant fibre system elicits an all-or-nothing maximal escape response. Each of the third-order giant fibres arises by the union of the axons of large numbers of small neurones, and are therefore syncytial. The rate of conduction of nerve impulses in decapods is proportional to the square root of the diameter of the fibre and the giant fibre system minimises the lapse of time between the occurrence of some disturbing event and the initiation of the escape response.

The statocysts of the Dibranchiata are large and complex as compared with those of other molluscan classes. The statocysts arise as ectodermal invaginations which become enclosed in hard cartilaginous capsules lying postero-ventral to the eyes. Within the statocyst the macula and its associated statolith record the position of the body in relation to gravity; this information is supplied directly to the eyes and controls their orientation. In the Decapoda the macular nerve probably innervates the first-order giant fibres. In the statocyst the crista is a continuous ridge arranged in three planes on its inner surface; two limbs of the crista lie in the horizontal plane and the third lies in the vertical plane. All limbs of the crista possess hair cells which detect movements in the fluid filling the statocyst, and by means of the crista the animal can detect angular accelerations in different directions. This capacity is possessed by only one other animal group, viz. the Vertebrata. Damage to both statocysts prevents the correct alignment of the eyes and discrimination between horizontal and vertical surfaces, and interferes with the ability of the animal to move about.

The eyes of the dibranchiate Cephalopoda are highly advanced as compared with those of gastropods in the series: *Patella, Trochus, Lambis,* which are briefly described. There are striking parallelisms between the eyes of cephalopods and those of vertebrates, but there are major differences—the cephalopod eye is an ectodermal invagination, not an outgrowth of the brain; the lens is a rigid spherical secretion, and is of fixed focal length, not a mass of cells capable of plastic deformation; there are numerous small optic nerves, no fovea, and no blind spot. Another striking difference is that the cephalopod eye is controlled by the statocysts and is held in a constant relation to gravity, the slit-shaped pupils always being horizontal. In the octopus the retina comprises a more or less rectangular array of rhabdomes, nerve fibres pass from the retina into the optic lobe where they synapse with neurones whose dendritic pick-up fields are primarily oriented along horizontal or along vertical axes, and it is clear that the horizontal and the vertical extents of objects seen are of great significance to an octopus. Our own vision is largely dependent on our personal experience and we cannot assume that an octopus sees in the same way that we do. Our own vision is largely dependent on processes of analysis of visual information and we are not conscious of these analytical processes. We have to consider how visual information may be analysed in the brain of an octopus. It seems probable that an object seen is scanned and classified according to its horizontal and its vertical extent, but other methods of analysis probably also occur. An octopus can be taught to discriminate between contrasted situations, e.g. a horizontal and a vertical rectangle, by rewards for attacking a "favourable" situation and by punishments for attacking an "unfavourable" situation. The octopus can remember such a discrimination for a considerable time. The methods of investigating the learning process, the mechanisms whereby retinal images may be analysed, and the neurophysiological processes involved in learning and memory, are described in detail.

The introduction of a crab plus a white plate into an aquarium presents a complex visual problem to an octopus which has previously learned by punishment to avoid such a situation. The sight of the crab is associated with previous gustatory pleasures, but the sight of the white square is associated with previous punishments by electric shock. The central nervous system comprises immense numbers of neurones including some which serve as analysers and summate the visual information according to one or more special attributes, e.g. the horizontal versus the vertical extent of the objects observed; persistent records of previous experiences are utilised in this summation. The output from these analysers will be in one of two possible neural channels—to attack or to refrain from attacking—and the channel used will depend on the result of a statistical summation of all relevant information received, including representations of previous experiences. The neurones in the optic lobes are connected in a random way with one another and with neurones supplying gustatory and tactile information, all this information being widely distributed throughout the optic lobes. The representation in an optic lobe of a visual situation must be retained for a sufficient length of time to permit its association with subsequent signals of pleasure (gustatory) or of pain (electric shock, or stings); this constitutes a short-term memory which may be largely housed in the optic lobes alone. Severe lesions to the supra-oesophageal brain centres, leaving the optic lobes undamaged, will destroy any long-term memories but will leave undamaged the capacity to carry short-term memories. A short-term memory presumably comprises self-excitatory reverberations in cyclical series of neurones in the optic lobes, while a long-term memory may become established if by repetition of the situation these reverberating circuits are extended to include parts of the supra-oesophageal brain mass, especially the vertical lobe.

The learning process concerns the imprinting of changes in the probability of attack under given circumstances. These changes cannot concern a single neurone or a single series of neurones, but must be widely distributed through the substance of the brain; the more widely is the change distributed in the brain, the more firmly is the memory established. The long-term memory probably involves rhythmical reinforcement of the cyclical reverberations passing through neurones in the vertical, subvertical, and optic lobes. Although there are major structural differences between the brains of cephalopods and of vertebrates, it is possible that there may be basic similarities in their modes of functioning; this possibility is indicated by the occurrence of common features in the mechanisms of analysis of visual situations by octopuses, goldfish, rats, and monkeys.

The Octopoda diverged from their pelagic ancestry, adopted a benthic mode of life, began to use their arms for tactile exploration, and developed a capacity for tactile learning and memory which is not possessed by the Decapoda. This tactile sense is capable of discriminating between smooth and rough surfaces according to the degree of distortion of the rims of the suckers, but is not capable of discrimination on account of the shapes of objects since the arms lack a jointed skeleton and there is no basis for computing the relative positions of different sections of the arms. Proprioceptive information is not passed to the supra-oesophageal brain mass and the octopus cannot discriminate between objects which differ only in weight. Tactile information is passed up the arm in question via the brachial ganglion to the inferior frontal and sub-frontal lobes of the brain, and the tactile memory system has developed on much the same lines as the visual memory system described above. However, a tactile memory acquired by one arm will only be acquired by the other arms after an appreciable interval of time during which the record is spreading through the neural tissue. A special lobe of the brain, the sub-frontal lobe, which is not present in the brain of decapods, has developed in the Octopoda to house the neural tissue required for maintaining the tactile learning and memory.

Various gastropods have been shown to respond automatically to the presence of certain chemicals, or food substances in the vicinity. In the Cephalopoda, on the other hand, chemo-tactile sense organs widely distributed in the skin of the body and arms transmit information to the central nervous system on the basis of which an octopus can learn to distinguish by touch objects which only differ chemically. It is not known whether octopuses can discriminate a number of distinct tastes, like ourselves, or whether the same sense cells are stimulated to differing extents by different chemicals.

A neurosecretory system of unknown function has been found in association with the vena cava in *Sepia*, *Eledone*, and *Octopus*; the system is briefly described. It is thought to liberate a hormone into the blood stream. In the Octopoda the stellate ganglion bears an epistellar body which is innervated by a branch from the pallial nerve. The epistellar body contains neurosecretory cells, and it is thought possible that it supplies a hormone which maintains the tonus of the body. The epistellar body is briefly described.

The functions of the various parts of the brain have been investigated by observation of the effects of stimulation by the application of electrodes at various points on the brain, and by observation of the behaviour of specimens after brain lesions of known extent. The functions of all the parts of the brain, determined as indicated above, are described in turn.

The dibranchiate Cephalopoda differ from all other major taxonomic groups in the Mollusca in that their primary adaptation was to suit the lives of free-swimming predacious carnivores. This basic adaptation set an ineradicable imprint on the form of the body and on its mode of organisation; cephalisation in particular has been carried to an extremely advanced condition, and the Dibranchiata unquestionably stand at the summit of invertebrate evolution. The mode of life of a free-swimming predacious carnivore has arisen independently in numerous gastropod lineages, but these individually represent secondary, or perhaps even tertiary steps in the adaptive radiation of the class; the bodily form and mode of organisation was already at least partly determined, and the evolutionary potential of each of these various gastropod lineages was strictly limited. In contrast the Cephalopoda have probably always been water-borne and the lineage culminating in the modern dibranchiate forms has specialised in various interrelated ways to suit the lives of pelagic hunters: retention of a primary bilateral symmetry; lightening of the body by reduction of the shell; locomotion partly by an unique form of jet propulsion; cephalisation, with the development of unusually powerful and well-organised eyes, and with a truly remarkable central nervous system capable of making important discriminations based on information obtained from these distance receptors. It is true that certain of the Octopoda are benthic hunters, but there can be no doubt that this is the result of a comparatively recent, and secondary, adaptation.

The central nervous system in the Dibranchiata comprises supra-oesophageal and infra-oesophageal ganglionic masses joined by short, stout lateral connectives which also receive the optic stalks from the bases of the large, paired optic ganglia. To this circum-oesophageal brain mass are attached the large paired stellate ganglia which lie lateral to the visceral mass near the mouth of the mantle cavity; superior and inferior buccal ganglia which lie closely apposed to the base of the buccal mass; and a gastric ganglion which lies on the "ventral" surface of the stomach. A particularly important feature of the nervous system of the Decapoda is the presence of a system of giant fibres which administer the retractor muscles of the head and funnel, and the pallial muscles.

Respiration and Locomotion

A respiratory water current generated by ciliary mechanisms on the ctenidia would be quite inadequate for the maintenance of large and active animals such as the present-day *Loligo*, *Sepia*, and *Octopus*, etc. One of the earliest steps in the emergence of the dibranchiate cephalopods must have been the elaboration of an alternative system of maintenance of the respiratory current, and this was provided by the contraction and relaxation of pallial muscle fibres. This system comprises two series of intermingled fibres, which together transform the mantle into a thick and powerful muscular wall quite unlike the mantle in any other molluscan group. In fact the system is unique. Closely packed circular fibres contract and so reduce the volume of the mantle cavity and force an exhalant water stream out through the funnel. Radial muscle fibres pass transversely through the mantle wall, and when these fibres contract the thickness of the mantle wall is reduced, the volume of the mantle cavity is increased, and an inhalant stream of water is drawn into the mantle cavity through the space between the base of the funnel and the rim of the mantle wall.

These pallial muscle fibres are activated by nerve fibres which originate in the stellate ganglion and which pass out to the pallial muscles in a regular series of stellar nerves. The nerve fibres in any one stellar nerve are responsible for innervation of all the muscle fibres in the relevant sector of the mantle wall. Examination of sections of the stellar nerves and of the mantle wall shows that there are vastly more muscle fibres than nerve fibres, and it follows that individual nerve fibres innervate very large numbers of muscle fibres. Co-ordinating mechanisms in the central nervous system ensure that the muscle fibres in all sectors of the mantle wall, on both left and right sides of the body, contract and relax synchronously, thereby producing gentle rhythmic inhalant and exhalant respiratory movements.

The buoyancy of a cuttle-fish or of a squid is almost identical with that of the surrounding water, the body is streamlined, and very little energy is required to set the animal in motion. This energy is partly supplied by movements of the lateral fins, and partly by the exhalant respiratory jets of water from the funnel. The funnel is flexible and is adequately supplied with muscles by means of which it can be moved so that the exhalant jet of water can be played in one direction or another, and the direction of movement of the animal will coincide with the resultant of the forces supplied by the fins and by the exhalant jets. The same system therefore serves both for respiration and for locomotion.

In the Decapoda—but not in the Octopoda—each stellar nerve contains one giant nerve fibre in addition to large numbers of normal fibres. It is apparent that when the animal is undisturbed the gentle pallial movements associated with respiration and with slow locomotion are mediated by the large numbers of small nerve fibres, and the giant fibres are inactive. The giant fibres are only utilised in emergency, when the animal has been alarmed (Young, 1938). If a squid is disturbed by some startling event nearby it will execute a violent escape movement by ejecting a powerful jet of water from its funnel, which causes it to shoot backwards away from the source of alarm. This escape movement is produced by intervention of the whole giant fibre system, which ensures maximal response from the circular pallial muscle fibres and also causes retraction of the head and of the funnel. To be effective such an escape reflex requires to be kept under very effective, yet delicate control. It should be triggered off "instantaneously" and maximally on all appropriate occasions, yet it should not be fired unnecessarily by sudden visions or events which do not represent a hazard for the animal. The structure and function of the giant fibre system adequately meets these delicate requirements.

In order to understand the mode of action and biological significance of the giant fibre system, it is necessary to describe it in some considerable detail. The system is rather complicated, and it will be helpful to check each step in the description by reference to Fig. 131. The following account is of *Loligo pealii* and is drawn from the detailed analysis of Young (1939).

The giant nerve cells and their fibres, which are arranged in much the same way in sepioid and in teuthoid decapods, comprise an hierarchical series commencing with two giant cells of the first order housed in a special lobe of the central nervous system, the lobus magnocellularis (see Fig. 131). These two giant cells lie above the statocysts, in the vicinity of the meeting points of the cerebral, optic, pedal, and pallio-visceral ganglia. The two cell-bodies lie in the ventral part of the lobus magnocellularis and their dendrites pass forwards and spread through the whole of the lobe. The ventralmost dendrites are in close association with the afferent nerve fibres from the statocyst—and in particular with the fibres from the maculae. The surfaces of the nerve-cell bodies, and of their dendrites, are closely covered with end-feet which are not separated from the cell by any form of sheath, but which press closely against the cell surface. It seems probable that the two giant cells of the first order can be activated through these end-feet by *visual stimuli* transmitted via the optic lobes; by *tactile stimuli* from the skin of the arms, which are transmitted inwards via the brachial ganglia; and by *gravitational data* detected chiefly by the maculae in the statocysts. It is possible that these giant cells may also be activated from the higher centres of command sited in the supra-oesophageal ganglia. It seems probable that these two giant cells of the first order cannot be activated by only very few afferent impulses—otherwise the animal would be in a constant state of flight. The giant cells may respond only to massive stimulation from any one receptor system, and they may also respond to special co-ordinated patterns of stimulation from diverse sources. The axons of the two giant cells of the first order pass backwards from their respective nerve-cell bodies, and they meet and fuse in a median inter-axonic bridge sited in the neuropil of the pallio-visceral ganglia. Posterior to this unique point of axonic fusion the two axons separate, pass further backwards, and divide into a number of branches. Each of these branches makes synapse with one of the giant nerve fibres of the second order. It seems obvious that the fusion of the axons of the two first-order giant cells serves to ensure that unilateral stimulation of the animal will not fail to fire the entire escape mechanism, on both sides of the body. Vigorous unilateral contraction of the circular pallial muscle fibre system would be less effective as the basis for an escape movement.

The number of giant cells of the second order varies from one decapod to another; in *L. pealii* there are seven on each side, of which one passes to

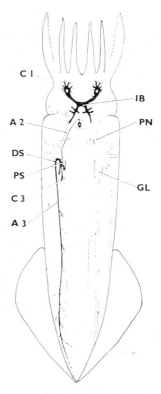

Fig. 131. Diagrammatic representation of the giant fibre system in the nervous system of the decapod *Loligo*. The giant fibres of the first and third orders are shown in solid black and the central nervous system and principal nerves are shown in outline. The giant fibres of the first order have their cell bodies (C1) in the magnocellular lobe; their axons fuse at the inter-axonic bridge (IB). One of the giant fibres of the second order (A2) is shown passing in the pallial nerve (PN) to the stellate ganglion. One of the giant fibres of the third order (A3) is shown arising from a group of nerve-cell bodies (C3) which are gathered together in the giant fibre lobe (GL) of the stellate ganglion. For interpretation of other lettering, see p. 448. (Originally published in Young, 1936, *Cold Spring Harbor Symposia on Quantitative Biology*, **4**, 4, fig. 6.)

the stellate ganglion, one passes to the intrinsic muscles of the funnel, and the remainder pass to the retractor muscles of the head and of the funnel. Each of these giant nerve fibres of the second order arises from a single giant nerve-cell body situated in the pallio-visceral ganglion. At the synapse between the giant fibres of the first order and of the second order a branch from the fibre of the first order lies alongside the axon of the second-order fibre, the two being separated from each other by a sheath. Many collaterals from the second-order fibre penetrate this sheath and make synaptic contact with the surface of the branch of the first-order fibre. When there is a discharge of the first-order giant cells there will be a very extensive area of excitation of the second-order fibre at this synapse. Thus a single discharge of the first-order giant cells is adequate to ensure activation of all the second-order giant cells. This synapse is known in each case as the *distal synapse*. In addition to this each second-order giant fibre also possesses a *proximal synapse* with many smaller nerve fibres of uncertain origin. It is possible that these proximal synapses provide machinery for the activation of the escape response under certain circumstances when the external stimuli do not suffice to attain the threshold of response of the first order giant cells.

The single second-order giant fibre which enters the stellate ganglion subdivides into several branches, each of which accompanies one of the third-order giant fibres and makes a *distal synapse* with it. The two are separated by their connective tissue sheaths, but large numbers of collaterals of the third-order fibre pass through holes in the intervening sheath and press closely against the surface of the branch of the second-order giant fibre. Once again, these very extensive synaptic areas ensure that a single discharge of the second-order fibre will activate all the third-order giant fibres.

In addition to the single second-order giant fibre which passes from the pallio-visceral lobe to the stellate ganglion there is also an accessory giant-fibre which synapses in the pallio-visceral lobe with many small neurones and not with the first-order giant fibres. When this accessory giant fibre reaches the stellate ganglion it breaks down into innumerable small branches, each of which makes a *proximal synapse* with one of the very many small neurones sited there. This accessory giant fibre may be another safety measure built in to the system of neurones which administers the escape mechanism; it may be activated in the pallio-visceral ganglion by impulses which were not directed to the second-order giant fibres arising there.

The series of third-order giant fibres completes the efferent motor pathway to the circular pallial muscle fibres. The number of these giant fibres of the third order differs in different decapods, there being one in each stellar nerve. Where the stellar nerve branches, so does the giant fibre within, and the third-order giant fibre innervates all the circular muscle fibres in the sector of mantle wall served by that stellar nerve. Each third-order giant fibre is a syncytium, for it arises by union of large numbers of axons of the small neurones mentioned above as participating in the proximal synapses with the accessory giant fibre.

To summarise, any sudden external event in the vicinity of a decapod —e.g. the appearance of a potential predator, or a major physical disturbance of the water—will probably evoke an escape reaction by the decapod. The reaction may be initiated by a visual or a tactile disturbance or by a disturbance in the animal's stability, and is transmitted by an integrated system of giant nerve fibres which ensures "instantaneous", maximal, bilateral response even to unilateral stimulation. The first pair of giant fibre cells doubtless require fairly massive stimulation to ensure that the escape reaction is not invoked unnecessarily. The interaxonic bridge between these two cells ensures that even a unilateral esternal disturbance evokes a full bilateral escape response. Withdrawal of the head and of the funnel will compress the mantle cavity and so augment the escape reaction, and there is only one synapse on the giant fibre pathway which mediates this withdrawal. The escape reaction depends on close co-ordination of contraction of all the circular muscle fibres of the mantle; this is achieved by the branching of the third-order giant fibres, so that no more than ten or eleven of these fibres command the entire array of circular pallial muscle fibres. There are only two synapses on the giant fibre pathway to the circular pallial muscle fibres; there is minimal consumption of time in transmission of impulses through synapses, while the very extensive nature of the synaptic contacts ensures that there will be no failure to excite the neurones distal to the synapse.

The giant fibre system aids the escape reaction in one further respect, which is especially worthy of mention. It has been shown that the rate of conduction of a nerve impulse varies according to the diameter of the nerve fibre, conduction being faster in the larger fibres. The rate of conduction increases approximately in proportion to the square root of the diameter of the fibre (Pumphrey and Young, 1938). The presence of giant fibres, and the utilisation of these when the escape reaction is invoked, results in a great saving in "reaction time". Due to possession of a giant fibre system an alarmed decapod can get under way far more rapidly than it could otherwise. Furthermore, it has been shown that in the series of third-order giant fibres, the longer stellar nerves contain giant fibres of greater diameter, and the shorter stellar nerves contain giant fibres of lesser diameter. In consequence there is a more rapid delivery for impulses which have a greater distance to travel, and the time of contraction of the pallial muscles is a little more closely co-ordinated than would be the case if all nerve impulses travelled at a uniform speed. The exhalant jet of water will be expelled with slightly greater vigour, and the animal will dart backwards at a slightly higher speed than would otherwise be the case.

Structure and Function of the Statocysts

The statocysts of gastropods and of bivalves are comparatively small and simple organs which may contain one single statolith or alternatively several smaller statoliths. Such statocysts are adequate to the needs of these sedentary or relatively slow-moving animals. In the dibranchiate Cephalopoda, which are highly specialised as fast-swimming predacious carnivores, the statocysts differ in being relatively large and very much more complicated in structure. The statocysts in the Dibranchiata evolved in response to the requirements of nektonic animals and there was no significant alteration in their basic mechanism or in their utilisation in the Octopoda when these diverged and adopted a benthic mode of life. This explains certain limitations in the organisation and the capabilities of the octopods as revealed by experimental investigation. Attention is directed here to the structure and mode of utilisation of the statocysts in *Octopus vulgaris* (see Figs. 132 and 133).

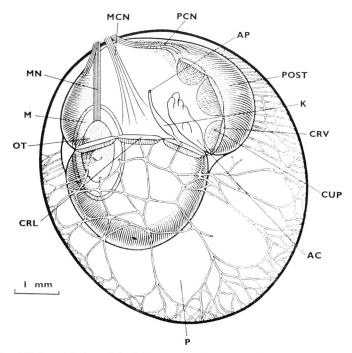

FIG. 132. Lateral view of the left statocyst of *Octopus vulgaris*. The longitudinal (CRL) and the vertical (CRV) limbs of the crista possess three sections each; they are served by the middle and posterior crista nerves (MCN, PCN) respectively. The otolith (OT) lies in the macula (M) which is served by the macular nerve (MN). The statocyst lies suspended in a wide space filled with perilymph (P). For interpretation of other lettering, see p. 448. (Originally published in Young, 1960, *Proc. roy. Soc. B*, **152**, 5, fig. 2.)

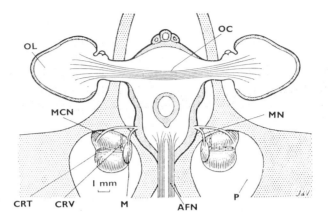

FIG. 133. Transverse section through the head of an *Octopus*, seen from the anterior side, showing the positions of the statocysts in relation to the horizontal plane—which is indicated by the optic commissure (OC) joining the two optic lobes (OL). For interpretation of other lettering, see p. 448. (Originally published in Young, 1960, *Proc. roy. Soc.* B, **152**, 5, fig. 1.)

The statocysts arise as ectodermal invaginations which become enclosed in hard, cartilaginous capsules, and in a moderately large *Octopus* this capsule can be felt as a hard lump beneath the posterior border of each eye. The original ectodermal invagination persists in the adult as a slender blind canal, known as Kolliker's canal, which no longer communicates with the surface. In decapods the statocyst lies in contact with the surrounding cartilaginous capsule, but in octopods the sac is suspended from the cartilage in a space which is filled with perilymph. The wall of the octopod statocyst is mainly thin and membranous, but it is supported in places by cartilaginous thickenings; its shape is maintained by the action of scattered muscle fibres in the wall of the sac against the internal fluid pressure. Scattered fibres and blood vessels cross the perilymph to support the statocyst. The sac is firmly attached to the cartilaginous capsule only at one point, at which a number of separate nerves pass from the statocyst to the brain.

There are three types of sense organ in the statocyst: the crista, which serves to record angular accelerations in different planes; the macula, which records the bodily position in relation to gravity; scattered "hair" cells which may detect movements in the endolymph which fills the lumen of the statocyst.

The crista is a continuous ridge which is arranged in three planes on the inner surface of the statocyst. The statocyst wall is supported externally by a system of cartilages which form a continuous framework external to the crista and the macula. The transverse limb of the crista corresponds with the horizontal plane joining the two eyes; another limb of the crista is also horizontal, but at right angles to the first, while the third limb of the crista lies vertically and parallel with the sagittal plane. In comparison with this arrangement, in the Vertebrata the inner ear possesses two semi-circular canals in vertical planes at right angles to each other but both set obliquely with reference to the principal axes of the animal, and only one semicircular canal in the horizontal plane. The Cephalopoda and the Vertebrata are the only animal groups to have developed their stato-cysts as organs for the resolution of angular accelerations in different directions (Young, 1960).

The sense organ which comprises the crista is not continuous, but it consists of nine units of which three occur in each limb of the crista. These nine units are not identical, there being two types which differ slightly and which are arranged alternately. Each limb of the crista therefore possesses two units of one kind and one of another kind. Each of these units consists in principal of either a double or a single row of sensory "hairs" which together form a flap projecting into the lumen of the statocyst. Each limb of the crista possesses its own nerve, by means of which afferent impulses are transmitted to the brain (see Fig. 134).

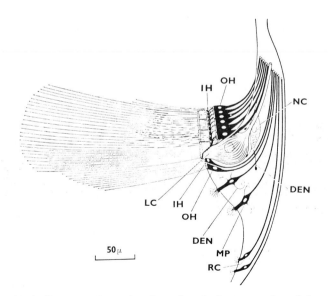

FIG. 134. A diagrammatic section through a single-row portion of the crista of *Octopus vulgaris*. In such a portion of the crista a row of large hair cells (LC) is flanked both above and below by inner and outer rows of hair cells (IH, OH). Other receptors scattered over the inner surface of the wall of the statocyst are probably associated with multipolar cells (MP), as shown. For interpretation of other lettering, see p. 448. (Originally published in Young, 1960, *Proc. roy. Soc.* B, **152**, 17, fig. 9.)

It is a reasonable inference that the crista serves to detect changes in momentum in the three planes, the hairs in one limb of the crista being stimulated differently by upward and by downward movements in the endolymph which fills the statocyst; similarly, in the vertical segment of the crista movements to right or to left will be separately detected by the sensory hairs. Receipt of this information in the brain permits computation and the issue of motor instructions to keep the animal on the correct course to its target (e.g. "homing" on to a swimming crab).

Opposite the crista a flexible plate, the anticrista, projects into the lumen of the statocyst. When the octopus attacks a crab it swims with one eye in a leading position, and the anticrista probably protects the crista from movements in the endolymph caused by acceleration.

Removal of one or both statocysts interferes with the ability of the animal to move about. After unilateral removal of the statocyst, Boycott found that the animal would only swim backwards. After bilateral statocyst removal the animal would not swim unless forced to do so, and was unable to maintain direction; it swam in spirals, sometimes performing somersaults, and it showed rolling, pitching, and yawing movements. After bilateral statocyst removal the animal seemed unable to distinguish between horizontal and vertical surfaces, and it would move over vertical surfaces with its head down, or sideways, and with the visceral mass displaced by gravity—the normal animal always moves with the head uppermost (Boycott, 1960; Young, 1960) (see Fig. 135).

There is no evidence to suggest that *Octopus* can hear, and any responses to vibrations can be made equally well by individuals which have had both statocysts removed (Hubbard, 1960).

The macula is an oval area on the median wall of the statocyst, the long axis of the macula lying exactly in the vertical plane. Associated with the macula there is a single, large, oval statolith which rests against the sensory "hairs" of the macula. The sensory epithelium includes concentric rings of neurosensory hair cells, the hairs of which may be attached to the statolith. Multipolar nerve cells pass from the sensory epithelium to the macular nerve. The Octopoda do not possess a giant fibre system, but in the Decapoda the macular nerve passes into the magnocellular lobe, and its fibres lie in close association with the dendrites of the first-order giant nerve cells. Disturbances in the water which upset the equilibrium of a decapod will be detected by the maculae and afferent impulses passing in the macular nerves may result in firing the first-order giant cells and so invoke the rapid escape reaction.

Probably the most important function served by the statocysts is the maintenance of constant orientation of the eyes, with the slit-shaped pupils in a horizontal position regardless of the degree of inclination of the substratum and the bodily posture of the octopus. This function is probably dependent upon information supplied by the maculae. Correct orientation of both eyes is maintained even after the loss of one statocyst, nerve fibres from the statocysts presumably decussating in the pedal lobes. Information from the statocysts regarding the posture of the body in relation to gravity is not passed to the "higher" centres of the brain and is not itself available for direct integration with information supplied to these "higher" centres by the eyes. On the contrary, the information supplied by the statocysts is passed to a nerve centre in the lateral pedal lobes which directly controls the movements of the pupil of the eye and

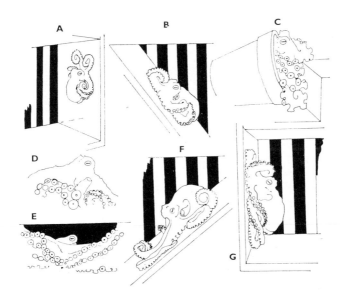

FIG. 135. The orientation of the eyes of *Octopus vulgaris* is by reflex action, under the influence of the statocysts. As long as at least one of the statocysts is intact, and there is no interference with its input to the central nervous system, the two eyes are always held with the slit-shaped pupil in the horizontal plane, regardless of the posture of the octopus (A–E). After removal of both statocysts this reflex no longer occurs, and the orientation of the eyes depends on the position of the octopus (F, G). (Originally published in Wells, 1960, *J. exper. Biol.* **37**, 492, fig. 1.)

thereby maintains constant orientation of the retina with reference to gravity. Compensatory eye movements depend on the integrity of the statocysts, and correct interpretation of visual information is wholly dependent on the proper functioning of at least one of the statocysts. This is an understandable mechanism for a phylogeny of nektonic animals swimming in middle waters where there is no horizon to observe and where detection of gravity is the only available aid in maintenance of stability. It might at first sight appear disadvantageous to a benthic hunter such as *Octopus* to be unable to pass information from the statocysts to the "higher" brain centres for direct integration with data from other sense organs. But as long as the eyes are correctly oriented with reference to gravity all the requirements of a hunter can be satisfied, and it is extremely unlikely that an octopus could survive any natural event that caused bilateral destruction of the statocysts. After experimental removal of both statocysts the eye is no longer correctly oriented with reference to gravity and this makes it impossible for the animal to discriminate efficiently between certain figures presented to it experimentally—the information passed from the retina to the brain is interpreted there on the basis that the retina is correctly oriented in relation to gravity, which may or may not be the case (see also pp. 416 and 422).

Vision and Visual Discrimination

The cephalic eyes of gastropods and cephalopods exhibit a wide range of structural complexity. Among the simplest are those of the prosobranch limpet *Patella* in which the eye consists of nothing more than a shallow pit lined by a retinal epithelium served by nerve fibres. In *Trochus* spp. the eye is more advanced in structure, there being a deep spherical pit which is lined by the retina and which is filled by a spherical crystalline lens. The pit remains open at the surface, and light is admitted at this point and is concentrated on the retina by the lens. In many other prosobranchs, e.g. *Lambis lambis*, the cephalic eyes are more highly advanced in structure. The pit has become closed from the surface, and the epithelium overlying the lens is in the form of a transparent cornea; the spherical chamber lined by the retina is appreciably larger than the lens, and in consequence there is a space between the lens and the retina, and this space is filled by a vitreous body. Eyes of similar complexity occur in the Pulmonata. In all these cases it has often been assumed that no clear image is cast on the retina; the animals probably react only to sudden changes in the intensity of illumination, e.g. a shadow cast by a passing animal, or objects moving rapidly through a wide angle of arc in the immediate vicinity of the animal. However, Newell (1965) has found that the lens of *Littorina littorea* throws a clear, undistorted image at 155 μ from the centre of the lens. He concluded that when underwater visual acuity is limited by the dioptrics rather than by the fineness of the retinal mosaic. When the animal is in air the front surface of the eye will also cause convergence, and visual acuity should be greater. The eye is housed in a bulge at the base of the tentacles and is surrounded by a fluid-filled perioptic sinus; contraction of muscle fibres in the ocular bulge may provide accommodation for vision of near objects. Movements of the tentacles when palpating foreign objects will not interfere with the position and orientation of the eyes (see Fig. 136).

In the dibranchiate Cephalopoda the eyes are very much more highly

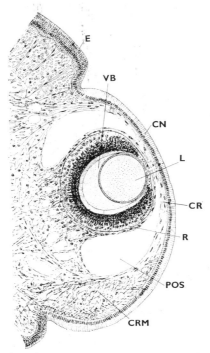

FIG. 136. Vertical section through the eye of *Littorina littorea* showing the conjunctiva (CN), the cornea (CR), the lens (L), the vitreous body (VB), the retina (R), and the perioptic sinus (POS). For interpretation of other lettering, see p. 448. (Originally published in Newell, 1965, *Proc. zool. Soc. Lond.* **144,** 78, fig. 1.)

evolved, and in gross morphology show marked parallel evolution with those of the vertebrates. Thus in the Dibranchiata the eye is a globular mass which lies in an orbit and is capable of partial rotation by a series of extrinsic muscles. An image is cast on the retina, the eye is capable of accommodation, and the animal can perceive objects which enter its field of view. Cannon (1958, pp. 112–14) seemed to have difficulty in envisaging how the vertebrate eye could have evolved on mendelian lines.*

* "Now random mutation, he says, could not lead to the evolution of the whole co-ordinated structure, for the chances of all the mutations occurring in the same direction at the appropriate time are so remote as to be negligible. A truer statement was never made by any geneticist." But homeostasis is a fundamental property of all living systems and by virtue of this property living systems necessarily generate a high degree of improbability. We have no reason to doubt that the human eye evolved on orthodox neo-darwinian lines just because such a process may seem statistically improbable. Moreover, Cannon was unreasonable in requiring all the relevant mutations to occur "in the same direction"; progressive mutations would presumably acquire dominance, while retrogressive mutations would become recessive and so would not hinder the general advance in efficiency of the organ.

Although the molluscan examples mentioned above do not themselves constitute an evolutionary series, and there are many fundamental differences between them and the eyes of vertebrates, nevertheless they indicate a somewhat comparable sequence of structural and functional stages through which the vertebrate eye must have passed on orthodox neo-darwinian evolutionary lines.

In gross morphology there are striking resemblances in structure and function between the eyes of dibranchiate cephalopods and of vertebrates, but closer investigation shows that these are analogous, not homologous features. The differences between the two types of eye are at least as striking as the similarities. Whereas the vertebrate eye is formed as an outgrowth of the brain, that of the cephalopod is formed by invagination of the ectoderm. The lens in the vertebrate eye is cellular in structure and is capable of plastic deformation by contraction of the ciliary muscles for accommodation to near objects. In contrast the lens in the cephalopod eye is a rigid, spherical secretion which is formed by two planoconvex components secreted by and attached to the epithelial body sandwiched between them. The lens of the cephalopod eye is of fixed focal length, and when the eye is at rest it is focused on comparatively near objects. Accommodation of the eye for distant vision is effected by contraction of the ciliary muscle, which draws the lens closer to the retina. Unlike the vertebrate eye, that of the Dibranchiata receives a large number of small optic nerves which enter from behind the eye. The retina contains only rod-like sensory elements, and there is no blind spot and no fovea. There is only one fluid-filled chamber to the cephalopod eye, that lying between the lens and the retina. In the Dibranchiata the eye lies in a fluid-filled orbit, the lumen of which communicates with the exterior by a narrow duct opening under the anterior end of the eyelid. A highly contractile iris lies external to the lens, and is perforated by a horizontal, slit-like pupil.

The wall of the eyeball is supported by cartilages, but the eyeball is not as firm as that of a vertebrate. The eyes can be rotated a certain amount by extrinsic muscles of which there are thirteen in the cuttle-fish, *Sepia*. These extrinsic muscles of the eye are innervated by branches from four motor nerves (Tompsett, 1939). The great number of these muscles may be a necessary consequence of the lack of rigidity of the eyeball.

The most important feature in which the eyes of the Dibranchiata differ from those of the Vertebrata is their constant orientation in relation to gravity, which is achieved by a reflex association between the statocysts and the eyes. The slit-shaped pupils are always maintained in a horizontal position, and the information supplied to the brain is interpreted there on the basis of correct alignment of the eyes. The Octopoda secondarily adopted a benthic mode of life, but they retained this reflex system of alignment of the eyes as a heritage from their long pelagic ancestry. In *Octopus* the retina contains a more or less regular system of rhabdomes each of which is built up from four rhabdomeres. The rhabdomeres contain piles of tubules oriented in parallel and in two of them the tubules are oriented in the horizontal plane while in the other two rhabdomeres the tubules are oriented at right angles, in the vertical plane. Adjacent rhabdomes are separated by slender pigment cells (Young, 1962) (see Fig. 137).

Nerve fibres pass inwards from similarly oriented pairs of rhabdomeres into the optic lobes where they synapse with neurones whose dendritic fields may also be primarily oriented along horizontal or along vertical axes. It seems clear that the rectangular orientation of units in the visual

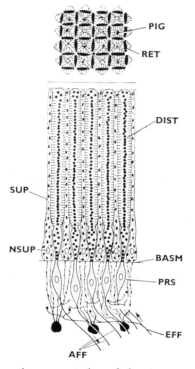

FIG. 137. Diagrammatic representation of the structure of the retina of *Octopus*, above in tangential section, and below in radial section. Each retinal cell comprises a proximal, nucleate portion (PRS) and a distal rhabdome-carrying segment (DIST). The rhabdomes are separated by supporting cells (SUP) the nuclei of which (NSUP) lie close to the basal membrane (BASM). In tangential section (above) the retinal cells (RET) are seen to be arranged in a rectilinear fashion to form rhabdomes, with granules of pigment (PIG) in the centre of the rhabdome. Afferent fibres pass from the bases of the retinal cells to the optic nerves, and efferent axons (EFF) supply the retina. For interpretation of other lettering, see p. 448. (Originally published in Young, 1962 *Phil. Trans.* B, **245**, 2, fig. 1.)

system is of great importance to the animal in the interpretation of the visual data received. *Octopus* can be taught to discriminate between certain geometrical figures, e.g. between two identical rectangles one of which is disposed vertically while the other is placed horizontally. In contrast, *Octopus* cannot distinguish between the same two rectangles when they are placed obliquely, one leaning to the right and the other leaning to the left. These, and other capabilities and limitations in perception by *Octopus*, will be discussed in more detail in another section of this chapter in connection with the ability of *Octopus* to learn and to remember lessons. For the present it is necessary to ponder over the limitations in the ability of *Octopus* to discriminate between figures, as indicated above, and to seek to rationalise these.

From our own subjective experience we appreciate that vision does not depend solely on the degree of perfection of the optical system, but also involves experience and judgement. Due to lack of experience, of vocabulary, of concepts, and of powers of judgement, a very young child could not be expected to extract as much information regarding its environment as it could years later as an adult. An artist, an engineer, and a naturalist, all having good sight and all looking at the same scene, would all presumably *see* the same things, but would almost certainly *perceive* some different things. By virtue of their specialist training and personal interests their eyes would pick out, and their minds would classify, certain items not noticed by other observers. Using both eyes we experience the benefits of stereoscopic vision and can judge the shapes of objects and their relative distances. If we mask one eye we still seem to possess the same powers when viewing familiar scenes, due to judgement borne of long experience, but this is illusory as can easily be demonstrated by attempting to pour a fluid into a small vessel while one eye is masked. By analogy with a simple camera we can readily imagine the lens of the eye causing a real, inverted image of the environment to be formed on the retina at the back of the eye. We know that the retina contains immense numbers of visual elements very closely packed together. By analogy with the stippled illustrations seen in the daily papers we can envisage how the inverted image could be recorded, each visual element contributing a single light or dark dot, the innumerable very closely packed dots blending to build a coherent picture or representation of the external scene. Acuity of vision would depend on the optical accuracy of the lens and closeness of spacing of the retinal cells. It is probable that the eyes of octopods and decapods operate optically in much the same way, though the accuracy of their vision through water is doubtless inferior to our own vision in air.

However, we do not see the inverted images which we believe to fall on our retinas. We become aware of the nature of our surroundings by virtue of analysis in the brain of the information supplied to it by the retina, and as we have seen above, our awareness of our surroundings depends very much on the nature of our personal experience and training. Our visual perception is enriched in proportion to the relevant enrichment of our vocabulary and concepts. When we are shown a crab or a rectangular figure, we recognise it conceptually as well as seeing it visually. Needless to say, a very small child would recognise neither, and an illiterate savage would certainly not recognise the rectangle.

When we wish to consider what an octopus or a cuttlefish may perceive in nature, or when confronted with some special experimental situation, we obviously have to go behind the inverted image on the retina and try to discover what resources the animal may have for analysing this image and what criteria may be used in the analysis. Considerable success has been gained in experiments designed to provide information on the ability of octopuses to discriminate visually between contrasted shapes presented in their field of view, on the preferences they may have for one shape as compared with another, and on the principles on which these discriminations and these preferences may depend. This line of investigation will be discussed at a later stage in this chapter.

Familiar members of the Dibranchiata are raptorial carnivores which attack their prey at sight. Such an attack may be initiated by an inherited visual releaser mechanism, e.g. an automatic response to movements of small objects in the immediate vicinity. In addition, however, cephalopods such as *Octopus* and *Sepia* are capable of learning, i.e. of modifying their behaviour as the result of previous experience. *Octopus* is a convenient animal for experimental investigation of feeding behaviour; if a few bricks are suitably arranged at one end of an aquarium the octopus will adopt the space under the bricks as a den and will occupy this when not foraging; experimental material can then be presented at the opposite end of the aquarium tank at a constant distance from the octopus. Again, when fed upon crabs an octopus is apparently almost insatiable, and no irregularities in behaviour during experiments need arise as a result of fluctuations between hunger and satiety.

After a few days of captivity in an aquarium, during which time crabs are supplied regularly as food, in most cases an octopus adapts itself to these conditions and will attack the crab within 5 seconds of presentation. Such an octopus is then in a suitable condition for experimentation; it can be trained to recognise either of two contrasting situations, one favourable and the other unfavourable, and to react appropriately to each of these situations. In the earlier experiments of this nature the favourable situation was the presentation of a crab on the end of a piece of thread. If the octopus attacked the crab within a certain interval of time it was rewarded by being allowed to eat the crab. Alternatively, the octopus was presented with an unfavourable situation, namely a crab and a small white square plate. If the octopus attacked the crab + plate it was given an electric shock—8–10 volts (a.c.)—which was sufficiently strong to serve as a punishment, but not strong enough to damage the octopus. The shock was comparable with the stings an octopus would receive from an anemone if it were to attack a hermit crab with a commensal anemone (Boycott and Young, 1955). The octopus was presented with either of these situations, in random order, at regular intervals. Records were kept as to whether the octopus attacked the situation presented and, if so, how many seconds elapsed between presentation and attack. The results of these investigations showed that the octopuses could distinguish between the two situations, and quickly learned to adopt the correct behaviour, i.e. to attack the crab when this was presented alone, and to refrain from attacking the crab + white plate. In other words, an octopus can remember and is able to learn to modify its behaviour according to previous experience.

In a second type of experiment the octopuses were presented with two contrasted abstract situations, i.e. situations neither of which included a food object. The two situations presented to the octopus might be a geometrical figure such as a rectangle, presented in the vertical or in the horizontal position; alternatively the octopus might be presented with two contrasted geometrical figures such as a circle and a square (see Fig. 138).

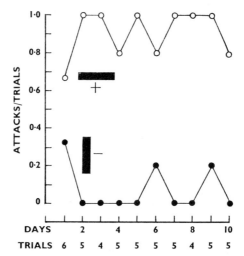

FIG. 138. *Octopus vulgaris* can discriminate between contrasted figures such as a horizontal rectangle and a vertical rectangle. The octopus has been trained to attack the "positive" figure (horizontal rectangle) by being rewarded with food after each attack, and it has learned not to attack the "negative" figure (vertical rectangle) as a result of punishment with an electric shock on each occasion that it attacks the negative figure. The graphs record the behaviour of a single octopus, the number of attacks divided by the number of trials being plotted for the positive figure (○) and for the negative figure (●). (Originally published in Boycott and Young, 1956, *Proc. zool. Soc. Lond.* **126,** 523, fig. 15.)

There is greater prospect of attack on the figures if these are kept moving, and under these conditions the octopuses are likely to learn more quickly than if the figures are stationary. In each case one situation was treated as "positive" and the octopus was trained to attack the positive figure. This was done by rewarding the octopus with a crab or a small piece of fish every time it attacked the positive figure within a specified period after presentation. Conversely the other situation was "negative" and the octopus was punished with an electric shock each time it attacked the negative figure. In many cases the octopuses were able to discriminate efficiently between the positive and the negative situations, and to adopt appropriate action. The octopus learned to associate the positive figure with a subsequent reward, and the negative figure with a subsequent punishment. This is a good example of a conditioned response. Such experiments have been the bases for investigations on two contrasted lines. One line of investigation has been towards an understanding of the mechanism or mechanisms whereby the octopus analyses images formed on the retina, and thereby discriminates between different situations. The other line of investigation has been with a view to interpreting the neurophysiological processes involved in learning and memory, and with a view to identifying the functions of the various lobes of the brain. Both lines of investigation have been rewarding.

If an octopus is presented at random with either of two abstract situations, as described above, and is rewarded for attacking one situation (with a piece of food) and is punished for attacking the other situation (with an electric shock) the octopus may be able to learn to discriminate between the two situations. If so, the octopus will tend to attack the favourable, or "positive" situation, and will tend to refrain from attacking the "negative" situation. The animal is deemed to have performed "correctly" if it attacks the positive situation within 20 seconds, and also if it fails to attack the negative situation for the same time interval. The behaviour of the octopus during these trials can be recorded objectively as the percentage of correct responses. If the octopus is presented with only one situation at a time, either the positive or the negative situation, in random order, then it is liable to improve its performance gradually until it reaches a certain limit, beyond which the percentage of correct responses will not increase with further training. The measure of this asymptote of performance may be taken as an indication of the ease or difficulty with which the octopus discriminates between the two situations. If the discrimination is an easy one the octopus may achieve 80 per cent correct responses, or even more. If the discrimination is a very difficult one, the percentage of correct responses may not differ significantly from that expected by chance, i.e. 50 per cent, in which case there is no evidence to suggest that the octopus can distinguish between the two situations. The powers of visual discrimination of octopus can be determined by testing fairly large numbers of octopus in this way, with a variety of geometrical shapes. As a result of such investigations Sutherland (1957) has put forward a theory as to the nature of the mechanisms by which *Octopus* may analyse visual data and on the basis of which it may be able to discriminate between contrasted situations.

It is suggested that the image falling on the retina is analysed according to the degree of excitation of the dendritic fields of a variety of neurones in the optic lobes. Each of these neurones is deemed capable of firing in response to stimulation of an area of retinal receptors by an image with certain spatial attributes. A group of such neurones with identical functions could be termed an "analyser".* There may be a number of different types of analyser, each capable of recording the degree of occurrence of one particular type of feature in the visual field, e.g. vertical extent, horizontal extent, extent of perimeter in relation to area, degree of brightness, etc. The first stage in the process of discrimination may be the selection of the appropriate analyser. The concept of the selection and "switching-in" of the correct analyser can explain some of the observed paradoxes in the processes of learning in animals (Sutherland, 1964, 1964a).

As mentioned above, due to the reflex association between the statocysts and the eyes the latter are constantly orientated with the slit-shaped pupils in the horizontal plane; the rhabdomeres in the retina are predominantly arranged on a rectangular grid, either horizontally or vertically; furthermore it has been shown that behind the retina there are neurones whose dendritic pickup fields are elliptical in shape, and the long axes of their dendritic fields tend to lie either vertically or horizontally, with a preponderance in the latter plane (Young, 1960a) (see Fig. 139). It seems clear that recognition of the horizontal plane is of the greatest importance to the animal. On the basis of these functional and anatomical facts, Sutherland (1957, 1959) suggested that there is an initial analysis of the images falling on the retina according to the degree of horizontal extent and of vertical extent of these images. This could be achieved by a rectilinear arrangement of sensory cells in the retina and a backing of neurones, each of which is specific to a horizontal row of retinal cells, or to a vertical column of retinal cells, and which serves to count the number of sensory cells along that row which have received excitation. The degree of excitation of these analysing neurones would be in proportion to the extent of the image in the horizontal and in the vertical planes, and on the basis of this information the animal could perhaps succeed in discriminating between two contrasted shapes presented in its visual field. Although more recent work may suggest some errors of detail, and some imperfections in prediction from this theory, the postulation retains much of its original interest: on the basis of this theory it has

---

* It is not suggested that such an analyser is a recognisable morphological unit. An analyser should be conceived as an array of neurones of identical function which are dispersed widely through the substance of the optic lobes and which are capable of responding to the same kind of stimulus received at many different parts of the retina.

been possible to make accurate predictions as to the ability of octopuses to discriminate with ease, or with difficulty, between selected pairs of shapes. The theory has therefore been of value in leading to further advances in knowledge and understanding of visual perception in animals. According to this original theory the image of a horizontal rectangular figure would

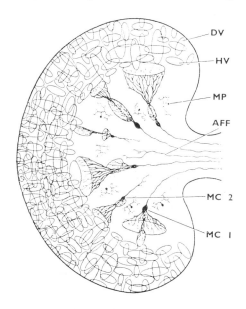

FIG. 139. Diagrammatic thick horizontal section through the optic lobe of *Octopus*, indicating the probable arrangement of dendritic fields of some bipolar cells in single layers of the plexiform zone. Some dendritic fields are oriented vertically (DV) while others are oriented horizontally (HV). Afferent fibres (AFF) pass out from the optic lobe via the optic tract. For interpretation of other lettering, see p. 448. (Originally published in Young, 1961, *Biol. Rev.* **36**, 60, fig. 15.)

provide a high output over a narrow sector of neurones surveying the horizontal rows of retinal sensory cells, and it would provide a low output over a wide sector of neurones surveying the vertical columns of retinal sensory cells (see Fig. 140). This theory of visual perception was capable of being tested, and, when tested, generally favourable results were obtained.

Geometrical shapes were devised which were identical in area but which differed in horizontal and vertical extent, and it was found that octopuses were able to discriminate between them; clearly in such cases the area of the shape was irrelevant (Sutherland, 1961). Similarly, octopuses could discriminate between shapes which had the same total length of outline and the same area, but which differed in horizontal and in vertical extent. Various other possible bases of discrimination, e.g. the presence or absence of

**RATE OF FIRING
FOR OUTPUTS FROM COLUMNS**

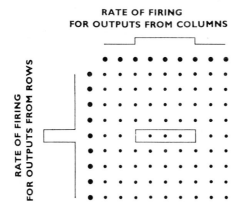

**RATE OF FIRING
FOR OUTPUTS FROM ROWS**

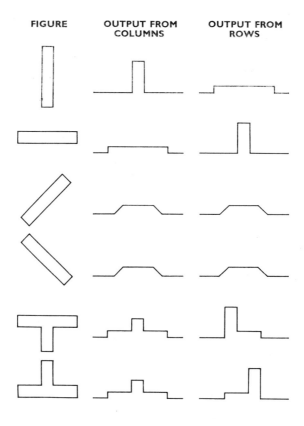

| FIGURE | OUTPUT FROM COLUMNS | OUTPUT FROM ROWS |

straight edges, or of angles, the ratio between the square root of the area and the length of the circumference, etc., had also been found irrelevant to the ability of octopuses to discriminate between shapes (Sutherland, 1959a, 1963a).

Sutherland devised a number of geometrical shapes of similar area and determined the horizontal projection and the vertical projection of each shape. These shapes fell into two contrasted groups, all of the shapes within each group having identical horizontal and vertical projections. The theory predicts that the octopus should be able to discriminate between any two shapes which belong to different groups, because these shapes would on analysis be found to possess different horizontal and vertical projections. This prediction was found to be correct. According to the theory octopuses should be unable to discriminate between any two shapes in the same group, since they would on analysis be found to have the same horizontal and vertical projections. This prediction was found to be generally correct, but the octopuses *were* found to be capable of discriminating—with considerable difficulty—between shape ⌢⌢ and shape ⌢ (Sutherland, 1959). Again, octopuses were able to discriminate easily between patterns of reduplicated alternating black and white rectangles placed either horizontally or vertically, in spite of the fact that the two patterns were so arranged as to have identical horizontal and vertical projections (Sutherland *et al.*, 1963). In such cases there obviously must be some secondary basis of discrimination.

Young (1960a) found that among the neurones in the optic lobes which have elongated dendritic pickup fields, a slightly larger number had the long axis of the dendritic field corresponding to the visual horizontal plane than to any other plane. On the basis of this neuro-anatomical observation

FIG. 140. Diagrammatic representation of a mechanism whereby *Octopus* may be able to differentiate between different objects in its field of view. Above, Analysis of the image of a horizontal rectangle on the retina of an octopus. The retinal elements are arranged in a rectilinear grid, as indicated by the columns and rows of dots. Circles along the upper border represent cells each of which is concerned with data from a single column of retinal cells, and the rate of firing of such cells is indicated by the figure above. Similarly, the row of circles on the left represent cells each of which is specific to a single horizontal row, and the figure on the left indicates the rate of firing of these cells by the image falling on the retina. Below, A number of geometrical figures are shown in the left column, and in the central and right-hand columns are shown the outputs from the cells analysing the vertical columns and the horizontal rows of retinal cells respectively. In some cases it is seen that on this hypothesis two related figures are easily discriminable on this theory, e.g. the horizontal and the vertical rectangles; in other cases the two figures are indiscriminable according to this theory, e.g. the rectangles inclined to left and to right. (Originally published in Sutherland, 1957, *Nature, Lond.* **179**, 12, figs. 1 and 2.)

Sutherland predicted that it should be easier for octopuses to discriminate differences in length in the horizontal plane than similar differences disposed in the vertical plane. Experiments were designed to test this and the results upheld the prediction (Sutherland, 1960, 1961). Sutherland and Muntz (1959) found that it was very difficult to train octopuses to discriminate between stationary shapes, and that octopuses were particularly attracted by objects which moved in the direction of their long axes. These observations are obviously in close accord with the normal experiences of the animals, for their prey will move in this way. This is even true of crabs, for these will appear to the octopus to be elongated transversely, and they scuttle or swim sideways.

The capabilities of octopuses in visual discrimination between different geometrical shapes have also been investigated in another type of experiment with the use of transfer tests. In such an experiment groups of octopuses are trained to discriminate between contrasted shapes, for example a square and a cross of equal area, half of the animals being trained with the square as the positive figure and the other half with the cross as the positive figure. The training having reached asymptote of performance with these figures, other geometrical shapes of similar area are substituted to see whether they are treated like the positive or like the negative figure. When tested with these transfer figures the octopuses were neither rewarded nor punished. As the result of these tests it was possible to arrange all the geometrical transfer shapes used in a fairly consistent order between the two training shapes—the cross and the square—at the two ends of the series. To generalise, at one end of the series the shape has a relatively large perimeter in relation to its area, and Sutherland described this as an "open" shape. At the other end of the series the shape has a perimeter which is small in relation to its area, and Sutherland describes this as a "closed" shape. The series of shapes studied have therefore been arranged serially along a dimension which Sutherland termed "open–closed", and it appears that octopuses exhibit natural preferences in respect to the position a shape may occupy along this dimension. In short, octopuses tend to prefer "open" shapes such as upright or oblique crosses to "closed" shapes such as circles and squares. This corresponds with their tendency to feed on crabs, which exhibit many projecting angles (legs). The existence of an hierarchical system of preferences for different shapes may provide an indication of a further quality on the basis of which images on the retina may be analysed, but the present investigations seem to rule out differences in area, differences in the length of the perimeter, and differences in brightness of the object (Sutherland, 1962).

Octopuses were able to discriminate between diamond-shaped plaques which were either a uniform grey colour or carried horizontal or vertical black stripes. By varying the thickness of the black stripes Sutherland (1963b) was able to determine the narrowest stripe which the octopus could discriminate from the uniform grey plaques, and the distance from the plaques from which the octopus could launch an attack. By this means he was able to make an estimate of the visual acuity of the octopus; he judged the minimum separable for vertical striations to be about 17′ of arc, whilst Young (1960a) had estimated the size of a single retinal unit in the octopus to correspond to a visual angle of 1′18″ of arc. There is a striking difference between these two values, and it seems probable that the minimum separable is not related to the size of the sensory cells of the retina but to the requirements for firing the neurones with orientated dendritic pickup fields in the optic lobes. In contrast the minimum separable figure for humans, about 30″ of arc, is of the same order as the size of the retinal receptor cells.

There should not necessarily be only one basis for analysis of visual data; there may be a number of alternative types of analyser, and the processing of visual data may take place in two stages: firstly, the selection of the appropriate analyser; secondly, analysis of the outputs of these analysers. On the basis of analysis according to the nature of the horizontal and vertical projections, certain geometrical figures should not be discriminable by the octopus. However, it has been shown that some such shapes are in fact easily discriminated in spite of the fact that their horizontal and/or vertical projections are identical. The simplest solution is that there are alternative systems on which visual data can be analysed.

Hubel and Wiesel (1959) have shown that in the cat there are neurones in the striate cortex which have elliptical receptive fields. In some of these the receptive field comprises a central excitatory strip and paired lateral inhibitory strips. Such a neurone is only activated maximally by a visual image which is so orientated that it is superimposed on the central excitatory strip, and misses the flanking inhibitory areas. It is not impossible that the orientated dendritic pickup fields in the optic lobes of the octopus may operate on a comparable basis. It might seem improbable that cephalopods and mammals should resemble one another in this way, yet it has already been shown that goldfish and rats have behaved comparably to octopuses in the ordering of transfer shapes (Sutherland, 1963).

Visual Learning

Experiments involving the responses of octopuses to contrasted visual situations have been employed in two complementary lines of investigation, the one concerned with the mechanisms whereby the octopus analyses images formed on the retina, and the other with the processes involved in learning and memory. We have already considered the former line of investigation, and we now turn to the latter.

Let us consider an octopus which has been trained under experimental conditions in the laboratory to discriminate between two contrasted visual situations, and which is now resting in its den, passively observing the scene. An image of the contents of the aquarium is formed on the retina of each eye, and information is being passed from the majority of retinal receptors into the optic lobes. In the absence of any significant movement there is no feature in the scene which arouses the attention of the octopus, and only "background" information is being passed from the optic lobes into the centres of correlation in the brain, which comprise the vertical, superior frontal, and subvertical lobes.

Now suppose that into the aquarium there is introduced a crab and a white square, a situation which the octopus has previously been trained to leave alone. The movement will affect very many of the retinal receptors, the occurrence of movement will be recorded by many of the widely dispersed deep-seated neurones which are classifiers for the occurrence of movement, and information will be passed to the correlation centres of the brain that there is a movement in the immediate vicinity of the octopus. In response to this information the octopus now "shows attention", and is no longer surveying the scene passively. The crab is a familiar object and a natural source of food, whilst the white square is a relatively unfamiliar object recently found to be obnoxious and best left alone. Classifying neurones in the optic lobes will collectively detect the principal attributes of this total visual situation. These classifiers must be individually capable of either of two outputs; one such output promotes attack while the other suppresses attack. Young (1965) has suggested that the probable mechanism in the selection of the appropriate neural pathway is the blockage of the inappropriate channel; this may be effected by metabolic changes in the adjacent small neurones leading to inhibition and thereby blockage of the unwanted neural pathway.

The optic lobes comprise an elaborate meshwork of neurones which permit interaction of information from different parts of the retina, thereby enabling the identification of any object seen according to its principal visual attributes. The optic lobes also receive tactile and gustatory information from other brain centres, and are the site of association between earlier visual stimuli and subsequent tactile and/or gustatory information (Boycott and Young, 1955). The association of a visual situation with either a reward or a punishment is "read in" to the memory by the blockage of the inappropriate neural channels in the manner indicated above, and this promotes the adoption of the "correct" behaviour on a subsequent similar occasion. Classifiers in the optic lobes will therefore report the visual situation on the basis of previous total experience, in terms of "to attack" or "not to attack". According to the relative strengths of these two contrasted types of coded information the situation is summed up in the light of previous experience and a command is sent to the higher motor centres, either endorsing the natural tendency to attack the crab or countermanding this tendency.

The first phase in the learning process must therefore involve detection of which type of classifier is providing the appropriate discrimination. It is then necessary to "switch in" this type of classifier, i.e. to discount information submitted by all other types of classifier. This explains the initial fall in efficiency when an octopus trained on one discrimination is given a new problem requiring discrimination on a different basis.

The principal parts of the brain of the octopus comprise astronomically large numbers of neurones; the action taken by an octopus in solving problems is sometimes correct and sometimes incorrect, the "decision" to attack or not to attack being a matter of statistical probability (Young, 1955). The learning process concerns the imprinting of changes in the probability of attack under given circumstances. When an octopus attacks an unfamiliar object and as a result of that attack it obtains benefit in the form of a meal, there is an increased probability of the same kind of object being attacked when it is next encountered. It is necessary to conclude that some kind of record of the experience has been made in the central nervous system, and that this is available for comparison at some later time. This record underlies the ability to learn and to remember (which, of course, are two aspects of the same general function). The question before us is: What is the nature of this record?

The first point of importance to be established is that the record of an event cannot be confined to a single neurone or even to a number of neurones arranged in a single series, but must be quite widely diffused through the substance of the brain. The reason for this conclusion is simple—the octopus attacks an object with one eye in a leading position, and there will be an image of the object attacked in that eye only. Yet on a subsequent occasion the animal is likely to behave in an appropriate manner even if it attacks with the other eye in the leading position. The two optic lobes are placed in communication by a large commissure, and what is learned with one eye is able to spread to the neural tissues of both sides of the brain. Again, when an object such as a swimming crab moves into the field of view of an octopus, and is then attacked by the octopus, the image of the object will move across the retina and will grow rapidly in size so that a very large number of retinal receptors will be stimulated by it. We have to conclude that the record of a visual event must involve a very large number of neural pathways through the brain, not a single pathway. The more widely is the record diffused through the brain as the result of repeated experience, the more firmly is the memory established, and the more certain is the adoption of the appropriate response on subsequent occasions. When an octopus under training has reached asymptote of performance on a particular problem it may reasonably be assumed that all available neural pathways have acquired the record.

The octopus can express itself in the main in one of only two ways in the situations to which it is exposed experimentally—it can attack, or refrain from attacking the object seen. An attack is launched in response to commands from the higher motor centres of the brain, but these higher motor centres are themselves subordinate to the correlation centres of the brain and the latter may permit or may inhibit an attack. This suggests that the record of a visual event concerns facilitation of one or other of two general pathways via the correlation centres to the motor centres. One such pathway permits and even reinforces the command to attack on what has in the past proved to be a profitable situation. The other general pathway inhibits the launching of an attack.

If an octopus were to attack anything and everything on the slightest provocation it would certainly waste an unnecessary amount of energy, and would probably soon get into serious difficulties. Conversely, the animal would never get any food if it were to recoil from the very slightest discomfort. To strike a happy medium the attack is presumably not launched unless the higher motor centres are receiving considerable and repetitive favourable information. Likewise a pending attack will only be withheld if an impressive barrage of danger signals is being received. The intensity of the barrage of signals to the higher motor centres will depend on two factors—the number of retinal receptors involved, which is a measure of the size and proximity of the object seen, and the frequency of the signals. The latter depends primarily upon persistence of the vision, but it could be further augmented if there were a device in the brain causing repetition of the signals. Such multiplication of the signals could be achieved by the presence of cyclical arrangements of neurones within which excitation would tend to reverberate and from which collateral fibres would carry the frequently repeated signals to their destination. It has been suggested, therefore, that a record can be established in a memory store in the octopus brain when reverberating excitations have been set up in a sufficient number of neural pathways. It is implied that as these reverberating circuits gradually fade so will the memory fade with them, but reinforcement of the memory can be effected by further repetition of the relevant experience.

A normal octopus which has been trained to discriminate between two visual situations, one "positive" and the other "negative", can remember the lesson for many days and will behave appropriately after an appreciable interval during which it has not seen either situation. This long memory seems to be destroyed if the animal then suffers an operation in which the greater part of the vertical lobe is removed (Boycott and Young, 1955, 1955a) or if deep incisions are made which effectively isolate the vertical lobe from the optic lobes (Figs. 141 and 142). This indicates that the vertical lobe is deeply involved either in the learning process or in the storage of the memory, or in "reading-out" from the memory store. An operation involving removal of most of the vertical lobe does not affect the general behaviour of the animal. If such an operated animal is shown the negative situation, e.g. the crab plus white square, it will fling itself upon the crab in spite of all previous experience and will receive punishment by electric shock. Having received this punishment the operated octopus will withdraw, continuing to watch the crab plus white square warily. As long as it retains

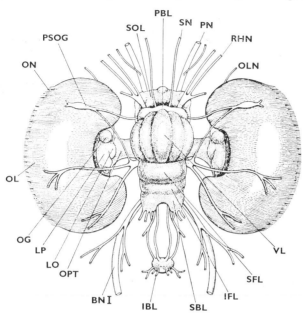

FIG. 141. Dorsal view of the brain and optic lobes of *Octopus vulgaris*. The anterior end of the brain is lowermost in the figure. Note the optic gland (OG), which lies above the distal end of the optic tract (OPT), and which is concerned with the maturation of the gonad. For interpretation of other lettering, see p. 448. (Originally published in Young, 1961, *Biol. Rev.* **36,** 36, fig. 1.)

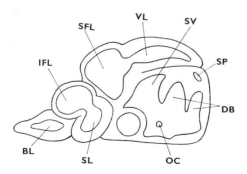

FIG. 142. Sagittal section through the brain of an octopus, indicating the relative positions of the various lobes. For interpretation of lettering, see p. 448. (Originally published in Wells, 1964, *Quart. J. exper. Psychol.* **16,** 161, fig. 3.)

the image of the negative situation continuously on its retina the octopus may refrain from further attack; it may even carry the memory not to attack this particular situation for a few minutes after removal of crab plus white square, up to a maximum of about 30 minutes, as may be demonstrated by re-presenting the negative situation at intervals of 3–5 minutes (Boycott and Young, 1955, 1955a). The crab is a very familiar item of food and the memory of this is too deeply imprinted to be dislodged by training experiments which associate the crab with pain. However, the operated octopus can still discriminate between two abstract visual situations after removal of most of the vertical lobe, particularly if no rewards are given. If the operated animal is rewarded after correct response, this encourages the animal to attack at the next presentation, regardless of the situation seen. If the operated animal views both the positive and the negative situation *simultaneously*, and no rewards are given, then it tends still to show correct responses. The vertical lobe, therefore, cannot comprise the whole of the memory store. It seems that the operated animal still retains its memory, but that in the absence of the vertical lobe it is no longer easy for the animal to "read out" information from the memory (Boycott and Young, 1957; Young, 1965.)

As mentioned earlier, it is necessary for the animal to be able to bridge the gap between the initial receipt of visual information and the subsequent arrival of gustatory or pain signals, in order to make the necessary correlations between these. This need is met by the existence of a short-term memory system in the tissues of the optic lobes. Such short-term memories fade rapidly unless reinforced through the mediation of neurone circuits passing through the vertical lobe.

There is a disadvantage in the type of experiment in which "crab" and "crab plus white square" are the contrasted visual situations, in that the crab is a component of both situations. It has been shown that octopuses are also able to discriminate between two abstract situations, e.g. horizontal rectangle versus vertical rectangle, in which they receive their reward in the form of a piece of fish subsequent to attacking the "positive" figure and the usual electric shock as punishment for attacking the negative figure. There is an apparent anomaly in the behaviour of octopuses which have been trained to make this discrimination and which have then suffered severe lesions to the vertical lobe of the brain. Such animals are under no compulsion to attack the figure which they had previously learned to treat as "positive", whereas octopuses trained with crabs invariably attacked the crab when this was presented to them after operation. There seem to be two possible alternative explanations for this anomaly.

1. The crab is a natural source of food for octopuses, and it is possible that octopuses inherit a reflex mechanism initiating attack on crabs of a certain size relative to that of the octopus. If this were true, the octopus

does not have to learn to attack crabs, but can be taught, by painful experience, to refrain from attacking crabs under certain circumstances. Octopuses are very small when they hatch from the egg, they pass the first part of their life in the plankton, and there would be great technical difficulty in rearing octopus hatchlings to the point where such a proposition could be tested.

2. The octopus has fed successfully upon crabs for such a very long time that the memory ensuring attack upon a crab has become consolidated in some way in the central nervous system to such an extent that the animal has acquired in its own lifetime a visual releaser mechanism ensuring attack upon a crab at sight.

So far as the mature octopus is concerned there is no significant difference between the two possibilities; in either event the attack upon the crab is automatic. There is no equivalent compulsion to attack an abstract figure which the octopus has been taught to treat as "positive".

It remains a matter for speculation whether in the octopus all well-established memories must remain based indefinitely upon continuous reverberating neurone circuits or whether important memories, once firmly established, can be transferred to a simpler and more economical long-term storage mechanism, thereby releasing large numbers of neurones for the learning of new lessons?

The brains of cephalopods and of vertebrates are built on very different plans, but there are certain points of similarity. Thus in the higher vertebrates the brain comprises astronomical numbers of neurones, there are correlation centres where information from different sources can be associated, and the linking of neurones at random permits diffusion of information through large volumes of brain substance, and also permits infinite diversification of association of information received. The results of experiments have suggested that in such diverse animals as octopuses, goldfish, rats, and monkeys, there are common features in the mechanisms of analysis of visual situations (Sutherland, 1963). This suggests the possibility that what is learned of the mode of functioning of the cephalopod brain may be at least partly relevant to that of the mammalian, and of the human brain. Although there are basic architectural differences between the brains of cephalopods and of mammals, in both cases these brains are computer systems that have evolved independently to solve the same kinds of problems. There may well be close analogies between the two systems. J. Z. Young (1956) has drawn attention to the benefit that may be gained from the further analogy between the modes of functioning of such brains and of man-made computers. The use of computers has necessitated the development of new technical terms, the use of which is necessary for the proper understanding of the mechanisms of computer systems. The application of the terminology of this tool-language to the study of the central nervous systems of animals may help to clarify neurophysiological problems and lead to their solution (no serious attempt has been made to use this tool-language extensively in the present chapter, as this would have necessitated lengthy explanatory digressions).

Tactile Discrimination and Learning in the Octopoda

When the Octopoda diverged from their pelagic ancestry and adopted a benthic mode of life they began to use their arms for tactile exploration, and this naturally threw an additional burden on the central nervous system. Development of the capacity for tactile learning involved the association of representations of tactile stimuli with those of subsequent reward (food) or pain. Enlargement of the relevant region of the supra-oesophageal part of the brain supplied the additional nervous tissue required, and thereby led to the formation of the sub-frontal lobe. It is notable that such a lobe has not developed in the Decapoda, which do not indulge in the same way in tactile exploration and which therefore do not require such a specialised section of the brain (Wells, 1959a).

The arms of octopuses are highly distensible and, since they lack joints, they are capable of infinite diversification of orientation. Accordingly there can be no frame of reference in the central nervous system of an octopus by means of which the animal could compute the relative positions of particular suckers, or even of certain sections of the arms, and so judge the shapes of objects handled. Thus an octopus is unable to discriminate by means of its tactile sense between a sphere and a plane surface, provided that the radius of curvature of the sphere is not too small and the rims of the suckers of the octopus are not distorted. When an octopus picks up an object the arm will be stretched to greater or to lesser extent according to the weight of the object. Stretch receptors sited in the muscles of the arm supply the proprioceptive information required locally to correct undue stretching of the arm, but this information is not passed up the arm to the brain. Thus the octopus has been found to be incapable of discriminating between two perspex cylinders of identical dimensions, one of which had been heavily weighted internally (Wells, 1961).

The situation in an octopod must be radically different from that in an arthropod, or in a tetrapod, in which the possession of jointed limbs makes the development of a proprioceptive information supply to the central nervous system a meaningful possibility. The ability of man to manipulate tools, to use a typewriter, and to play the piano, depends upon the possession of jointed limbs and of a finely developed proprioceptor system. This manipulative ability, in turn, has played a significant part in the development of intelligence in man. In the octopus, on the other hand, since the arms lack articulated skeletal supports, proprioceptive information is solely of local utility; it is not transmitted to the higher centres of the brain, and so it cannot be utilised by the octopus in discriminating between objects that are inspected by touch. The general plan of the nervous system of the Dibranchiata was laid down to meet the needs of a pelagic mode of life in which there was no need for proprioceptive feed-back to the higher centres of the brain; this condition persists in the benthic octopods (Wells, 1963). Although the octopus has a remarkably well developed brain, due to the lack of proprioceptive feed-back there is no possibility of the development of manipulative skills by octopods, and this sets a limit to the potential development of intelligence in the group.

Since in the octopus proprioceptive information can only be utilised locally within the individual arms, it follows that tactile discrimination is necessarily limited to textural features which can be detected by the suckers. The suckers on the arms of octopuses contain a variety of sensory cells some of which, lying in the epithelium, may perhaps serve as chemoreceptors. Other sense organs, lying at the base of the epithelium of the rim of the sucker, may detect distortions caused by irregularities in the surfaces of objects grasped by the sucker (see Fig. 143). By means of such tactile receptor organs the octopus can discriminate between objects which differ sufficiently in surface texture. This capacity to discriminate by touch is obviously of the greatest importance to a benthic hunter which may seek out its prey in situations of poor illumination in sub-littoral recesses or grottoes. Experiments on octopuses which have been blinded have indicated the types of discrimination which octopuses can make solely by touch, and the limitations of this system of information. These experiments have also shown that such tactile information is passed up the arms to the brachial ganglia and thence to the supra-oesophageal lobes of the brain, and that there is a tactile memory system which is basically comparable with the visual memory system which has already been described above.

It has been shown that octopuses can discriminate solely by touch between a smooth and a grooved perspex cylinder of identical dimensions. They can also discriminate between cylinders which differ only in the proportion of grooved to smooth surface, e.g. they can distinguish a cylinder the surface of which bears 30 per cent grooving from one which only bears 14 per cent grooving. However, they cannot discriminate on the basis of orientation of grooving, e.g. between cylinders which bear longitudinal grooves and cylinders with transverse, annular grooves, nor can they distinguish from either of these a cylinder which is grooved both longitudinally and transversely (Wells and Wells, 1957). This failure to distinguish orientation of pattern is entirely due to the absence of any proprioceptive input to the central nervous system. Inability to distinguish orientation, or pattern, of surface texture is of no disadvantage to the animal; when an exploring tentacle encounters a prospective food object in some gloomy recess in the rocks, and passes it under the inter-brachial web for oral inspection, or when the animal flings itself upon some passing crab, the suckers are attached at random and without any attempt at uniform orientation. Orientation of pattern of surface texture of foreign bodies is therefore devoid of significance to the octopus. The

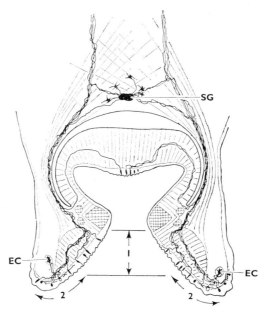

FIG. 143. Section through a sucker of an octopus, showing the site of the sub-acetabular ganglion (SG) and the regions in which various sense organs are found. Two types of sense organ are most abundant in the inner part of the sucker disc (region 1); three other types of sense organ are most abundant on the rim of the sucker disc (region 2). Comparatively small numbers of large encapsulated sense cells lie deep-seated on the outer side of the sucker rim (EC). (Originally published in Wells, 1964a, *J. exper. Biol.* **41**, 443, fig. 6.)

only significant information is the degree to which the rims of the suckers are distorted by irregularities in the surfaces of objects grasped. The octopus can distinguish between a sphere and a cube only because when grasping the latter some of the some of the suckers are likely to be distort-ed by contact with an edge or a corner of the cube. The animal is less efficient in discriminating a sphere from a cube the corners of which have been rounded off, while it treats a slender rod as equivalent to a cube and different from a sphere since the rod causes maximal distortion of the attached suckers (Wells, 1964, 1964a).

When an octopus has been blinded it takes up a position with its arms well spread out, and it is possible to present an experimental object repeatedly to the same arm. Initially all such objects are passed under the interbrachial web for oral inspection, but the animal quickly learns to reject useless objects without such oral inspection. Under such experimental conditions the octopus learns within three or four trials to reject a "negative" object associated with punishment by electric shock. When an octopus has learned to give the correct response with one arm it may give the wrong response if tested on another arm, and it has to learn all over again with a different arm. This is particularly true when the training trials are only 3 minutes apart. When training trials are 20 minutes apart, and after the elapse of some time, other arms are likely to have acquired the lesson and may give the correct response (Wells, 1959). The learning process, which concerns the setting up of an association between tactile experience and subsequent pain, is brought about in a part of the brain which is directly concerned with one particular arm. Once the lesson has been learned by that arm, the record gradually spreads to the equivalent sections of neural tissue which individually stand in comparable relation with all the remaining arms. What seems curious is that the process of diffusion of the record should take so long —a matter of several hours. Wells (1959) considers whether this indicates that the diffusion of the record is based on processes of growth of the neurones involved, rather than the spread of nerve impulses through an existing network.

The majority of sensory nerve fibres pass without synapse through the lower centres of the brain, and continue to the supra-oesophageal lobes. Experiments with blinded octopuses, which have also suffered various lesions to the brain, have shown that the tactile memory system is located in the inferior frontal and sub-frontal lobes; when these lobes are extensively damaged the capacity for tactile learning and memory is correspondingly reduced. The tissue of the sub-frontal lobe is clearly divided into left and right halves, between which there are extensive connections, and there may also be further subdivisions into fields which administer the individual arms, though such postulated "arm-fields" have not yet been discovered (Wells, 1959, 1959a). The inferior frontal lobe is divided into three parts, of which the median lobe is not essential to the process of learning, but in its absence there can be no transfer of the record from one arm to another (Young, 1965).

The Olfactory Sense in Molluscs and Olfactory Learning in the Octopoda

Various gastropods have been shown to possess a very acutely developed olfactory sense, e.g. the limpet *Patella vulgata* can distinguish differences in the salinity of water with which it is splashed, and responds appropriately (Arnold, 1957); the larvae of the dorid *Adalaria proxima* will only settle and metamorphose in the presence of *Electra pilosa*, and evidently respond to some chemical emanation from this polyzoan (Thompson, 1958); experiments with choice chambers have shown that various herbivorous gastropods can detect the presence of suitable seaweeds and move in the correct direction to feed on these; the piscivorous species of *Conus* undoubtedly detect the presence of their prey by virtue of a well-developed olfactory sense, while the opisthobranch *Bullia laevissima* responds quickly to the presence of carrion in the vicinity and has been shown to react positively to the presence of the chemicals trimethylamine and tetra-methyl ammonium, which are the natural products of decomposition of food substances (Brown, 1961; Brown and Noble, 1960). In some cases it has been demonstrated that this olfactory sense is exercised by a well-defined sense organ, the osphradium; in *Patella* the detection of salinity differences is effected by strips of sensory epithelium on the sides of the foot, but in very many cases little or nothing is known of the site of the olfactory organs. The recent literature has been reviewed by Kohn (1961). In all of these examples, as far as is known at present, the behaviour of the animal is an automatic response to a specific stimulus, i.e. it is a reflex action. The animal does not learn to make the correct response; it cannot be taught to respond in a particular way under experimental conditions; it has no memory store in its central nervous system (see Figs. 144–6).

Fig. 144. When touched by the tube feet of the starfish *Asterias rubens*, the prosobranch *Nassarius reticulatus* makes violent leaps with its long and muscular foot in order to escape. (Original: Dr. Gunnar Thorson's Christmas card for 1951.)

The Cephalopoda also possess a well-developed olfactory sense. In addition to a special olfactory pit, olfactory or gustatory sense cells are widely distributed in the skin of the body and of the arms, and an octopus can discriminate between objects which only differ chemically, when grasping these objects with its suckers. If an octopus is presented, on the one hand, with a living bivalve and, on the other hand, with the empty shell valves of the same species, closed and filled with wax, the octopus has no difficulty

Fig. 145. When touched by the tube feet of the starfish *Asterias rubens*, the bivalve *Cardium echinatum* escapes by convulsive movements of its enormous foot. (Original: Dr. Gunnar Thorson's Christmas card for 1959.)

in selecting the former and rejecting the latter. Wells (1963) experimented on blinded octopuses to determine their chemo-tactile capabilities, and found that they could detect substances in solution in sea water at concentrations 100 times as dilute as those detectable in distilled water by man. He showed that the chemo-tactile information obtained by sense cells in the epithelium of the suckers was passed to the central nervous system and formed the basis of a system of learning, and memory. In this respect *Octopus*, and presumably other dibranchiate cephalopods also, has a far more highly advanced nervous system than that found in any other molluscan class.

Wells (1963) experimented with octopuses which had been blinded by section of the optic nerves on both sides of the body. These octopuses were presented with perspex cylinders which had been covered with thick porous paper stuck down with perspex cement and which had then been immersed in one or other of various test solutions. As in experiments on visual discrimination described elsewhere in this chapter, in these chemo-tactile tests the octopuses were presented successively with two alternative objects, one "positive" and the other "negative". The octopuses were trained to accept the perspex cylinder bearing the positive solution, by being rewarded with a small piece of fish each time they attacked it. Attacks on the cylinder with the negative solution resulted in punishment by the application of a weak electric shock. If the octopus was able to learn to make the correct responses it was obviously able to discriminate between the two solutions under test. It was first established that the octopuses could discri-

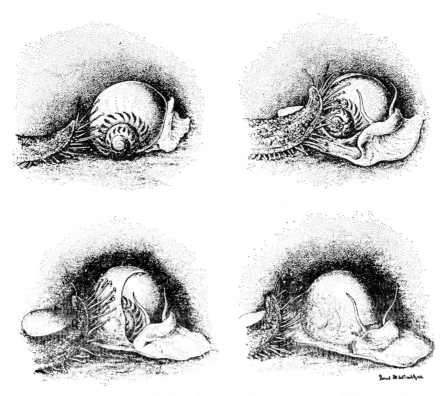

FIG. 146. Even when withdrawn into its shell, *Natica catena* quickly responds to contact by the starfish *Asterias rubens;* the foot of the snail is expanded, water is taken into the tissues, and a fold of flesh spreads forwards over the shell, covering it completely. The suckers on the tube feet of the starfish are unable to grip the mucoid surface of the thin sheath of flesh covering the shell, and the snail is immune from attack. (Original: Dr. Gunnar Thorson's Christmas card for 1953.)

minate between plain sea water and a 7 per cent solution of sucrose and could still make this discrimination when the sucrose solution had been diluted with sea water by 10^{-3}. Similarly, octopuses could discriminate between plain sea water and 7% HCl diluted by 10^{-5}, and a 3 per cent solution of quinine sulphate diluted by 10^{-7}. It was also found that octopuses could distinguish between equimolar solutions of any pair of these chemicals, and that they could distinguish between sea waters of different salinities. It is not known whether, like ourselves, octopus can discriminate a number of distinct tastes or whether the test chemicals differed only in the degree to which they stimulated the same epithelial sense cells. In either event, the chemotactile sense is of great importance to the animal, which is very well endowed in this respect.

Postulated Neurosecretory Structures in the Dibranchiata

Alexandrowicz (1964, 1965) has described a supposed neurosecretory system associated with the vena cava which is substantially the same in the decapod *Sepia officinalis* and in the two octopods *Eledone cirrosa* and *Octopus vulgaris*. The system comprises an extraordinarily large number of small neurones, particularly in *Eledone* in which the NSV—as the system has been named by Alexandrowicz—contains over 2 million neurones. Anteriorly there is a superficial layer of small nerve-cell bodies in the visceral lobe of the sub-oesophageal brain mass, and in addition to this there are two pairs of ganglionated nerve trunks which pass backwards to the vena cava. The axons of these neurones are uniformly slender, they pass through the muscular wall of the vena cava and ramify extensively under the epithelium of the vena cava and the bases of its principal tributaries, forming a thick neuropil there. It is assumed that the function of the system is the liberation of some hormone—as yet undetected—into the blood stream. No other function seems in the least probable.

In *Sepia* and in the oegopsid Decapoda the cell bodies of the third-order giant fibres are dispersed through the stellate ganglion, but in the myopsid Decapoda, e.g. *Loligo* and *Sepiola*, these cell bodies are aggregated in a special giant fibre lobe which lies at the posterior end of the stellate ganglion. In the Octopoda there is no giant fibre system and in the place of the above-mentioned giant fibre lobe there is alternatively a sac-shaped epistellar body which lies within the thick connective tissue sheath which encloses the stellate ganglion. It has been suggested that the Octopoda arose from a myopsid decapod ancestry, the loss of the giant fibre system being accompanied by transformation of the giant fibre lobe of the stellate ganglion into an epistellar body the function of which is presumed to be neurosecretion. The epistellar body has been found in species of *Eledone* and of *Octopus*. In *Octopus* it is separated from the stellate ganglion by a thin sheet of connective tissue which is penetrated at one point by a branch from the pallial nerve to the epistellar body. The wall of the epistellar body consists externally of a connective tissue sheath, then a layer of neurosecretory cells each of which sends one long process in between the cells of the endothelium which lines the central cavity of the organ. The branch from the pallial nerve leads to terminal knobs on the outer sides of the neurosecretory cells, which might be interpreted as modified bipolar neurones.

The epistellar body exhibits considerable variation in appearance and content; at times it is stretched and its interior is filled with an optically homogeneous acidophile substance and a few amoebocytes. Alternatively, the walls of the epistellar body may be relaxed, the quantity of acidophile material is reduced and large quantities of amoebocytes are present; intermediate conditions have been observed. Excision of the epistellar bodies from a number of specimens of *E. moschata* led to a state of great depression which lasted for about a week, after which the animals apparently recovered completely. During the period of dejection the chromatophores were all contracted, which indicates a state of low tonus, and it has been suggested that the function of the epistellar body is the liberation of a hormone the function of which is to maintain the tonus of the body. If this is true it is hard to understand how the experimental animals recovered so quickly from the operation. There was no indication of regeneration of the epistellar body on subsequent re-examination (Young, 1936).

The Functional Organisation of the Brain in the Dibranchiata

In view of the high degree of cephalisation and the considerable elaboration of the central nervous system in the Dibranchiata, it is desirable now to review briefly the functions of the various lobes of the brain in the Decapoda and the Octopoda. The central nervous system is constructed on the same general plan in the two groups, and the following account is based on the comprehensive review of the functional organisation of the brain of the decapod *Sepia* (Boycott, 1961), noting differences in structure and function of the brain of *Octopus* wherever necessary. The functions served by the individual lobes of the brain have been investigated in two contrasted ways: (1) By observation of the behaviour of normal and of operated specimens it is possible to determine the limitations imposed on the activities of the animal by brain lesions of known extent. Much work on these lines has been carried out on the octopus. (2) Animals are operated upon to expose the brain, electrodes are inserted into the brain at selected points, and responses to faradic stimulation are noted. Taken together these two methods of investigation have yielded a considerable body of information on the functioning of the various lobes of the brain in the Dibranchiata.

The parts of the brain may be assembled into three main groups: the sub-oesophageal mass, the higher motor centres of the supra-oesophageal mass, and the optic lobes and correlation centres of the supra-oesophageal mass; these will now be considered in turn.

The Sub-oesophageal Mass

The sub-oesophageal brain mass consists of both lower and intermediate motor centres. The anterior and posterior chromatophore lobes are lower motor centres in that they issue nerve fibres which pass directly to the effector organs and low level mechanical stimulation of these lobes may cause very local responses by the chromatophores. The remainder of the sub-oesophageal mass does not function in this manner and is deemed to comprise intermediate motor centres.

At the anterior end of the sub-oesophageal mass the brachial lobe is penetrated by tracts of fibres which pass from the anterior pedal lobe into the arms. By avoiding these nerve tracts and stimulating the substance of the brachial lobe itself no responses were obtained either from the arms or from the chromatophores. Stimulation of the anterior pedal lobe resulted in movements of the arms and tentacles, some of the responses being recognisable as components of normal behaviour, e.g. raising of the first pair of arms to the position of attention. Stimulation of the dorsal part of the anterior pedal lobe caused simultaneous *contraction* of the chromatophores over the head, arms, and mantle. Stimulation of the posterior pedal lobe resulted in retraction of the head and movements of the funnel. The lateral pedal lobes administer the extrinsic and intrinsic muscles of the eyes, and stimulation of these lobes causes movements of both eyes or of the ipsi-lateral eye. The anterior chromatophore lobe lies in front of the anterior pedal lobe and when it is stimulated this brings about *expansion* of the chromatophores on the front of the head and on the arms, movements of the skin, and erection of the skin papillae of the head and arms.

The pallio-visceral lobe functions both as a lower and as an intermediate motor centre. Superficial stimulation of the lobe at various points may cause a violent expiratory spasm, retraction of the head, movements of the collar, movements of the funnel, or ejection of ink. Stimulation of the neuropil of the lobe may cause any or all of these actions, but without coordination. The posterior chromatophore lobe lies latero-dorsal to the pallio-visceral lobe; it functions as a lower motor centre. Threshold stimulation of the lobe causes *expansion* of chromatophores of the ipsi-lateral side. There is a commissure between left and right lobes and stimulation of one of the lobes at higher voltage results in expansion of the chromatophores bilaterally, due presumably to diffusion of nerve impulses across the commissure. Mechanical stimulation of the lobe, e.g. by prodding it at various points with the electrode, indicated that anterior, middle, and posterior parts of the lobe are responsible for local groups of chromatophores in comparable regions of the mantle.

The fin lobe lies latero-dorsal to the pallio-visceral lobe and in front of the posterior chromatophore lobe. From an anatomical point of view the fin lobe is a lower motor centre, but it does not behave as such in response to faradic stimulation for this elicits only unnatural movements of the whole fin. There is no commissure between the left and right fin lobes, and stimulation of one lobe never produces bilateral responses of the fins. The characteristic undulating movements of the fins, and the diverse forms of co-ordination of left and right fins must be controlled from the supra-oesophageal brain mass, presumably by nerve tracts which pass indirectly to the fin lobes via the posterior pedal lobes.

The magno-cellular lobe contains the cell bodies of the two first-order giant fibres in the Decapoda. Stimulation of this lobe in *Sepia* causes powerful expiratory action, doubtless due to the firing of the giant fibre system. In the Octopoda there is no giant fibre system, and in *Octopus* stimulation of the posterior part of the magno-cellular lobe causes *inspiration*, while *expiration* is caused by stimulation of the pallio-visceral lobe. These two lobes, therefore, together control the respiratory rhythm In octopods.

The Higher Motor Centres of the Supra-oesophageal Mass

The functions of breathing, swimming, turning, attacking prey, and the adoption of particular colour patterns, cannot be administered by the lower and intermediate motor centres, but are dependent on the co-ordinating actions provided by the higher motor centres, which issue the necessarily complex instructions to the effector organs via the intermediate and lower motor centres. These functions are not dependent in any way upon the correlation centres, and an animal which has suffered the total loss of the correlation centres by experimental lesion can function normally in all respects so long as the higher motor centres remain undamaged. Loss of the correlation centres only upsets the capacity of the animal to learn and to remember.

Many of the activities of the body are highly complex and are built up by co-ordination of activity of separate effector organs. This is true, for example, of respiration, swimming, and feeding. Each component of the activity may be controlled by one particular lobe, or perhaps by two lobes of the brain acting in conjunction. In some cases, e.g. the medial basal lobe and the anterior basal lobe, two lobes may possess numerous interconnections and it may be experimentally impossible to isolate the activities of one from those of the other, or to determine the extent to which they act jointly. With these precautionary remarks it is now possible to consider the various higher motor centres of the supra-oesophageal brain mass in turn, and to itemise their functions as far as these have been indicated by the results of electric stimulation.

The medial basal lobe is a higher motor centre which controls the sub-oesophageal centre concerned with the movements of the mantle, collar, and funnel for the purposes of respiration and swimming. Stimulation of the medial basal lobe may produce gentle respiratory movements or alternatively the giant fibre system may be fired, thereby invoking the rapid escape reaction and perhaps also the ejection of ink. Stimulation of the medial basal lobe may cause protraction or retraction of the head, and the latter, of course, is a component of the escape reaction. The medial basal lobe and the anterior basal lobe are both responsible for the production of co-ordinated movements of the fins and stimulation of either of these lobes may produce any of the different co-ordinated bilateral patterns of undulation of the fins which occur in normal life, i.e. waves commencing anteriorly in the ipsi-lateral fin and posteriorly in the contra-lateral fin, and vice versa; waves commencing anteriorly in both fins; waves commencing posteriorly in both fins.

The anterior basal lobe, as mentioned above, is capable of invoking any of the patterns of fin movement that occur in normal life; another function which it evidently shares with the medial basal lobe is the protraction and retraction of the head, but in addition the anterior basal lobe is able to cause the rotation of the head. Stimulation of the anterior basal lobe can result in producing all the components of change of direction, such as movement of the funnel and associated movements of the head, eyes, and arms, reproducing exactly the normal behaviour of a cuttle-fish when it turns to attack a prawn. The anterior basal lobe also controls the arm movements that immediately precede an attack, i.e. raising the first and second pairs of arms to a position of attention, and then recurving the arms backwards around the head, but this lobe is not responsible for the launching of the attack itself.

The interbasal lobe alone is responsible for the final protrusion of the tentacles to grasp the prey; stimulation of this lobe has elicited movements of the tentacles equivalent to the initial gentle protrusion of the tips of the tentacles out of the tentacle pits, the sudden violent protrusion of the entire tentacles to grasp the prey, and their subsequent retraction. There must be extensive interaction between this and other lobes of the supra-oesophageal mass for the purpose of attacking, grasping, biting, and swallowing the prey.

The lateral basal lobes are the higher motor centres responsible for expansion of the chromatophores and the erection of skin papillae on the head, arms, and mantle. Efferent motor tracts pass from the lateral basal lobe to the anterior and posterior chromatophore lobes on the same side of the body. When the lateral basal lobe is stimulated there may be ipsi-lateral, or bilateral, expansion of chromatophores and erection of skin papillae.

The sub-pedunculate lobe plays a major part in the sexual cycle of the female *Octopus* by inhibiting the development of the optic gland and thereby delaying the onset of sexual maturity until a body weight of over 1000 g has been reached. It seems probable that the development of the testis in the male is delayed in a similar manner, but not to so great an extent. It is possible that the subpedunculate lobes exercise the same control over maturation of the gonads in the Decapoda as well as in the Octopoda (Wells and Wells, 1959; Wells, 1964).

Stimulation of the superior and inferior buccal ganglia has elicited biting movements by the jaws. The whole mechanism of eating is undoubtedly complex, and co-ordination of the component actions will depend on interaction of these ganglia with the medial basal lobe and with the inferior frontal lobe.

Stimulation of 'the peduncle and olfactory lobes, which lie closely apposed, caused general expansion of the chromatophores all over the body.

The Optic Lobes and the Correlation Centres of the Supra-oesophageal Mass

Little information is gained by electrical stimulation of the optic lobes, or of the optic tracts, for this elicits any of the responses which may be obtained more directly by stimulation of the various higher motor centres of the brain. The most useful information regarding the functions of the optic lobes has been obtained in experiments on visual discrimination and visual learning, using normal octopuses and octopuses which have suffered various brain lesions. This work has been described in detail in an earlier section of this chapter, and the principal functions of the optic lobes may be summarised as follows.

Neurones in the substance of the optic lobes tend to have their dendritic pick-up fields oriented in planes corresponding to the horizontal or to the vertical plane in the visual field. Such neurones tend to be fired by objects in the visual field which are oriented in particular ways. In this and perhaps in other ways objects seen can be classified according to certain attributes, such as, for example, the amount of vertical extent as compared with horizontal extent, etc. This capacity of the optic lobes to classify objects seen, lays the foundations for the capacity of the animal to discriminate between such objects.

The second important function of the optic lobes is to maintain a representation of an object seen until the object is handled and bitten, when tactile and gustatory stimuli are also received. This maintenance of the visual representation may be effected by cyclical reverberation in closed circuits of neurones within the optic lobes. The information is widely dispersed through the substance of the optic lobe and is passed via the optic commissure from one optic lobe to the other.

A third function of the optic lobes is to receive both tactile representations from the skin and gustatory representations from the mouth, to allow these representations to become dispersed through the substance of the optic lobes and to become associated with the visual representation received earlier. Classifying neurones summate the information received from these diverse sources on the basis of previous experience, and in terms of "good to attack" or "not good to attack", and the results of the analysis are passed to the higher motor centres sited in the basal lobes of the supra-oesophageal mass. This association in the optic lobes between an initial visual input, a subsequent tactile input and a final gustatory input, provides a short-term learning and memory system on the basis of which the animal is able to benefit from its own previous experiences.

The system of visual learning and memory is maintained by dual circuits passing from the optic lobes through the supra-oesophageal mass and back to the optic lobes again. The lower of these circuits passes through the sub-vertical lobe, while the upper circuits passes from the optic lobe to the superior frontal lobe, thence via five major tracts into the vertical lobe, and finally through the sub-vertical lobe and back to the optic lobe again. One of the functions served by these circuits is the general amplification of the signal being generated in the optic lobes so that this signal reaches the threshold necessary to direct the higher motor centres appropriately. Serious interference with these circuits impairs the visual learning and memory system, and it is not easy to execute experiments which will define more closely the functions of the individual lobes of the brain which are involved (Boycott and Young, 1955, 1956, 1957; Young, 1965).

Tactile exploration by the arms, and the development of a system of tactile learning and memory, seems to be confined to the Octopodidae. To meet these new requirements a special lobe of the brain, the sub-frontal lobe is developed in the Octopodidae. There is no such lobe in the Decapoda. The tactile learning and memory system also comprises dual circuits through the supra-oesophageal brain mass, and there is considerable parallelism between the tactile and the visual memory systems. The sub-frontal lobe is divided internally into left and right halves, between which there are numerous transverse connections; there may be further sub-divisions into arm fields, but there is no visible histological evidence of this. The inferior frontal system is divided into three parts, there being lateral inferior lobes on either side of the sub-frontal lobe, and a median inferior frontal lobe above it. The median inferior frontal lobe plays no part in the tactile learning process, but it is involved in the distribution of tactile representations and in its absence learning is confined to the individual arm concerned.

The *vertical lobe* is not essential to the tactile learning process, but in its absence tactile learning is much more slow. It appears that as for the visual learning system here also the vertical lobe provides general amplification of the signal (Wells, 1959, 1959a; Young, 1965).

Key to the Lettering on the Figures

A2	Giant fibre of the second order.	M	Macula.
A3	Giant fibre of the third order.	MC1 ⎱	Cells of the medulla with deep and superficial dendrites oriented in different directions and spreading to different extents.
AC	Anti-crista.	MC2 ⎰	
AFF	Afferent nerve fibres.		
AFN	Anterior funnel nerve.		
AP	Internal aperture of Kölliker's canal.	MCN	Middle crista nerve.
		MN	Macular nerve.
BASM	Basal membrane of retina.	MP	Multipolar cells.
BN1	First brachial nerve.	NC	Nerve cell of crista.
BL	Buccal lobe of brain.	NSUP	Nuclei of supporting cells of retina.
C1	Cell body of first-order giant fibre.		
		OC	Optic commissure.
C3	Cell body of third-order giant fibre.	OG	Optic gland.
		OH	Outer row of hair cells.
CN	Conjunctiva.	OL	Optic lobe of brain.
CR	Cornea.	OLN	Olfactory nerve.
CRL	Longitudinal crista.	ON	Optic nerves.
CRM	Muscles of cornea.	OPT	Optic tract.
CRT	Transverse crista.	OT	Otolith.
CRV	Vertical crista.	P	Perilymph.
CUP	Cupula of crista.	PBL	Posterior basal lobe of brain.
DB	Dorsal basal lobes of brain.	PCN	Posterior crista nerve.
DEN	Dendrite.	PIG	Granules of pigment.
DIST	Distal, rhabdome-carrying segment of retinal cell.	PN	Pallial nerve.
		POS	Perioptic sinus.
DS	Distal synapse between collateral of second-order giant fibre and the third-order giant fibre.	POST	Posterior sac.
		PRS	Proximal, nucleate section of retinal cell.
		PS	Proximal synapse between the accessory giant fibre and the neurones contributing to the third-order giant fibres.
DV	Vertically oriented dendritic fields in the optic lobe.		
E	Epidermis.		
EC	Large encapsulated sense organ.	PSOG	Posterior superior ophthalmic ganglion.
EFF	Efferent axon to the retina.	R	Retina.
GL	Giant fibre lobe of the stellate ganglion.	RC	Receptor cells of plexus.
		RET	Retinal cells.
HV	Horizontally oriented dendritic field in the optic lobe.	RHN	Retractor nerve of head.
		SBL	Superior buccal lobe of brain.
IB	Interaxonic bridge between the two first-order giant fibres.	SFL	Superior frontal lobe of brain.
		SG	Sub-acetabular ganglion.
IBL	Inferior buccal lobe of brain.	SL	Sub-frontal lobe of brain.
IFL	Inferior frontal lobe of brain.	SN	Sympathetic nerve.
IH	Inner row of hair cells.	SOL	Sub-oesophageal lobe of brain.
K	Kölliker's canal.	SP	Sub-pedunculate lobe of brain.
L	Lens..	SUP	Supporting cells of retina.
LC	Row of large hair cells.	SV	Sub-vertical lobe of brain.
LO	Olfactory lobe of brain.	VB	Vitreous body.
LP	Peduncle lobe of brain.	VL	Vertical lobe.

Reference List

ALEXANDROWICZ, J. S. (1964) The neurosecretory system of the vena cava in Cephalopoda. I. *Eledone cirrosa*, *J. mar. biol. Ass. U.K.* **44**, 111–32.

ALEXANDROWICZ, J. S. (1965) The neurosecretory system of the vena cava in the Cephalopoda. II. *Sepia officinalis* and *Octopus vulgaris*, *J. mar. biol. Ass. U.K.* **45**, 209–28.

ARNOLD, D. C. (1957) The response of the limpet *Patella vulgata* L. to waters of different salinities, *J. mar. biol. Ass. U.K.* **36**, 121–8.

BOYCOTT, B. B. (1960) The functioning of the statocysts of *Octopus vulgaris*, *Proc. roy. Soc.* B, **152**, 78–87.

BOYCOTT, B. B. (1961) The functional organisation of the brain of the cuttlefish *Sepia officinalis*, *Proc. roy. Soc.* B, **153**, 503–34.

BOYCOTT, B. B. and YOUNG, J. Z. (1955) A memory system in *Octopus vulgaris* Lamarck, *Proc. roy. Soc.* B, **143**, 449–80.

BOYCOTT, B. B. and YOUNG, J. Z. (1955a) Memories controlling attacks on food objects by *Octopus vulgaris* Lamarck, *Publ. Staz. zool. Napoli.* **27**, 232–49.

BOYCOTT, B. B. and YOUNG, J. Z. (1956) Reactions to shape in *Octopus vulgaris* Lamarck, *Proc. zool. Soc. Lond.* **126**, 491–547.

BOYCOTT, B. B. and YOUNG, J. Z. (1957) Effects of interference with the vertical lobe on visual discrimination in *Octopus vulgaris* Lamarck, *Proc. roy. Soc.* B, **146**, 439–59.

BOYCOTT, B. B. and YOUNG, J. Z. (1958) Reversal of learned responses in *Octopus vulgaris* Lamarck, *J. Anim. Behav.* **6**, 45–52.

BOYCOTT, B. B. and YOUNG, J. Z. (1960) The comparative study of learning, *Symp. Soc. exp. Biol.* **4**, 432–53.

BROWN, A. C. (1961) Chemo-reception in the sandy-beach snail *Bullia*, *S. Afr. J. Lab clin. Med.* **7**, 160.

BROWN, A. C. and NOBLE, R. G. (1960) Function of the osphradium in *Bullia* (Gastropoda), *Nature, Lond.* **188**, 1045.

CANNON, H. G. (1958) *The Evolution of Living Things*, Manchester University Press.

DEUTSCH, J. A. and SUTHERLAND, N. S. (1960) Theories of shape discrimination in *Octopus*, *Nature, Lond.* **188**, 1090–4.

DILLY, P. N. (1963) Delayed responses in *Octopus*, *J. exp. Biol.* **40**, 393–401.

HUBBARD, S. J. (1960) Hearing and the octopus statocyst, *J. exp. Biol.* **37**, 845–53.

HUBEL, D. H. and WIESEL, T. N. (1959) Receptive fields of single neurones in the cat's striate cortex, *J. Physiol.* **148**, 574–91.

HUBEL, D. H. and WIESEL, T. N. (1962) Receptive fields, binocular interaction and functional architecture in the cat's visual cortex, *J. Physiol.* **160**, 106–54.

KOHN, A. J. (1961) Chemo-reception in gastropod molluscs, *Amer. Zoologist* **1**, 291–308.

MOODY, M. F. and PARRISS, R. J. (1960) The discrimination of polarised light by *Octopus*, *Nature, Lond.* **186**, 839–40.

NEWELL, G. E. (1965) The eye of *Littorina littorea*, *Proc. zool. Soc. Lond.* **144**, 75–86.

PUMPHREY, R. J. and YOUNG, J. Z. (1938) The rates of conduction of nerve fibres of various diameters in cephalopods, *J. exp. Biol.* **15**, 453–66.

SUTHERLAND, N. S. (1957) Visual discrimination of orientation and shape by the octopus, *Nature, Lond.* **179**, 11–13.

SUTHERLAND, N. S. (1957a) Visual discrimination of orientation by octopus, *Brit. J. Psychol.* **48**, 55–71.

SUTHERLAND, N. S. (1959) A test of a theory of shape discrimination in *Octopus vulgaris* Lamarck, *J. comp. physiol. Psychol.* **52**, 135–41.

SUTHERLAND, N. S. (1959a) Visual discrimination of shape by *Octopus*: circles and squares, and circles and triangles, *Quart. J. exp. Psychol.* **11**, 24–32.

SUTHERLAND, N. S. (1960) Theories of shape discrimination in *Octopus*, *Nature, Lond.* **186**, 840–4.

SUTHERLAND, N. S. (1961) Discrimination of horizontal and vertical extents by *Octopus*, *J. comp. physiol. Psychol.* **54**, 43–48.

SUTHERLAND, N. S. (1962) Visual discrimination of shape by *Octopus*: squares and crosses, *J. comp. physiol. Psychol.* **55**, 939–43.

SUTHERLAND, N. S. (1963) Shape discrimination and receptive fields, *Nature, Lond.* **197**, 118–22.

SUTHERLAND, N. S. (1963a) The shape-discrimination of stationary shapes by octopuses, *Amer. J. Psychol.* **76**, 177–90.

SUTHERLAND, N. S. (1963b) Visual acuity and discrimination of stripe widths in *Octopus vulgaris* Lamarck, *Pubbl. staz. zool. Napoli* **33**, 92–109.

SUTHERLAND, N. S. (1964) Visual discrimination in animals, *Brit. med. Bull.* **20**, 54–59.

SUTHERLAND, N. S. (1964a) The learning of discrimination by animals, *Endeavour* **23**, 148–52.

SUTHERLAND, N. S., MACKINTOSH, J. and MACKINTOSH, N. J. (1963) The visual discrimination of reduplicated patterns by *Octopus*, *Anim. Behaviour.* **11**, 106–10.

SUTHERLAND, N. S. and MUNTZ, W. R. A. (1959) Simultaneous discrimination training and preferred directions of motion in visual discrimination of shape in *Octopus vulgaris* Lamarck, *Pubbl. Staz. zool. Napoli* **31**, 109–26.

THOMPSON, T. E. (1958) The natural history, embryology, larval biology and post-larval development of *Adalaria proxima* (Alder and Hancock) (Gastropoda, Opisthobranchia), *Phil. Trans.* B, **242**, 1–58.

TOMPSETT, D. H. (1939) *Sepia*, L.M.B.C. Memoir No. 32, 1–184. University Press, Liverpool.

WELLS, M. J. (1959) Functional evidence for neurone fields representing the individual arms within the central nervous system of *Octopus*, *J. exp. Biol.* **36**, 501–11.

WELLS, M. J. (1959a) A touch learning centre in *Octopus*, *J. exp. Biol.* **36**, 590–612.

WELLS, M. J. (1960) Proprioception and visual discrimination of orientation in *Octopus*, *J. exp. Biol.* **37**, 489–99.

WELLS, M. J. (1961) Weight discrimination by *Octopus*, *J. exp. Biol.* **38**, 127–33.

WELLS, M. J. (1963) Taste by touch: some experiments with *Octopus*, *J. exp. Biol.* **40**, 187–93.

WELLS, M. J. (1963a) The orientation of *Octopus*, *Ergebn. Biol.* **26**, 40–54.

WELLS, M. J. (1964) Tactile discrimination of shape by *Octopus*, *Quart. J. exp. Psychol.* **16**, 156–62.

WELLS, M. J. (1964a) Tactile discrimination of surface curvature and shape by the octopus, *J. exp. Biol.* **41**, 435–45.

WELLS, M. J. and WELLS, J. (1957) The function of the brain of *Octopus* in tactile discrimination, *J. exp. Biol.* **34**, 131–42.

YOUNG, J. Z. (1935/6) The giant nerve fibres and epistellar body of cephalopods, *Quart. J. micr. Sci.* **78**, 367–86.

YOUNG, J. Z. (1938) The functioning of the giant nerve fibres of the squid, *J. exp. Biol.* **15**, 170–85.

YOUNG, J. Z. (1939) Fused neurons and synaptic contacts in the giant nerve fibres of cephalopods, *Phil. Trans.* B, **229**, 465–503.

YOUNG, J. Z. (1955) Memory, *Mem. Proc. Manchr. lit. phil. Soc.* **96**, 1–8.

YOUNG, J. Z. (1956) The evolution of homeostatic control, *4th Wallace Memorial Lect. Univ. Malaya, Singapore*, pp. 1–11.

YOUNG, J. Z. (1960) The statocysts of *Octopus vulgaris*, *Proc. roy. Soc. Lond.* B, **152**, 3–29.

YOUNG, J. Z. (1960a) Structure of the optic lobe of *Octopus*, *Nature, Lond.* **186**, 836–9.

YOUNG, J. Z. (1961) Learning and discrimination in the octopus, *Biol. Rev.* **36**, 32–96.

YOUNG, J. Z. (1962) The retina of cephalopods and its degeneration after optic nerve section, *Phil. Trans.* B, **245**, 1–18.

YOUNG, J. Z. (1962a) The optic lobes of *Octopus vulgaris*, *Phil. Trans.* B, **245**, 19–58.

YOUNG, J. Z. (1965) Two memory stores in one brain, *Endeavour* **24**, 13–20.

APPENDIX A

THE FOUR MINOR CLASSES

Monoplacophora

This small class, which has only recently been recognised, comprises a small group of Cambro-Silurian fossils in the family Tryblidiacea, and the single modern genus *Neopilina*. Ten living specimens and three empty shells of *Neopilina galatheae* were collected by the Galathea expedition, at a depth of 3570 m, from a bottom of dark, muddy clay off the west coast of Mexico (Lemche, 1957). Later, four specimens of *N. ewingi* were collected in a little over 3000 fathoms from the Peru–Chile trench (Clarke and Menzies, 1959). It was only on the examination of the soft parts of *Neopilina* (Lemche, 1957) that it was recognised that a new class would have to be erected for this genus and also for the fossil genera *Pilina*, *Scenella*, *Stenothecoides*, *Tryblidium*, *Archaeophiala*, *Drahomira*, *Proplina*, and *Bipulvina* (Lemche and Wingstrand, 1959). The reason for this remarkable decision was the undoubted occurrence in *Neopilina* of the segmental repetition of parts in various organ systems including the gills, the muscles, the nervous system, the nephridia, and the gonads (see Fig. 147). On the basis of the detailed information of the anatomy of *Neopilina* (Lemche, 1957; Lemche and Wingstrand, 1959) it was clear that the numerous muscle scars on the shells in the fossil genera must also be interpreted as homologous, segmentally repeated structures.

Hitherto the phylum Mollusca has sometimes been designated as an assemblage of unsegmented animals (Borradaile *et al.*, 1958; Parker and Haswell, 1943). Pelseneer (1906) more cautiously concluded that "signs of primitive segmentation are no longer evident" in the Mollusca. It will be seen from the following brief account of the anatomy of *Neopilina* that it is undoubtedly a Mollusc, possessing numerous features characteristic of the phylum. It unquestionably shows segmental repetition of parts, and it will therefore be necessary to revise our definition of the basic structure of the body in the Mollusca.

This fascinating find of yet one more "living fossil" calls for a new study of various important issues. Firstly, we shall have to reshape our ideas concerning the form of the "hypothetical ancestral molluscan stock" from which all modern molluscan forms can be derived by descent with modification. Secondly, we shall have to consider whether the discovery of segmentation in a Mollusc will throw any new light on the phylogenetic origins and relationships of the Mollusca. Finally, we shall have to have

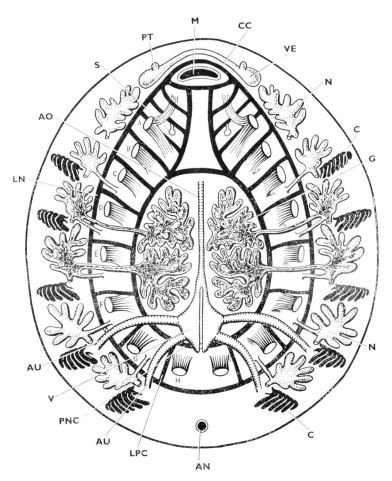

Fig. 147. *Neopilina galatheae*, diagrammatic representation of the serial repetition of various organ systems. The mouth (M) lies anteriorly and the anus (AN) lies posteriorly in the median line. The muscular system includes three anterior segments (not represented here) in the vicinity of the head, and eight body segments (A–H). Nephridia (N) are present in six body segments; gonads (G) are present in two body segments and their gonad ducts communicate to the exterior via the third and fourth nephridia. Ctenidia (C) are present in five body segments. The nervous system comprises a lateral nerve cord (LN) and a pedal nerve cord (PNC) on each side of the body, these two nerve trunks being placed in communication by a regular series of ten connectives (LPC). For interpretation of other lettering, see pp. 492–3. (Originally published in Lemche and Wingstrand, 1959, *Galathea Rept.*, vol. 3, plate 54, fig. 165.)

another look at the serially repeated organs in other molluscan classes, e.g. the shell and the ctenidia in the Polyplacophora, the ctenidia in *Nautilus* (another "living fossil"), and perhaps also the ganglia of the central nervous system in the Scaphopoda, Gastropoda, and Bivalvia. Can these various cases continue to be dismissed as secondary multiplications, or can some of them now fall into perspective as being further examples of the retention of an archaic segmental structure in the Mollusca?

It is clear, therefore, that the interest which attaches to the discovery of *Neopilina* is not so much the proof derived from anatomical considerations that it is a Mollusc, or the inescapable logic that demands for it the creation of a new class, or the interest which may be derived by comparing its organ systems with those of the other molluscan classes; on the contrary it is the fresh opportunity which it affords to delve into the origins of the phylum and the phylogenetic relationships of the classes of Mollusca. Nevertheless, we must begin by setting out a brief account of the anatomy of the animal, drawing heavily on the most excellent monograph by Lemche and Wingstrand (1959).

The Mantle and the Shell

The shell consists of a single depressed, limpet-shaped valve, the apex of which lies almost vertically above the anterior margin, in the median line (see Fig. 148). The margin of the shell is almost circular, being slightly longer in the sagittal plane (maximum measurements $37 \times 35 \times 13$ mm high). The larval shell, or protoconch, is dextrally coiled, and may be retained on the apex of the adult shell. The adult shell lacks any coiling and, apart from the protoconch, the adult is bilaterally symmetrical in all respects. There has been no process of torsion. The shell is relatively thin and comprises three layers, an outer periostracum which arises by secretion in the periostracal groove in the margin of the mantle, a middle prismatic layer which increases in thickness towards the margin of the shell, and an inner nacreous layer which remains of approximately equal thickness over different zones of the shell. Thus the shell becomes slightly thicker towards the margin. The prismatic layer is secreted by the outer side of the outermost lobe of the mantle margin, and the nacreous layer is secreted by the epithelium of the upper surface of the remainder of the mantle lining the shell. The structure of the shell, and its method of formation, are therefore basically the same as in, for example, the Bivalvia.

The shell of *Neopilina* is relatively thin and fragile, and its inner surface does not show any scars at the points of insertion of the main muscles of the body. In fossil material, on the other hand, numerous muscle scars may be clearly shown on the inner surface of the shell. The interpretation of these scars had hitherto been problematical, and a true appraisal of their significance had been hindered by the general view that the Mollusca were unsegmented animals. It is now clear that the serially repeated muscle scars on the inner surface of the shell of the fossil *Pilina unguis*, for example, are comparable in disposition with the muscles of the living *N. galatheae*, and are to be interpreted as evidence of metameric segmentation (see Fig. 149). The antiquity of the class Monoplacophora and of its segmental body structure is therefore well established.

The mantle may be considered as consisting of two parts. Firstly, there is the epithelium covering the body of the animal and lining the greater

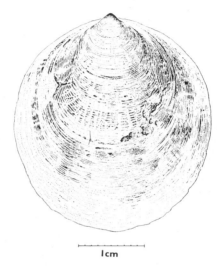

I cm

FIG. 148. *Neopilina galatheae*, view of the dorsal surface of the shell, showing the anteriorly directed apex. (Originally published in Lemche and Wingstrand, 1959, *Galathea Rept.*, vol. 3, plate 1, fig. 2.)

part of the interior of the shell. This part of the mantle is incapable of movement, and it is responsible for secreting the innermost, nacreous layer of the shell as mentioned above. Secondly, there is a marginal extension of the mantle, which surrounds the entire animal like a skirt, and which in life is presumably closely pressed against the under surface of the periphery of the shell and will probably extend to the very margin of the shell when the animal is undisturbed. Due to the presence of radial muscle fibres inserted into the shell along a continuous, circular pallial line, this marginal part of the mantle is capable of retraction in response to disturbance, and it would presumably be extended again to its former position by blood pressure.

The margin of the mantle is subdivided into three lobes, the epithelium on the outer surface of the outer lobe being responsible for the secretion of the prismatic layer of the shell. As the animal grows in size there is a comparable increase in the extent of this outer lobe of the mantle margin, and accordingly the prismatic layer of the shell becomes thicker towards the margin of the shell. There is a groove between the extensive outer lobe and the more compact middle lobe of the mantle margin, and the epithelium lining this groove comprises the periostracal gland. A flexible sheet of periostracum arises in this groove, passes outwards ventral to

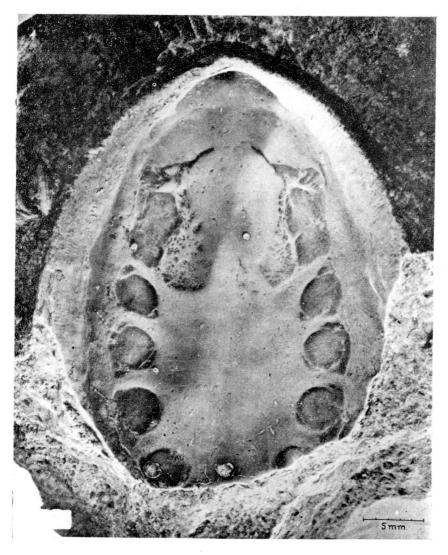

Fɪɢ. 149. *Pilina unguis*, a silurian fossil. The inner side of the shell, showing metamerically repeated muscle scars. (Originally published in Lemche and Wingstrand, 1959, *Galathea Rept.*, vol. 3, plate 42, fig. 134.)

the outer lobe of the mantle margin, and bends round the margin of the shell in continuity with the periostracal coat of the shell. Growth of the shell is effected by extension of the mantle tissues so that a portion of the newly formed, and still flexible, periostracum provides a linear projection of the shell, beyond its present margin. This flimsy periostracal extension is then invested and strengthened by a deposition of the prismatic layer of the shell, by the outer side of the outer lobe of the mantle margin. Some time later the inner surface of the prismatic layer is invested by a deposition of nacre. During this growth process the insertions of muscles on to the ventral surface of the shell will gradually migrate outwards, so that the proportions and the dispositions of the parts remain broadly unchanged. It is not known how these migrations are effected, but it may be as a result of a slow resorbtion of muscle fibres on the median side of muscles and a more rapid production of new muscle fibres on the outer sides of the same muscles. In this way the position of the muscle insertion would slowly move radially outwards, while the muscle would slowly increase in girth in proportion to the overall growth of the animal.

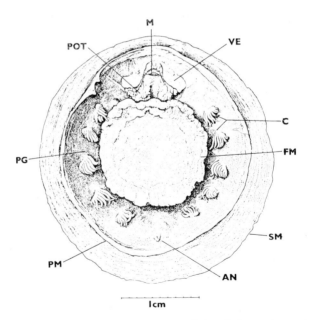

FIG. 150. *Neopilina galatheae*, ventral view of the whole specimen, showing the organs in the mantle cavity. Five pairs of ctenidia (C) lie in the laterally sited pallial groove (PG). The mantle margin (PM) has shrunk away from the periphery of the shell (SM). The mouth (M) is flanked by a velum (VE) on each side, and a series of post-oral tentacles (POT) lie behind the velar lobes. For interpretation of other lettering, see pp. 492–3. (Originally published in Lemche and Wingstrand, 1959, *Galathea Rept.*, vol. 3, plate 1, fig. 1.)

The Mantle Cavity

The mantle cavity is in the form of a shallow gutter which surrounds the entire animal, being delimited internally by the walls of the foot and laterally by the pallial fold which underlies the margin of the shell. Anteriorly the mantle cavity contains the mouth, which is surrounded by an anterior velar ridge which passes backwards into a pair of lateral velar

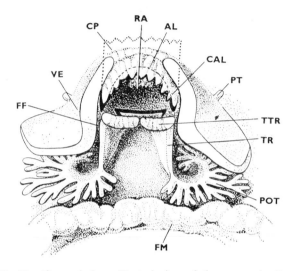

FIG. 151. *Neopilina galatheae*. Ventral view of the organs in the anterior part of the mantle cavity, diagrammatic interpretation. The radula (RA) is seen lying within the mouth. The anterior lip of the mouth (AL) bears a cuticular plate (CP). Posterior to the mouth there is a transverse part of the tentacle ridge (TTR). Food probably enters the mouth via the feeding furrow (FF) which lies between the velum (VE) and the tentacle ridge (TR). The post oral tentacles (POT) may assist in the feeding process. For interpretation of other lettering, see pp. 492–3. (Originally published in Lemche and Wingstrand, 1959, *Galathea Rept.*, vol. 3, plate 19, fig. 66.)

folds, and a pair of posterior tentacle ridges which arise transversely just behind the mouth and then pass backwards and outwards between the foot and the velar folds. The outer parts of the tentacle ridges are loosely coiled and bear about a dozen dichotomously branched post-oral tentacles, and are reminiscent of the spirally coiled lophophore of Brachiopoda (see Figs. 150 and 151). The tentacles appear to be poorly ciliated and are unlikely to participate actively in the transport of food to the mouth. In *N. ewingi* the post-oral tentacles are more numerous than in *N. galatheae* (Clarke and Menzies, 1959). Two short pre-oral tentacles lie on either side of the mouth, lateral to the velar folds. Between the lateral velar folds and the tentacle ridges there is a groove, which has been designated the "feeding furrow". If the velar folds and the tentacle ridges could be equated with the outer and the inner labial palps of a bivalve mollusc, the feeding groove could be compared with the lateral oral groove of the latter. The function of the feeding groove, of course, has not yet been observed.

Laterally the mantle cavity of *N. galatheae* contains five pairs of gills, each of which is suspended from the roof of the mantle cavity by a slender base. From this slender base the gill hangs downwards in the mantle

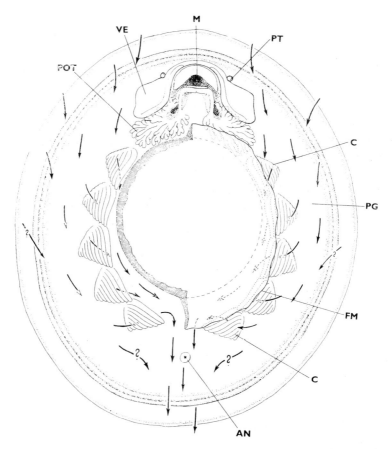

FIG. 152. *Neopilina galatheae*, diagrammatic representation of the ventral surface of the animal, showing the mouth (M), the velum (VE), the post oral tentacles (POT), the ctenidia (C), and the anus (AN). On one side of the diagram the margin of the foot (FM) is cut away to expose the ctenidia fully. For interpretation of other lettering see pp. 492–3. (Originally published in Lemche and Wingstrand, 1959, *Galathea Rept.*, vol. 3, plate 15, fig. 57.)

cavity, and is seen to consist of a stem which bears one series of seven or eight lamellae on one side only. The proximal lamella is longest, and the remainder decrease regularly in size towards the distal end of the stem. Due to a twist in the stem of each gill the lamellae come to lie on the posterior side of the stem (see Figs. 152 and 153). There is a second, rudimentary series of lamellae on the opposite side of the stem of the gill in certain instances, and on this basis it is thought reasonable to accept the structure as being a ctenidium and homologous with the ctenidia of other classes of Mollusca. Too little is known about this alleged second

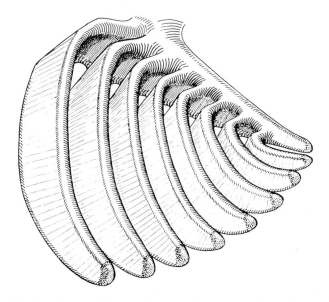

FIG. 153. *Neopilina galatheae*, diagrammatic representation of one cteni-dium, as seen from the ventral side. The ctenidial axis, which is attached to the visceral mass at its base, bears a single series of long gill lamellae which decrease in length as they approach the tip of the ctenidial axis. (Originally published in Lemche and Wingstrand, 1959, *Galathea Rept.*, vol. 3, plate 16, fig. 58.)

series of lamellae to comment on them with any confidence. Almost the whole of the ctenidium, including the stem as well as the lamellae, is covered with one of three main types of ciliated epithelium. The epithe-lium on the flattened sides of the lamellae may include secretory cells, and its ciliated cells bear relatively long cilia—about 10 μ long. The very tips of the lamellae bear an epithelium of dense, strongly staining colum-nar cells with long and powerful cilia—at least 20 μ long. The dorsal and ventral edges of the lamellae and the surfaces of the stem of the ctenidium are covered with a third type of epithelium which includes many globular

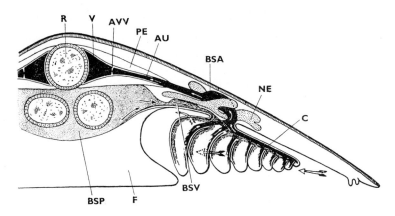

FIG. 154. *Neopilina galatheae*, diagrammatic transverse section through the body in the region of the heart. Large arrows indicate the presumed course of water between the ctenidial leaflets, and smaller arrows indicate the course of circulation of the blood. Blood in the peri-intestinal blood space (BSP) passes towards the ctenidium (C) via an afferent vein (BSV). Arterial blood passes from the ctenidium in the arterial blood sinus (BSA) to the auricle (AU) and thence, through the auriculo-ventricular valve (AVV) to the paired ventricles (V) on either side of the rectum (R). The nephridium (N) discharges into the mantle cavity close to the base of the ctenidium. For interpretation of other lettering, see pp. 492–3. (Originally published in Lemche and Wingstrand, 1959, *Galathea Rept.*, vol. 3, plate 46, fig. 143.)

mucous cells and ciliated cells which carry comparatively short cilia. There are no skeletal supports within the ctenidia. *N. ewingi* differs in having six pairs of ctenidia (Clarke and Menzies, 1959).

The excretory organs open into the mantle cavity by six nephridiopores of which the first pair lie between the velar fold and the first ctenidium, while the remaining five pairs open at the bases of the five pairs of ctenidia. In both sexes the third and fourth nephridia serve as gonoducts, for liberation of the gametes into the mantle cavity. There are no accessory genital structures.

The anus lies at the apex of a low papilla which lies in the median line posteriorly in the mantle cavity (Fig. 152).

The Foot

The foot is a short circular column, the peripheral walls of which are muscular due to the presence of a zone of circular muscle fibres, and also to the presence of dorso-ventral fibres of the pedal retractor muscles. The flat, ventral wall of the foot, however, is thin, transparent, and is relatively lacking in muscular support; it is covered by a ciliated epithelium and forms a creeping sole. A pedal gland, lying along the anterior border of the foot, may aid creeping movements by supplying mucus. The foot probably cannot act as a sucker, due to the lack of central or sub-central elevator muscle fibres.

Movements of the foot are presumably effected by interplay of muscular action on the hydro-skeleton provided by blood in the peri-intestinal blood sinus. Thus contraction of circular elements would constrict the column and so extend the foot, while contraction of dorso-ventral elements would have the reverse effect.

The Alimentary Canal

The epithelium of the upper lip of the mouth is covered by a strong and continuous cuticle which thickens as it passes into the mouth to form a cuticular plate with lateral, jaw-like projections (Fig. 155). The epithelium of the lower lip is similarly covered with a thin cuticle which enters the oral cavity and passes into a sub-radular sac.

The mouth points downwards and backwards, and in consequence the "anterior" border of the mouth forms a backwardly directed floor, or "ventral" wall to the mouth, and the radula occupies a central position on the "roof" of the mouth.

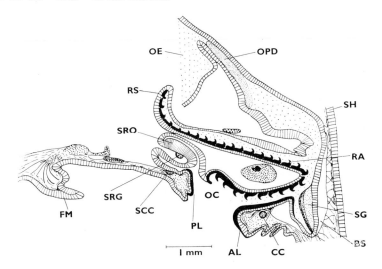

FIG. 155. *Neopilina galatheae*, sagittal section through the oral region. The radula (RA) lies in the radular sac (RS). On the posterior side of the mouth there is a sub-radular organ (SRO) with an associated sub-radular ganglion (SRG). The pharynx receives a pair of pharyngeal diverticula (OPD) laterally, and then passes back into the oesophagus (OE). For interpretation of other lettering, see pp. 492–3. (Originally published in Lemche and Wingstrand, 1959, *Galathea Rept.*, vol. 3, plate 23, fig. 81.)

The sub-radular organ is a blind sac which opens into the oral cavity by a narrow slit on its posterior wall. The sac is innervated by a sub-radular ganglion, and it may serve as a sense organ. The pharynx runs forwards from the mouth, then upwards, and finally backwards to enter the oesophagus. A radula-sac opens on to the upper (morphologically posterior) wall of the first, horizontal part of the oesophagus. When the distal end of the radula is protruded, it can therefore be seen on the posterior roof of the oral cavity. The radula-sac is a long, posteriorly directed diverticulum, which carries the radula on its ventral wall. The radula consists of about forty-five V-shaped rows of teeth with the tip of each row lying anteriormost. Each row contains eleven teeth, the radular formula being 5, 1, 5. Feeding seems to be effected by the anteriormost ten rows of teeth, which show signs of wear. These teeth presumably work against the thickened cuticular lining of the ventral (morphologically anterior) wall of the oral cavity.

In each row of teeth the median tooth, and the first lateral tooth on each side of it, are relatively small; the second lateral and the third lateral tooth on each side are very much more powerfully built and each terminates in a blunt distal hook; the fourth lateral tooth on each side is membranous in form, and its distal border is delicately pectinate; the fifth lateral tooth on each side is similar to the second and third lateral teeth but is less robust (see Fig. 156). These transverse rows of teeth are borne on a radular membrane, they are secreted at the blind end of the radula-sac and gradually pass forwards to replace wornout teeth at the anterior end of the radula. The anterior part of the radula is presumably protruded, when the pectinate fourth lateral teeth may diverge. On withdrawal of the radula these pectinate teeth may close together and grasp small objects presented to them, carrying them backwards into the pharynx. Within the pharynx such food matter will probably be triturated between the radula and the anterior cuticular lining of the pharynx before being passed to the oesophagus. Examination of the contents of the stomach indicates that *Neopilina* is probably a "detritus" feeder, living on radiolarians and debris collected from the substratum (but see Fig. 127, p. 378).

An unpaired anterior salivary gland lies in the mantle anterior to the mouth, and may be homologous with the salivary gland of chitons. Just before opening into the oesophagus, the pharynx bears a pair of large, lobed pharyngeal diverticula, of unknown function. The oesophagus passes backwards to the stomach, lying below the pharyngeal diverticula, and compressed between the digestive diverticula. The oesophagus is lined by a ciliated columnar epithelium. The stomach lies above the anterior part of the foot, it receives the oesophagus anteriorly and discharges into the mid-gut posteriorly. Laterally the stomach bears a pair of digestive diverticula, each of which opens into the stomach by a single,

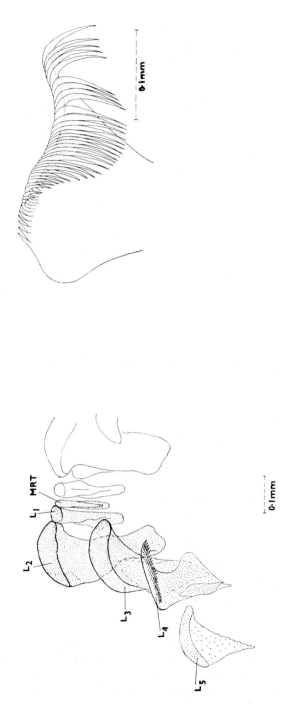

FIG. 156. *Neopilina galatheae*; left, one oblique row of radular teeth, incomplete on the right side of the figure. There is a small median tooth (MRT) and five lateral teeth (L_1–L_5), of which the fourth lateral tooth bears a finely denticulate margin. Right, enlarged view of the fourth lateral tooth. (Originally published in Lemche and Wingstrand, 1959, *Galathea Rept.*, vol. 3, plate 25, figs. 87 and 88.)

long, slit-shaped opening. The functions served by the digestive diverticula are not known.

A short, blunt style-sac enters the stomach dorsal to the opening of the mid-gut, the cavity of the style-sac being separate from that of the mid-gut. A well-defined crystalline style projects from the style-sac across the stomach towards the oesophageal orifice; it is probably rotated by the strongly ciliated epithelium of the style-sac and, if so, it will help to mix the contents of the stomach as in the case of bivalves. The crystalline style is constructed of superimposed concentric hollow cones, the posterior-most of these being held in the style-sac. This suggests that the style is secreted in the style-sac and pushed forward into the stomach, as in the majority of bivalves.

The mid-gut consists of six concentric turns (anti-clockwise as seen from above). There are no folds or typhlosoles in the mid-gut. The anterior part of the mid-gut possesses tall epithelial cells, comparable with those lining the wall of the stomach. These high cells are probably ciliated. Further posteriorly the mid-gut epithelial cells become lower and broader, and are definitely ciliated. In the last two coils of the mid-gut granules are found in the epithelial cells. In the rectum the epithelium is distinctly ciliated, and these granules are numerous. The anus opens on a low papilla in the mantle cavity, posterior to the foot.

The Vascular System

The heart possesses a pair of ventricles which lie on either side of the rectum, and each passes into a lateral anterior aorta. These lateral aortae unite anteriorly above the rectum to discharge blood into a single median anterior aorta. There are two pairs of auricles, the anterior pair draining blood from the first four pairs of gills, while the posterior pair of auricles drain blood only from the fifth pair of gills (see Fig. 157).

Arterial blood from the anterior aorta passes into a large haemocoele surrounding the radular apparatus; a peri-intestinal haemocoele in which lie the stomach, digestive diverticula, mid-gut and gonad; and a haemocoele in the pallial fold, which supplies venous blood to the five pairs of gills. The ventricles and auricles lie within a pericardial chamber. Each auricle communicates with the appropriate ventricle via an auriculo-ventricular valve.

The Excretory System

There are six pairs of nephridia, which lie in the proximal half of the pallial fold. The first pair of nephridia do not communicate internally with the coelome, but the cavity of each of these nephridia is continuous with that of the adjacent second nephridium. The next three pairs of nephridia all open internally by a nephrostome into the dorsal coelome.

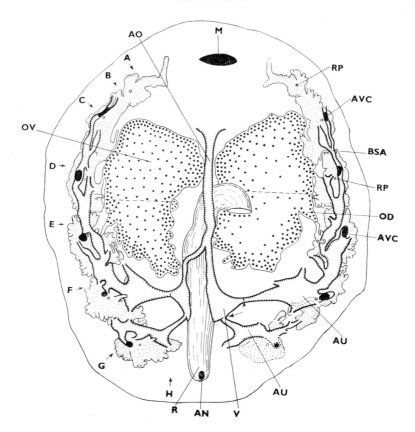

Fig. 157. *Neopilina galatheae*, diagrammatic representation of the vascular system and the excretory system. The six nephridia open into the lateral mantle cavity each by its own renopore (RP). There are two pairs of gonads (ovary, OV), and these discharge by gonad ducts (oviduct, OD) via the third and fourth nephridia. Arterial blood drained from the ctenidia is collected into two pairs of auricles (AU), the first four ctenidia being drained into the first pair of auricles and the last ctenidium into the posterior auricles. The two pairs of auricles open into paired ventricles (V) which lie lateral to the rectum (R). For interpretation of other lettering, see pp. 492–3. (Originally published in Lemche and Wingstrand, 1959, *Galathea Rept.*, vol. 3, plate 46, fig. 144.)

All six pairs of nephridia open to the exterior by renopores. The renopores of the second to sixth nephridia lie at the bases of the first to fifth gills, respectively, on their median sides. The renopore of the first nephridium opens further anteriorly, where there is no related gill. The third and fourth nephridia on each side each receive a gonad duct, and transmit gametes to the exterior via their renopores. Details of the fine structure of the nephrostomes and the ducts from these into the glandular part of the nephridium are not available, but the presence of basal granules in the epithelial cells suggest that the nephrostomes are ciliated.

The occurrence of nephrostomes connecting the individual excretory organs with the coelome have been definitely identified for the third and fourth nephridia, and were strongly indicated for the second, but no communication with the coelome could be established for the anterior-most nephridial mass. In the case of the last two pairs of nephridia cellular strands were seen to pass from the nephridia to the pericardium, which replaces the coelome in this part of the body. These strands may be homologous with nephridial ducts, but no actual duct could be discerned.

The Reproductive System

Serial sections of only two specimens indicate that there are separate sexes, with no signs of change of sex. In the male specimen there were clearly two pairs of gonads, which opened by separate gonad ducts into the third and the fourth nephridia respectively. In the female the gonads extend over a comparable area, but it was not possible to determine for certain that two separate pairs of gonads were present. Nevertheless, in the female also two pairs of gonad ducts discharged into the third and fourth nephridia respectively.

There are no copulatory organs, no indications of sex dimorphism, and no traces of sperm in the female reproductive system. Eggs found in the lumen of the nephridia close to the renopore were without shells or secondary membranes. This strongly suggests that the genital products are discharged into the water, and that fertilisation is random in the sea water.

The gonads lie in the peri-intestinal blood sinus, the ovary in the female being confined to the bottom of this sinus, while in the male lobes of the testis extend upwards among the loops of the intestine.

There are two genital segments in *Neopilina* and it is suggested that these can be directly compared with such segments in certain Polychaeta, the gonad duct being a coelomoduct and the excretory organ a true metanephridium.

The Coelomic System

The coelome consists of three parts: the dorsal coelome, the pericardium, and the lumina of the gonads. Each of these is paired, and there is no trace of any intercommunication between the cavities of the left and right sides.

The dorsal coelome. This consists of a pair of broad, flattened sacs, one on each side of the body, extending from the side of the median dorsal aorta outwards to the bases of the pedal retractor muscles. The coelome lies immediately below the dorsal body wall. Due perhaps to collapse at the time of preservation, the lumen of the dorsal coelome is almost obliterated in places, and it cannot be seen whether it is subdivided by segmental septa. Antero-laterally each dorsal coelome bears an extension which sweeps forwards and inwards; these two arms of the left and right dorsal coelomes meet anteriorly in the median line in front of the pharynx, but their cavities remain separated from each other by a median septum.

Each dorsal coelome probably receives three nephrostomes. No nephrostomial opening could be demonstrated for the anteriormost nephridial mass; a nephrostomial opening was "strongly indicated" for the second nephridium, while two nephrostomes were clearly shown for the third and fourth nephridia (which also communicate by gonad ducts with the two pairs of gonads).

The genital coelome. The lumen of each gonad is considered to be coelomic in origin. There are two pairs of these in the male, and this is probably also true of the female. The anterior and posterior gonads communicate respectively with the third and the fourth nephridium by a gonoduct.

The pericardial sacs. These sacs lie to left and right of the rectum. Each encloses the ventricle and the two auricles on that side of the body, there being a pair of lateral extensions to encompass the auricles, and also an anterior extension of each pericardial sac lateral to the paired dorsal aortae. The pericardial sacs receive a cellular strand from each of the last two (fifth and sixth) nephridia; these may be regarded as homologous with nephrostomes, but are not functional links between the lumina of the pericardial sacs and of the nephridia.

The Muscular System

A series of dorso-ventral muscles, inserted into the shell dorsally and passing downwards towards the mouth and towards the foot, is clearly arranged on a metamerically segmental plan.

The anterior series of muscles. A complex series of muscles inserted into the shell anteriorly on either side of the mouth serve to move the lips of the mouth, the velar lobes, the post-oral tentacles, and the radular apparatus. Contraction of individual muscles will cause retraction of the region concerned, and the extension of the organ in question will presumably be effected by blood pressure brought about by the contraction of other muscles in this general region. These muscles have been grouped according to the region of shell on which they are inserted, and may perhaps be taken to indicate that the head region was derived from three or more segments. The number of segments involved is far from clear, and the relation between such possible "head" segments and "body" segments is uncertain. These muscles have therefore been designated as X, Y, and Z, respectively, instead of being listed numerically (see Fig. 158).

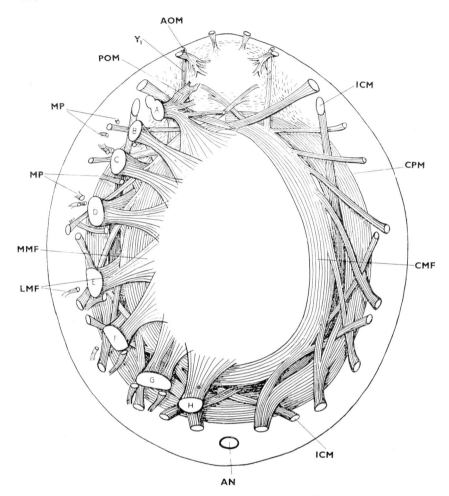

Fig. 158. *Neopilina galatheae*, diagrammatic representation of the musculature of the head and body. There are three groups of muscles in the
region of the head, X, Y, and Z, not all of which are shown in this figure;
there are eight groups of muscles in the body region, A–H. For interpretation of the lettering, see pp. 492–3. (Originally published in Lemche and
Wingstrand, 1959, *Galathea Rept.* vol. 3, plate 35, fig. 121.)

The posterior series of muscles. Here the serial repetition of parts is undoubted, and eight groups of muscles can be recognised. These have been
designated A, B, C, D, E, F, G, and H, signifying the presence of eight
"body" segments. Each of these muscle groups is complex, but each group
is built on the same common plan, and this system provides clear evidence
of metameric segmentation.

In each group the chief muscle is a retractor muscle of the foot, the fibres of which diverge into a median and a lateral bundle. These two parts are homologous with comparable muscles in chitons. The lateral bundle passes downwards into the muscular lateral wall of the foot, while the median bundle passes inwards under the floor of the peri-intestinal sinus. In addition to these powerful muscle bundles there are in each group a pair of smaller anterior and posterior retractor muscles, which run obliquely forwards, and obliquely backwards, from their insertions into the shell to join the circular muscle of the foot. This circular muscle lies peripherally in the base of the foot, and its contraction will cause the column of the foot to contract radially and thus extend in height.

There is an additional system of circular muscle fibres below the epithelium at the base of the foot and below the pallial groove. This system is not strictly circular since it is not continuous transversely at the anterior end of the animal. Fibres of this system are inserted into the shell at four points, two antero-laterally near muscle group A, and two postero-laterally near muscle group H (see Fig. 158, ICM).

In the gill-bearing "segments" the groups of dorso-ventral muscles each include an internal and an external retractor muscle of the gill. Small and irregularly disposed pallial muscles pass from their insertions into the shell, radially outwards into the margin of the mantle.

Below the radula there is a transverse muscle bundle which joins a pair of short "cartilages". The radular membrane can be made to slide forwards and backwards over this support. Laterally these "cartilages" are each attached to a radular vesicle. These two vesicles are presumably rigid in life, due to distention with fluid. The lateral border of each vesicle is attached to the ventral body wall by a group of six muscle bundles. The purpose of these is presumably to establish each radular vesicle as a fixed base upon which movements of the radula may be effected by movements of other muscles. Two of these muscles serve as protractors of the vesicles, and these pass forwards from the vesicles to their insertions on to the shell, anteriorly. The remaining four muscles radiate downwards into the ventral body wall and serve to stabilise the lateral borders of the vesicles. The single, large, retractor muscle of the radula passes backwards from the posterior and lateral borders of the radular diverticulum, towards the lateral tips of the radular vesicles. Contraction of this muscle will draw the radula backwards. A powerful protractor muscle of the radula passes forwards from the radular diverticulum, on each side of the body, to be inserted on to the shell anteriorly in the X area. Contraction of this muscle will draw the radula forwards into the mouth, while contraction of the radular retractor muscle—mentioned above—will return the radula into its sheath. Another muscle pair serves to retract the entire radular apparatus backwards into the body.

In general it may be concluded that the mode of support of the radula on a system of "cartilages" and fluid-filled vesicles, and the mode of action of the muscles operating the whole radular apparatus, are comparable with the dispositions found in the Polyplacophora.

Due perhaps to the thinness of the shell, the areas of insertion of the various muscles on to the shell do not show as muscle scars. In the known fossil genera now assigned to this class the shell is thicker, and muscle scars are present. The arrangement, and the size of these muscle scars are closely comparable with the arrangement and the size of the various muscles found in the modern *Neopilina*, and this fully justifies placing these genera in the Monoplacophora. The thicker shells of these fossils suggests that they lived in comparatively shallow waters, and, indeed, they were recovered from strata laid down in shallow water.

The serially repeated dorso-ventral muscles, the basic supporting structures for the radula, and the muscles operating the radula in *Neopilina* are directly comparable and homologous with those in the Polyplacophora.

The Nervous System

A pair of large, but ill-defined cerebral ganglia lie on either side of the mouth, and communicate with each other by a pre-oral and a post-oral commissure. A pallial nerve arises from the lateral border of each cerebral ganglion and these nerves unite ventral to the rectum to form a circum-pallial nerve system which innervates the margin of the mantle, the gills, and the renopores. The circum-pallial nerve possesses an outer layer of nerve-cell bodies, and no ganglionic swellings. Two branchial nerves, one anterior and one posterior, serve each gill (see Fig. 159).

A pair of pedal nerves arise from the post-oral cerebral commissure, close to the postero-median border of the cerebral ganglia. These pedal nerves form a circum-pedal nerve system, since the two nerves unite posteriorly, and are joined anteriorly to each other by an inter-pedal commissure. This circum-pedal nerve cord bears a superficial layer of nerve-cell bodies, and has no ganglionic swellings; it lies in a narrow peri-neural blood sinus. The pedal nerve cord supplies the muscles of the foot with median and lateral pedal nerves.

A particularly interesting feature of the peripheral nervous system is the presence of a series of ten latero-pedal connectives, which provide regular intercommunications between the pedal and the pallial ring nerves. Two small nerves leave the anterior part of the pedal nerve cords and supply the post-oral tentacle ridges. The second latero-pedal connective supplies a nerve to the statocyst, and also three small nerves to the post-oral tentacles.

The pre-oral cerebral commissure is devoid of nerve-cell bodies; it supplies nerves to the anterior lip of the mouth and also to the anterior part of the mantle margin. In addition to the post-oral commissure, the postero-median corners of the cerebral ganglia give rise to two more commissures. The first of these is the sub-radular commissure which bears a single, median sub-radular ganglion; the second is the buccal commissure, which bears paired buccal ganglia on its course, laterally.

The cerebral ganglia are triradiate, consisting of swellings on the bases of the cerebral commissure, the pallial nerve, and the pedal nerve cord. This ganglion may be comparable with the cerebropleural ganglion of bivalves.

The presence of nerve-cell bodies along the course of the lateral and

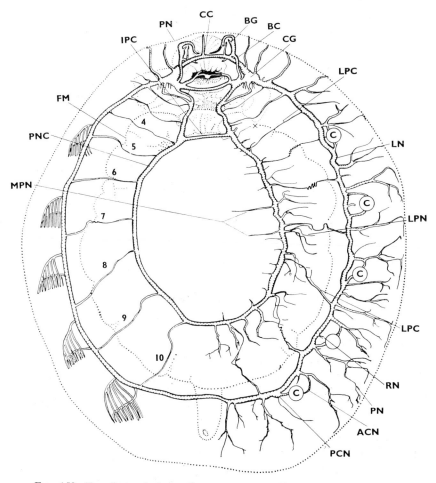

FIG. 159. *Neopilina galatheae*, diagrammatic representation of the nervous system. The lateral nerve cord (LN) and the pedal nerve cord (PNC) are linked together by a regular series of ten lateropedal connectives (LPC, 1–10). Anteriorly the lateral nerve cords swell to form the cerebral ganglia (CG). For interpretation of other lettering, see pp. 492–3. (Originally published in Lemche and Wingstrand, 1959, *Galathea Rept.*, vol. 3, plate 43, fig. 135.)

pedal nerve cords, excepting the anterior parts of both, may be regarded as a primitive feature, comparable with that in the chitons. The chitons are even more primitive in lacking a differentiated cerebral ganglion pair.

Sense Organs

Pre-oral tentacles. With their rich innervation these may be chemo-receptors or perhaps they are tactile organs.

The sub-radular organ. Innervated from the sub-radular ganglion, this pit in the posterior border of the oral cavity may be a chemo-receptor.

The statocysts. There are two statocysts, situated behind the post-oral tentacles. Each is a vesicle formed by invagination of the epithelium of the mantle cavity, with which it may remain in communication via a slender duct. (It was not possible to decide whether the duct opened into the mantle cavity.) The form of the statoliths—if any were present—is not known, but the statocysts presumably respond to gravity. The ventral epithelium of the vesicle appears to be ciliated, nerve fibres appear to originate here, and these fibres pass along the latero-pedal connective and forwards into the pallial nerve cord.

Aesthetes, cephalic eyes, and osphradia are all absent.

Polyplacophora

The Chitons, or "coat-of-mail shells" are wholly marine organisms. The body is dorso-ventrally compressed, elongated, bilaterally symmetrical, with an anterior mouth and a posterior anus. They are to be found on the sea shore, and on the sea bed down to considerable depths, and are particularly adapted in their body form for crawling over irregular rocky surfaces.

The mantle covers the whole dorsal surface of the animal including the head, which consequently lacks eyes and other cephalic sense organs. Dorsally the mantle typically bears a series of eight overlapping shell plates. Lateral to these shell plates the mantle forms a tough girdle. The latero-ventral margin of the mantle may be provided with a fringe of fine spicules. Ventrally, and separated from the head by a narrow transverse groove, there is a broad, elongated, suctorial foot. Between the foot and the margin of the mantle there is a peripheral groove, which contains a series of ctenidia, and which comprises the mantle cavity. The margin of the mantle can be raised locally to admit a respiratory stream of sea water at any point. The animal glides very slowly over rocky surfaces, being able to mould its body to the contour of the rock, due to the articulations between adjacent shell plates. The animal can therefore move over wave beaten rocks on the most exposed shores, adequately protected against dislodgement by wave impact at all times. The suctorial action of the foot is very powerful, as also are the muscles in the margin of the girdle, so it is almost impossible to detach a large chiton which has been alarmed, or which is stationary on an exposed rock.

The Polyplacophora are of ancient origin, fossils having been found in Ordovician rocks. In these early forms the eight shell plates lacked articulations. The class includes 6 families with 43 genera, of which 7 genera occur in British waters.

The Mantle and the Shell

Typically the exoskeleton consists of a series of eight transverse shell plates, each of which overlaps the anterior border of the one behind, all of the plates being visible on the dorsal surface. The anterior border of the first plate, and the posterior border of the eighth plate are approximately semicircular; the remaining six plates are roughly rectangular in shape.

Each shell plate consists of two contrasted calcareous layers. The upper layer, or "tegmentum" is perforated by numerous parallel canals which accommodate pallial outgrowths which bear apical sense organs (aesthetes, and micraesthetes, to be discussed below). The external surface of the tegmentum commonly bears a median beak, or keel, and its surface area may be subdivided into distinct areas by a variety of ornamentation which may be of aid in specific identification.

The lower layer of the shell, or "articulamentum", is formed of a more compact substance. In more primitive Polyplacophora the articulamentum is coextensive with the tegmentum, but in more advanced forms the articulamentum projects anteriorly as a shelf which underlies the posterior border of the shell plate which lies in front. The articulamentum also projects laterally and forms an insertion plate which is deeply embedded in the encircling mantle, or "girdle" as it is called. These projections of the articulamentum may bear one or more marginal notches, the number and disposition of which is of aid in specific identification. Such articular projections of the articulamentum are not found in the earliest polyplacophoran fossils from the Ordovician.

The eight shell plates articulate one with another, and form a compact skeletal structure which is manipulated by metamerically repeated muscles. The shell plates are embedded in, and are secreted by the mantle, which surrounds the shell in the form of a tough, marginal girdle. The dorsal surface of the girdle is rough to the touch, due to the presence of tubercles, irregular protuberant chitinous or calcareous spicules, and sometimes a series of compact tufts of bristles, e.g. in *Acanthochitona*.

The tegmentum is said to have no homologue in the shells of other molluscan classes; it is secreted by a lobe of the girdle which overlaps the border of the articulamentum, and the pallial outgrowths that penetrate the tegmentum and bear aesthetes arise from this lobe.

In *Cryptoplax* this reduplication of the girdle has extended over the dorsal surface of the animal, completely enclosing and hiding the shell plates from view. An unusual condition is found in *Schizoplax*, for here the shell valves are divided into left and right halves joined by a median ligament (Lemche and Wingstrand, 1959).

The Mantle Cavity

The mantle cavity is in the form of an oval gutter which encircles the animal; its median wall is supplied by the foot, and its lateral wall by the mantle, or girdle as it is called in this class. Anteriorly the mantle cavity is largely occluded by the proboscis, with the mouth in its centre. Two labial palps lie one on each side of the proboscis. The anus opens into the mantle cavity posteriorly on a small median papilla. A variable number of ctenidia lie in the mantle cavity postero-laterally, the number of ctenidia increasing

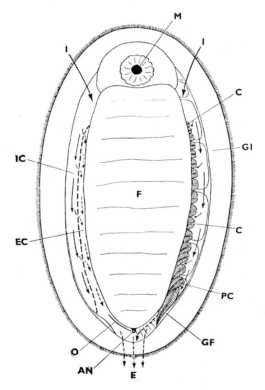

FIG. 160. *Lepidochitona cinereus*, ventral view. On the right side of the figure the approximate positions of the shell plates are shown. On the left the ctenidia are omitted in order to depict the inhalant (IC) and exhalant (EC) chambers and the ciliary currents therein. For interpretation of the lettering, see pp. 492–3. (Originally published in Yonge, 1939, *Quart. J. micr. Sci.* **81**, 371, fig. 1.)

as the animal develops. The ctenidia serve principally for respiration, but in a few examples they also afford protection for the early stages of development of eggs which are fertilised and retained in the mantle cavity (see Fig. 160). Two excretory pores open into the mantle cavity posteriorly, one on each side. The ctenidium which lies immediately behind this pore is known as the "post-renal" ctenidium, and is either the largest or at least one of the largest of the ctenidia. This is the first ctenidium to be formed (Pelseneer, 1898, 1899). Two gonad ducts open into the mantle cavity posteriorly, one on each side and a little anterior to the excretory pore.

Sensory organs, termed "osphradia", lie in the posterior part of the mantle cavity close to the anus, and are innervated by the pallial nerve. The function of these sensory patches is not known for certain.

The Ctenidia

Each ctenidium consists of an axis which hangs downwards from its attachment to the roof of the mantle cavity (see Fig. 161). Within the inner margin of the axis an afferent branchial blood vessel brings blood to the ctenidium for oxygenation; within the outer margin of the axis there

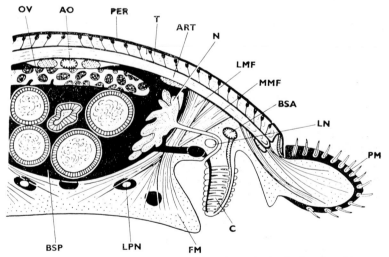

FIG. 161. Diagrammatic transverse section through the body of a chiton. The mantle margin, or girdle (PM) extends lateral to the shell plates. The ctenidia (C) lie in a lateral mantle cavity. The nephridia (N) discharge into the mantle cavity. The gut lies in a peri-intestinal blood sinus (BSP) below the gonad (ovary, OV). The venous blood supply to the ctenidia is shown in black; arterial blood passes from the ctenidia via the arterial blood sinus (BSA) to the heart. For interpretation of other lettering, see pp. 492–3. (Originally published in Lemche and Wingstrand, 1959, *Galathea Rept.*, vol. 3, plate 55, fig. 166.)

is an efferent blood vessel which drains oxygenated blood from the ctenidium. Adjacent to each of these vessels there is a bundle of longitudinal
muscle fibres, by means of which the ctenidium may be contracted, or
may be gently moved to one side. Extension of the ctenidium will be
effected by relaxation of the muscle fibres, together with the inflow of
blood. A pair of branchial nerves lie below the epithelium, external to the
afferent and the efferent vessels respectively. The ctenidial axis bears two
series of alternating short, broad ctenidial filaments, one series on the
anterior side of the axis, and the other series on the posterior side.

The upper and the lower surface of each ctenidial filament can be divided
into three zones. Of these, the outer zone bears very few cilia, the middle
zone bears a dense cover of long, lateral cilia and at the margin, a bunch
of even longer attaching cilia, while the innermost zone bears scattered
cilia. The lateral cilia, and to a lesser extent the scattered cilia of the innermost zone, create a respiratory current which drives water from the outer
part of the mantle chamber, through the spaces between adjacent ctenidial
filaments, and so into the inner, exhalant part of the mantle cavity. The
flow of blood in each ctenidial filament is from the afferent to the efferent
vessel, which is in the opposite direction to the respiratory stream of
water, and this is considered to make the most of the opportunity for
oxygenation of the blood. The long attaching cilia on the margin of each
ctenidial filament will interlock with corresponding cilia on the filaments
of the adjacent ctenidium, and in this way the whole series of ctenidia
become temporarily bound together to form a perforated partition between the inhalant and the exhalant water streams in the mantle cavity.

Along the afferent and efferent borders of the ctenidial axis, and also
on either side of the series of bunches of attaching cilia, there are cilia
which beat towards the tip of the ctenidium. These serve to collect particles suspended in the inhalant current, and to direct them to the tip of
the ctenidium. Here the particles, consolidated in a mucoid secretion,
are passed to the cleansing ciliated tracts in the exhalant chamber, and
are voided posteriorly.

The ctenidial filaments are short and broad. Their basic structure,
and ciliation are comparable with those of primitive gastropods, such as
Diodora or *Haliotis*. These similarities necessitate the conclusion that the
ctenidia of Polyplacophora are true ctenidia, and homologous with those
of other molluscan classes. The only significant difference is the absence,
in ctenidial filaments of the Polyplacophora, of internal chitinous supporting rods along the side of the filament facing the inhalant water
current (Yonge, 1947). This absence may be correlated with the shortness
of the filaments, the weakness of the inhalant current, and the consequent
lack of need for such an internal support. Whether the shortness of the
filaments is primitive, or is a secondary feature brought about due to the
increased number of ctenidia, is a matter for debate (Yonge, 1939).

The ctenidia of Polyplacophora are true ctenidia, which have become secondarily increased in number. The series of ctenidia is generally regarded as being non-segmental, as the number and disposition of the ctenidia do not correspond with those of other serial structures such as the shell valves. This view, however, was formulated prior to the discovery of the metamerically segmented *Neopilina* (Monoplacophora), and may be in part a rationalisation of the situation to fit the class to a phylum which was supposed to be wholly lacking in segmental structure. Perhaps the speculation is now permissible that in the Polyplacophora there is a number of body segments, and that perhaps more than one of these bears true, segmental ctenidia, but that the segmental arrangement of the ctenidia has been obscured through a secondary increase in numbers.

The Foot

The foot is a broad, creeping sole which is admirably adapted for progression over irregular rocky surfaces through the passage along the foot of waves of muscular contraction; it is also well adapted to gripping the rock surface suctorially, giving more than adequate protection against dislodgement by wave impact.

The Alimentary Canal

The proboscis lies antero-ventrally in front of the foot, from which it is separated by a narrow transverse groove. The mouth lies in the centre of the proboscis, and leads vertically into a buccal cavity, the walls of which are invested by a cuticular sheath (see Fig. 162). There are no jaws. The posterior wall of the buccal cavity bears a blind sac, the walls of which are also lined by a sheath of chitin. On the dorsal wall of this sac there is a sensory sub-radular organ, which is not covered by chitin. The posterior half of this organ is glandular, while the anterior half is sensory, and is innervated from the sub-radular ganglion which lies immediately above. Antero-dorsally the buccal cavity bears a pair of salivary glands, which may be homologous with those of the monoplacophoran, *Neopilina.*

Postero-dorsally the buccal cavity communicates with the oesophagus above, and the radula sac below. The radula sac is a blind cylindrical tube underlying the oesophagus, and containing the radula. The radula consists of a flexible chitinous radular membrane, which bears a series of transverse rows of chitinous teeth, with seventeen teeth in each row. (The radular formula is 8.1.8.) In each row there are three comparatively small teeth centrally, on either side of which there is a single large tooth and then a set of six smaller, polygonal teeth. The third of these six lateral teeth may be enlarged, and its distal end may be pectinate. A somewhat

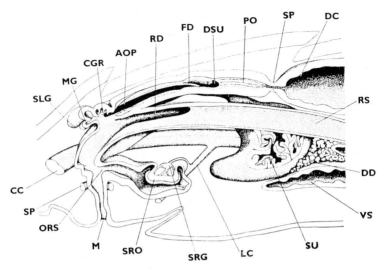

Fɪɢ. 162. *Lepidochitona cinereus*, sagittal section through the anterior part of the alimentary canal. The sub-radular organ (SRO) lies in a sac which opens into the buccal cavity. The radular sac (RS) lies below the oesophagus and also opens into the buccal cavity. Sphincter muscles (SP) control the mouth (M) and the passage between the posterior oesophagus (PO) and the dorsal food channel (DC). Salivary glands (SLG) open into the buccal cavity. For interpretation of other lettering, see pp. 492–3. (Originally published in Fretter, 1937, *Trans. roy. Soc. Edinb.* **59**, 122, fig. 2.)

similar condition is found in *Neopilina* where, however, there are fewer lateral teeth.

The processes of feeding and digestion have been studied by Fretter (1937) on *Lepidochitona cinereus* and *Acanthochitona crinitus*. In these examples the radula is supported by a pair of fluid-filled vesicles which lie laterally and are themselves supported by four sub-median "cartilages". The radula is moved over this supporting framework by a series of muscles, so that the anterior end of the radula protrudes through the mouth and is then withdrawn again. In general, structure and function are closely similar with *Neopilina*. The above-named chitons feed by rasping filamentous, and encrusting algae, and debris, from the surface of the rocks over which they glide. They may also rasp bits off larger seaweeds. The prospective feeding area is first tested by extrusion of the sub-radular organ through the mouth and placing it against the substrate; the radula is then protruded and withdrawn, the teeth of the radula rasping food particles off the substrate in the withdrawal movement. The substrate is again tested by the sub-radular organ prior to the next "bite". As the radula is withdrawn it becomes curled to form a shallow gutter, with the lateral teeth pointing inwards and backwards. Small particles of food material on the radula are therefore caught between the lateral teeth and are drawn up into the buccal cavity. Mucus from the salivary glands will lubricate the radula and prevent it from scraping the walls of the buccal cavity and the mouth. The radula serves to supply the stomach with minute particles of plant material and organic debris, bound together with mucus secreted by glands in the walls of the buccal cavity, and this is regarded as being the type of feeding mechanism possessed by the most primitive ancestral molluscs.

Oesophageal glands secrete a diastase and a glycogenase on to the ingested food material. The digestive diverticula open by two ducts into the posterior end of the stomach, which lacks a crystalline style. The orifices of the ducts from the digestive diverticula are each protected by a sphincter muscle. The digestive diverticula are arranged in two masses, which fill the spaces between the coils of the mid-gut and the stomach. Each lobe consists of a very large number of blind tubules. The walls of the tubules of the digestive diverticula consist of two types of cells, one of which secretes an extracellular protease which passes into the lumen of the stomach. The other type of cell contains aggregations of lime the functions of which are not known for certain. The contents of the stomach and of the anterior part of the mid-gut are highly acid. Digestion is principally extracellular, though some wandering phagocytes may ingest particulate matter and then return through the wall of the stomach. Muscular contractions of the stomach wall play an important part in mixing the food material with the extracellular enzymes. Since no cellulase is present the cell walls of plant material are indigestible, and large quantities of plant material may pass through the gut undigested. The soluble products of digestion are absorbed by the tall secretory cells of the tubules of the digestive diverticula. In the Polyplacophora the hind part of the mid-gut is remarkably long and this may be related to the importance of consolidation of the faeces into firm pellets which will not foul the posteriormost ctenidia. The posterior part of the mid-gut and the short hind-gut are lined by a ciliated epithelium, action of which imparts a spiral backward motion to the faecal pellets. Amoebocytes in the lumen of these parts of the gut can pass freely through the epithelium from the lumen of the gut into the haemocoele.

In contrast to *Lepidochitona* and *Acanthochitona*, which have been shown to feed on small particles of plant material and of detritus, some of the Polyplacophora are carnivorous. Barnawell (1959) has shown that the gut contents of five species of *Mopalia* contain from 15 per cent to 60 per cent of animal matter, this including sponges, hydroids, bryozoans, annelids, barnacles, cyprid larvae, and small specimens of *Mytilus*. *Mopalia lignosa* and *M. porifera* were stated to be sufficiently massive "to rasp entire barnacles along with their shells from their places of attachment". Since Polyplacophorans rasp their food from rock surfaces it is to be expected that at least some of their food material will be of animal origin.

On the other hand, McLean (1962) has shown that *Placiphorella*, another member of the family Mopalidae, is a predacious carnivore. *P. velata* possesses a pre-cephalic mantle lobe which bears tentacle-like projections, and also a large anterior head flap, an extension of the girdle which can be raised above the substratum (see Fig. 163). *P. velata* feeds

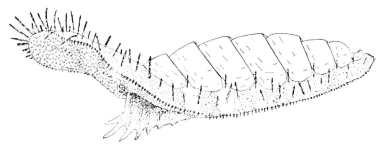

FIG. 163. Lateral view of *Placiphorella velata* in the feeding position, the head-flap being raised and the pre-cephalic tentacles being pressed against the substratum. When small crustaceans wander under the raised head-flap this clamps down suddenly and traps the prey, which is then ingested. (Originally published in McLean, 1962, *Proc. malac. Soc. Lond.* **35,** 24, fig. 2.)

upon small amphipods and other comparably small organisms which are trapped under the head flap. The red-coloured undersides of the head flap project conspicuously from crevices in the rocks and when its under side is touched by a small moving organism, such as an amphipod, it clamps down on the rock quickly, thus trapping the prey. The movement is astonishingly rapid for an animal belonging to a class which is generally regarded as being rather sluggish. Specimens studied in aquaria fed upon amphipods up to 6 mm long, on small shrimps, polychaet worms, and even on small crabs. Crustaceans over 2 mm long were apparently torn up by the radula and were swallowed piece by piece. *P. velata* was also found to be able to feed on plant material.

The Vascular System

A large pericardium lies dorsally in the posterior half of the body, beneath the last two shell plates. Blood is collected from the ctenidia by two auricles, which discharge into a single, median ventricle by one or more auriculo-ventricular valves on each side of the body. There is one such valve in the Lepidopleuridae, the Mopalidae, and certain of the Ischnochitonidae; there are typically two such valves on each side in the Acanthochitonidae, Cryptoplacidae, and Chitonidae, while there are three in *Chiton squamosus* and four in *C. goodalli.* It may be questioned whether this could be regarded as evidence of a primitive segmental structure.

The left and right auricles intercommunicate posterior to the ventricle. The ventricle is elongated, and discharges blood forwards into a single anterior aorta which serves the various organs, and passes blood into a peri-visceral blood sinus. Venous blood passes from this to a longitudinal blood sinus which lies alongside the mantle cavity, and which supplies blood to the afferent branchial blood vessels.

In the very large *Ischnochiton magdalenensis* a special visceral arterial system is developed in association with the gut, and the thick walls of these vessels are largely composed of storage cells, in which droplets of fat and spherules of protein are stored. This development may be related to the large size of the animal, for such a storage tissue is not known in smaller species of chiton (Fretter, 1937).

The Excretory Organs

There are two nephridia, which lie symmetrically on either side of the body. Each is a U-shaped tube of which one end opens posteriorly into the pericardium by a short, slender, ciliated funnel. In some cases the two limbs of the excretory organ have fused to form a single forwardly directed tube, with reno-pericardial, and excretory orifices at its posterior end, e.g. *Nuttalochiton.* The other end of the tube opens posteriorly into the exhalant part of the mantle cavity immediately in front of the post-renal ctenidium, at the excretory pore. The mass of the nephridium presses forwards in the peri-visceral blood sinus, and bears numerous median and lateral branches which increase the area of its walls, which are all excretory. In some genera the left and right nephridia are adjacent, and may show various degrees of fusion.

The nephridia do not play any part in the liberation of ova or spermatozoa; the gonad has its own duct. It is possible therefore, as has been suggested by Lemche and Wingstrand (1959), that the single pair of nephridia are derived from a "pericardial segment" and have come to serve the whole body, the gonad and the genital duct being derived from a "pre-pericardial segment". If this is true, although the nephridia and the gonads do not themselves display metameric segmentation, it could be concluded that in the Polyplacophora there are at least two "body segments", one of which has become specialised for excretory purposes and the other for reproduction.

The Reproductive System

The sexes are separate; the testis, as also the ovary, is an unpaired median sac anterior to the pericardium, which communicates with the lateral pallial groove by a pair of gonad ducts, one on each side. These gonad ducts do not communicate with the nephridia. The eggs are liberated into the pallial groove where in some cases they are fertilised and undergo early development. In other examples the fertilised eggs may be laid singly or in strings. The gonad lies in the perivisceral blood sinus, dorsal to the viscera.

The Nervous System

The nervous system generally resembles that of *Neopilina* (Monoplacophora) in consisting mainly of a circum-oral nerve ring, and paired pallial and pedal nerve cords. It is even more primitive than that of the Monoplacophora, for there are no cerebral ganglia, and nerve-cell bodies are distributed over the course of all the larger nerve cords.

The circumpallial nerve cord circumnavigates the body, lying in the mantle above the mantle cavity, fusing with its counterpart anterior to the mouth, and dorsal to the rectum. A ventral commissure unites the two lateral pallial cords posterior to the mouth. This circumpallial nerve ring supplies the margin of the mantle, the ctenidia, and the osphradia, with an irregular series of nerves, and it also innervates the gonads, heart and excretory organs. In addition to the anterior cerebral commissure and the ventral commissure, two pairs of nerves pass inwards anteriorly from the pallial cord. One pair passes to the paired buccal ganglia, which serve the complex of muscles which operate the radula, while the other pair pass to the paired sub-radular ganglia which serve the sensory epithelium of the sub-radular organ.

The pedal nerve cords pass backwards from the head region in a blood sinus, through the muscular substance of the foot. The pedal cords supply nerves irregularly to the muscles of the foot, both medially and laterally. Irregular transverse nerves join the pedal and the pallial nerve cords.

There is no trace of segmental structure in the nervous system in the Polyplacophora.

The structure of the nervous system in the Polyplacophora suggests that the buccal ganglia were the first ganglia to become elaborated in the Mollusca, this development being necessary at a very early stage to give adequate control of the complex of muscles operating the radula.

Sense Organs

Mention has already been made of the sub-radular sense organ, by means of which potential food is tested prior to each rasping action. Mention has also been made of the osphradia and the sensory patches at various points on the epithelium of the mantle cavity, which may serve to test the inhalant stream of water for quantity of suspended matter, or for the presence of sperm or ova of the same species. There are no statocysts in the Polyplacophora (Lemche and Wingstrand, 1959).

There is a lack of macroscopic cephalic sense organs, and this may well be a primitive condition.

Pallial eyes of rather a peculiar form may be present; the shell plates, particularly the anteriormost plate, are penetrated by pallial outgrowths which terminate just below the surface of the shell in a visual organ which possesses a pigment layer, a retina, and sometimes also a lens. These visual organs are innervated by fibres of the pallial nerve cords, and are presumably photo-receptors; they are commonly known as "aesthetes". These aesthetes are usually arranged in rows which run diagonally from the centre of each valve, outwards to its lateral border.

The pallial papilla which terminates in an aesthete also gives off laterally a number of slender processes which also penetrate the shell and terminate in small tubercles at the surface of the shell. Each of these smaller organs is known as a "micraesthete". They are presumably sense organs, but their function is not yet known.

Behaviour

Evans (1951) has studied the movements of *Lepidochitona cinereus* and found that this species exhibited a negative photo-orthokinesis, tending to move at random and to accumulate in regions of deepest shade under experimental conditions. When exposed in air on damp surfaces, *L. cinereus* moved downwards, exhibiting positive geotaxism under these conditions.

Aplacophora

The Aplacophora are a small group of worm-like, sub-littoral marine molluscs, which are partly primitive and partly specialised in form. They have been the subject of comparatively little recent investigation.

In the majority of genera there is a median ventral groove with a central longitudinal ridge which represents the foot, and the mantle covers the dorsal, lateral, and the greater part of the ventral side of the animal. A large mucus gland opens into the groove anteriorly. In *Chaetoderma*, the ventral groove and the pedal ridge are absent and the mantle forms a continuous cylindrical external surface. In all cases there is no shell, the exo-skeleton being represented only by a cuticular layer which bears a variety of forms of spicule. At the posterior end of the body there is a chamber which can be equated with the mantle cavity of other molluscan classes; the anus and the urino-genital apertures open into this mantle cavity, which also contains the gills. In *Chaetoderma* the two gills are bi-pectinate and are comparable with those of the Polyplacophora; in *Neomenia* and in several other genera there is a circlet of laminar gills in the mantle cavity, whilst in some genera there are no gills.

Specialised features include the reduction or loss of the foot, the loss of the shell, the great diversity or sometimes the loss of the radula, and the modification of the nephridia in certain genera to form accessory sexual organs.

The possession of well-defined cerebral and pleural ganglia indicates that the Aplacophora are more advanced than the Polyplacophora at least in this respect, and this suggests that the simple body form of the Aplacophora is probably attributable to secondary simplification rather than to primitive simplicity.

Pelseneer (1906) regarded the Aplacophora as degenerate forms, derived from a polyplacophoran ancestry.

The Alimentary Canal

The anterior mouth opens into a muscular pharynx which is lined by a thick cuticle, and which typically receives one, or two pairs of salivary glands and the radula sac. In some genera the salivary glands are absent, e.g. *Neomenia*. The radula is highly variable in form; there may be several teeth, two teeth, or even only one tooth in each transverse row, whilst in a number of genera the radula is absent.

The animals are all carnivorous, feeding on Protozoa, or corals, or hydroids, etc. The alimentary canal is short and straight. The oesophagus is short, and leads into a cylindrical stomach which may bear symmetrical lateral caeca, and the walls of which constitute a secretory and, presumably, an absorptive surface. In *Chaetoderma* a comparatively large "hepatic caecum" passes backwards ventral to the mid-gut. In some genera, on the other hand, a caecum may pass forwards dorsal to the oesophagus. The mid-gut is short and straight, and is lined by a ciliated epithelium; the anus opens into the mantle cavity.

The Vascular System

The pericardium lies at the posterior end of the body, adjacent to the mantle cavity and dorsal to the hind-gut. In those species in which gills are present blood passes from a haemocoele into the gills, and from the gills via a pair of "auricular ducts" into the ventricle. The ventricle is contractile and pumps blood forwards into a dorsal sinus. Blood passes from this into a peri-visceral blood space and thence to the gills for oxygenation before being returned to the heart. The blood contains haemoglobin in red blood corpuscles. Where gills are absent, oxygenation occurs in the epithelium of the inner wall of the mantle cavity, and of the foot.

The Reproductive and Excretory Systems

The majority of Aplacophora are hermaphrodite, having paired gonads which discharge their products into the pericardium; each gonad is an ovo-testis, producing ova on its median wall and spermatozoa on its lateral wall. The paired nephridia are U-shaped, with inner openings into the pericardium. The proximal limb of each nephridium may bear one or two caeca, which serve as sperm reservoirs. In most cases the distal limbs of the nephridia fuse in their terminal portions to form a median pouch the walls of which are glandular and serve to secrete the egg-shells. In some species copulation occurs, and spicules mounted laterally to the single urino-genital aperture apparently act as copulatory, or as excitatory organs.

In *Chaetoderma*, which is uni-sexual, there is a single median gonad; the nephridia are more typically molluscan in form, with no modifications to serve the reproductive system.

The Nervous System

For such small and comparatively simple worm-like animals the nervous system is moderately well developed, showing a pronounced advance as compared with the Polyplacophora in the possession of well-defined cerebral and pleural ganglia. A pair of cerebral ganglia lie antero-dorsal to the pharynx and give rise to two pairs of longitudinal nerve trunks which are homologous with the pedal and the pleural nerve cords of the Polyplacophora and the Monoplacophora. The anterior ends of the pedal cords are developed into a pair of communicating pedal ganglia, and pleural ganglia are situated at the anterior ends of the pleural nerve cords. The pleural cords unite posteriorly by an anastomosis dorsal to the rectum. The pedal cords are connected to each other, and also to the pleural cords, by a regular series of transverse nerve fibres. There are ganglionic swellings at the bases of these transverse fibres, on the pedal trunks.

In addition to the above, the cerebral ganglia give rise to two connectives which pass to a pair of stomatogastric, or buccal ganglia. These ganglia lie below the anterior end of the oesophagus, and serve to operate the radula where this is present.

The serial ganglionic swellings and transverse nerve fibres leaving the pedal trunks are suggestive of metameric segmentation, but this indication is not reflected in any other organ system.

Sense Organs

As would be expected in a group of small animals which live in comparatively dark surroundings in water of at least moderate depth, there are no
organs of special sense comparable with the cephalic sense organs found in
the more advanced molluscan classes, i.e. there are no eyes, tentacles, or
statocysts. The only known sensory organs are club-shaped epithelial pro-
tuberances covered by a thick cuticular layer, and also a median invagin-
able sensory papilla which lies postero-dorsally above the hind-gut. Numerous papillae surrounding the mouth in some genera may be sensory.

Scaphopoda

This is a small, but distinctive class, of which the fossil record extends
back as far as the middle Silurian (Pelseneer, 1906). Pelseneer stated that
there are about 150 living and 275 fossil species, and he arranged these in
two families each of which comprised five genera. Thiele (1934), on the
other hand, recognised only two families of one and of three genera respectively. Winckworth (1932) records two species of *Dentalium*, two of *Sipho-
nodentalium*, and one of *Cadulus*, as occurring in British marine waters.
Recognition of the status of the group dates from 1857 when Lacaze
Duthiers placed *Dentalium* in an independent group, the Solenoconcha
(=Scaphopoda) on the basis of its special anatomical features. Studies of
the living animal are apparently limited to the works of Yonge (1937) and
of Morton (1959) on *Dentalium entalis*, and of Dinamani (1964) on *D.
conspicuum* and references will be made to these in the following account.

The Mantle and the Shell

The shell is a gently curved, tapering cylinder of which the concave side
is dorsal. It is about $1\frac{1}{2}$ in. in length. The shell is lined internally by a tubular mantle which is perforate at both extremities, the margin of the mantle
at the larger, anterior, orifice being the site of increments to the shell during
growth. *D. entalis* occurs in offshore waters, and is found burrowing at
moderate depths (15 to 100 fathoms) in fine or medium sand, shell gravel,
or even in gravel (Marine Biological Association, U.K., 1957).

Observations on the Living Animal

Observation of *D. entalis* in aquaria shows that it lies three-quarters
buried in the substratum with the dorsal concave surface uppermost and with
the small posterior orifice of the shell protruding from the sand. Burrowing
in the sand is effected by the muscular foot. Protrusion of the foot is by
blood pressure, and the firm, pointed tip of the foot easily penetrates the

substratum. On either side of the foot there is an erectile fold of skin which is retracted and pressed closely against the sides of the foot when this is thrust into the sand. When the foot is embedded in the substratum the lateral fold is inflated with blood to form a wide flange which anchors the foot firmly. Muscular contraction in the basal half of the foot then causes the shell to be drawn forwards through the soil. Periodic sudden contractions of the foot are probably responsible for the violent expulsion of water

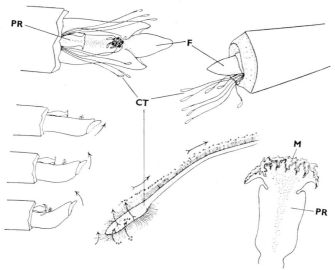

FIG. 164. A ciliary feeding mechanism employed by *Dentalium conspicuum*. The captacula (CT) are extended into the substratum and small particles are passed backwards along a ciliary pathway on the stalk of the captaculum to an area above the base of the foot (F), where the material can be ingested by the proboscis (PR). For interpretation of other lettering, see pp. 492–3. (Originally published in Dinamani, 1964, *Proc. malac. Soc. Lond.* **36**, 2, fig. 1.)

from the apex of the shell, which has been observed at intervals of about 10 minutes; these contractions of the foot may aid in the extrusion of faeces from the anus, and the faeces will then be carried out of the shell via the posterior orifice in the accompanying powerful exhalant jet of water. The gradual expansion of the foot may be partly responsible for the slow intake of water through the same posterior orifice, as the ciliary mechanisms in the mantle cavity are not thought to be sufficiently powerful alone to generate the observed slow inhalant current.

The only observations ever to have been made on the processes of feeding of a scaphopod appear to be those of Dinamani (1964) on *D. conspicuum*. In this species the swollen tips of the captacula are ciliated and a ciliated tract leads up the captaculum to its base (see Fig. 164). Dinamani observed

small particles being collected from the substratum on the tips of the captacula, and being conveyed up the ciliated tracts to accumulate in a furrow
on the dorsal surface of the foot just in front of the frilly lips of the proboscis. Periodical upward movements of the foot then cause the accumulation
of small particles to be moved closer to the mouth. From this point the
mass of small particles, bound in mucus, was probably ingested by muscular movements of the lips, but this was not actually seen to occur.

The Mantle Cavity

The mantle cavity of *D. entalis* is very simple in form. There are no
ctenidia, the respiratory exchanges of such a small animal being adequately performed by the mantle. There is a weak ciliation on the roof and
sides of the posterior half of the mantle cavity, particles falling in this area
being conveyed forwards rather slowly towards a system of seventeen oblique ciliated ridges that lie on the sides and floor of the mantle cavity just
in front of the anus. These ridges bear long and active cilia which doubtless
contribute to the slow inhalant current, and which direct particles falling
on them towards the base of the foot.

The mouth is a horizontal slit with thickened and lobed upper and lower
lips. It lies at the end of a muscular proboscis which arises dorsally at the
base of the foot and which overlies the proximal half of the foot. On either
side of the base of the proboscis there is a broad lobe from which arise a
large number of prehensile filaments which are known as captacula. On
each side of the proboscis there may be nearly a gross of these captacula of
various sizes (see Fig. 165). The smaller captacula are probably of the most
recent formation, and they will grow to take the place of others which may
have been lost by autotomy. The captacula are capable of great elongation
when they are protruded by pressure of blood; they are passed out through
the anterior mantle pore and are thrust into the substratum in all directions,
and attach themselves to foraminifera lying in the sand; they draw these
foraminifera back by the contraction of longitudinal muscles, and pass them
into the mouth. The process has apparently never been directly observed,
since the animals are rather "shy", but it is considered probable that the
captacula become attached to the foraminiferan by suction, the tip of each
captaculum being expanded to form a small bulb with a terminal ciliated
depresion. Probably a number of captacula become attached to a single
foraminiferan and jointly drag it back to the mouth for ingestion. This not
only presupposes a suctorial grip by the end of the captaculum, but also
the existence of chemo-receptors at the tips of the captacula for the recognition of suitable food material. Both these functions remain to be clearly
demonstrated. There is a small nerve ganglion near the tip of each captaculum.

In *D. entalis* the food ingested consists of various species of foraminifera including *Elphidium* (=*Polystomella*), *Bulimina*, *Quinqueloculina*, and *Discorbis*. It has been suggested that the captacula serve to remove particulate material which accumulates at the base of the foot due to the action of ciliary mechanisms in the mantle cavity and to convey this material to the mouth, where it is ingested. If this occurs regularly in nature it could be

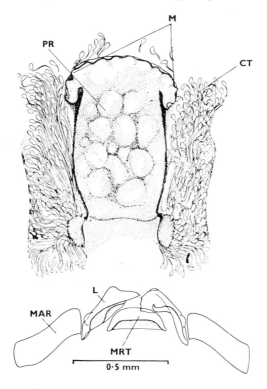

FIG. 165. *Dentalium entalis*, dorsal view of the anterior end after removal of the shell and the dorsal mantle wall. The captacula (CT) lie retracted at the sides of the proboscis (PR), which is seen to be packed with ingested foraminiferans. A single transverse row of radular teeth is shown below. For interpretation of other lettering, see pp. 492–3. (Originally published in Morton, 1959, *J. mar. biol. Ass. U.K.* **38**, 233, fig. 4.)

claimed that *Dentalium* feeds partly on organic debris, but the most recent information is that *Dentalium entalis* is mainly a highly selective feeder on foraminifera.

The Alimentary Canal

Behind the proboscis there is a buccal mass into which the ingested material is passed, perhaps partly by peristaltic contraction of the proboscis and partly by the action of the radula. The radula is housed in a radula sac. It bears approximately eighteen transverse rows of teeth, with five teeth in each row. The radula tooth formula is 1.1.1.1.1, the central tooth being a crescentic transverse plate without cusps. On either side of the central tooth there is a powerful curved lateral tooth which is flanked in turn by a simple marginal plate. Protrusion of the radula doubtless causes the two lateral teeth to diverge, and on withdrawal of the radula these teeth will come together and will grasp any object of suitable size which lies in reach. In this way foraminifera lying in the proboscis will be grasped and drawn back individually into the buccal mass. Such a radula, of relatively immense size, which is clearly highly specialised in function as also in form, cannot be regarded as being a primitive feature.

A median chitinous jaw lies antero-dorsally in the buccal mass, and possibly foraminiferan skeletons are broken by being pressed against this jaw while they are firmly grasped between the lateral teeth of the radula, and before they are passed into the oesophagus. The oesophagus bears a pair of oesophageal pouches, which are judged to be homologous with those of the Polyplacophora and of the Archaeogastropoda. The oesophagus opens into the stomach on its right side. The globular stomach is thin-walled, yet it is strongly contractile, and serves chiefly as a gizzard. Paired digestive diverticula open into the stomach posteriorly by very wide apertures. The mass of the left and right lobes of the digestive diverticula lies posterior and lateral to the stomach, and ventral to the gonad.

A series of ciliated ridges extends over the posterior wall of the stomach, some of these ridges reaching the orifices of the ducts from the digestive diverticula, while others pass up to the roof of the stomach. Ciliary currents pass along the grooves between these ridges, and direct waste material into the mid-gut, which arises on the left anterior wall of the stomach. The floor, and part of the sides of the interior of the stomach are lined by a cuticular sheath which is raised at one point to form a little spur. This can be regarded as homologous with the gastric shield of the Bivalves.

Waste material from the stomach is periodically squeezed by peristalsis into the mid-gut, which describes a double loop anterior to the stomach before descending as the hind-gut to the anus. The faeces, which are not formed into compact pellets, are extruded periodically due to muscular contraction of the foot.

The Vascular System

The vascular system is extremely simple, there being no pericardium, no heart, and no vessels differentiated into arteries and veins; there is simply a spacious haemocoele with extensions into the mantle, the foot, and the captacula. This system of blood spaces has no endothelial lining. A blood space near the anus bears a contractile wall; this space is in communication with the mantle cavity via two pores through which blood may be emitted during extreme contractions of the viscera (Pelseneer, 1906). Movement of the blood through this system of sinuses is probably chiefly effected by contractions and expansions of the foot.

The Excretory Organs

There are two excretory organs, which lie antero-ventral to the gonads. Each is a fairly wide sac with pleated walls, and opens into the mantle cavity lateral to the anus. There is no internal aperture, and the two excretory organs do not intercommunicate.

The Reproductive System

The sexes are separate; there is a single, unpaired gonad in the roof of the mantle cavity, posterior to the digestive diverticula. The anterior end of the gonad leads into a single duct which opens into the right excretory organ (Pelseneer, 1906). The gametes are liberated into the water, fertilisation of the ova and development of the larvae being external.

Embryology and Larval Development

Knowledge of the larval development of *Dentalium* apparently rests mainly on the classical study by Kowalewsky (1883). The eggs are laid singly and undergo irregular development to produce a blastula with six micromeres and one macromere. Gastrulation is by invagination and the embryo elongates and acquires an apical tuft of cilia and a velum. The young larva is a trochophore. The shell gland, which is formed on the dorsal surface, extends over the right and left sides of the larva so producing symmetrical mantle lobes which eventually fuse in the mid-ventral line. After fusion of the mantle lobes the shell assumes a tubular form. At the age of 5 or 6 days the veliger larva leaves the plankton, the velum atrophies, and the young individual begins to crawl on the substratum. Finer details of the results of experimental study of the development of *Dentalium* are to be found in Raven (1958).

The Nervous System

The nervous system consists of four pairs of nerve ganglia, joined together by connectives. The cerebral ganglia are fused into a single mass and lie dorsal to the oesophagus. They serve the proboscis and the captacula. The pleural ganglia, which innervate the mantle, lie closely behind the cerebral ganglia to which they are connected. Cerebro-pedal and pleuro-pedal connectives pass downwards from the cerebral and the pleural ganglia respectively. On each side of the body these connectives fuse to form a single connective which passes forwards to the pedal gang-lion on that side of the body. The two pedal ganglia lie close together in the middle of the foot, which they innervate. Two pleuro-visceral com-missures pass backwards, one from each pleural ganglion, to the visceral ganglia which lie on either side of the anus and which are interconnected by a short commissure anterior to the rectum.

In addition to the above ganglia and commissures, which follow essentially the same plan as in the Gastropoda and the Bivalvia, there is also a stomato-gastric complex which finds its counterpart in the Poly-placophora, Gastropoda, and Cephalopoda. This nerve complex arises from the cerebral ganglia and consists of a number of small ganglia and nerves which serve the oesophagus and the buccal mass.

Sense Organs

Three types of sense organs are known in the Scaphopoda. There are rod-like end-organs between the epithelial cells at the distal ends of the captacula. These are, perhaps, the chemo-receptors required for recogni-tion of suitable food material, and they are in communication with the ganglion which lies within the captaculum, near its tip. There is a sub-radular organ, which is a ciliated ridge on the ventral side of the buccal mass. Nerve end-cells occur in the epithelium of this ridge, and they are served by a pair of nerves which communicate with the cerebral ganglia. Two statocysts, each of which contains a number of statoliths, lie in the substance of the foot posterior to the pedal ganglia. The statocysts are innervated from the cerebral, and not from the adjacent pedal ganglia.

Key to the Lettering on the Figures

ACN	Anterior ctenidial nerve.	CAL	Cuticularised corner of ante-rior lip of mouth.
AL	Anterior lip of mouth.		
AM	Anterior muscle of the mouth.	CC	Cerebral commissure.
AN	Anus.	CG	Cerebral ganglion.
AO	Aorta.	CGR	Ciliated groove.
AOP	Anterior oesophageal pouch.	CMF	Circular muscle of the foot.
ART	Articulamentum.	CP	Cuticular plate on the anterior lip and in the ventral wall of the pharynx.
AU	Auricles.		
AVC	Entrance of arterial vessel from the ctenidium.		
		CPM	Circular pallial muscle.
AVV	Auriculo-ventricular valve.	CT	Captacula.
BC	Buccal connective.	DC	Dorsal food channel.
BG	Buccal ganglion.	DD	Digestive diverticula.
BS	Blood sinus.	DSU	Duct of sugar gland.
BSA	Arterial blood sinus.	E	Exhalant water stream.
BSP	Peri-intestinal blood sinus.	EC	Exhalant chamber.
BSV	Venous blood sinus.	F	Foot.
C	Ctenidia.	FD	Fold dividing posterior oeso-phageal pouch.

FF	Feeding furrow.	PE	Pericardium.
FM	Margin of foot.	PER	Periostracum.
G	Gonad.	PG	Pallial groove.
GF	Girdle fold.	PL	Posterior lip of mouth.
GI	Girdle.	PM	Pallial margin.
I	Inhalant water stream.	PN	Pallial nerve.
IC	Inhalant chamber.	PNC	Pedal nerve cord.
ICM	Intermediate circular muscle.	PO	Post-oesophagus.
IPC	Inter-pedal commissure.	POM	Posterior muscle of mouth.
L	Lateral tooth of radula.	POT	Post-oral tentacles.
L_1-L_5	First to fifth lateral radular teeth.	PR	Proboscis.
		PT	Pre-oral tentacle.
LC	Labial commissure.	R	Rectum.
LMF	Lateral muscle of the foot.	RA	Radula.
LN	Lateral nerve cord.	RD	Diverticulum of radular sac.
LPC	Latero-pedal connective.	RN	Nerve to renopore.
LPN	Lateral pedal nerves.	RP	Renopore.
M	Mouth.	RS	Radular sac.
MAR	Marginal tooth of radula.	S	Statocyst.
MG	Mucous gutter.	SCC	Sub-cerebral commissure.
MMF	Median muscle of the foot.	SG	"Anterior salivary gland."
MP	Pallial muscles.	SH	Shell.
MPN	Median pedal nerves.	SLG	Salivary gland.
MRT	Median tooth of the radula.	SM	Margin of shell.
N	Nephridia.	SP	Sphincter.
O	Osphradium.	SRG	Sub-radular ganglion.
OC	Oral cavity.	SRO	Sub-radular organ.
OD	Oviduct.	SU	"Sugar gland."
OE	Oesophagus.	T	Tegmentum.
OPD	Opening into pharyngeal diverticula.	TR	Tentacle ridge.
		TTR	Transverse part of tentacle ridge.
ORS	Opening of radular sac into buccal cavity.	V	Ventricle.
OV	Ovary.	VE	Velum.
PC	Post-renal ctenidium.	VS	Ventral sac of stomach.
PCN	Posterior ctenidial nerve.	VVE	Ventral edge of velum.

Reference List

BARNAWELL, E. B. (1959) The carnivorous habit among the Polyplacophora, *Veliger* **2**, 85–88.

BORRADAILE, L. A., POTTS, F. A., EASTHAM., L. E. S. and SAUNDERS, J. T. (1958) *The Invertebrata*, 3rd edn., Cambridge University Press.

CLARKE, A. H. and MENZIES, R. J. (1959) *Neopilina* (*Vema*) *ewingi*, a second living species of the palaeozoic class Monoplacophora, *Science* **129**, 1026–7.

DINAMANI, P. (1964) Feeding in *Dentalium conspicuum*, *Proc. malac. Soc. Lond.* **36**, 1–5.

EVANS, F. G. C. (1951) An analysis of the behaviour of *Lepidochitona cinereus* in response to certain physical features of the environment, *J. Anim. Ecol.* **20**, 1–10.

FRETTER, V. (1937) The structure and function of the alimentary canal of some species of Polyplacophora (Mollusca), *Trans. roy. Soc. Edinb.* **59**, 119–64.

KOWALEWSKY, A. (1883) Étude sur l'embryogénie du Dentale, *Ann. Mus. Hist. nat. Marseille* **1**, No. 7.

LEMCHE, H. (1957) A new living deep-sea mollusc of the Cambro-Devonian class Monoplacophora, *Nature, Lond.* **179**, 413–6.

LEMCHE, H. and WINGSTRAND, K. G. (1959) The anatomy of *Neopilina galatheae* Lemche, 1957 (Mollusca, Tryblidiacea), *Galathea Rep.* **3**, 1–63, 56 pl.

MARINE BIOLOGICAL ASSOCIATION, U.K. (1957) *Plymouth Marine Fauna.*

MATTHEWS, G. (1953) A key for use in the identification of British chitons, *Proc. malac. Soc. Lond.* **29**, 241–8.

MCLEAN, J. H. (1962) Feeding behaviour of the chiton *Placiphorella, Proc. malac. Soc. Lond.* **35**, 23–26.

MORTON, J. E. (1958) *Molluscs*, Hutchinson University Library, London.

MORTON, J. E. (1959) The habits and feeding organs of *Dentalium entalis*, *J. mar. biol. Ass. U.K.* **38**, 2, 225–38.

PARKER, T. J. and HASWELL, W. A. (1943) *A text-book of zoology*, 6th ed., revised by O. Lowenstein, Macmillan, London.

PELSENEER, P. (1898) Morphologie des branchies et des orifices rénaux et génitaux des Chitons, *Bull. sci. Fr. Belg.* **31**, 23–30.

PELSENEER, P. (1899) Recherches morphologiques et phylogénétiques sur les Mollusques archaïques, *Mém. cour. Acad. R. Belg.* **57**.

PELSENEER, P. (1906) *A Treatise on Zoology*, V, *Mollusca*, ed. E. Ray Lankester, A. and C. Black, London.

RAVEN, C. P. (1958) *Morphogenesis: The Analysis of Molluscan Development*, London, Pergamon Press.

THIELE, J. (1934) *Handbuch der systematischen Weichtierkunde*, Fischer, Jena.

WINCKWORTH, R. (1932) The British Marine Mollusca, *J. Conch.* **19** (7), 211–52.

YONGE, C. M. (1937) Circulation of water in the mantle cavity of *Dentalium entalis*, *Proc. malac. Soc. Lond.* **22** (6), 333–6.

YONGE, C. M. (1939) On the mantle cavity and its contained organs in the Loricata (Placophora), *Quart. J. micr. Sci.* **81**, 367–90.

YONGE, C. M. (1947) The pallial organs in the Aspidobranch Gastropoda, and their evolution throughout the Mollusca, *Phil. Trans.* B, **232**, 443–518.

YONGE, C. M. (1957) Reflexions on the Monoplacophoran *Neopilina galatheae* Lemche, *Nature, Lond.* **179**, 672–3.

YONGE, C. M. (1957a) *Neopilina*: survival from the Palaeozoic. *Discovery*, **18**, 255–6.

THE THREE MAJOR CLASSES

Bivalvia

Dissection of an Eulamellibranch (Aspatharia)

The following account is of the anatomy of the West African *Aspatharia brumpti*, a member of the family Mutelidae. The account is closely applicable also to the European *Anodonta cygnea* which belongs to the related family Unionidae. Examine the specimen provided and check the validity of the account, sentence by sentence and draft suitable amendments, wherever necessary, to make the account fully applicable to *A. cygnea*. Any differences noted are attributable to the differences between the two genera, while the very extensive similarities are due to the close affinity between the Mutelidae and the Unionidae, both of which are assigned to the order Unionacea.

In a similar way, but with a greater degree of adaptation, the account could be applied to the investigation of bivalves in other orders and families, for in many respects the account concerns the general features of the majority of bivalves.

External features. The body is protected by left and right shell valves which are approximately oval in outline, and are attached to each other at a hinge on the dorsal margin. The shell valves are similar in shape and size, i.e. they are equivalve. Growth of the shell valves is by accretion at their margins and occurs irregularly, the size and shape of the shell valves at earlier stages in the life of the animal being indicated by growth lines which were caused by some interruption in the growth of the shell at that time. The greater part of the shell is covered with a layer of periostracum, which is dark in colour in large and old specimens. At the oldest part of the shell the periostracum may have been worn away, the underlying prismatic layer of calcareous material may also have been eroded, exposing the innermost nacreous layer of the shell. This area of erosion occurs around the umbo, a small backwardly pointing beak situated near the dorsal edge of the shell. Draw the shell, represent the antero-posterior axis of the animal by drawing a horizontal line indicating the greatest length of the shell. Drop a perpendicular to this line from the umbo, and note that it divides the shell into a smaller anterior part and a larger posterior part. The shell valves are therefore inequilateral.

Examine specimens lying undisturbed in an aquarium and note the median foot which may be thrust out antero-ventrally between the shell valves. Note also the two siphons at the posterior end of the shell; the inhalant siphon is ventral and the exhalant siphon is dorsal (see Fig. 166).

General dissection. Hold the animal with its posterior end pointing to the left, and with its right shell valve uppermost. Insert a wedge mid-

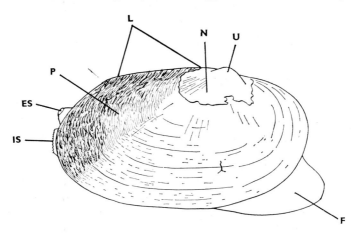

Fig. 166. *Aspatharia brumpti*, external view of the living animal, seen from the right side. The foot (F), the inhalant siphon (IS), and the exhalant siphon (ES) are protruded. The extent of the opisthodetic ligament (L) is indicated. In an extensive area around the umbo (U) the periostracum (P) has been worn away and the nacreous layer (N) is exposed.

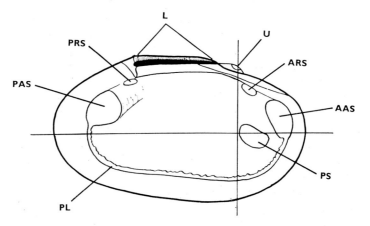

Fig. 167. *Aspatharia brumpti*, interior of the left shell valve, showing scars left by the anterior adductor (AAS), posterior adductor (PAS), the anterior retractor (ARS), the posterior retractor (PRS), and the protractor (PS) muscles. The pallial line (PL) is well developed. For interpretation of other lettering, see p. 508.

ventrally between the shell valves. Using a sharp scalpel, separate the margin of the mantle from the interior of the right, uppermost shell valve. Continue this process until you reach the two adductor muscles which unite the two shell valves anteriorly and posteriorly, close to the dorsal borders of the valves. Cut through these muscles as close to the shell as possible and then raise the right shell valve, noting that it is secured along its dorsal border to that of the left shell valve by a tough ligament. The ligament has to be broken in order to part the valves widely. Note that the ligament lies wholly posterior to the umbo, a condition described as opisthodetic. In life the elasticity of the ligament tends to cause the shell valves to gape slightly along their free ventral borders, and the adductor muscles operate against the ligament to close the shell valves tightly when this is necessary.

Remove the right shell valve, clean and dry its interior surface, and examine this to see the scars which mark the areas of insertion of the anterior and posterior adductor muscles. Postero-ventral to that of the anterior adductor muscle, there is a conspicuous scar formed by the protractor muscle which serves to draw the foot forwards. Close to the umbo there is a small oval scar formed by the anterior retractor muscle. Another small oval scar lies antero-dorsal to that of the posterior adductor, and immediately below the posterior end of the ligament; this is the site of insertion of the posterior retractor muscle. The mantle lobe is secured to the shell a short distance from its free margin by a continuous band of muscle fibres which produce a scar on the shell known as the pallial line, which extends as a continuous line joining the ventral border of the anterior adductor with that of the posterior adductor. Examine the surface of the mantle and find the cut surfaces of these muscles.

The mantle cavity. Make a vertical cut in the mid-ventral border of the right mantle lobe. Raise the two parts of the right mantle, make such other cuts as may be necessary, and pin these flaps to the dish above the dorsal margin of the shell. The space between the left and right lobes of the mantle is the infra-branchial chamber; it contains the following organs:—In the median line there is the large visceral mass dorsally, above the mantle cavity, with the much contracted, and orange coloured foot hanging from its lower border. Anteriorly, two labial palps hang downwards into the infrabranchial chamber just below the cut surface of the protractor muscle. The uppermost of these palps is known as the outer labial palp, and the other as the inner labial palp. (see Figs. 168 and 170.) Reflect the outer palp and note that the opposed surfaces of the two palps are covered with fine vertical corrugations. These corrugations, or folds, are covered with a ciliated epithelium, the function of the cilia being to sort particulate material and to pass particles of suitable size towards the fine groove between the bases of the two palps. Unsuitable material is rejected and passed to the ciliary cleansing currents on the

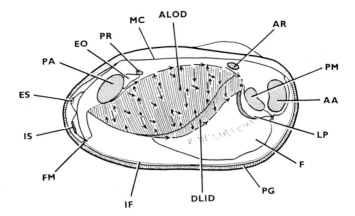

FIG. 168. *Aspatharia brumpti*, the organs in the mantle cavity, seen from the right side after removal of the right shell valve and the right mantle lobe. Ciliary feeding currents are indicated by arrows on the ascending lamella of the outer demibranch (ALOD) and on the ventral border of the descending lamella of the inner demibranch (DLID). The labial palps (LP) extend from the anterior border of the ctenidium towards the mouth. For interpretation of other lettering, see p. 508.

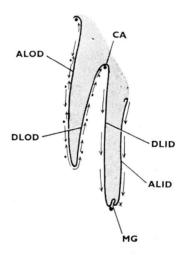

FIG. 169. *Aspatharia brumpti*, diagrammatic transverse section through the ctenidium, showing the course of ciliary feeding currents. The ctenidium is attached to the visceral mass at the ctenidial axis (CA). There is a marginal food groove (MG) at the free ventral border of the inner demibranch only. Note that, as also shown in Fig. 168, the ascending lamella of the outer demibranch (ALOD) possesses both dorsalward and ventralward ciliary currents. ● represents oralward currents at three different positions on the ctenidium. For interpretation of other lettering, see p. 508.

surface of the mantle. In the lateral oral groove particles are directed forwards, towards and into the mouth. The outer labial palp continues forwards and joins with that of the other side of the body forming, as it does so, an anterior lip to the mouth. Similarly the two inner palps join

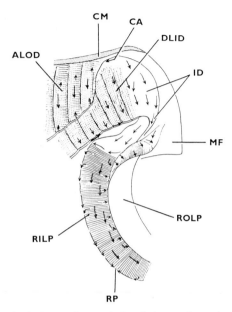

FIG. 170. *Aspatharia brumpti*, association between the anterior end of the ctenidium and the labial palps on the right side. Ciliary currents are indicated by arrows. The anterior end of the inner demibranch (ID) extends between the posterior ends of the inner (RILP) and outer (ROLP) labial palps. The opposed surfaces of the two palps are ridged and grooved and these folded surfaces bear complex patterns of ciliary activity, as is shown in greater detail in Fig. 171. Unacceptable material is prevented from reaching the mouth and is rejected from the free ventral margin of the palps at the point where anteriorward and posteriorward marginal currents meet (RP), the waste material being transferred to the ciliary cleansing mechanisms of the visceral mass and of the mantle. For interpretation of other lettering, see p. 508.

to form a posterior lip to the mouth. The mouth, bounded by these two lips, lies above the anterior border of the foot, and just below the lower surface of the anterior adductor muscle. Trace the lateral oral groove forwards, and find the mouth, which may be probed gently with a seeker. (See Fig. 170.)

The ctenidia lie on either side of the visceral mass, one ctenidium on each side of the body. Each ctenidium consists of two demibranchs, that lying superficially in the dissection being the outer demibranch. Anteriorly, the two demibranchs reach forwards to the posterior ends of the palps.

The ctenidia extend backwards behind the visceral mass to the base of the siphons. Raise the two demibranchs of the right ctenidium, and note that anteriorly the inner demibranch lies alongside but is not attached to the side of the visceral mass, while posteriorly it is attached to the inner demibranch of the left ctenidium.

Examine the mantle at the posterior end of the body, and note that there is a short area of fusion between left and right mantle lobes, forming a ventral wall to the inhalant siphon. Water is drawn into the infrabranchial chamber at this point. Examine the inhalant and the exhalant apertures, and note that they are complete, but very short tubes, with smooth rims lacking tentacles. Both surfaces of the outer demibranch

DORSAL

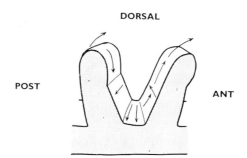

POST

ANT

Fig. 171. *Aspatharia brumpti*, diagrammatic representation of the ciliary sorting mechanisms on the folded, opposed surfaces of the labial palps. The figure represents a thick section cut through two adjacent folds, and the arrows represent the direction of effective beat of cilia in a number of separate ciliated tracts. The cumulative effect of these various ciliated tracts is to reject large particles, and large quantities of small particles, and to permit only few, small, particles to pass dorsalwards up the grooves between the folds, into the proximal oral groove, and finally to enter the mouth.

are divided into a very regular series of folds, this condition being described as plicate. Examine the surface of the demibranch under the high power of a binocular dissecting microscope, and note that it is composed of fine vertical filaments, there being about twenty similar filaments to each plica. Now examine the inner demibranch; are both surfaces plicate here also, or is the innermost surface smooth? The filaments are joined to each other by tissue junctions which run at right angles to the filaments, and so the surface of the ctenidium is divided to form a very fine meshwork. Water passes through this meshwork, from the infra- to the supra-branchial chamber. Examine the free margins of the two demibranchs, and note that there is a deep marginal food groove along the edge of the inner demibranch, but not along that of the outer demibranch. Small particles which are suspended in the inhalant water current are filtered from the water as this passes between the ctenidial filaments. Such particles are passed by ciliary action along the ctenidial filaments, and bound into masses with mucus which are carried by ciliary action forwards towards the mouth in two main streams: (1) along the groove between the bases of the inner and the outer demibranchs, which is known as the ctenidial axis; (2) within the marginal groove at the free border of the inner demi-branch. This material collected by the ctenidia, after exposure to selection and rejection mechanisms at the labial palps, is passed to the mouth for acceptance as food.

Supra-branchial chambers lying, as their name suggests, above the ctenidia, convey the water from the ctenidia to the exhalant siphon at the posterior end of the animal. Faeces are discharged into this exhalant stream, as also are the waste products from the excretory organs. In contrast, particulate material which is rejected from the ctenidia and labial palps, together with any material collected by ciliary cleansing currents on the visceral mass, the foot, and the inner side of the mantle, are bound together with mucus into pseudofaeces and are rejected oc-casionally, by sudden closure of the shell valves, either at the inhalant siphon, or at the ventral gape.

The pericardium. Remove the right lobe of the mantle completely. The pericardium lies above the dorsal border of the ctenidia, immediately below the anterior end of the ligament. Make an incision in the pericardial wall, taking care not to damage the heart within. Remove the side wall of the pericardium, exposing the heart fully. Blood is collected from the left and right ctenidia by a pair of triangular, thin-walled auricles, which discharge into the single median ventricle (see Fig. 172). The auriculo-ventricular aperture is a conspicuous horizontal slit on the side of the ventricle. The ventricle is wrapped around the rectum in such a way that the rectum seems to penetrate the lumen of the ventricle (but in fact does not do so). The ventricle sends blood forwards in an anterior aorta, which lies above the rectum, and which serves the anterior part of the visceral mass. A posterior aorta passes backwards from the ventricle under the rectum, to serve the posterior end of the body. A conspicuous opaque bulge on the side of the ventricle, posteriorly, probably serves as a safety device, preventing disturbance to the action of the ventricle if blood is suddenly driven back towards the heart due to contraction of the posterior part of the body. The excretory organs lie in a triangular space between the dorsal border of the ctenidium, the front of the posterior adductor muscle, and the posterior retractor muscle. The internal opening of the excretory organ is on the floor of the pericardium, while the opening to the exterior is on the side of the visceral mass, in the supra-branchial chamber. (See Fig. 172.)

The nervous system. Examine the surface of the visceral mass just behind the anterior end of the labial palps, and just above the line of attachment of the outer labial palp. Scrape the mantle tissues away gently here,

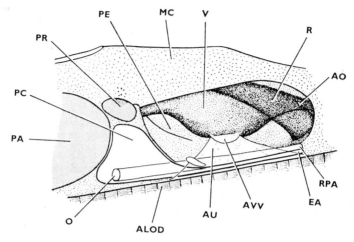

FIG. 172. *Aspatharia brumpti*, view of the pericardium from the right side. The rectum (R) passes upwards from the visceral mass, into the pericardium (PE) and backwards dorsal to the posterior adductor muscle (PA). Blood is collected from the ctenidium into the right auricle (AU), and through the auriculo-ventricular valve (AVV) into the ventricle (V). The ventricle pumps arterial blood to the body via the anterior aorta (AO) and a posterior aorta. The ventricle is wrapped round that part of the rectum which lies in the pericardium. The excretory organ is a U shaped tube which opens internally by the reno-pericardial aperture (RPA) into the pericardium, and externally into the supra-branchial space by the excretory aperture (EA). The proximal limb of the excretory organ opens by a wide oval aperture (O) into a posterior dilatation (PC) of the distal limb. For interpretation of other lettering, see p. 508.

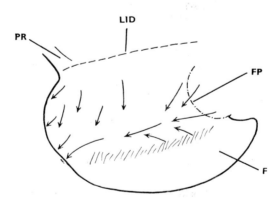

FIG. 173. *Aspatharia brumpti*, ciliary cleansing mechanisms, represented by arrows, on the right side of the visceral mass. − − − represents the line of attachment (LID) of the dorsal margin of the ascending lamella of the inner demibranch to the visceral mass. −·−·− represents the position of the free posterior margins of the labial palps (FP). For interpretation of other lettering, see p. 508.

and search for the right cerebral ganglion, which lies just below the epithelium, and which is rendered conspicuous by its orange colour. The right and left cerebral ganglia are united by a commissure which passes anterior to the oesophagus (do not attempt to search for this unless you have spare time at the end of the exercise).

Remove the right ctenidium. The transverse septum formed by the united bases of the ctenidia extends backwards to join the ventral border of the exhalant siphon. The axis of the left ctenidium is supported by a membrane which reaches backwards as far as the middle of the posterior

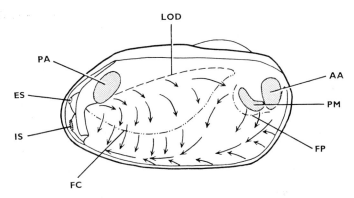

FIG. 174. *Aspatharia brumpti*, ciliary cleansing mechanisms, represented by arrows, on the inner surface of the mantle; the dissection shows the left mantle lobe, lining the left shell valve. — — — represents the line of attachment (LOD) of the dorsal margin of the ascending lamella of the outer demibranch to the mantle. — ·· — ·· — ·· represents the position of the free ventral margin (FC) of the ctenidium. For interpretation of other lettering, see p. 508.

adductor muscle, at which point a blood vessel passes downwards and backwards towards the hind end of the ctenidium.

Examine the ventral surface of the posterior adductor opposite the point of entry of this vein into the suspensory membrane of the ctenidium. Separate the epithelium of the under side of the posterior adductor at this point, and find the conspicuous, orange coloured right visceral ganglion. The left visceral ganglion lies closely against it. The visceral ganglion is connected to the cerebral ganglion of the same side by a cerebro-visceral connective which passes through the digestive gland and ventral to the pericardium, but there will probably be insufficient time to trace the whole course of the connective.

The left and right pedal ganglia are fused to form a transverse oval bar which lies embedded in the visceral mass in the median line, a little dorsal to the limit of the muscular substance of the foot. Make a median longitudinal incision in the foot and extend this upwards until the substance of the digestive gland is reached. Tease the glands with a needle, washing repeatedly with a pipette to remove fragments, until the ganglia are found. Once exposed, their orange colour renders them conspicuous. The pedal ganglia are connected to the cerebral ganglia of the same side by cerebro-pedal connectives but, once again, there is not likely to be sufficient time to trace these connectives.

The alimentary canal. Pass a seeker through the mouth and up the oesophagus. Trace the oesophagus into the stomach and slit this open along its right side. The digestive diverticula open into the stomach by numerous orifices which are mostly arranged in two groups on the left and on the right of the anterior floor of the stomach. The midgut leaves the stomach posteriorly. If the animal was freshly collected before dissection, a crystalline style may project from the combined style-sac and mid-gut into the stomach. It rotates slowly, thereby stirring the stomach contents; it dissolves slowly, liberating an amylase into the stomach, thereby initiating extra-cellular digestion of carbohydrate materials.

After several turns in the visceral mass, the mid-gut passes into the rectum, which traverses the ventricle in the pericardium, as already noted, passes over the dorsal surface of the posterior adductor muscle, and finally opens into the supra-branchial chamber at the anus. Slit open the posterior part of the supra-branchial chamber and find the terminal part of the rectum and the anal papilla on the dorsal and posterior surfaces of the posterior adductor muscle.

The reproductive system. The individual is unisexual; the tissues of the gonad pack the spaces between the coils of the mid-gut and hind-gut, behind the mass of digestive diverticula. The gonad duct opens on the side of the visceral mass, into the supra-branchial chamber. The genital and excretory apertures lie close together, and may be seen with the aid of a binocular dissecting microscope.

Diversity of Form in the Bivalvia

The following questionnaire indicates the diversity of form which may be encountered in the class Bivalvia, and this diversity will be most fully appreciated by examining in turn a number of bivalve genera such as *Mytilus, Cardium, Scrobicularia, Barnea, Mya, Ostrea*, etc., and completing the questionnaire separately for each genus studied.

Shell valves: equivalve/inequivalve
           equilateral/inequilateral
Hinge line: straight/arcuate
Axis of movement of valves: longitudinal/vertical/both
                                   longitudinal and vertical

Outline of valves: orbicular/oval/triangular/emarginated anteriorly

External surface: glossy/dull
self-coloured/radial streaks

Sculpture: radial ridges/concentric ridges/decussations/spines/rows of fine teeth

Ligament: external/internal
if external, prosodetic/opisthodetic

Mantle:

Fusions: mantle isthmus only/also below exhalant opening/also short fusion below inhalant siphon/fusion below inhalant siphon extends far forwards

Nature of mantle fusion: inner mantle fold only/inner and middle mantle folds/inner, middle and outer folds all involved

Cruciform muscle: present/absent

Sensory elements: pallial tentacles present/absent
pallial eyes present/absent

Siphonal orifices: not raised upon siphons
two short separate siphons
two long siphons united at base only
siphons united throughout

Inhalant siphonal orifice: with incurved straining tentacles/without straining tentacles

Surface of siphons, where present: periostracal sheath present/absent

Muscles:

Adductor muscles: isomyarian/anisomyarian/monomyarian
if isomyarian, muscles contracting simultaneously/muscles contracting alternately

Retractor muscles: anterior retractor muscle present/absent
anterior protractor muscle present/absent
posterior retractor muscle present/absent
byssal retractor muscle present/absent

Orbicular muscle: pallial sinus absent/shallow/very deep

Ctenidia: flat/plicate
homorhabdic/heterorhabdic

Structure and ciliation of ctenidial surfaces: Atkins type A–G (see Fig. 49, p. 117)

Labial palps: large, with ridged ciliary sorting surface/greatly reduced

Foot: Wedge shaped/tongue shaped/cylindrical with terminal sucker/
    rudimentary
  Byssus apparatus: present, active in adult
                     present, inactive in adult
                     absent

Alimentary canal:

  Style-sac: separate from mid-gut/conjoined with mid-gut

  Crystalline style: present, firm
                      present, soft and reduced
                      absent
  Stomach: internal structure type I, II, III, IV, V (see Fig. 81. p. 226)

Pericardium: pericardial glands: on auricles
                                     on pericardial walls
                                     on auricular and pericardial walls
  posterior aorta: aortic bulb present/absent

Excretory organs: reno-pericardial apertures
                     median fenestra between left and right distal limbs
                     present/absent

  excretory apertures: separate from genital aperture/united with genital
                       aperture

Gonads: male/female/hermaphrodite
  condition of gametes: spermatozoa active/inactive/not found
                         ova abundant, ripe/otherwise
  modification of demibranchs to form brood pouches:
                 inner demibranch modified/not modified
                 outer demibranch modified/not modified
  results of artificial fertilisation:

Nervous system: cerebral ganglia
                   pedal ganglia
                   visceral ganglia
  date of investigation
  number of specimens studied

TABLE 16. *Scheme of classification of the bivalvia*

Only a few orders, families, and genera are listed, these having been chosen to illustrate the main lines of diversification of the class. Only the principal references in the text are cited.

Sub-class	Order	Family	Genus	Pages
Protobranchia	*Nuculacea*	Nuculidae	*Nucula*	107, 219
		Nuculanidae	*Nuculana*	107–8
		Solenomyidae	*Solenomya*	222–3
Septibranchia	*Poromyacea*	Cuspidariidae	*Cuspidaria*	110, 225
Polysyringia	Mytilacea	Mytilidae	*Mytilus*	154
(filter-feeding			*Lithophaga*	180
bivalves)		Pinnidae	*Pinna*	138, 166–7
	Pectinacea	Pectinidae	*Pecten*	189–92
		Limidae	*Lima*	173
	Anomiacea	Anomiidae	*Anomia*	174
	Ostreacea	Ostreidae	*Ostrea*	178, 235–8
	Unionacea	Unionidae	*Anodonta*	313
		Mutelidae	*Aspatharia*	495–504
	Sphaeriacea	Sphaeriidae	*Sphaerium*	312, 341
			Pisidium	341
	Dreissenacea	Dreissenidae	*Dreissena*	388
	Erycinacea	Montacutidae	*Montacuta*	197
	Cardiacea	Cardiidae	*Cardium*	439
		Tridacnidae	*Tridacna*	132, 175
			Hippopus	339
	Veneracea	Veneridae	*Venus*	57, 159
			Paphia	
		Petricolidae	*Petricola*	388
	Mactracea	Mactridae	*Mactra*	157–9
			Lutraria	162
	Tellinacea	Donacidae	*Donax*	160
			Egeria	342
		Semelidae	*Scrobicularia*	9
		Tellinidae	*Tellina*	140, 379
			Macoma	378
	Solenacea	Solenidae	*Solen*	164
			Ensis	155, 163
	Saxicavacea	Saxicavidae	*Hiatella*	292
	Myacea	Myidae	*Mya*	162
		Aloididae	*Aloidis*	192
	Gastro-chaenacea	Gastro-chaenidae	*Rocellaria*	184
	Adesmacea	Pholadidae	*Pholas*	182–6
			Barnea	139
			Martesia	120
		Xylophaginidae	*Xylophaga*	306, 314
		Teredinidae	*Teredo*	188, 241
			Bankia	
	Pandoracea	Thraciidae	*Thracia*	
	Clavagellacea	Clavagellidae	*Brechites*	169

Key to the Lettering on the Figures

AA	Anterior adductor muscle.	LP	Labial palps.
AAS	Scar formed by anterior adductor muscle.	MC	Median mantle crest.
		MF	Part of the right mantle lobe, folded forwards.
ALID	Ascending lamella of inner demibranch.		
		MG	Marginal food groove.
ALOD	Ascending lamella of outer demibranch.	N	Nacreous layer of the shell, exposed by erosion of the periostracum and the prismatic layer.
AO	Anterior aorta.		
AR	Anterior retractor muscle.		
ARS	Scar formed by anterior retractor muscle.		
		O	Opening of the median limb into the posterior chamber of the excretory organ.
AU	Auricle.		
AVV	Auriculo-ventricular valve.		
CA	Ctenidial axis.	P	Periostracum, which is thick and wrinkled on the postero-dorsal sector of the shell.
CM	Cut edge of mantle.		
DLID	Descending lamella of inner demibranch.		
		PA	Posterior adductor muscle.
DLOD	Descending lamella of outer demibranch.	PAS	Scar formed by the posterior adductor muscle.
EA	Excretory aperture.	PC	Posterior chamber of the excretory organ.
EO	Excretory organ.		
ES	Exhalant siphon.	PE	Pericardium.
F	Foot.	PG	Periostracum, arising in the periostracal groove.
FC	Position of the free ventral margin of the ctenidium.		
		PL	Pallial line.
FM	Anterior limit of fusion of the left and right mantle lobes.	PM	Protractor muscle.
		PR	Posterior retractor muscle.
FP	Position of the free margins of the labial palps.	PRS	Scar formed by the posterior retractor muscle.
ID	Fleshy, non-filamentar extension of the inner demibranch.	PS	Scar formed by the protractor muscle.
IF	Inner fold of the mantle margin.	R	Rectum, passing through the ventricle.
IS	Inhalant siphon.	RILP	Right inner labial palp.
L	Ligament.	ROLP	Right outer labial palp.
LID	Line of attachment of the upper margin of the inner demibranch.	RP	Rejection point on the labial palps.
		RPA	Reno-pericardial aperture.
LOD	Line of attachment of the upper margin of the outer demibranch.	U	Umbo.
		V	Ventricle.

Gastropoda

The Gastropoda comprise three sub-classes of which the Prosobranchia are the most primitive, and this subclass is judged to be ancestral to the remainder of the Gastropoda. It will be convenient to treat the three sub-classes separately, commencing with the Prosobranchia.

Sub-class Prosobranchia

Due to the adoption of a wide variety of habitats, modes of life, and methods of feeding, this sub-class has exhibited a great diversity of form, and it is impossible to select any example which may be regarded as characteristic of the sub-class. Nevertheless, it is convenient to study the form and function of one particular type in order to comprehend the diversities of form and function exhibited throughout the sub-class. The type which has been chosen for detailed examination is the whelk, *Buccinum*, the reasons for its choice being that it is easily available at a low price, it is of convenient size for dissection, and it does not exhibit any extreme of anatomical modification to suit its mode of life. Other genera might serve equally well. There is a good illustrated account of the anatomy of *Buccinum* in:

DAKIN, W. J. (1912) *Buccinum*, Liverpool Marine Biology Committee, Memoir No. 20, Williams and Norgate, London.

Dissection of Buccinum undatum

The shell. The heavily built shell is spirally coiled around a central columella, the direction of coiling being dextral i.e. clockwise as seen from the apex of the shell. The mouth of the shell can be closed by an operculum.

External features. For convenience the animal should be narcotised and removed from its shell. Place the animal in a large dish with a wax bottom, and cover it with water. Note the large muscular foot the ventral surface of which forms a broad creeping sole. Posteriorly the foot bears an oval chitinous operculum on its dorsal surface. When the animal contracts into its shell the tail end of the foot is the last part to be withdrawn, and the operculum closes the mouth of the shell.

Place the specimen with the creeping sole lowermost, and secure it with pins passing through the sides of the foot. The dorso-ventrally flattened head lies above the anterior end of the foot; the "mouth" is a median slit on the ventral surface of the head; the head bears a pair of tentacles antero-laterally, and there is an eye lateral to the base of each tentacle. If the animal is a male an enormous, muscular penis is borne on the right side of the body, above and behind the head. The penis is compressed, and bears a sub-terminal papilla at the apex of which the vas deferens opens.

The mantle cavity lies above and behind the head, the roof of the mantle cavity being formed by the mantle. The margin of the mantle is thickened, and muscular, and on the left side it is developed to form an incomplete tube, the siphon. The siphon is protruded beyond the margin of the shell to receive the inhalant water stream.

Posterior to the mantle cavity the visceral mass is spirally coiled to fit the interior of the shell. On the right side of the body, and underlying the coiled visceral mass, note the inner end of the columella muscle. This is normally attached to the columella of the shell, and serves to draw the animal back into its shell when it is disturbed.

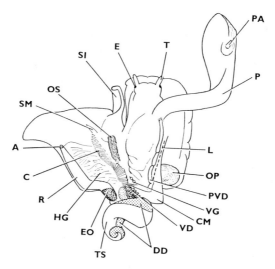

Fɪɢ. 175. *Buccinum undatum*. A longitudinal cut was made in the wall of the mantle cavity along its right side (— — — L), immediately below the rectum (R), and the roof of the mantle cavity was pinned out on the left side of the dissection. Note the osphradium (OS) and the ctenidium (C) suspended from the roof of the mantle cavity, and the hypobranchial gland (HG) lying between the rectum and the ctenidium. The pallial vas deferens (PVD) passes forwards in the floor of the mantle cavity on its right side, and then passes more deeply into the substance of the penis (P). Note that the visceral ganglia (VG) can be seen through the thin floor of the mantle cavity, close to its posterior end. For interpretation of other lettering, see pp. 533–4.

Dissection of the mantle cavity (see Fig. 175). Cut through the mantle on the right side close to the body and the columella muscle. Clean the interior of the mantle cavity with a pipette and continue the cut backwards to the posterior end of the mantle cavity. Pin out the flap of mantle on the left side of the dissection.

(a) *Male.* A muscular thickening which passes across the floor of the mantle cavity from the posterior end, close to the side of the columella muscle, and thence to the base of the penis, is the pallial vas deferens.

(b) *Female.* There is a conspicuous thickening on the right hand side of the roof of the mantle cavity, close to the margin of the columella muscle, and this thickening terminates abruptly just within the mantle cavity. This thickening is the pallial oviduct, the major part of which serves as a capsule gland. When the mantle cavity has been opened as directed above, this pallial oviduct should lie along the left hand margin of the reflected roof of the mantle cavity.

Rinse the dissection well and remove as much mucus as is possible. Then examine the roof of the mantle cavity and note the ctenidium, which consists of a single series of short, broadly triangular ctenidial leaflets. The shape of these leaflets indicates that they serve solely for respiratory purposes. In life, the ctenidium is suspended from the roof of the mantle cavity. The osphradium is a sense organ which lies lateral to the anterior half of the ctenidium, in the path of the inhalant water stream from the siphon. The osphradium consists of a longitudinal thickening which bears two series of triangular leaflets laterally; the whole is covered with a sensory epithelium. In some gastropods the osphradium is considered to measure the quantity of suspended matter contaminating the inhalant water stream, whilst it is claimed to serve as a chemo-receptor in certain genera. The roof of the mantle cavity bears a large hypobranchial gland which secretes much mucus. This serves to immobilise small particles which settle from the respiratory current, the mucus-bound particles then being cleared from the mantle cavity by ciliary action. The rectum lies on the right-hand side of the mantle cavity under normal circumstances, but should have been reflected to the extreme left of the dissection, and should be seen running along the side of the hypobranchial gland. The anus opens within the mantle cavity, and the well compacted faeces will be discharged from the mantle cavity via the exhalant water stream on the right side of the head. In the female the rectum runs along the side of the much larger pallial oviduct, between this and the hypobranchial gland.

The mantle cavity has come to occupy this anterior position above the head as the result of a process of torsion whereby the visceral mass was rotated through 180° of arc anti-clockwise as seen from above. This process of torsion occurred in the remote past at the origin of the class Gastropoda, and is recapitulated in larval development of each individual prosobranch. As a result of torsion the alimentary canal is twisted, and in the nervous system the visceral commissures are crossed, and it is important to bear these facts in mind when turning to the dissection of the internal organs. Note that torsion has nothing to do with the spiral coiling of the visceral mass.

The alimentary canal. Make a median incision through the body wall only in the floor of the mantle cavity, and extend this cut forwards towards the head. This exposes the long proboscis, lying in a muscular sheath from which muscle bundles radiate outwards to the body wall. The true mouth lies at the tip of the proboscis, which can be protruded through the orifice on the head. From the mouth the pharynx runs backwards through the substance of the proboscis and enters the oesophagus, which can be seen emerging from the posterior end of the proboscis. The oesophagus passes

forwards along the ventral surface of the proboscis sheath until it reaches
the level of the opaque creamy yellow salivary glands. Here the oesophagus
passes through the concentrated circum-oesophageal ring of nerve ganglia
and commissures, and then travels backwards along the floor of the body
space. The left salivary gland is larger than the right, and the two glands
hide the mass of ganglia from view. A salivary duct leaves each gland and
accompanies the oesophagus backwards to the posterior end of the proboscis sheath, and then enters the proboscis sheath together with the oesophagus. The salivary ducts lie external to the ganglionic ring. The posterior
part of the oesophagus is partly hidden from view by the pale brown gland
of Leiblein, which can be traced forwards and is found to open by a short
duct into the oesophagus (see Fig. 176).

Trace the oesophagus backwards into the mass of digestive diverticula
which make up the bulk of the visceral mass; in so doing, note the presence
of two nerve ganglia immediately under the body wall and above the

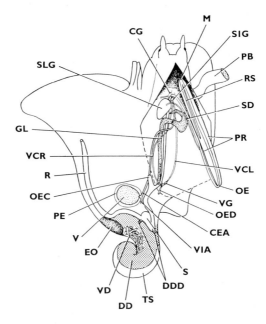

FIG. 176. *Buccinum undatum*. General dissection of a male specimen. Details
of the organs in the mantle cavity are omitted. The body cavity has been opened by a median incision in the floor of the mantle cavity, and the proboscis
(PR), which has been pressed slightly to one side, has been slit open to expose
the radula sac (RS) and the proximal part of the oesophagus (OE). The specimen is laid out to display the detailed structure of the alimentary canal. Note
that, as the result of torsion, the left visceral connective (VCL) passes backwards towards the visceral ganglia (VG) along the right side of the body, whilst
the right visceral connective (VCR) crosses over from the right to the left side.
For interpretation of other lettering, see pp. 533–4.

oesophagus, at the posterior end of the body space. The oesophagus bears a short caecum on its left side and then continues backwards to enter the stomach. The stomach lies on one side of the visceral mass, close to the surface. The intestine passes directly from the stomach to the rectum which opens by the anus on the right side of the mantle cavity.

The nervous system. Press the proboscis sheath over to the right in order to expose the salivary glands below. At this point the oesophagus was previously seen to bend back sharply on itself and then to pass posteriorly on the floor of a perivisceral space which is the great cephalic sinus. Immediately after bending backwards, the oesophagus passes through a circum-oesophageal complex of ganglia which lies just in front of the anterior margin of the salivary glands. Rinse the cephalic sinus for a few minutes with 70 per cent alcohol, which helps the dissection by making the nerves opaque white and more readily visible. Now remove the left salivary gland, and find the circum-oesophageal ganglionic ring.

The dorsal part of the circum-oesophageal ring is formed by a broad, white cerebral commissure, which joins the left and right cerebral ganglia; cut through this commissure. Identify and remove the oesophagus, and the large cephalic aorta which also passes through the ring. The circum-oesophageal complex can now be studied from within. Antero-ventrally a large pair of pedal ganglia are fused medially, and each sends about eight nerves, some of which are very large, forwards into the foot. Two very short commissures leave the pedal ganglia: laterally a cerebro-pedal commissure passes to the cerebral ganglion on each side; posteriorly a pleuropedal commissure passes to a pleural ganglion. The left and right pleural ganglia are not symmetrically arranged, that on the right being produced backwards into a conspicuous supra-intestinal ganglion and a smaller sub-intestinal ganglion. Each cerebral ganglion bears a small buccal ganglion on its anterior side, and the two buccal ganglia are joined to each other by a long slender buccal commissure. The buccal ganglia send nerves to the odontophore and associated muscles which operate the radula. The cerebral ganglia send nerves to sense organs including the tentacles, the eyes and the statocysts.

The left pleural ganglion supplies nerves to the siphon, to the columella muscle and the osphradial ganglion.

The supra-intestinal ganglion, on the right side of the body, is produced far backwards and two nerves leave its posterior extremity, the large pallial nerve and the smaller *right visceral connective*. The sub-intestinal ganglion, which also lies on the right side (but which lies median to the supra-intestinal ganglion), supplies a number of pallial nerves and also a small *left visceral connective*.

The right visceral connective passes over to the left side of the body, crossing above the oesophagus and above the left visceral connective, this crossing of the two connectives being due to torsion. The two connectives pass backwards to the two visceral ganglia, which lie posterior to the cephalic sinus, above the oesophagus and immediately below the body wall. If the specimen is reasonably fresh, these two ganglia should be visible through the body wall at a point immediately in front of the darkly coloured digestive gland. The visceral ganglia are joined to each other by a short commissure. That on the right side is larger, and is joined by the left visceral connective to the sub-intestinal ganglion. The smaller ganglion on the left side is joined by the right visceral connective to the supra-intestinal ganglion. Trace these two visceral connectives backwards from their origins in the circum-oesophageal complex, along the floor of the cephalic sinus, under the gland of Leiblein, to the two visceral ganglia noting that the connectives cross as they pass backwards.

The reproductive system. The sexes are separate.

(a) *Female.* The ovary lies in the coiled part of the visceral mass, and is visible at the surface. The oviduct passes forwards along the surface of the visceral mass close to the columella of the shell. At the posterior end of the mantle cavity the oviduct opens into the much broader pallial oviduct. The proximal part of the pallial oviduct is an albumen gland which surrounds the fertilised eggs with a layer of albumen. The distal three quarters of the pallial oviduct is a capsule gland, where the eggs are provided with a protective capsule. The pallial oviduct runs alongside the rectum, and opens into the mantle cavity.

(b) *Male.* The testis lies in the coiled part of the visceral mass, and is visible at the surface. The vas deferens passes forwards at the surface of the visceral mass, close to the columella of the shell. Near the anterior end of the digestive diverticula the vas deferens is swollen and greatly convoluted, and this region serves as a seminal vesicle. The vas deferens then passes into the pallial vas deferens, a muscular tube which passes across the floor of the mantle cavity and into the base of the penis. The pallial vas deferens is visible as a thick ridge on the right floor of the mantle cavity. The vas deferens passes up the interior of the penis and opens at the tip of a small sub-terminal papilla.

The vascular system. The heart lies in the pericardium close to the posterior end of the ctenidium. It can be seen in surface view. The pericardial wall is very thin, and the large globular ventricle can be seen by transparency. Cut through the pericardial wall near the posterior border of the ventricle, and reflect the pericardial wall forwards in order to see the single auricle, which is flimsy and inconspicuous. The auricle receives venous blood dorsally from the buff coloured renal organ which lies adjacent. The auricle also receives blood antero-ventrally from the posterior end of the ctenidium. A single aorta leaves the ventral corner of the ventricle, and this aorta divides almost immediately into an anterior cephalic artery and a posterior visceral artery. The cephalic artery supplies a branch to the anterior border of the mantle, and another which accompanies the posterior oesophagus in the floor of the cephalic sinus; this last branch of the cephalic artery passes through the circumoesophageal nerve ring, and then branches to the foot and to the proboscis.

The proboscis. Remove the proboscis from the proboscis sheath and pin it down securely with the dorsal side uppermost. Slit the proboscis mid-dorsally, and note the oesophagus which passes forwards within, to its tip. Slit open the oesophagus mid-dorsally and note the anteroventral orifice through which the tip of the radula can be seen. Slit open the radula sac and extract the radula. Note that the radula consists of a broad transparent membrane anteriorly, which bears a long series of teeth along the median line. The anterior teeth are dark brown and are very powerful, while further posteriorly the teeth are pale and less robust. The young teeth first appear at the posterior end of the radula sac. There are three teeth in each transverse tooth row. Examine the radula under the monocular microscope and draw. This type of radula is described as "rachiglossan".

The era of comparative anatomy is giving way to a new era of functional morphology and comparative physiology, and we cannot afford to continue to be hindered by obscurities of terminology. If "Taenioglossa" is strictly equivalent to "Mesogastropoda", and if "Stenoglossa" is similarly equivalent to "Neogastropoda", then in each case one of these synonyms should be expunged from our vocabulary. If the Neritacea are to be withdrawn from the Archaeogastropoda in recognition of important structural and functional considerations, as recommended by Wilbur and Yonge, then the residue of the Archaeogastropoda should be given a new name so that no ambiguity should attach to the use of the term Archaeogastropoda, as is now the unfortunate case.

Should the Prosobranchia be divided into its constituent orders on the basis of the structure of the ctenidia according to Pelseneer, on the structure of the heart according to Fretter and Graham, or otherwise? Taxonomic categories above the species level are to a certain extent abstractions on which opinions may differ according to the weight placed on different pieces of evidence and if we are striving towards a natural, i.e. a phylogenetic classification we may never have the final answer. If, alternatively, we can be content with an approximation to a natural classification which at the same time is efficient from a clerical point of view, we could settle on a generally agreed system of classification and relieve students of malacology of an unnecessary burden of synonymous and ambiguous terms. It would be very helpful if an international convention of specialists were to meet to produce an agreed classification and scheme of nomenclature.

The classification adopted here is based on the classification of Wilbur and Yonge, except that the Neritacea have not been abstracted from the Archaeogastropoda.

TABLE 17. *Schemes of classification of the sub-class Prosobranchia*

Pelseneer, P. (1906)	Thiele, J. (1931)	Fretter, V. and Graham, A. (1962)	Wilbur, K. M. and Yonge, C. M. (1964)
Order 1 Aspidobranchia	Order 1 Archaeogastropoda	Order 1 Diotocardia (= Archaeogastropoda)	Order 1 Archaeogastropoda
Sub-order 1 Docoglossa Sub-order 2 Rhipidoglossa			Order 1a Neritacea
Order 2 Pectinibranchia Sub-order 1 Taenioglossa	Order 2 Mesogastropoda	Order 2 Monotocardia Sub-order 1 Taenioglossa (= Mesogastropoda)	Order 2 Mesogastropoda
Sub-order 2 Stenoglossa	Order 3 Stenoglossa	Sub-order 2 Stenoglossa (= Neogastropoda)	Order 3 Neogastropoda

TABLE 18. *Scheme of classification of the Prosobranchia adopted in the present work*

Only a few orders, families, and genera are listed, these having been chosen to indicate the main lines of diversification of the sub-class. Only the principal references in the text are cited

Order	Sub-order	Family	Genus	Pages
Archaeogastropoda	Zeugobran-chia	Haliotidae	*Haliotis*	17–18
		Fissurellidae	*Diodora*	18, 366
			Fissurella	367
	Patellacea	Patellidae	*Patella*	368–75
			Patina	380
		Acmaeidae	*Acmaea*	365
	Trochacea	Trochidae	*Calliostoma*	19
			Gibbula	363–4
			Monodonta	363–4
			Trochus	365
		Turbinidae	*Turbo*	365
	Neritacea*	Neritidae	*Nerita*	46, 281
			Theodoxus	46
		Hydrocenidae	*Hydrocena*	46
• Mesogastropoda	Architaenio-glossa	Cyclophoridae	*Cyclophorus*	26
			Alycaeus	62
			Opisthosto-ma	26, 348–50
			Diplomma-tina	26, 350, 354
		Viviparidae	*Viviparus*	53
		Ampullariidae	*Pila*	28
			Ampullaria	
	Littorinacea	Littorinidae	*Littorina*	359–65, 414–15
	Rissoacea	Hydrobiidae	*Hydrobia*	378
		Rissoidae	*Rissoa*	
	Cerithiacea	Turritellidae	*Turritella*	52
		Vermetidae	*Vermetus*	53–5
		Melaniidae	*Melania*	
		Cerithiidae	*Cerithium*	
	Ptenoglossa	Janthinidae	*Ianthina*	57–8, 279
	Calyptraeacea	Capulidae	*Capulus*	51
		Calyptraeidae	*Calyptraea*	51
			Crepidula	49–51, 285–7
	Strombacea	Aporrhaidae	*Aporrhais*	52

* Wilbur and Yonge (1964) isolate the Neritacea from the Archaeogastropoda as a separate order yet recognise that the Archaeogastropoda and Neritacea are very closely related. See pp. 280–2, 515.

TABLE 18 (*cont.*)

Order	Sub-order	Family	Genus	Pages
Mesogastropoda (*cont.*)		Strombidae	*Strombus*	47
			Pterocera	47
	Heteropoda	Atlantidae	*Atlanta*	58, 87
		Carinariidae	*Carinaria*	58
		Pterotracheidae	*Pterotrachea*	58
	Naticacea	Naticidae	*Natica*	57, 379
	Lamellariacea	Lamellariidae	*Velutina*	284
			Lamellaria	
	Cypraeacea	Cypraeidae	*Trivia*	56, 283
			Cypraea	
Neogastropoda	Muricacea	Muricidae	*Murex*	64
			Drupa	365
			Purpura	67
			Thais	367
			Nucella	67
			Urosalpinx	64–5
	Buccinacea	Buccinidae	*Buccinum*	62, 509–15
			Busycon	64
	Volutacea	Olividae	*Oliva*	67
		Volutidae	*Voluta*	
	Toxoglossa	Conidae	*Turris*	
			Philbertia	67
			Conus	67–71

Sub-class Opisthobranchia

The Opisthobranchia were derived from a prosobranch ancestry, and from a monotocardian stock as is indicated by the presence of only one ctenidium, and only one auricle. The opisthobranchiate condition was achieved by detorsion, which presumably served to meet some major functional problem. It seems probable that detorsion occurred on more than one occasion, there being parallel—or convergent—evolution in this respect in a number of independent phylogenies. If this is true the Opisthobranchia are polyphyletic, and this would account for the very considerable diversity of form in the sub-class.

So great is the anatomical diversity in the Opisthobranchia no one type can be chosen the anatomy of which is broadly applicable throughout the sub-class. To understand the Opisthobranchia thoroughly it would be necessary to study several different examples, one for each major taxonomic group. This is beyond the scope of the present book, and the detailed examination of only one example is suggested. For this the genus *Aplysia* is chosen, on account of its large size, world-wide distribution, generally easy availability, and ease of dissection.

The provision of satisfactory material for dissection depends on due care being taken to achieve complete relaxation before the specimens are put in the fixative, and the use of a considerable excess of fixative. The specimens should be moved frequently while lying in fixative to ensure that they are adequately exposed to the preservative. It is useless to try to preserve large numbers of specimens in the same vessel as they will weigh each other down and cause distortion and imperfect preservation. Above all, the material should be reasonably fresh when dissected, otherwise the internal organs will be unduly brittle and the dissection will be unsatisfactory.

The following account is of *A. winneba*, from the West coast of Africa. There is some anatomical variation from species to species within the genus, but this should cause no difficulty. Principal features of variation are mentioned in the text at appropriate points. It will be of interest to check the account critically against the specimen being dissected, and to draft suitable amendments to the text where necessary, to fit the account to the species under investigation.

The following references will be invaluable:

EALES, N. B. (1960) Revision of the world species of *Aplysia* (Gastropoda, Opisthobranchia), *Bull. Brit. Mus. (nat. Hist.) Zoology* **5**, No. 10.
EALES, N. B. (1921) *Aplysia*, Liverpool Marine Biology Committee, Memoir No. 24, University Press, Liverpool.

General dissection of Aplysia

External features. Note the extensive creeping sole by means of which the animal can crawl over the surfaces of rocks and boulders. Contraction of muscle fibres in the substance of the foot may give the creeping sole a corrugated surface or, under extreme conditions, may cause marked contortions of the sole. The lateral border of the creeping sole is usually clearly defined by a low longitudinal ridge, and the sides of the visceral mass are usually comparatively smooth. Anteriorly a transverse groove separates the foot from the ventral surface of the head. Dorsally the visceral mass bears a pair of conspicuous epipodia; in life these are considerably larger, are very mobile, and by co-ordinated slow graceful flapping can effect a gentle swimming action. At their posterior ends the two epipodia may unite or may remain separate, according to the species.

The mantle cavity lies between the epipodia, on the right-hand side of the large oval visceral hump, which is protected by a single shell valve of comparable size and shape (see Fig. 177). The shell is much reduced in form and is very light; it consists mainly of periostracum, with a thin layer of calcareous material below this. The shell is originally an exo-skeleton, but it is obscured from view by a thin lamellar extension of the mantle. In most species there remains a small central perforation in the reflected layer of mantle through which a small portion of the shell can be seen, and the shell is still strictly an external structure. In some species, however, the reflected layer of mantle completely covers the shell, which has thereby become wholly internal.

At the posterior border of the visceral hump, on the right side, the mantle margin is raised and coiled to form an anal funnel. The anus opens within this funnel on its posterior wall. A single ctenidium lies in the mantle cavity, overhung by the right border of the visceral hump, and with its posterior tip curling upwards into the anal funnel. Raise the ctenidium and see

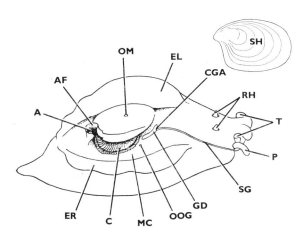

FIG. 177. *Aplysia* spp. View of the animal from the right side, to display the external features. As the result of partial detorsion the mantle cavity (MC) lies on the right side. There is only one ctenidium (C). The anus (A) opens into an anal funnel (AF) in the posterior part of the mantle cavity. The shell (SH) is almost completely covered by a reflection of mantle tissues in which a small central aperture (OM) is present in some species. A seminal groove (SG) leads from the genital aperture (CGA), down the right side of the body to the penis (P) which is here shown extruded. For interpretation of other lettering, see pp. 533–4.

that it consists of two series of complex leaflets arranged on either side of a longitudinal axis, and that the mantle cavity and ctenidium extend far underneath the visceral hump.

A short ridge passes forwards on the anterior floor of the mantle cavity, close to the right border of the visceral hump. This is the genital duct, which opens into the mantle cavity by the common genital aperture. A deep seminal groove passes forwards from the common genital aperture down the right side of the body towards the head, to the base of the penis which lies retracted—or protruded to greater or lesser extent—on the right side of the head. Examine the seminal groove and probe the common genital aperture with a seeker.

An opaline gland underlies the anterior floor of the mantle cavity and will be seen later, during dissection. This gland opens into the mantle cavity either by a number of small ducts, or in some species, by one large duct. A purple gland lies embedded in the right side of the visceral hump, in the shelf of mantle tissue which overlies the ctenidium. When the living animal is disturbed on the sea shore, it may emit a liberal secretion of a rich purple colour from this gland. The osphradium lies on the floor of the mantle cavity, just in front of the base of the ctenidium, but it can only be seen clearly in the living animal.

The head bears a deep vertical cleft on its anterior surface, and the mouth lies within this cleft. Lateral to the mouth the head bears a pair of ridges which curl outwards and backwards to join the anterior pair of tentacles, which are deeply grooved on their outer sides. Posterior to these tentacles the head bears a pair of rhinophores. Examine the skin a short distance in front of the rhinophores and find the small and inconspicuous eyes.

Remove the layer of mantle which is reflected over the shell and the visceral hump, and examine this layer with a hand lens to see the centrally placed mantle aperture. Remove and examine the shell. Examine the tissues underlying the shell; this area can be divided by an oblique line into two halves, that on the left and anterior side being occupied by the pericardium and the excretory organ, while the mantle cavity occupies the part on the posterior and right side. The purple gland lies along the right side in the roof of the mantle cavity.

Remove the purple gland and the thin roof of the mantle cavity, exposing the ctenidium fully. The base of the ctenidium is attached to the posterior floor of the mantle cavity; the tip of the ctenidium is directed backwards, towards the anal funnel. An afferent ctenidial vein supplies blood to the base and posterior border of the ctenidium. An efferent vein runs along the anterior and right border of the ctenidium and passes blood from the ctenidium to the single auricle of the heart. Examine the free border of the ctenidium, on its right side, and note the position of the efferent ctenidial vein in a groove with irregular ctenidial leaflets on each side.

The pericardium. The excretory organ is an opaque white mass of tissue which lies on the left side of the visceral hump, and which runs obliquely forwards and to the right in front of the ctenidium. The excretory organ lies on the left and posterior wall of the pericardium. Cut through the roof of the pericardium in front of the excretory organ, part the flaps of the cut and see the heart lying within. Remove the roof of the pericardium to expose the heart fully. The globular auricle lies at the right side of the pericardium and receives blood from the efferent ctenidial vein on the anterior border of the ctenidium. The auricle opens into the ventricle by an auriculo-ventricular valve which has two horizontal lips. The ventricle discharges ventrally, near its left extremity into a wide aorta which immediately divides into three. Immediately below the

ventricle the anterior aorta passes to the right across the floor of the pericardium. On its anterior and posterior sides the anterior aorta bears
sacculations known as the crista aortae, of unknown function. In some
species these sacculations are developed into conspicuous flaps, the
cavities of which are confluent with that of the aorta. In such cases the
crista aortae may expand and take up blood driven back towards the
heart when the animal contracts suddenly, and so protect the ventricle
from interference. Slit open the auricle, ventricle and anterior aorta and
examine their internal structure as described above.

At the base of the anterior aorta, near the left border of the pericardium,
a gastro-oesophageal artery passes immediately downwards to serve
the crop and the gizzard. Immediately behind the origin of the gastrointestinal artery, an abdominal artery passes backwards under the ventricle
and excretory organ, to serve the posterior end of the visceral mass. The
further courses of these three arteries can be followed in the course of the
general dissection.

General dissection. Cut through the floor of the mantle cavity on the
right side of the common genital aperture and the orifice of the opaline

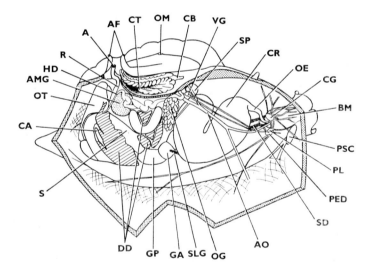

FIG. 178. *Aplysia* spp. General dissection, seen from the right side. The specimen has been opened by a longitudinal incision in the dorsal body wall, leading forwards from the floor of the mantle cavity towards the head and leading backwards to the right of the anus. The right side of the body wall has
been cut to reduce tension, and has been pinned down. The dissection displays
the principal features of the alimentary, reproductive, and nervous systems.
Due to detorsion, the visceral connectives are uncrossed; however, in this
view of the internal organs the visceral ganglia (VG) have been turned upwards, and this makes it appear that the connectives cross just in front of the
crop (CR). For interpretation of lettering, see pp. 533–4.

gland; extend the cut backwards along the right-hand side of the mantle cavity, to the right of the anal funnel, and back towards the tail; extend the cut forwards obliquely to the mid-dorsal line between the rhinophores and the anterior tentacles, taking care not to damage the underlying structures.

Note the complex lattice of muscle fibres in the body wall, and the numerous slender white nerves which pass through the body cavity towards their endings in the muscle layers of the epipodia, and the sides of the body. In particular, note two slightly thicker nerves which pass round the right side of the gut and terminate in two ganglia just below the common genital aperture. These are the cerebro-visceral connectives. A transparent and colourless blood vessel crosses the body cavity on the right side of the gut, sending one large branch forwards and outwards to the body wall, and another branch forwards and downwards towards the buccal mass. This artery, which enters the visceral cavity just behind the above mentioned ganglia, is the anterior aorta.

Place the animal on its left side in a wax-bottomed dish, cover it with water, and pin it down securely. Remove portions of the right epipodium and body wall, make vertical incisions in the remaining part of the right body wall without damaging the above mentioned nerves and arteries, and pin back the flaps of body wall so as to expose the viscera as fully as possible. (See Fig. 178.)

The alimentary canal. The buccal mass is a compact globular body which lies in the head region. It is partly obscured from view by numerous stout nerves which pass forwards from the large cerebral ganglia towards the penis, and towards the sense organs of the head. A slender oesophagus leaves the posterior face of the buccal mass, passing through the circumoesophageal complex of nerve ganglia and connectives. Raise the oesophagus and find the slender ducts of the two salivary glands, which open into the buccal mass laterally. The oesophagus passes backwards into the large, capacious, and thin-walled crop, which is coiled once. Raise the crop and find the next section of the gut, the anterior gizzard. This is a short section of gut which is surrounded by a thick band of circular muscle fibres. The inner surface of the anterior gizzard is lined by a number of large pyramidal teeth which serve to compress, and perhaps to triturate, the contents of the gut. The salivary glands are attached at the sides of the anterior gizzard, and their ducts pass forwards under the crop towards the buccal mass. The thin-walled posterior gizzard leads back to the stomach, which is surrounded by lobes of the digestive diverticula and parts of the reproductive system, the whole being enclosed in a transparent membrane to form a globular visceral mass.

Remove the membrane covering the visceral mass and trace the route taken through this by the intestine as far as is possible by surface inspection. A long, slender caecum, which arises from the posterior side of the stomach, can be seen on the right side of the visceral mass, posteriorly. The rectum passes dorsalwards on the left side of the visceral mass, to terminate at the anus, which has already been found on the posterior wall of the anal funnel. Examination of the interior of the various parts of the alimentary canal will be deferred until the end of the dissection.

Make an enlarged scale drawing of the dissection, showing all parts of the alimentary canal and such other anatomical features as have already been identified.

The reproductive system. The ovo-testis is a large, pale structure on the postero-dorsal surface of the visceral mass. Separate the ovo-testis from the adjacent tissues and press it backwards slightly, to reveal the slender, little hermaphrodite duct which passes forwards from its anterior surface. Make an enlarged scale drawing of the complex reproductive system, identifying the parts by reference to Fig. 179. An albumen gland and a

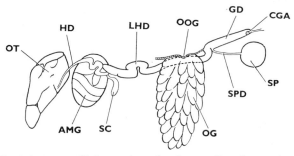

Fig. 179. *Aplysia* spp. If the specimen has been collected recently, and has been skilfully narcotised and preserved, it may be possible to loosen the viscera extensively, and to see all the details of the reproductive system shown in the figure. Specimens which have been preserved for a long time tend to become brittle, and it is no longer possible to separate the various organs. Comparison of such a specimen with the present figure may make it possible to discern the principal features of the reproductive system. For interpretation of the lettering, see pp. 533–4.

mucus gland together make up a compact ovoid mass which communicates with the common genital aperture via the large hermaphrodite duct. During copulation, spermatozoa produced by each animal are passed down the hermaphrodite duct, out of the common genital aperture, and along the seminal groove to the base of the penis. Foreign spermatozoa, together with a certain amount of debris, are introduced into the large hermaphrodite duct and thence into a small blind spermatocyst on the anterior side of the albumen and mucus gland. The spermatozoa are stored in the spermatocyst until used, and the debris is withdrawn and accumulated in the spermatheca, a globular vesicle lying close to the common genital aperture, and communicating with the anterior end of the large hermaphrodite duct by a slender spermathecal duct. Examine the inner surface of the body wall on the right side of the buccal mass, and see the penis sheath, into which the penis is normally withdrawn, and the retractor muscles of the penis, which are inserted into the body wall a little further posterior. Slit open the penis sheath and examine the penis.

The nervous system. A ring of ganglia and commissures surrounds the gut on the posterior side of the buccal mass. This consists of paired cerebral, pleural, and pedal ganglia. The cerebral ganglia lie dorsally and may be found as two distinct ganglia connected by a cerebral commissure or may be fused to form a single ganglionic mass (according to the species). The pleural ganglia lie latero-ventrally and are connected to the cerebral ganglia by lateral cerebro-pleural commissures. The very large pedal ganglia lie immediately below the pleural ganglia, against which they are very closely pressed and are connected to them by very short commissures. The pedal ganglia are connected to each other by the ventral pedal commissure and to the cerebral ganglia by the lateral cerebro-pedal commissures.

The cerebral ganglia give rise to a number of nerves which serve the skin on the front and sides of the snout, the anterior tentacles, and the rhinophores; they also give rise to a pair of extremely slender optic nerves, which run parallel with and close to the nerves to the rhinophores. A pair of commissures leave the ventral surface of the cerebral ganglia and pass to the large stomatogastric ganglia, which lie on the posterior surface of the buccal mass, immediately below the origin of the oesophagus. Cut through the oesophagus and the two salivary ducts, and draw the circum-oesophageal ganglionic complex backwards to expose the stomatogastric ganglia and the commissures which unite them to the cerebral ganglia. The stomatogastric ganglia supply one nerve to the oesophagus, and three nerves to the buccal mass, on each side. Replace the circum-oesophageal ganglionic complex in its normal position.

The pleural ganglia give rise to one very small nerve to the body wall, and to one long and thick pleuro-visceral connective on each side of the body (see Fig. 180). These connectives, which were noted at the beginning of the dissection, pass to a pair of ganglia which lie adjacent to the spermatheca. The connective from the right pleural ganglion joins the parietal ganglion, and the other member of the ganglion pair, which is joined to the left pleural ganglion by the other connective, is known as the visceral ganglion. The parietal ganglion (on the right side) and the visceral ganglion (on the left side) lie closely pressed against each other.

A stout nerve leaves the posterior surface of the parietal ganglion and runs backwards to the base of the ctenidium, which it serves, and sends a small branch to serve the purple gland. Three nerves leave the posterior and lateral border of the visceral ganglion to serve the anal siphon and anus, the accessory genital glands, and the floor of the mantle cavity.

Since the animal has been dissected from the right side, the parietal and visceral ganglia are viewed from below, and the pleuro-visceral connectives appear to be crossed. In their normal relations, however, these connectives are not crossed since detorsion has occurred in the Opisthobranchia, with a secondary return to a euthyneurous condition.

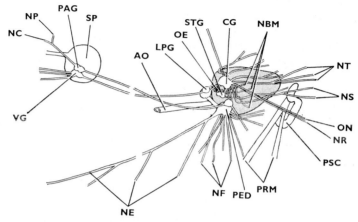

FIG. 180. *Aplysia* spp. Enlarged drawing of the nervous system, taken from the same dissection as that used for Fig. 178. Note the stomatogastric ganglia (STG) which lie on the posterior side of the buccal mass, below the origin of the oesophagus. Once again it is emphasised that the visceral connectives only appear to be crossed as the result of the upward displacement of the visceral (VG) and parietal ganglia (PAG) in the course of the dissection. For interpretation of other lettering, see pp. 533–4.

The pedal ganglia give rise to a large number of nerves which pass out to the muscles of the foot, the body wall, and to the epipodia.

Make an enlarged scale drawing of the nervous system seen from the right side, including the buccal mass, penis sheath and retractor muscles of the penis.

The internal structure of the alimentary canal. Cut the buccal mass in half longitudinally in the median line, examine the cut surfaces and see the radular membrane with numerous rows of chitinous teeth in each tooth row. Open the crop and note the comparatively large pieces of green and of red weeds which are ingested. Slit open the anterior and posterior gizzards and see the large pyramidal teeth which line the wall of the former (these teeth readily become detached from their insertions in preserved material). In the posterior gizzard there may be a small number of more slender, pointed teeth or, in some species, there may be a number of small rounded bosses protecting the entrance into the stomach. Slit open the stomach and see the series of folds on its floor, which constitute a ciliated sorting area; note also the orifices by which the digestive diverticula open into the stomach, and the orifice of the caecum.

TABLE 19. *Scheme of classification of the Opisthobranchia adopted in the present work*

Only a few orders, families, and genera are listed, these having been chosen to indicate the main lines of diversification of the sub-class. The Pyramidellidae, the Succineidae and the Oncidiidae have recently been judged to be opisthobranchs and have accordingly been included in the following table, but no attempt has been made to determine their precise positions in the classification of the Opisthobranchia. Only the principal references in the text are cited.

Sub-class	Order	Family	Genus	Pages
		Pyramidellidae	*Odostomia*	85–6
			Pyramidella	
	Succineacea	Succineidae	*Succinea*	75–6
Pleurocoela	Cephalaspi-dea			
		Actaeonidae	*Actaeon*	72
		Akeridae	*Akera*	74
		Scaphandridae	*Cylichna*	317
			Scaphander	83
		Philinidae	*Philine*	81, 257–8
	Aplysiacea	Aplysiidae	*Aplysia*	247, 519–27
Pteropoda	Euthecoso-mata			
		Cavoliniidae	*Creseis*	76–7
			Cavolinia	77
			Clio	
	Pseudothe-cosomata	Cymbuliidae	*Cymbulia*	77
	Gymnosomata	Clionidae	*Clione*	83–4
Sacoglossa		Stiligeridae	*Hermaea*	248
			Alderia	283
		Elysiidae	*Elysia*	248, 283
		Limapontiidae	*Actaeonia*	248
			Limapontia	248, 282–3
	Oncidiacea	Oncidiidae	*Oncidella*	247
Acoela Notaspidea		Pleurobranchi-dae	*Pleurobran-chus*	78–9
Nudibran-chia	Doridacea	Polyceridae	*Acanthodoris*	
			Adalaria	78, 315
			Onchidoris	78
		Dorididae	*Doris*	
			Archidoris	254
			Jorunna	78, 254–5
	Aeolidiacea	Dendronotidae	*Dendronotus*	
		Phyllirrhoidae	*Phyllirrhoe*	
		Tethyidae	*Melibe*	82–3
			Tethys	
		Calmidae	*Calma*	257
		Aeolididae	*Aeolidia*	79, 256
			Facelina	256

Sub-class Pulmonata

Not all the air-breathing snails are pulmonates—some are prosobranchs, while one family of opisthobranchs has adopted a terrestrial, air-breathing mode of life. It is more than possible that the sub-class Pulmonata is polyphyletic, the terrestrial and air-breathing mode of life having been evolved a number of times in different phylogenies. Conversely, not all pulmonates are air breathing—some inhabit fresh waters and others inhabit the sea shore and have retained the habit of breathing water, but with the development of secondary gills.

As is indicated below, the Pulmonata are divided into two major groups, the Basommatophora with eyes at the base of the cephalic tentacles, and the Stylommatophora with eyes at the tip of the oculiferous cephalic tentacles. This is probably a major phylogenetic cleavage of the sub-class. The most convenient types for investigation by dissection fall into the latter category, in the families Achatinidae and Helicidae. The choice of a suitable example for anatomical investigation is largely one of convenience on the basis of easy local availability and relatively large size. The Achatinidae are readily available over a wide area in the humid tropics. Anatomical differences between genera in the Helicidae and Achatinidae are not so great as to cause difficulty, while the recognition of such differences can be a matter of considerable interest.

In the Achatinidae, dissection of the genus *Archachatina* is a little more difficult than that of the related *Achatina fulica*, for two reasons: in the former the sole of the foot is less broad, the stem which raises the visceral mass above the foot is more slender, and it is not easy to secure the animal ventral surface lowermost in a dissecting dish, but it can be conveniently dissected when secured on its left side. Further, I have found that the internal organs, especially the reproductive system, of *Archachatina* to be bound together by tough strands of connective tissue, which call for meticulous care in dissection whereas in *Achatina fulica*, as in *Helix pomatia*, it is extremely simple to separate the components of the reproductive system. *A. fulica* has a broad pedal sole and can be dissected with ease while secured with the ventral surface lowermost, as is true also for *H. pomatia*.

For these reasons I have chosen to provide an account of the mode of investigation of the slightly more difficult example, *Archachatina*, and have given appropriate supplementary notes and figures illustrating significant differences in *Achatina fulica* and in *H. pomatia*.

General dissection of Archachatina

External features. Examine living specimens moving and feeding in a vivarium, and note the following features. The spirally coiled shell is up to 3 in. in length. The coiling is dextral, i.e. when viewed from the apex of the shell the twist is clockwise. In all there are six or seven turns in a large shell. The sides of the head and foot are irregularly marked with grooves. The foot has a flat creeping sole, which is marked off from the head anteriorly by a transverse groove in the centre of which the mucus gland opens.

The mouth is anterior, and is bordered laterally by a pair of broad lobes. The dorsal surface of the head bears two pairs of tentacles, which may be partly introverted. Of these, the posterior pair are larger, and bear eyes terminally. The genital aperture lies a little posterior to the right oculiferous tentacle.

The thickened edge of the mantle is fused to the body wall above and behind the head, there being a small aperture on the right side, the pneumostome, which leads into the mantle cavity.

The mantle cavity. Remove the shell piece by piece, with bone forceps, and pin the animal securely in a dish with the sole of the foot lowermost. The pins can be passed through the sides of the foot. Cut through the line of fusion of the mantle edge with the body wall *without perforating the body wall,* and extend the cut round to the right side, *ventral to the pneumostome.* Now cut along the right wall of the mantle cavity *ventral to the rectum,* and turn the roof of the mantle cavity over to the left side of the dissection.

Remove the pins and re-fix the animal on its left side, and pin out the roof of the mantle cavity so that it is fully exposed. The rectum runs along its outer border. Anterior to the rectum lies the large excretory organ, and the excretory duct runs along the side of the rectum which is nearest to the excretory organ. Slit open the terminal part of the rectum, and wash out its contents; slit open the terminal part of the excretory duct, probe it with a seeker and find the excretory aperture adjacent to the anus.

The lung occupies the roof of the mantle cavity, and is especially well developed anteriorly. It receives blood from the blood spaces of the visceral mass via a vein which runs along the left and anterior borders of the mantle cavity. Blood is drained from the lung by a pulmonary vein which runs into the single auricle of the heart. The pericardium lies in front of the excretory organ. Open the pericardium and see the single small auricle and the large ventricle. Blood is passed from the ventricle to the body via a single aorta.

General dissection. Remove the pins, and re-secure the animal with the sole of the foot lowermost. Make a longitudinal incision through the body wall in the mid-dorsal line, exposing the crop and other viscera. Pin down the flaps of the body wall, and extend the cut into the region of the coiled digestive diverticula, gently liberating the various organs.

The reproductive system. The ovotestis lies embedded in the digestive diverticula on the inner side of the coil. It may be large or small according to the state of the individual. A slender coiled hermaphrodite duct passes forwards from the gonad along the inner surface of the twisted visceral mass, and joins a smooth oval albumen gland, the size of which is variable (see Fig. 181). Anterior to the albumen gland, the hermaphrodite duct swells out to form a very thin walled "uterus" into which open a large number of columnar prostate glands. If the uterus contains no eggs, this part of the duct appears quite inconspicuous against the bulk of the prostate glands which open into it. Within the uterus there is a longitudinal fold covering a groove along which spermatozoa travel.

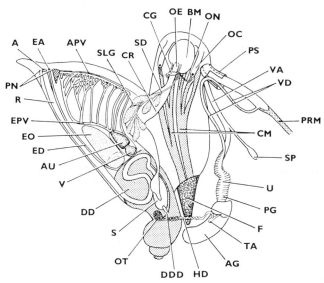

Fig. 181. *Archachatina* sp., general dissection. The mantle cavity has been opened by a longitudinal incision along its right side, below the rectum, and the roof of the mantle cavity and the rectum have been reflected to the left side of the dissection. The reproductive system has been loosened from the remainder of the viscera, and has been spread out on the right side of the dissection. For interpretation of the lettering, see pp. 533–4.

At the anterior end of the uterus, the sperm groove passes into a separate vas deferens, which runs alongside the oviduct. A third duct, a posterior-ward diverticulum of the oviduct, passes backwards to the spermatheca, and all three ducts are bound together by connective tissue. The oviduct runs directly forwards to the vagina, whilst the vas deferens turns backwards to the apex of the penis. The apex of the penis is secured by a retractor penis muscle. The base of the penis is surrounded by a penial sheath, which also surrounds a small part of the length of the vas deferens, where this duct is closely pressed against the penis.

Turn the reproductive organs over to the right side of the dissection.

The alimentary canal. The mouth leads into a hard globular buccal mass, from the posterior end of which the oesophagus arises. The anterior end of the oesophagus is surrounded by a nerve ring consisting of the fused cerebral ganglia mid-dorsally, pedal and visceral ganglia ventrally, and cerebro-visceral and cerebro-pedal connectives laterally. The oesophagus dilates into a thin-walled crop on the sides of which are the salivary glands. The salivary ducts pass forwards, within the nerve ring, and open into the posterior end of the buccal mass. The crop passes backwards into a globular stomach which is embedded in the digestive diverticula. The intestine leaves the stomach, and passes forwards to the right-hand side of the mantle cavity. Here the rectum passes forwards to the anus. Slit open the stomach and find the two ducts leading into the lobes of the digestive diverticula.

The muscular system. A broad, flat, muscle bundle passes backwards from each side of the buccal mass, and these are joined by muscles passing backwards from the oculiferous tentacles, from the penis, and from the sole of the foot. These all join to form a columella muscle which enters the visceral mass and is inserted into the inner wall of the shell.

The nervous system. The ganglia and nerves are surrounded by a tough layer of connective tissue. The cerebral ganglia lie dorsal to the oesophagus and send a stout nerve laterally to each eye. A slender branch leaves the base of this nerve and travels forwards to the skin of the head. Cerebro-pedal and cerebro-visceral connectives pass transversely round the oesophagus to a compact ganglionic mass ventral to the oesophagus. This mass includes both pedal and visceral ganglia. Although there has been no detorsion, the system is euthyneurous, due to the concentration of the nervous system and the shortening of the visceral loop. The ventral nerve mass gives rise to a large number of nerves which pass to the muscles of the foot and to the walls of the visceral mass.

Cut between the cerebral ganglia mid-dorsally, and examine the internal surface of the nerve ring. On each side three thick nerves pass forwards from the cerebral ganglion to the snout in the region of the mouth. Ventral to these, a slender connective passes to the posterior surface of the buccal mass ventral to the origin of the oesophagus, where it joins one of a pair of stomatogastric ganglia. The stomatogastric ganglia are united to each other by a short commissure, and they send a number of nerves to the buccal mass and to the ducts of the salivary glands.

Slit open the ventral ganglionic mass and see that it is composed of an antero-ventral pedal ganglion and a postero-dorsal visceral ganglion. A blood vessel passes between the two, and carries blood to the buccal mass.

Find the slender, tubular mucus gland beneath the pedal ganglia. Remove the buccal mass intact, pin it down and bisect it in the sagittal plane, and find the jaw dorsally and the radula borne on an odontophore ventrally. Remove the radula and examine it under the monocular microscope; it is covered by close set rows of innumerable fine teeth.

Achatina fulica. This has a broad, flat pedal sole and can be secured with this ventralmost, and can be dissected from the dorsal aspect. Due to the lack of extensive connective tissue adhesions the internal organs can be separated with comparative ease. The main difference from *Archachatina*, in my experience, is that the vas deferens is not bound by a collar to the base of the penis sheath but passes directly to the inner apex of the penis.

Helix pomatia. The principal anatomical differences between this species and the two members of the Achatinidae described above concern the reproductive system, as follows:

(1) The female duct does not bear any columnar prostate glands.

(2) There are a conspicuous dart sac and a dichotomously branching mucus gland, both of which open into the common genital aperture. (See Fig. 182.)

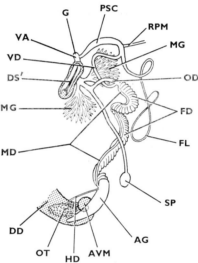

FIG. 182. *Helix pomatia*. The structure of the reproductive system, showing the dart sac (DS′) and the branching mucous glands (MG) which do not occur in *Achatina* or in *Archachatina*. The dart sac is shown cut open longitudinally to expose the long, slender, sharp, calcareous dart within. For interpretation of other lettering, see pp. 533–4.

Reference to the principal paper on the anatomy of the Achatinidae:

MEAD, A. R. (1950) Comparative genital anatomy of some African Achatinidae (Pulmonata), *Bull. Mus. comp. Zool.* **105** (2), 533–4.

TABLE 20. *Scheme of classification of the Pulmonata adopted in the present work*

Only a few orders, families, and genera are listed, these having been chosen to summarise the more important references to the Pulmonata in the text

Sub-class	Order	Family	Genus	Pages
Basommatophora	Actophila	Ellobiidae	*Ellobium*	90, 252
	Patelliformia	Gadiniidae	*Gadinia*	32
		Siphonariidae	*Siphonaria*	31, 92, 365
		Lymnaeidae	*Lymnaea*	29, 90, 249
		Planorbidae	*Planorbis*	89
		Ancylidae	*Ancylus*	30
Stylommatophora	Vertiginacea	Vertiginiidae	*Boysidia*	349
			Gyliotra-chela	349
	Achatinacea	Achatinidae	*Achatina*	383–5
			Archachatina	528–31
	Zonitacea	Arionidae	*Arion*	90, 253
		Limacidae	*Limax*	90
	Bulimulacea	Bulimulidae	*Partula*	343–7
	Helicacea	Pleurodontidae	*Amphidro-mus*	356
		Helicidae	*Cepaea*	283, 390–4
			Helix	249, 532
	Streptaxacea	Streptaxidae	*Streptaxis*	90
			Gonaxis	90, 386

Key to the Lettering on the Figures

A	Anus.
AF	Anal funnel.
AG	Albumen gland.
AMG	Albumen and mucous glands.
AO	Anterior aorta.
APV	Afferent pulmonary vein.
AU	Auricle.
AVM	Apex of visceral mass.
BM	Buccal mass.
C	Ctenidium.
CA	Caecum of stomach.
CB	Base of ctenidium.
CEA	Cephalic artery.
CG	Cerebral ganglia.
CGA	Common genital aperture.
CM	Parts of columellar muscle.
CR	Crop.
CT	Tip of ctenidium, passing to

	the left under the visceral hump.
DD	Digestive diverticula.
DDD	Duct from the digestive diverticula.
DS	Dart sac.
DS'	Dart sac, cut open to show the dart.
E	Eye.
EA	Excretory aperture.
ED	Excretory duct.
EL	Left epipodium.
EO	Excretory organ.
ER	Right epipodium.
EPV	Efferent pulmonary vein.
F	Hind end of foot.
FD	Female part of genital duct.
FL	Flagellum.

G	Genital aperture.	OT	Ovotestis.
GA	Anterior gizzard.	P	Penis.
GD	Genital duct, seen as a thickening in the floor of the mantle cavity.	PA	Papilla.
		PAG	Parietal ganglion.
		PB	Base of penis.
GL	Gland of Leiblein.	PE	Pericardium.
GOA	Gastro-oesophageal artery.	PED	Right pedal ganglion.
GP	Posterior gizzard.	PG	Columnar prostate glands.
HD	Hermaphrodite duct.	PL	Right pleural ganglion.
HG	Hypobranchial gland.	PN	Pneumostome, cut open.
L	Line along which the mantle cavity was cut open.	PR	Proboscis.
		PRM	Retractor muscle of the penis.
LHD	Large hermaphrodite duct.	PS	Penial sheath, partly cut away.
LPG	Left pleural ganglion.	PSC	Penis sac.
M	Mouth.	PVD	Pallial vas deferens.
MC	Mantle cavity.	R	Rectum.
MD	Male part of the genital duct.	RH	Rhinophore.
MG	Mucous gland.	RPM	Retractor muscle of penis.
NC	Nerve to ctenidium.	RS	Radular sac.
NBM	Nerves from the right stomatogastric ganglion to the buccal mass.	S	Stomach.
		SC	Spermatocyst.
		SD	Salivary duct.
NF	Nerves to the foot and body wall.	SG	Seminal groove.
		SH	Shell.
NP	Nerve to the purple gland.	SI	Siphon.
NR	Nerve to the rhinophore.	SIG	Supra-intestinal ganglion.
NRE	Nerves to the right epipodium.	SLG	Salivary gland.
NS	Nerves to the skin on the side of the snout.	SM	Suspensory membrane of ctenidium.
NT	Nerves to the tentacles and skin of the snout.	SP	Spermatheca.
		SPD	Spermathecal duct.
		STG	Stomatogastric ganglia.
OC	Oculiferous tentacle, retracted.	T	Tentacles.
		TA	Talon.
OD	Oviduct.	TS	Testis.
OE	Oesophagus.	U	"Uterus".
OEC	Caecum of oesophagus.	V	Ventricle.
OED	Diverticulum of oesophagus.	VA	Vagina.
OG	Opaline gland.	VCL	Left visceral commissure.
OM	Orifice in the layer of mantle reflected over the shell.	VCR	Right visceral commissure.
		VD	Vas deferens.
ON	Optic nerve.	VG	Visceral ganglion.
OOG	Orifice of opaline gland.	VIA	Visceral artery.
OP	Operculum.		
OS	Osphradium.		

Cephalopoda

Sub-class Dibranchiata

No consideration has been given in this book to fossil representatives of the Cephalopoda in the Nautiloidea, Ammonoidea, or Belemnoidea, nor to the Pearly Nautilus, the one remaining living representative of the Nautilidae, which is not readily available for investigation by students.

The remainder of the recent cephalopods comprises the subclass Dibranchiata, which evolved in adaptation to the pelagic mode of life and as free-swimming predacious carnivores. There has been an extensive adaptive radiation in the sub-class, meeting the special requirements of habitats such as the open ocean, the abyss, and shallow coastal seas, and exploiting a variety of modes of life. Thus the octopods are thought to have been derived from a free-swimming decapod ancestry and to have become adapted to a benthic mode of life. No attempt has been made here to describe the adaptive radiation of the Dibranchiata, due to inadequate personal contact with the materials.

The squid, *Loligo*, and the cuttle-fish, *Sepia*, are very suitable examples for morphological investigation, both being very widely distributed in the seas and oceans. The following account is of the cuttle-fish, and, so far as I am aware from dissections carried out in Britain, West Africa, and Malaysia, no difficulties should arise in the application of this account to locally collected material.

A very full and most carefully illustrated account of the anatomy of *Sepia* is given in:

TOMPSETT, D. H. (1939) *Sepia*, Liverpool Marine Biology Committee, Memoir No. 32, University Press, Liverpool.

Dissection of the cuttle-fish, Sepia

External features. The head bears eight short tapering arms which bear four rows of suckers on their ventral surfaces. The suckers are stalked, and have horny rims. The bases of the arms are united by webs. In addition there are two long tentacles which arise from pits lying latero-ventral to the mouth. The tentacles are smooth and cylindrical, and terminate in a broad flap bearing numerous very large suckers. The tentacles are longer than the long axis of the body. The mouth is situated in the centre between the arms. It is surrounded by a circular flap of flesh, the rim of which bears seven short prolongations. Within the mouth can be seen the dorsal and ventral jaws.

Two large eyes lie laterally on the head, behind the bases of the arms. At the anterior border of each eye there is a duct which leads into the cavity superficial to the lens.

Ventral to the head there is a semicircular flap of flesh which bears a large conical funnel. This flap bears a pair of sockets latero-ventrally, which interlock with a pair of knobs in a corresponding position on the inner surface of the wall of the mantle cavity. Note the wide entrance to the mantle cavity. The body, which is supported internally by a massive but light shell, bears a pair of lateral fins which meet at the posterior end of the body.

The mantle cavity (see Fig. 183). Place the animal on the dissecting board
with the physiologically ventral surface uppermost, and open the mantle
cavity by a median longitudinal incision. Towards the hind end of the
mantle cavity the viscera are connected to the ventral body wall by a mesentery. Cut this mesentery and nail out the two flaps of the mantle laterally,
taking great care not to damage the ink sac which lies in the apex of the
visceral mass. Observe the following features in the mantle cavity: two
ctenidia, each secured to the mantle wall by a latero-dorsal suspensory

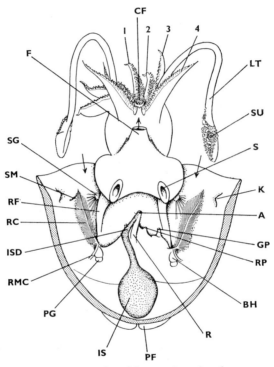

Fig. 183. *Sepia officinalis*. Interior of the mantle cavity of a young male specimen, greatly reduced in size. Note the stellate ganglion (SG), the two ctenidia (RC), the anus (A), the genital papilla (GP), the two renal papillae (RP),
and the retractor muscles of the funnel (RF). Some structures in the visceral
mass can already be seen due to the transparency of the epithelium, e.g. the
ink sac (IS) and the ink sac duct (ISD), the branchial hearts (BH), and the pericardial glands (PG). The directions of inhalant and exhalant water currents
are indicated by arrows. For interpretation of other lettering, see pp. 542–3.

membrane; a pair of branchial hearts which lie in the body cavity can be
seen at this stage, one at the base of each ctenidium; the rectum and the
ink sac duct passing forwards to the anus, which is a wide aperture with
a pair of lateral flaps; powerful muscles running to the funnel, and to the
head; the stellate ganglia at the base of the funnel and lateral to the above

muscles; the ink sac, blue in colour, covering the posterior half of the visceral mass.

(a) *The female*. Two large white nidamental glands lying median to the bases of the ctenidia and having conspicuous anterior apertures; two cream-coloured accessory nidamental glands just anterior to the nidamental glands; the oviduct lies on the left side, and is rendered conspicuous by the large white oviducal glands. The renal papillae lie on each side of the rectum, about 1 in. anterior to the accessory nidamental glands in a large specimen.

(b) *The male*. The large male duct is on the left side (in the same position as the female duct), with a slightly expanded aperture around which spermatophores may be found. The outline of the spermatophore-producing apparatus can be seen by transparency of the superficial tissues.

Carefully dissect the epithelium away from the ink sac and the ink duct, *taking very great care not to damage the sac in any way*, as any leakage of ink over the dissection will cause considerable trouble. Remove the ink-sac and its duct intact. In doing this, note the very thin walls of the various coelomic spaces. In the case of the female, the nidamental and accessory nidamental glands should now be removed. While doing this, take care not to damage the underlying branchial hearts and pericardial glands.

The vascular system (see Fig. 184). Liberate the intestine opposite the bases of the ctenidia by cutting through the membrane on its right side and displace the intestine and rectum slightly to the left. A vein can now be seen running inwards and forwards from the right branchial heart, the course of the vein being obscured by renal tissue. Find the asymmetric ventricle beneath this vein without damaging the superficial tissues. Now find the flimsy left and right auricles which carry blood from the ctenidia to the stout ventricle. The ventricle pumps blood forwards in a cephalic artery which passes into the mass of the digestive glands, parallel with the oesophagus. Do not trace this artery far at present. Blood is pumped backwards in a posterior aorta, which divides into left and right posterior pallial arteries. A branch leaving the base of the posterior aorta serves the ink sac.

An anterior renal artery arises from the concave side of the ventricle and passes into the excretory organ. A genital artery arises at about the same point and passes back to the gonad.

Blood is collected from the anterior part of the body in a median cephalic vein, which runs close to the rectum, and this divides into a pair of venae cavae which pass through the excretory organ to the branchial hearts. In the cavity of the excretory organ these veins are invested with renal appendages, as was mentioned above. An anterior mantle vein drains blood from the anterior wall of the mantle cavity, it runs backwards parallel with and external to the ctenidium and joins the vena cava at the point where it enters the branchial heart. A posterior mantle vein approaches from behind the branchial heart and the base of this, where it joins the vena cava, is covered with renal appendages.

An ink sac vein and a genito-mesenteric vein empty into the right vena cava, and a mesenteric vein empties into the left vena cava.

Each branchial heart receives blood from the above three veins and pumps it into the ctenidium. On its posterior surface the branchial heart bears a large pericardial gland.

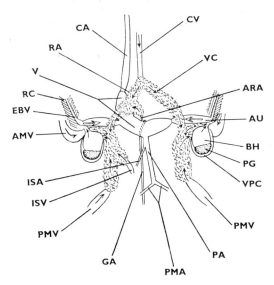

FIG. 184. *Sepia officinalis*. The principal features of the vascular system. Blood is returned from the head in the cephalic vein (CV) which opens into paired venae cavae (VC); blood is returned from the mantle in the paired anterior mantle veins (AMV) and paired posterior mantle veins (PMV); blood from these various sources is passed via the paired branchial hearts (BH) into the ctenidia (RC), the direction of flow of blood in the veins being indicated by solid black arrows. Oxygenated blood is drained from the ctenidia in the efferent branchial vein (EBV) into the paired auricles (AU) and thence into the asymmetrical ventricle (V). Blood leaves the ventricle in the cephalic artery (CA) which passes alongside the oesophagus towards the head; in the posterior aorta (PA); in the anterior renal artery (ARA), and in the genital artery (GA). The veins which lie in the cavities of the excretory organ are covered with renal appendages (RA). For interpretation of other lettering, see pp. 542-3.

The reproductive system. (a) *Female.* The single ovary lies at the posterior end of the visceral mass, it consists of numerous ova at various stages of development. The ova are discharged into the coelome, and are collected into a single oviduct on the left side of the body which arises close to the ovary and passes forwards to the female genital aperture anterior to the left ctenidium. The distal portion of the oviduct is swollen by a large oviducal gland which secretes an outer layer around the ova. The nidamental and accessory nidamental glands have already been seen.

(b) *Male.* The testis lies at the posterior end of the visceral mass. The gonad duct is much complicated by the presence of glands and by the necessity for the storage of spermatophores.

The alimentary canal (see Fig. 185). Remove the funnel; turn the rectum over to the right side. Skin the roof of the anterior part of the visceral mass, and expose the two large digestive glands. At about a quarter of their length forwards from their posterior extremities, these two glands discharge into two long wide ducts which open into the caecum of the stomach. These ducts are covered with lobed glandular masses, the "pancreas", which look somewhat similar to the excretory lobes on the vena cava. Having recognised this distinction, remove the venae cavae and the heart, and expose the underlying stomach. The right lobe of the digestive gland may now be wholly removed, thus exposing the course of the oesophagus and the cephalic artery.

In removing one of the digestive glands, note the position of the shell below, the course of the pallial nerve backwards from the visceral ganglion towards the stellate ganglion, and also the course of the visceral nerves in the median line, close to the cephalic vein. Cut through the cartilage of the head in the median ventral line, between the two ventral arms, thus exposing the buccal mass. Carry the cut backwards through the visceral ganglia to the anterior end of the digestive glands, thus exposing the two small white posterior salivary glands, which lie immediately anterior to the brown digestive glands. Ducts from the two posterior salivary glands unite and a median ventral duct passes forwards to the posterior end of the buccal mass.

Trace the oesophagus backwards to the point where it enters the stomach. The stomach is baggy, and lies on the right side of the body. Note the large gastric ganglion on the "ventral" surface of the stomach, opposite the termination of the oesophagus.

Trace the sympathetic nerves forwards from the gastric ganglion alongside the oesophagus towards the inferior buccal ganglia, from which they originate.

The stomach is comparatively thick walled; it opens on the left into a large thin-walled caecum. The wide intestine leaves the anterior side of the caecum, turns twice and then runs forwards to the anus. The ducts from the two lobes of the digestive gland run backwards side by side towards the caecum, they pass on either side of the base of the intestine, and open into the caecum together on the left side of the origin of the intestine.

Slit open the stomach and note the chitinous lining; examine the stomach contents and determine the nature of the food, if possible. Cut transversely through the stomach wall towards the base of the intestine.

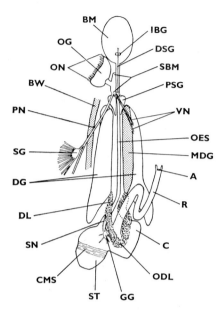

FIG. 185. *Sepia officinalis*. The alimentary canal. The jaws and radula lie within the elliptical buccal mass (BM) which has not been dissected open. The posterior salivary glands (PSG) lie antero-dorsal to the anterior ends of the large, paired digestive glands (DG); the two salivary ducts unite (DSG) and enter the postero-ventral surface of the buccal mass. The position of the inferior buccal ganglion (IBG) on the surface of the buccal mass is shown. The oesophagus passes alongside the cephalic artery in the median line to the muscular stomach (ST) whose walls are provided with circular muscle fibres (CMS) and which serves as a muscular gizzard. The large gastric ganglion (GG) lies on the ventral surface of the stomach, and is connected with the brain mass by two sympathetic nerves (SN) which accompany the oesophagus closely. The stomach communicates on its median side with a spiral caecum (C), by means of a small space, the vestibule, from which the intestine emerges. Coarse objects such as fish scales and bones, etc., pass directly from the stomach through the vestibule to the intestine and rectum, without entering the delicate caecum. The ducts of the two digestive glands (DL) unite posterior to the base of the intestine and open into the caecum at this point (ODL). Note that the duct from the left digestive gland passes underneath the intestine as seen in dissection. The two ducts from the digestive gland are covered with lobules of "pancreatic" tissue which is not unlike the renal appendages covering the surfaces of the large veins in this region. The left lobe of the digestive gland is shown pressed slightly to one side, and its median face is indicated by mechanical stipple. Note the two visceral nerves (VN) which emerge from the posterior end of the sub-oesophageal brain mass (SBM) and accompany the cephalic vein. The course of the pallial nerve (PN) to the stellate ganglion (SG) is indicated. For interpretation of other lettering, see pp. 542–3.

You will then expose a small chamber, the vestibule, lying between the stomach and the intestine. The caecum opens into the posterior wall of the vestibule. By adjustment of sphincter muscles, fragments of shell, etc., pass from the stomach through the vestibule and then into the intestine, and such fragments are prevented from entering the caecum, where only digestible material is received. Slit open the caecum, which is spiral in shape, and see the folded walls.

The nervous system. The gastric ganglion has already been seen, on the ventral surface of the stomach. Trace the pallial nerve through the muscles of the siphon to the stellate ganglion. A branch of the pallial nerve runs past the stellate ganglion on its median side, receives a nerve from the stellate ganglion, and then passes deep into the mantle muscles. Numerous pallial nerves radiate from the stellate ganglion to the muscles of the mantle, and to the fins.

Examine the ventral surface of the buccal mass, and find the pair of inferior buccal ganglia lying side by side near its posterior end. These are connected by a pair of transverse connectives to a single superior buccal ganglion which can be seen lying dorsal to the origin of the oesophagus by raising the buccal mass slightly.

Turn the animal over, so as to expose the "dorsal" surface; insert a seeker into the duct leading into the cavity between the cornea and the lens of one eye. Remove the cornea, cut away the skin around the eye so as to expose the lens and the whole of the iris. Remove the postero-dorsal roof of the orbit, and slice through the cartilaginous posterior wall of the orbit to expose the optic ganglion. Around the optic ganglion there is a large mass, the "white gland", of unknown function. Note the numerous short nerves passing from the optic nerve to the retina. Remove cartilage in the mid-dorsal line, and expose the cerebral ganglion. The brachial, pedal and visceral ganglia have been cut while exposing the anterior end of the oesophagus.

Remove the buccal mass, and find the brachial nerves by cutting the flesh transversely near the base of one of the arms. Trace the nerve for a short distance, and find a portion of the circular interbranchial connective which joins the eight brachial nerves just below the bases of the arms. Trace the nerves backwards to the brachial ganglion.

The buccal mass. Excise the buccal mass, noting carefully which is the "dorsal" surface. Keeping the dorsal surface uppermost, remove first the lower and then the upper jaw. Insert a seeker into the cut end of the oesophagus and find the mouth which lies posteriorly in the buccal mass between the bases of two conspicuous latero-dorsal lobes. These lobes contain the anterior salivary glands. Find the radula, which emerges anteriorly from the radula sac. Below the radula there is the "tongue".

Skeleton. Extract and examine the internal shell.

Eye. Slit this open horizontally, and examine the internal structure.

TABLE 21. *Scheme of classification of the Recent Cephalopoda adopted in the present work*

Only a few orders, families, and genera are listed, these having been chosen to summarise the more important references to the Cephalopoda in the text

Sub-class	Order	Family	Genus	Pages
Tetrabranchiata	Nautiloidea	Nautilidae	*Nautilus*	33
Dibranchiata				
Decapoda	Sepiacea	Spirulidae	*Spirula*	
		Sepiidae	*Sepia*	35, 260, 306, 325, 441–7
		Sepiolidae	*Sepiola*	
	Loliginacea	Loliginidae	*Loligo*	261–3, 324, 405–7
	Architeu-thacea	Architeuthidae	*Architeuthis*	
		Ommatostre-phidae	*Ommatostre-phes*	
		Cranchiidae	*Cranchia*	
Octopoda	Octopodacea	Octopodidae	*Octopus*	313, 318, 409, 417–37, 439
			Eledone	
	Argonauta-cea	Argonautidae	*Argonauta*	271–2, 306

Key to the Lettering on the Figures

A	Anus.	EBV	Efferent branchial vein.
AMV	Anterior mantle vein.	F	Funnel.
ARA	Anterior renal artery.	GA	Genital artery.
AU	Auricle.	GG	Gastric ganglion.
BH	Branchial heart.	GP	Genital papilla.
BM	Buccal mass.	IBG	Inferior buccal ganglion.
BW	Cut surface of body wall.	IS	Ink sac.
C	Spiral caecum of the stomach.	ISA	Ink sac artery.
CA	Cephalic artery.	ISD	Ink sac duct.
CF	Circular flap of skin surrounding the mouth.	ISV	Ink sac vein.
		K	Knob.
CMS	Circular muscle of the stomach.	LT	Left tentacle.
		MDG	Median face of left lobe of the digestive gland.
CV	Cephalic vein.		
DG	Lobes of digestive gland (="liver").	ODL	Opening of left and right ducts from the digestive glands into the spiral caecum of the stomach.
DL	Ducts of digestive gland, surrounded by "pancreatic" tissue.		
		OES	Oesophagus.
DSG	Duct of posterior salivary glands.	OG	Optic ganglion.
		ON	Optic nerves.

PA	Posterior aorta.	SBM	Sub-oesophageal brain mass.
PF	Posterior end of fin.	SG	Stellate ganglion.
PG	Pericardial gland.	SM	Suspensory membrane of cte-
PMA	Posterior mantle arteries.		nidium.
PMV	Posterior mantle vein.	SN	One of two sympathetic
PN	Pallial nerve.		nerves.
PSG	Posterior salivary glands.	ST	Stomach.
R	Rectum.	SU	Stalked suckers with chiti-
RA	Renal appendages of veins.		nous rims.
RC	Right ctenidium.	V	Ventricle.
RF	Retractor muscle of funnel.	VC	Vena cava.
RMC	Marginal part of retractor	VN	Visceral nerves.
	muscle of ctenidium.	VPC	Pocket of viscero-parietal coe-
RP	Renal papilla.		lome.
S	Socket.		

INDEX

Numerals in bold type indicate page references to figures and to many of the more important sections of the text